The Latin American Studies Book Series

The Latin American Studies Book Series promotes quality scientific research focusing on Latin American countries. The series accepts disciplinary and interdisciplinary titles related to geographical, environmental, cultural, economic, political, urban and health-related research dedicated to Latin America. The series publishes comprehensive monographs, edited volumes and textbooks refereed by a region or country expert specialized in Latin American studies.

The series aims to raise the profile of Latin American studies, showcasing important works developed focusing on the region. It is aimed at researchers, students, and everyone interested in Latin American topics.

Submit a proposal: Proposals for the series will be considered by the Series Advisory Board. A book proposal form can be obtained from the Publisher, Andrea Brody-Barre (andrea.brody-barre@springer.com).

Gustavo Federico Bonnat • María Clara Álvarez
Diana Leonis Mazzanti • María Paula Barros
Mariano Bonomo • Verónica Puente
Editors

Current Research in Archaeology of South American Pampas

Springer

Editors
Gustavo Federico Bonnat
INHUS, UNMDP-CONICET. Laboratorio de Arqueología Regional Bonaerense (LARBO), Facultad de Humanidades
Universidad Nacional de Mar del Plata (UNMDP)
Mar Del Plata, Buenos Aires, Argentina

Diana Leonis Mazzanti
Laboratorio de Arqueología Regional Bonaerense (LARBO)
Facultad de Humanidades
Universidad Nacional de Mar del Plata (UNMDP)
Mar Del Plata, Buenos Aires, Argentina

Mariano Bonomo
CONICET- División Arqueología
Facultad de Ciencias Naturales y Museo
Universidad Nacional de La Plata
Paseo del bosque s/n° (1900)
La Plata, Buenos Aires, Argentina

María Clara Álvarez
Grupo de Estudios Ambientales
IMASL-CONICET, UNSL
San Luis, Argentina

Facultad de Ciencias Sociales
Universidad Nacional del Centro de la Provincia de Buenos Aires
Olavarría, Argentina

María Paula Barros
INCUAPA, UNCPBA-CONICET
Universidad Nacional del Centro de la Provincia de Buenos Aires (UNCPBA)
Facultad de Ciencias Sociales
Olavarría, Buenos Aires, Argentina

Verónica Puente
Laboratorio de Arqueología Regional Bonaerense (LARBO)
Facultad de Humanidades
Universidad Nacional de Mar del Plata (UNMDP)
Mar Del Plata, Buenos Aires, Argentina

ISSN 2366-3421 ISSN 2366-343X (electronic)
The Latin American Studies Book Series
ISBN 978-3-031-55193-2 ISBN 978-3-031-55194-9 (eBook)
https://doi.org/10.1007/978-3-031-55194-9

This Springer imprint is published by the registered company Springer Nature Switzerland AG
The registered company address is: Gewerbestrasse 11, 6330 Cham, Switzerland

Paper in this product is recyclable.

Acknowledgments

We would like to thank all the people and institutions that made this book possible, namely, all the authors who sent their contributions; Springer/Nature editors, for allowing us to publish this volume, and particularly Jorge Rabassa, editor of the Latin American Studies series of Springer/Nature; the reviewers who agreed to evaluate the manuscripts; Universidad Nacional de Mar del Plata, Universidad Nacional del Centro de la Provincia de Buenos Aires, and Universidad Nacional de La Plata for their institutional support; and the Agencia Nacional de Promoción de la Investigación, el Desarrollo Tecnológico y la Innovación and the Consejo Nacional de Investigaciones Científicas y Técnicas, for providing funding for the development of the ninth CARPA.

Introduction: Current Research in the Archaeology of the South American Pampas

Abstract The South American Pampas is a huge biome that comprises the southern half of the state of Rio Grande do Sul in Brazil, the entire Uruguay, and central-eastern Argentina. In Argentina, the Pampas region constitutes the most important grassland ecosystem in the country (Mateucci, 2012). Based on a usually marked 600 mm isohyet (Burgos, 1968), this region is divided into two subregions: (1) the "Humid Pampas" or "Eastern Pampas" and (2) the "Dry Pampas" or "Western Pampas". In general, the Humid Pampas receives an average of 1000 mm precipitation per year and it is favored by the wetter winds from the Atlantic Ocean. In the Dry Pampas, the average annual rainfall is 400 mm. Today the typical vegetation of the region is grasslands, and one of the most relevant characteristics is the scarcity of trees (Sala et al., 2001). However, recent studies indicate the presence of xerophytic forests in some hilly sectors during the Pleistocene-Holocene transition (Brea et al., 2020). In terms of quantity and abundance of vegetal species, grasses are dominated by the Poaceae family. Currently, a large part of the native grasslands has been replaced by exotic seeded species, especially crops (Cabrera & Zardini, 1993).

These vast plains did not go unnoticed by different naturalists, travelers, and chroniclers who visited the Pampas from the beginning of the Spanish Conquest in the sixteenth century (e.g., Schmidel, 1945 [1563–1565]). The first sources mention the Querandí indigenous hunter-gatherer-fishers with high pedestrian mobility, who lived in the plains between the Central Hills of the province of Córdoba and the coast of the Río de la Plata. In 1581, Juan de Garay mentioned the presence of indigenous people in the Mar del Plata area, who wore woolen clothes brought from the Chilean Cordillera and had metallic objects (Vignati, 1960). References were more frequent at later times, especially in the eighteenth and nineteenth centuries (e.g., Cardiel, 1930 [1748]; Azara, 1943 [1796]; Miers, 1968 [1819]). The Andean influence on the Pampean plains became evident. The indigenous cultural dynamics intensified, as reflected in the different ethnonyms used in the chronicles: Puelches, Pampas, Serranos, and Aucas (=Mapuche), among others. An important milestone was Charles Darwin's visit to the Pampas in 1832–1833, when he was struck by the traditional customs of the so-called *gauchos*. Darwin's observations in the Pampas and Patagonia had a major influence on the subsequent development of his theory of evolution. Of particular importance were the fossils found in the towns of Punta Alta and Monte Hermoso, in Buenos Aires Province (Farinati et al., 2010). Another important milestone was Florentino Ameghino's postulation (1880–81, 1907, 1909, 1910a, b) of the Pampas region as the place where an independent process of hominization led to the origin of the genus *Homo* and different human species (Bonomo, 2002; Podgorny, 2021). Although this idea was erroneous, it generated interesting local and international debates at the time, and other ideas such as the coexistence between humans and Pleistocene mega mammals were accepted and vindicated 90 years later (Palanca et al., 1972; Politis & Gutiérrez, 1998).

In the twentieth century, studies on the Atlantic coast intensified amid an environment of strong controversy about the temporal ascription and authenticity of the new archeological discoveries of the marine ravines around Miramar (Daino, 1979; Bonomo, 2002). The recovered materials were mainly "bone tools", produced in fresh and fossil states, and lithic artifacts, some of which were embedded in the bones of extinct mega mammals. Some authors (Ameghino, 1915; Vignati, 1923; Frenguelli, 1936) accepted these findings, assigning them to the Tertiary or Early Quaternary periods. Other authors (Romero, 1918; Blanco, 1921; Boman, 1921) rejected the primary position of the materials or directly claimed that they were a scientific fraud. This latter idea prevailed and led to the abandonment of coastal archeological research for many years.

In the mid-twentieth century, Menghin (1957) and Bórmida (1960, 1969), representatives of the Austrian Culture-History School, established direct relationships between cultural artifacts and geographically and temporally bounded ethnic groups. From the repeated types of lithic tools, as well as the raw materials and manufacturing techniques used and the presence or absence of pottery, they defined different industries of the hunter-gatherers who inhabited the coast (e.g., "*Puntarubiense*"), the plains ("*Blancagrandense*" and "*Bolivarense*"), and the hills ("*Tandilense*"). These industries were integrated in "cultural circles" (= "*Kulturkreise*") defined for European prehistory. These cultural circles characterized groups of people with specific economics, technology, levels of cultural complexity, social structure, and religious beliefs.

In the following decades, some research reacted against the hegemonic cultural-historical approach, among which Madrazo's investigations (1973, 1979) stand out because he incorporated multidisciplinary studies in fieldwork and proposed a new interpretative model based on aspects of ecological and indigenous subsistence criteria.

In the 1980s, Pampean archaeology experienced considerable changes, including a new theoretical processual framework and a division into areas based on geographical, geological, and cultural aspects (Politis & Barros, 2006). It is from this moment on that systematic micro-regional research began to establish an early settlement of the region in ca. 14,000 cal. yr BP (Politis, 2008; Politis et al., 2016). The investigations showed that from the Late Pleistocene up to the Spanish Conquest, hunter-gatherer groups and equestrian headquarters occupied the Pampas. New archaeological research has been focused on studies from different approaches that incorporate sub-disciplines such as zooarchaeology, taphonomy, archaeobotany, and geoarchaeology to discuss fascinating topics including the early peopling, the mobility circuits in the past, the interaction between indigenous and colonial societies, processes of ethnogenesis and configuration of new ethnic identities (such as the Mapuche), and the perception of this complex past by the modern Pampean society, mainly composed of European immigrants and indigenous descendants (e.g., Politis, 2008; Archila et al., 2021).

In this context of growing research, the "*Congreso de Arqueología de la Región Pampeana Argentina*" (Conference of Archaeology of the Pampas Region, CARPA) became the main forum for the exchange and discussion of archaeological issues for the Pampas of the Southern Cone. The first CARPA was held in the city of Venado Tuerto (Santa Fe Province) in 1998. On that occasion, 68 papers were presented (Gradín & Oliva, 2004). Two years later, in 2000, the congress was held again in the city of Mar del Plata (Buenos Aires Province), where 83 presentations were delivered (Mazzanti et al., 2002). In 2002, the CARPA was held in the city of Olavarría (Buenos Aires Province) and 87 papers were presented at the conference (Martínez et al., 2004). In 2005, the congress was held in the city of Bahía Blanca (Buenos Aires Province) with a total of 91 presentations (Bayón et al., 2007). The city of Santa Rosa (La Pampa Province) hosted the congress in 2008, where 137 papers were presented (Berón et al., 2010). The sixth edition of the congress was held in the city of La Plata (Buenos Aires Province) in 2011, and 128 presentations were delivered (Bonomo et al., 2013). The seventh CARPA took place in 2014 in the city of Rosario (Santa Fe Province), where 99 studies were presented. The eighth edition of CARPA took place in the city of Luján (Buenos Aires Province) in 2017, and on that occasion, 98 works were presented.

The ninth CARPA was held online in August 2021, due to the sanitary conditions imposed at that time by the COVID-19 pandemic. It was organized by the Faculty of Humanities, Universidad Nacional de Mar del Plata, with the support of the Faculty of Social Sciences, Universidad Nacional del Centro de la Provincia de Buenos Aires, and the Faculty of Natural Sciences and Museum, Universidad Nacional de La Plata. Thus, three national universities joined forces to organize the conference. The objective was to guarantee its viability and to enhance the prestige of one of the most important scientific meetings of Argentine and South American archaeology. During the five days of the event, ten symposia, a general table, a poster table, a panel, three master conferences, and two special activities were held, and homage was paid to the late archaeologist Patricia Madrid, to whom the congress was also dedicated. The final number of presentations was 116. The choice of venue for the ninth CARPA was proposed by colleagues who emphasized the need to make visible the fact that Anthropology (sociocultural and archaeological orientations) had been taught at Universidad Nacional de Mar del Plata and had been closed down by the last military dictatorship (1976–1983) in 1977. New academic decisions that overcome anachronistic visions are working toward its prompt reopening as of 2024.

We wished to share the results of the ninth CARPA in a book that would bring together the main contributions and thematic axes addressed. Thus, the authors were invited to participate, resulting in this volume, which summarizes the main topics of the current research conducted in the Pampas. It comprises a selection that includes one of the master conferences and 15 chapters on pre-Hispanic and historical archaeology, as well as heritage studies. Chapter 1 corresponds to the master conference and is based on the results of years of research and work experience developed in the Salado River Depression (Buenos Aires Province). Here, González applied a variety of methodological approaches and summarized several lines of evidence to discuss the human strategies for the use of the resources in the landscape and the preparation and consumption of food.

Chapter 2 concerns the work of Scabuzzo, Costa Angrizani, and Bonomo, whose main goal was to study the diet of the pre-Hispanic populations that inhabited the Upper Delta of the Paraná River (Entre Ríos and Santa Fe Provinces, Argentina) during the Late Holocene. The authors present the isotopic data obtained from 15 individuals from two archaeological sites and discuss the importance of C_3 plans in the diet. In Chap. 3, Piccoli, Avila, and Barboza review the post-depositional history and paleopathological data of human skeletal remains found in the northern Pampas shallow lakes ("lagunas") landscapes (Santa Fe Province), Argentina. This research provided a better understanding of past biocultural processes and their social ramifications.

In Chap. 4, Bernardi, Arrieta, and Lynch Ianniello describe a paleodemographic study based on human burial remains from the Médano Petroquímica archaeological site (La Pampa Province, Argentina), dated to ca. 400 years BP. The authors discuss the dynamics of the prehistoric societies that inhabited the region at a time when important changes in the organization of these groups were taking place. In Chap. 5, Berón, Carrera Aizpitarte, Páez, Lucero, and Bedrán discuss late Holocene mobility circuits and interaction networks based on the archaeological record of The Tapera Moreira locality (La Pampa Province, Argentina). Thus, new data together with previous information allow the authors to evaluate the trajectories and social interactions of hunter-gatherer populations on a macro-regional scale for the last 5000 years.

In Chap. 6, Bonnat examines social interaction patterns in the hunter-gatherer societies that inhabited the Humid Pampas subregion (Buenos Aires Province, Argentina) since 14,000 cal. yr BP. The author discusses the lithic assemblages of 27 early archaeological sites to evaluate mobility ranges and procurement lithic strategies. In Chap. 7, Heider and Salgán use archaeological lithic assemblages to examine the use of landscape and hunter-gatherer mobility patterns in an extensive semi-arid area in the Western Pampas at the time of Hispanic-indigenous contact. To address these issues, the authors use the analysis of technological management based on the organization of technology, geochemical studies on obsidian, and the archaeology of internodal spaces on an archaeological scale.

In Chap. 8, Kaufmann, Álvarez, Blasi, and Barros discuss social interactions based on the archaeological record of The Hangar site, located in the northwest of the Interhills ranges ("Pampa Interserrana"; Buenos Aires Province, Argentina) and dated them to the late Holocene. The presence of silicified limestone allows the authors to study the circulation of goods and people across some sectors of the Pampas and neighboring regions.

In Chap. 9, Leipus presents the results of the technological and microwear analysis conducted on lithic assemblages from different archaeological sites located in the Paraná Delta (Entre Ríos Province, Argentina). The application of the microwear analysis made it possible to characterize the uses of the artifacts and to identify the work of various materials, such as wood, bone, and skin. In Chap. 10, Ghiani Echenique, Luengo, Fucks, Paleo, and Bernasconi describe the geomorphological processes of the Don Enrique archaeological site, located in the northern coastal plain of Samborombón bay (Buenos Aires Province, Argentina). This research represents the first archaeological record of Samborombón bay, where archaeological materials were found in situ in a buried paleo soil.

In Chap. 11, Tomassini and Eliges discuss the theoretical concept of "peri urban spaces" applied to historical archaeology in Argentina. The authors propose an archaeological approach to the study of a set of sites linked to the southern frontier of the Pampas and North Patagonia for the nineteenth century. In Chap. 12, Acevedo and Staropoli discuss various aspects involved in the management and protection of cultural heritage, in this case, the EB 1789 canyon from the eighteenth century in San Antonio de Areco (Buenos Aires Province, Argentina).

In Chap. 13, Levrand, Cantar, and Endere analyze the incorporation of heritage landscape categories and cultural itineraries in the internal legislation of South American countries. The authors note the lack of regulations for the protection of cultural heritage in some South American countries. On the other hand, they recognize the incorporation of the landscape and cultural itinerary categories into the Argentinian national legislation.

In Chap. 14, Sokol, Salerno, Escosteguy, Frère, and González discuss the importance of archaeological impact works for heritage conservation in the context of activities involving earthwork. These authors present their own experience in a large-scale project developed in the Salado River Basin (Buenos Aires Province, Argentina). In Chap. 15, Traba, Noya, and Camino discuss the heritage management of the Anchorena 942 archaeological site (San Martín theater) located in the city of Baradero (Buenos Aires Province, Argentina). The fortuitous finding of old structures beneath the theater's stage allowed the authors to discuss their inclusion in comprehensive long-term projects aimed at public use for educational, cultural, and tourist purposes.

Finally, we consider that the chapters mentioned above represent some of the various lines of research that are currently being developed in the Pampas region. We believe that these interesting investigations illustrate the effort of several years of field and laboratory work. This panorama shows that the future of archaeology in the Pampas is auspicious and will surely represent new challenges and pose questions that will be addressed in future conferences.

INHUS, UNMDP-CONICET. Laboratorio de Arqueología Regional Bonaerense (LARBO) Facultad de Humanidades Universidad Nacional de Mar del Plata (UNMDP) Mar del Plata, Buenos Aires, Argentina — Gustavo Federico Bonnat

Grupo de Estudios Ambientales IMASL-CONICET UNSL San Luis, Argentina

Facultad de Ciencias Sociales Universidad Nacional del Centro de la Provincia de Buenos Aires Olavarría, Argentina — María Clara Álvarez

Laboratorio de Arqueología Regional Bonaerense (LARBO) Facultad de Humanidades Universidad Nacional de Mar del Plata (UNMDP) Mar del Plata, Buenos Aires, Argentina — Diana Leonis Mazzanti

INCUAPA, UNCPBA-CONICET. Universidad Nacional del Centro de la Provincia de Buenos Aires (UNCPBA) Facultad de Ciencias Sociales Olavarría Buenos Aires, Argentina — María Paula Barros

CONICET- División Arqueología Facultad de Ciencias Naturales y Museo Universidad Nacional de La Plata Paseo del bosque s/n° (1900) La Plata, Buenos Aires, Argentina — Mariano Bonomo

Laboratorio de Arqueología Regional Bonaerense (LARBO) Facultad de Humanidades Universidad Nacional de Mar del Plata (UNMDP) Mar Del Plata, Buenos Aires, Argentina — Verónica Puente

References

Ameghino, F. (1880–81[1915]). *La antigüedad del hombre en el Plata*, Tomos I y II. Masson-Igon Hnos. París-Buenos Aires. In *Obras completas y correspondencia científica de Florentino Ameghino*, Tomo III, compiled by A. Torcelli. Taller de Impresiones Oficiales, La Plata.

Ameghino, F. (1907). Notas preliminares sobre el *Tetraprothomo argentinus* un precursor del hombre del Mioceno Superior de Monte Hermoso. *Anales del Museo Nacional de Buenos Aires, 9*, 1–82.

Ameghino, F. (1909). Le Diprothomo platensis. Un précurseur de l'homme du Pliocène inférieur de Buenos Aires. *Anales del Museo Nacional, serie 3, tomo, 12*, 107–210.

Ameghino, F. (1910a). Une nouvelle industrie lithique: L'industrie de la pierre fendue dans le Tertiaire de la région littorale au sud de Mar del Plata. *Anales del Museo Nacional de Buenos Aires, 20*(13), 189–204.

Ameghino, F. (1910b). *La industria de la piedra quebrada en el Mioceno superior de Monte Hermoso*. Separata del XVII.

Ameghino, C. (1915). El fémur de Miramar. Una prueba más de la presencia del hombre en el terciario de la República Argentina. Nota preliminar. *Anales del Museo Nacional de Historia Natural de Buenos Aires, 26*, 433–450.

Archila, S., Bonomo, M., & Hastorf, C. (2021). Introduction: South American archaeology's contributions to world archaeology. In M. Bonomo & S. Archila (Eds.), *South American contributions to world archaeology* (pp. 1–12). Springer.

Azara, F. (1943 [1796]). *Descripción e Historia del Paraguay y del Río de la Plata*. Editorial Bajel.

Bayón, C., Flegenheimer, N., González, M. I., Puppio, A., & Freire, M. (Eds.). (2007). *Arqueología de las Pampas, Tomos I y II*. EDIUNS.

Berón, M., Luna, L., Bonomo, M., Montalvo, C., Aranda, C., & Carrera Aizpitarte, M. (Eds.). (2010). *MamülMapu: Pasado y Presente desde la Arqueología Pampeana*. Editorial Libros del Espinillo.

Blanco, J. (1921). Las bolas de Parodi serán bolas....? *Estudios, 20*(1), 31–35.

Boman, E. (1921). Los vestigios de industria humana encontrados en Miramar (República Argentina) y atribuidos a la época terciaria. *Revista Chilena de Historia y Geografía, 39*, 330–352.

Bonomo, M. (2002). El hombre fósil de Miramar. *Intersecciones en Antropología, 3*, 69–85.

Bonomo, M., Prates, L., Scabuzzo, C., Madrid, P., Di Prado, V., Angrizani, R., Bagaloni, V., Matarrese, A., & Leon, C. (Eds.). (2013). *Revista del Museo de La Plata. Antropología, 13*(87). Special Volume.

Bórmida, M. (1960). Investigaciones paleoetnológicas en la región de Bolívar (Provincia de Buenos Aires). *Anales de la Comisión de Investigaciones Científicas de la Provincia de Buenos Aires, 1*, 190–283.

Bórmida, M. (1969). El Puntarrubiense. *Trabajos de Prehistoria, 26*(NS), 16–117.

Brea, M., Mazzanti, D., & Martínez, G. A. (2020). Xerophytic Forest Record of the Pleistocene/Holocene Transition and Use of Wood Resources by Early Human Groups in the Eastern Tandilia Range, Argentina. *Paleoamerica*. https://doi.org/10.1080/20555563.2019.1698189

Burgos, J. (1968). El Clima de la Provincia de Buenos Aires en relación con la vegetación natural y el suelo. In A. Cabrera (Ed.), *Flora de la Provincia de Buenos Aires, Tomo I* (pp. 33–39). Instituto Nacional de Tecnología Agropecuaria.

Cabrera, A. L., & Zardini, E. M. (1993). *Manual de la Flora de los Alrededores de la Provincia de Buenos Aires*. Editorial ACME.

Cardiel, J.S.J. (1930 [1748]). *Diario del Viaje y Misión al Río del Sauce. Realizado en 1748*. Editorial Coni.

Daino, L. (1979). Exégesis histórica de los hallazgos arqueológicos de la costa atlántica bonaerense. *Prehistoria Bonaerense*, 95–195. Olavarría.

Farinati, E. A., Manera, T., & Tomassini, R. L. (2010). La bahía que iluminó a Darwin. *Revista Española de Paleontología, 25*(1), 35–41. https://doi.org/10.7203/sjp.25.1.18906

Frenguelli, J. (1936). La serie geológica de la República Argentina en sus relaciones con la antigüedad del hombre. In R. Levene (Ed.), *Historia de la Nación Argentina (Desde los orígenes hasta la organización definitiva en 1862) I (1)* (pp. 97–120). Academia Nacional de la Historia.

Gradin, C., & Oliva, F. (Eds.). (2004). *Región Pampeana. Su Pasado Arqueológico*. Universidad Nacional de Rosario.

Madrazo, G. (1973). Síntesis de la Arqueología Pampeana. *Etnía, 17*, 13–25.

Madrazo, G. (1979). Los cazadores a la larga distancia de la Región Pampeana. *Prehistoria Bonaerense*, 11–67.

Martínez, G., Gutiérrez, M., Curtoni, R., Berón, M., & Madrid, P. (Eds.). (2004). *Aproximaciones Contemporáneas a la Arqueología Pampeana. Perspectivas Teóricas, Metodológicas,*

Analíticas y Casos de Estudio. Facultad de Ciencias Sociales, Universidad Nacional del Centro de la Provincia de Buenos Aires (UNICEN).

Mazzanti, D., Berón, M., & Oliva, F. (Eds.). (2002). *Del mar a los salitrales. Diez Mil Años de Historia Pampeana en el Umbral del Tercer Milenio*. UNMDP/LARBO and SAA.

Menghin, O. (1957). Das Protolitikum in Amerika. *Acta Praehistorica, 1*, 5–40.

Miers, J. (1968 [1819]). *Viaje al Plata. 1819–1824*. Solar-Hachette.

Palanca, F., Daino, L., & Benbassat, E. (1972). Yacimiento "Estancia La Moderna" (partido de Azul, Provincia de Buenos Aires). Nuevas perspectivas para la Arqueologìa de la Pampa Bonaerense. *Etnía, 15*, 19–27.

Podgorny, I. (2021). *Florentino Ameghino y Hermanos*. EDHASA.

Politis, G. (2008). The Pampas and the Campos. In H. Silverman & W. Isbell (Eds.), *Handbook of South American archaeology* (pp. 235–262). Springer.

Politis, G., & Barros, M. P. (2006). *La región pampeana como unidad espacial de análisis en la arqueología contemporánea* (Folia Histórica del Nordeste. N° 16: 51-74). Instituto de Investigaciones Neohistóricas- Conicet. Instituto de Historia – Facultad de Humanidades. UNNE. Resistencia, Province of Chaco, Argentina. ISSN: 0325-8238.

Politis, G., & Gutiérrez, M. (1998). Gliptodontes y cazadores-recolectores de la región pampeana (Argentina). *Latin American Antiquity, 9*(2), 111–134.

Politis, G., Gutiérrez, M. A., Rafuse, D. J., & Blasi, A. (2016). The arrival of Homo sapiens into the Southern Cone at 14,000 years ago. *Plos One, 11*(9), e0162870. https://doi.org/10.1371/journal.pone.0162870

Romero, A. (1918). El Homo pampaeus. Contribución al estudio de origen y antigüedad de la raza humana en Sudamérica según recientes descubrimientos. *Anales de la Sociedad Científica Argentina, 86*, 5–48.

Sala, O. E., Austin, A. T., & Vivanco, L. (2001). Temperate grassland and shrubland ecosystems. In S. Levin (Ed.), *Encyclopedia of Biodiversity 1, 5* (pp. 627–635). Academic.

Schmidel, U. (1945 [1563–1565]). Viaje al Río de la Plata. Emecé.

Vignati, M. A. (1923). Contribución al estudio de la Litotecnia Chapadmalense. *Physis, 6*, 238–247.

Vignati, M. A. (1960). El indigenado de la Provincia de Buenos Aires. *Anales de la Comisión de Investigaciones Científicas de la Provincia de Buenos Aires, 1*, 95–182.

Contents

Contributors and Reviewers

Contributors

Verónica J. Acevedo ArqueoLab-UBA, Instituto de Arqueología, Facultad de Filosofía y Letras, Universidad de Buenos Aires (UBA), Buenos Aires, Argentina

María Clara Álvarez Grupo de Estudios Ambientales IMASL-CONICET, San Luis, Argentina

Mario Arrieta CONICET-Laboratorio de Bioarqueología, Universidad Nacional de Río Cuarto (UNRC), Córdoba, Argentina

Juan D. Avila Centro de Estudios Interdisciplinarios en Antropología, Facultad de Humanidades y Arte, Universidad Nacional de Rosario (UNR), Santa Fé, Argentina

María C. Barboza Centro de Estudios Interdisciplinarios en Antropología, Facultad de Humanidades y Arte, Universidad Nacional de Rosario (UNR), Santa Fé, Argentina

María Paula Barros INCUAPA, UNCPBA-CONICET. Universidad Nacional del Centro de la Provincia de Buenos Aires (UNCPBA), Facultad de Ciencias Sociales, Olavarría, Buenos Aires, Argentina

Sofía Bedran Museo Etnográfico "Juan Bautista Ambrosetti", Facultad de Filosofía y Letras, Universidad de Buenos Aires (UBA), Buenos Aires, Argentina

Lila Bernardi CONICET-Laboratorio de Bioarqueología, Universidad Nacional de Río Cuarto (UNRC), Córdoba, Argentina

Emiliana Bernasconi CONICET-Instituto de Investigaciones en Biodiversidad y Medioambiente (INIBIOMA), Universidad Nacional del Comahue (UNCOMA), Neuquén, Río Negro, Argentina

Mónica Berón IDECU-CONICET. Museo Etnográfico "Juan Bautista Ambrosetti", Facultad de Filosofía y Letras, Universidad de Buenos Aires (UBA), Buenos Aires, Argentina

Facultad de Ciencias Sociales, Universidad Nacional del Centro de la Provincia de Buenos Aires (UNCPBA), Buenos Aires, Argentina

Gustavo Federico Bonnat INHUS, UNMDP-CONICET. Laboratorio de Arqueología Regional Bonaerense (LARBO), Facultad de Humanidades, Universidad Nacional de Mar del Plata (UNMDP), Mar del Plata, Buenos Aires, Argentina

Mariano Bonomo CONICET-División Arqueología del Museo de La Plata, Facultad de Ciencias Naturales y Museo, Universidad Nacional de La Plata (UNLP), La Plata, Buenos Aires, Argentina

Adriana M. Blasi CIC-División Mineralogía, Petrología y Sedimentología de la Facultad de Ciencias Naturales y Museo, Universidad Nacional de La Plata (UNLP), La Plata, Buenos Aires, Argentina

Ulises A. Camino Universidad de Buenos Aires (UBA), Facultad de Arquitectura, Diseño y Urbanismo, Instituto de Arte Americano, Centro de Arqueología Urbana, Buenos Aires, Argentina

Cooperativa de Trabajo Arqueocoop ltda, Buenos Aires, Argentina

Nahir M. Cantar CONICET, ILHAM- Facultad de Arquitectura, Urbanismo y Diseño, Universidad Nacional de Mar del Plata (UNMDP), Mar del Plata, Buenos Aires, Argentina

Manuel Carrera Aizpitarte IDECU-CONICET. Museo Etnográfico "Juan Bautista Ambrosetti", Facultad de Filosofía y Letras, Universidad de Buenos Aires (UBA), Buenos Aires, Argentina

Facultad de Ciencias Sociales, Universidad Nacional del Centro de la Provincia de Buenos Aires (UNCPBA), Olavarría, Buenos Aires, Argentina

Rodrigo Costa Angrizani CICYTTP-CONICET, UADER. Entre Ríos Argentina. División Arqueología del Museo de La Plata, Facultad de Ciencias Naturales y Museo, Universidad Nacional de La Plata (UNLP), La Plata, Buenos Aires, Argentina

Adolfo Eliges CONICET División Arqueología, Facultad de Ciencias Naturales y Museo-Universidad Nacional de La Plata (UNLP), La Plata, Buenos Aires, Argentina

María L. Endere INCUAPA, UNCPBA-CONICET. Universidad Nacional del Centro de la Provincia de Buenos Aires (UNCPBA), Facultad de Ciencias Sociales, Olavarría, Buenos Aires, Argentina

Paula Escosteguy CONICET; Instituto de Arqueología, Facultad de Filosofía y Letras, Universidad de Buenos Aires (UBA), Buenos Aires, Argentina

María M. Frère Instituto de Arqueología, Facultad de Filosofía y Letras, Universidad de de Buenos Aires (UBA), Buenos Aires, Argentina

Enrique E. Fucks Centro de Estudios Integrales de la Dinámica Exógena (CEIDE). Facultad de Ciencias Agrarias y Forestales y Facultad de Ciencias Naturales y Museo. Universidad Nacional de La Plata (UNLP), La Plata, Buenos Aires, Argentina

Naiquen Ghiani Echenique Laboratorio de Análisis Cerámico, Facultad de Ciencias Naturales y Museo, Universidad Nacional de La Plata (UNLP), La Plata, Buenos Aires, Argentina

María I. González Instituto de Arqueología, Facultad de Filosofía y Letras, Universidad de Buenos Aires (UBA), Buenos Aires, Argentina

Guillermo Heider CONICET, Departamento de Geología, Universidad Nacional de San Luis (UNSL), San Luis, Argentina

Cristian A. Kaufmann INCUAPA, UNCPBA-CONICET. Universidad Nacional del Centro de la Provincia de Buenos Aires (UNCPBA), Facultad de Ciencias Sociales, Olavarría, Buenos Aires, Argentina

Marcela Leipus División Arqueología, Facultad de Ciencias Naturales y Museo, Universidad Nacional de La Plata (UNLP), La Plata, Buenos Aires, Argentina

Norma E. Levrand CONICET, INES, Universidad Nacional de Entre Ríos (UNER), La Rioja, Entre Ríos, Argentina

Eliana N. Lucero ANPCyT. Museo Etnográfico "Juan Bautista Ambrosetti", Facultad de Filosofía y Letras, Universidad de Buenos Aires (UBA), Buenos Aires, Argentina

Mariel S. Luengo CONICET-Centro de Estudios Integrales de la Dinámica Exógena (CEIDE). Facultad de Ciencias Agrarias y Forestales y Facultad de Ciencias Naturales y Museo, Universidad Nacional de La Plata (UNLP), La Plata, Buenos Aires, Argentina

Ignacio Lynch Ilanniello Laboratorio de Bioarqueología, Universidad Nacional de Río Cuarto (UNRC), Córdoba, Argentina

Verónica Noya Dirección de Cultura, Secretaría de Cultura, Turismo, Educación y Deportes, Municipalidad de Baradero, Buenos Aires, Argentina

Florencia Páez IDECU-CONICET. Museo Etnográfico "Juan Bautista Ambrosetti", Facultad de Filosofía y Letras, Universidad de Buenos Aires (UBA), Buenos Aires, Argentina

María C. Paleo Laboratorio de Análisis Cerámico. Facultad de Ciencias Naturales y Museo. Universidad Nacional de La Plata (UNLP), La Plata, Buenos Aires, Argentina

Carolina V. Píccoli Centro de Estudios Interdisciplinarios en Antropología, Facultad de Humanidades y Arte, Universidad Nacional de Rosario (UNR), Santa Fé, Argentina

Virginia Salerno CONICET, Instituto de Arqueología, Facultad de Filosofía y Letras, Universidad de Buenos Aires (UBA), Buenos Aires, Argentina

María L. Salgán IDEVEA-CONICET, Universidad Tecnológica Nacional (UTN), San Rafael, Mendoza, Argentina

Clara Scabuzzo CICYTTP-CONICET, UADER. Entre Ríos, Argentina. División Arqueología del Museo de La Plata, Facultad de Ciencias Naturales y Museo, Universidad Nacional de La Plata (UNLP), La Plata, Buenos Aires, Argentina

Olivia Sokol CONICET. Instituto de Arqueología, Facultad de Filosofía y Letras, Universidad de Buenos Aires (UBA), Buenos Aires, Argentina

Laura Staropoli ArqueoLab-UBA, Instituto de Arqueología, Facultad de Filosofía y Letras, Universidad de Buenos Aires (UBA), Buenos Aires, Argentina

Aniela R. Traba Cooperativa de Trabajo Arqueocoop ltda, Buenos Aires, Argentina

Hernán Tomassini CONICET, Universidad Nacional del Sur (UNS), Bahía Blanca, Argentina

Reviewers

Luis Albornoz Ramos Facultad de Ciencias Sociales y Jurídicas, Universidad de Tarapacá (UTA), Arica, Chile

Marien Béguelin CONICET, Museo de La Plata. Universidad Nacional de La Plata (UNLP), La Plata, Buenos Aires, Argentina

Juan Bautista Belardi CONICET, Universidad Nacional de la Patagonia Austral (UNPA), Río Gallegos, Santa Cruz, Argentina

Gisela Cassiodoro Instituto Nacional de Antropología y Pensamiento Latinoamericano, CONICET, Universidad de Buenos Aires (UBA), Buenos Aires, Argentina

Silvina Castro CONICET, Laboratorio de Paleoecología Humana, Instituto Interdisciplinario de Ciencias Básicas (ICB), Universidad Nacional de Cuyo (UNCUYO), Mendoza, Argentina

Facultad de Filosofía y Letras, Universidad Nacional de Cuyo, Mendoza, Argentina

Luciana Catella División Arqueología, Facultad de Ciencias Naturales y Museo, Universidad Nacional de La Plata (UNLP), Buenos Aires, Argentina

CEAR, Facultad de Humanidades y Artes, Universidad Nacional de Rosario (UNR), Santa Fe, Argentina

María Gabriela Chaparro INCUAPA, UNCPBA-CONICET. Universidad Nacional del Centro de la Provincia de Buenos Aires (UNCPBA), Facultad de Ciencias Sociales, Olavarría, Buenos Aires, Argentina

Valeria Ciliberto CONICET- INHUS- CEHis Universidad Nacional de Mar del Plata (UNMDP), Facultad de Humanidades, Departamento de Historia, Mar del Plata, Buenos Aires, Argentina

Ignacio Clemente Conte Institució Milà i Fontanals (IMF-CSIC), Barcelona, España

Gabriel Cocco Museo Etnográfico y Colonial, Santa Fé, Argentina

Valeria Cortegoso CONICET, Instituto Interdisciplinario de Ciencias Básicas (ICB), Mendoza, Argentina

Hernán De Angelis CADIC-CONICET, GIATMA, Ushuaia, Tierra del Fuego, Argentina

Claudio G. De Francesco CONICET, Instituto de Investigaciones Marinas y Costeras (IIMyC), Universidad Nacional de Mar del Plata (UNMDP), Mar del Plata, Buenos Aires, Argentina

Mariana Fabra Instituto de Antropología de Córdoba (IDACOR), Facultad de Filosofía y Humanidades, Universidad Nacional de Córdoba (UNC), CONICET, Córdoba, Argentina

Cristian M. Favier Dubois INCUAPA, UNCPBA-CONICET. Universidad Nacional del Centro de la Provincia de Buenos Aires (UNCPBA), Facultad de Ciencias Sociales, Olavarría, Buenos Aires, Argentina

Anabel Feely IDECU-CONICET. Museo Etnográfico "Juan Bautista Ambrosetti", Facultad de Filosofía y Letras, Universidad de Buenos Aires (UBA), Buenos Aires, Argentina

Gonzalo Figueiro Departamento de Antropología Biológica, Facultad de Humanidades y Ciencias de la Educación, Universidad de la República (UDELAR), Montevideo, Uruguay

Gustavo Flensborg INCUAPA, UNCPBA-CONICET. Universidad Nacional del Centro de la Provincia de Buenos Aires (UNCPBA), Facultad de Ciencias Sociales, Olavarría, Buenos Aires, Argentina

Adolfo F. Gil Instituto de Evolución, Ecología Histórica y Ambiente (IDEVEA-CONICET-UTN), San Rafael, Mendoza, Argentina

Ana Gabriela Guraieb Instituto Nacional de Antropología y Pensamiento Latinoamericano, Ministerio de Cultura de la Nación, Buenos Aires, Argentina

María Isabel Hernandez Llosas CONICET, Instituto de Arqueología, Facultad de Filosofía y Letras, Universidad de Buenos Aires (UBA), Buenos Aires, Argentina

Salomon Hocsman CONICET, Grupo de Investigación en Arqueología Andina, Facultad de Ciencias Naturales e Instituto Miguel Lillo, Universidad Nacional de Tucumán (UNT), San Miguel de Tucumán, Tucumán, Argentina

Violeta Killian Galván CONICET, Universidad de Buenos Aires (UBA), Buenos Aires, Argentina

Virginia Pineau Instituto de Arqueología, Facultad de Filosofía y Letras, Universidad de Buenos Aires (UBA), Buenos Aires, Argentina

Diego D. Rindel CONICET, División Antropología, Facultad de Ciencias Naturales y Museo, Universidad Nacional de La Plata (UNLP), La Plata, Buenos Aires, Argentina

Jorge A. Suby INCUAPA, UNCPBA-CONICET. Universidad Nacional del Centro de la Provincia de Buenos Aires (UNCPBA), Facultad de Ciencias Sociales, Olavarría, Buenos Aires, Argentina

Simón Urbina Araya Escuela de Arqueología, Universidad Austral de Chile (UACh), Puerto Montt, Chile

Salado River Wetlands (Buenos Aires Province, Argentina): An Archaeological Perspective

María Isabel González

Abstract The Ninth Congress of Archaeology of the Argentine Pampas was held in August 2021. This chapter summarizes the experiences and reflections I presented in this congress, resulting from my long work experience in the Pampean region, particularly in the Salado River Depression. I present a consideration in dialogue with the data to address a history of at least 2400 years, and I use archaeological tools to approach that past, by combining theoretical and methodological positions. I emphasize my original contributions to the Salado River basin since I studied a region where there was no systematic research but only isolated works and data from collectors. It has not been easy, but I have worked intensively in this area of Buenos Aires Province where current human activity makes the archaeologist's work difficult. The abundance of archaeological remains indicates the intensive use of these sites: the presence of human remains, local manufacture of ceramics and lithic tools (using rocks transported from distant places), emphasis on the use of small fauna and avifauna closely related to internal aquatic environments, fishing gear, and the use of plants native to these environments. In addition, the location of the archaeological settlements in the river-lakes system of the Salado River basin shows the implementation of diversified strategies for the use of resources available in the landscape and for the preparation and consumption of food. It was a wetland with rich aquatic ecosystems that provided a reliable resource base for the hunter-gatherer-fisher (HGF) societies of the waterways.

Keywords Salado river · Research history · Hunter-gatherer-fisher

M. I. González (✉)
Universidad de Buenos Aires, Facultad de Filosofía y Letras, Instituto de Arqueología, Buenos Aires, Argentina
e-mail: igonzale@filo.uba.ar

G. F. Bonnat et al. (eds.), *Current Research in Archaeology of South American Pampas*, The Latin American Studies Book Series,
https://doi.org/10.1007/978-3-031-55194-9_1

Introduction

This chapter summarizes the presentation delivered during the Ninth Congress of Archaeology of the Argentine Pampas, held in August 2021. I would like to share, in this chapter, a series of experiences and reflections arising from my long practice in the Pampas region, particularly in the Salado River Depression. It is a review in dialogue with data to address a history of at least 2400 years. I will use archaeological tools to address that past, combining theoretical and methodological positions. As a strategy to address the research in this microregion, I have contributed by applying a variety of archaeological approaches that allowed me to recognize a way of life that had not been proposed for the Pampas region before.

The Hunter-Gatherer-Fisher Societies

Much research has been done on hunter-gatherer societies in the fields of both archaeology and ethnography. Thus, contributions have shown that this way of life was diverse, making it necessary to delve into the distinct characteristics of each space (Politis & Madrid, 2001; Politis & Barros, 2006; Barnard, 2014; Cobb, 2014, among others). Despite their variability, there is consensus that these societies had extensive knowledge about how to inhabit the territories, how to make good use of resources, and how to develop their technology. Although they had different degrees of mobility, they were organized in small groups, linked by kinship ties without differences in status, apart from sex, age, or individual abilities, to perform different social tasks. They were also connected to the broader community, an essential aspect that enabled them to establish marriage alliances, celebrate rituals, and exchange information and goods.

Particularly in this case study, the variety of approaches allowed us to define the Salado groups as hunter-gatherer-fisher societies, thus contributing to regional studies (González de Bonaveri, 2002; González, 2005). The recovered faunal remains revealed the use of different species linked to the wetland environment, such as coypu, overo lizard, rodents, fish, aquatic birds such as ducks and coots, as well as land birds such as the martineta. To a lesser extent, these populations complemented their diets with deer, Pampas deer, and terrestrial vegetables. Their food collection was expanded to include other resources, such as the mullet and black drum, estuarine fish that can live in waters with a wide range of salt concentrations (Escosteguy et al., 2014; González et al., 2016; Frère et al., 2016a). The role and scope in the consumption of coypu, fish, and other animals from this wetland has been one of the axes we used to discuss the intensification process, as the archaeofaunistic trend indicates the inclusion of a greater number of taxa with an increasing frequency of small animals, including fish (González, 2005).

The characteristics of the settlements allowed us to infer that the groups living there moved within the territory and occupied certain places, interacting with the

resources offered by the environment (Ceruti & González, 2007). The long-term settlements were usually located in the vicinity of permanent water courses, clearly showing a way of inhabiting internal aquatic places (Fig. 1). The location of the archaeological settlements in the fluvial and lacustrine or internal system of the Salado River basin in Buenos Aires accounts for the implementation of diversified strategies in the use of the resources available in the landscape, as well as the preparation and consumption of food. This wetland, with rich aquatic ecosystems, forms a reliable resource base for riverine hunter-gatherer-fisher societies. In this regard, Prates and Bonomo (2017) observed that throughout human development, aquatic environments have played an important role in attracting and concentrating human populations. These hunters, gatherers, and fishers from internal aquatic environments adapted to the environment where they lived by developing a specialized economy that used most of the available resources.

I herein emphasize my original contributions to the study of the Salado River basin because I initiated research on a space where there were no systematic investigations, only isolated works and data from collectors. It has not been easy, but I have worked intensively in this area of Buenos Aires Province, where human activity, the growth of cities and towns, and the advancement of agricultural activity hinder the work of archaeologists.

The Salado River basin of Buenos Aires Province is of great regional importance. It constitutes the backbone of the Pampas ecosystems, with numerous shallow lakes ("lagunas") and wetlands that host abundant and unique biodiversity and

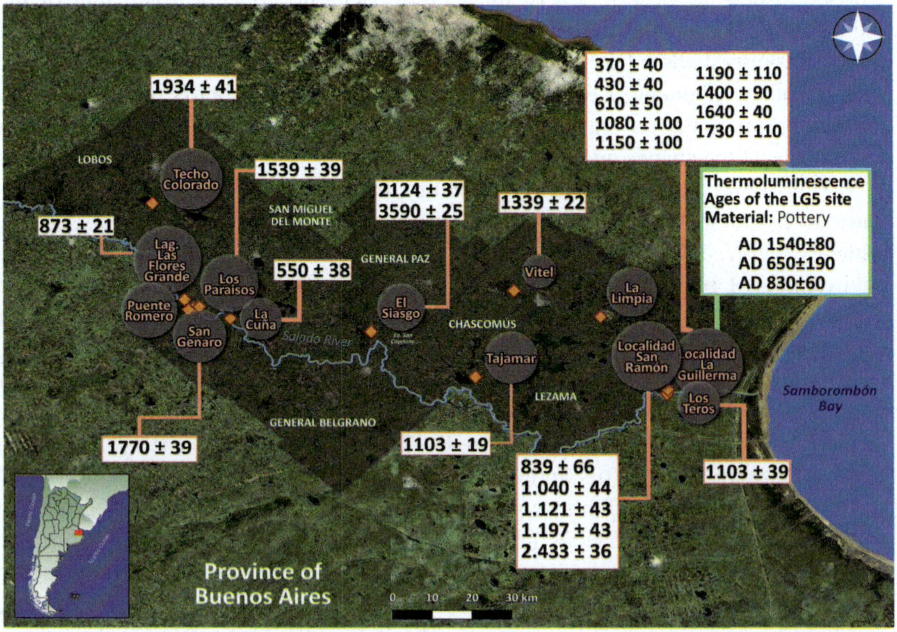

Fig. 1 Archaeological sites and chronology

closely interact with the agricultural production lands, associated with the cycles of the pluviometric regime. These are places where the presence of water concentrates a wide variety of biotic resources and constitutes communication nodes. This plain is characterized by the presence of two water bodies, the Samborombón and Salado Rivers, which flow into the Samborombón Bay among dunes and shell beds, a system open to the continental shelf. In this area, a great diversity of archaeological sites located near water bodies offers evidence of prolonged and recurrent human occupation starting in the Late Holocene. Here, the Salado River, large shallow lakes, hillocks, as well as tala, willow, and molle forests, constitute a wetland environment that represents an important supply of food, raw materials, and fuel.

The shallow lakes or "lagunas" that make up the Salado River system were used over long periods throughout the year. Its high topography prevented flooding and provided panoramic visibility and well-drained soils. They were located close to areas of the landscape that, during humid climatic intervals, offered biological wealth and enabled settlers to obtain faunal resources (Zárate, 2010). The abundance of archaeological findings indicates the intense and prolonged use of these sites: the presence of human remains, local pottery production, the manufacture of stone tools, fishing craft, and the use of plants typical of these environments.

The first archaeological studies in the chained shallow lakes of Chascomús began at the end of the nineteenth century and the beginning of the twentieth century. For example, Florentino Ameghino (1943) found pottery remains, arrowheads, and a spherical stone ball in the Vitel shallow lake. He considered that these technologies were very well developed. Other studies were conducted by Héctor Greslebin (1930), who published the description of the Ceferino Girado collection and that of his father, Emilio Greslebin, namely, materials collected in 1880–1881 on the coast of the Chascomús shallow lake, including the Manantiales, del Medio, and del Cacique shallow lakes. Clemente López Osornio (1942) studied potsherds recovered from the Samborombón River and the "loma de la laguna" in Chascomús and used them to reconstruct the shapes and designs of the pots.

In the second half of the twentieth century, I approached the study of the hunter-gatherer-fisher societies of the Salado by adopting a processualist perspective, which recognized the importance of adaptive relationships between humans and the environment. I began conducting fieldwork in the 1980s, within the interpretive frameworks of the new archaeology, in which these adaptive relationships played a central role. The processualist approach promoted, throughout the Pampas region, the study of spaces and topics that had been overlooked until then. New research questions were asked in the study of the hunter-gatherer groups that inhabited the region (Politis, 1984, 1988, 2000; Berón, 1997; Flegenheimer, 1986/1987; Salemme, 1997; Mazzanti, 1997). This approach also facilitated the creation of models and specific studies on the different types of recovered materials and promoted the use of interdisciplinary approaches (Zárate et al., 2002–2003). At the same time, it improved the recording and conservation techniques used during fieldwork. In our case study, these perspectives allowed us to gain new perspectives on the subject, highlighting, at the regional level, the peculiarities of the area related to the fishing

activity. This is how we proposed to refer to them as hunter-gatherer-fisher (HGF) societies and describe their mobility strategies.

In 1985, our first prospections and excavations revealed an abundant sample of pottery, lithic, and archaeofaunistic material and, to a lesser extent, isolated bone, human, plant, and charcoal remains. Over many years of work in this wetland, we gradually fulfilled different aspects of the research goals set. We carried out numerous prospecting works, particularly in the Salado River basin, in Chascomús district, and excavations in sites at La Limpia, Yalca, El Burro, Esquivel, and del Medio shallow lakes. In addition, we explored Manantiales, Las Flores Grande, and Los Cerrillos shallow lakes as well as El Siasgo stream, in the districts of San Miguel del Monte and General Belgrano. We also carried out systematic prospections in the Salado River basin in the Lobos and Roque Pérez districts.

We proceeded with different ceramic analyses and, at the same time, conducted experimental studies and applied archaeometric techniques (e.g., ceramic petrography and X-ray fluorescence) to compositionally characterize the ceramic and surface treatments, learn about the properties of the clayey raw materials used, correlate them with local outcrops, and infer clay preparation techniques and pottery-making methods.

What follows is a summary of our main contributions to the subject. This information contributed to the knowledge of the way of life of the HGF societies during the Late Holocene in this internal aquatic environment.

Chronology

We obtained the absolute ages for the microregion by analyzing different types of traces including dispersed charcoal from hearths, human bone remains, potsherds, fish bones, and sediment with organic matter from the matrix corresponding to the occupational levels. The analysis of mollusks made it possible to reconstruct the geological history of the site. Datings were performed using ^{14}C and the optically stimulated luminescence (OSL) methods (González, 2005). Calibrated radiocarbon ages indicate the presence of HGF for repeated periods (Frère et al., 2016b) from 2400 years BP. It is worth noting that the dating obtained for human remains from the El Siasgo stream shows the earliest evidence of human occupation of the area at 3500 years BP (Fig. 1).

Were there forests in the Buenos Aires plain? From an ecological point of view, the forests recognized by Parodi (1940a, b) occupied the north and northeast of Buenos Aires Province, reaching as far as the Salado River. He classified this forest land in the Salado area as undulating compact soil, covered with grasses and coarse subsoil. In pre-Hispanic times, at least 2000 years ago, the talar forest of the Salado Depression was used as a settlement site and the wood was used as fuel, as shown by anthracological studies of some charcoals recovered from the sites in the La Guillerma locality. The charcoal corresponded to wood from *Celtis tala*, *Schinus* sp., and *Salix* sp. (González, 2005; González & Frère, 2009). In the fluvial area of

Buenos Aires, the tala was considered a key element for the groups settled there (Páez et al., 1999; Pérez Meroni et al., 2010). Similarly, the anthracological studies carried out at the El Abra site in Tandilia (Brea et al., 2001) also showed that some charcoal was from tala (*Celtis* sp.), among other species, suggesting anthropic presence and use of plant resources during the Late Holocene.

In 1992, I undertook research from a geoarchaeological perspective, with valuable participation in the project of Dr. Marcelo Zárate. We adopted an interdisciplinary methodological approach focused on the methods and techniques used in Earth Sciences to solve archaeological problems. To date, the geomorphological and stratigraphic analyses have provided useful information to interpret the general localization of the La Guillerma sites and the San Ramón Río Salado site. Likewise, we have carried out datings of organic matter to analyze the results of the radiocarbon dates obtained and their archaeological and pedological significance. The organic matter samples from soil profiles provided information on the average residence time. We call them shallow sites because pedogenetic processes predominate in them, with intense bioturbation activity by microfauna and plant roots (Zárate et al., 2002–2003). These sites have a low degree of resolution since the soils represent time-transgressive surfaces that may contain diachronic cultural associations. For this reason, we had to control the evaluation of integrity and degree of resolution with archaeological indicators.

A special event was held at the La Guillerma archaeological site, as part of the fieldwork for a course on geoarchaeology delivered by Dr. Marcelo Zárate at the University of La Plata in 1993. This activity was intended to create a didactic experience and promote an interdisciplinary approach to the study of the archaeological locality (Salerno & González, 2014). The presence of experts, both archaeologists and geologists, favored the discussion in the field about the formation of the sites and the characteristics of the materials, particularly those related to ceramics (Fig. 2). The discussion on the landforms of these landscapes was especially interesting because it gave rise to the concept of "shallow site." This category was later discussed in a specific report (Zárate et al., 2002/2003). In addition, some of the archaeologists were researchers from Northwest Argentina with considerable experience in the analysis of ceramics from that region, which favored discussion about different aspects of the manufacture, finishing, and decoration of Pampean ceramic materials.

In our research, we decided to apply the conceptual framework of experimental archaeology because we considered that it was adequate for the study of archaeological artifacts. We applied it to the study of both the lithic and ceramic contexts (Vigna, 2010; Weitzel, 2012; Francese et al., 2011; González, 2005) and the bone remains (Escosteguy, 2020) recovered from the excavations. We also conducted actualistic studies in ceramic vessels on wild resources using chemical analysis (Frère et al., 2010). Particularly, the study of ceramics involved experimentation conducted as a work program since we were able to contact local ceramists with whom we proposed and contrasted hypotheses through the construction of current references. The long-term objective of this project is to find answers to the questions

Fig. 2 Experts, archaeologists, and geologists visited La Guillerma in 1993

we posed about the archaeological remains, tangible evidence that comes to us from the past social scene.

Human Remains

Given our interest in feeding practices, we examined the available isotopic studies on human bone remains and analyzed the isotopic values of the most abundant resources in the archaeological sites studied. The scarcity of data for the Salado River Depression led us to consider the isotopic values of archaeofauna from Paraná Delta sites that had a similar chronology (Escosteguy et al., 2017a). We took into account those samples for which information on δ13Ccol and δ15N was available. We selected six samples of pre-Hispanic human individuals obtained from donations made by residents of the area and from archaeological recovery. As a result of fieldwork in the microregion, three sites with human remains are known (La Guillerma Ñandú, La Guillerma 1, and La Guillerma 5), representing a minimum number of six individuals, all of them buried in isolation. We complement the preceding information with the bioarchaeological study of the skeleton recovered by Carlos Ameghino in the Siasgo stream at the beginning of the twentieth century. For these sites, four radiocarbon datings of human skeletons indicate that the burials took place in the Late Holocene (González, 2005; Scabuzzo & González, 2007; Frère et al., 2016b; Escosteguy et al., 2017b). In the microregion of the Salado River

Depression, paleodietary studies reflected temporal differences in feeding: the values of La Guillerma Ñandú, more depleted in δ13C, indicate the consumption of C3 plants and terrestrial herbivores that consume these vegetables, while the data from the later skeletons, more enriched in δ13C, would be consistent with the consumption of small herbivores (such as coypu) and diets based on C4 plants (Scabuzzo & González, 2007). The isotopic results obtained indicate that the diet of the individual from El Siasgo was mixed and based on herbivores that consumed C3 vegetables, freshwater fish, birds, and the complementary intake of marine food (see discussion in Escosteguy et al., 2017b).

We have recently begun studies on the content of plant micro-remains in human dental tartar from pre-Hispanic individuals. The first results of the analysis indicate that they are highly fertile, with a considerable amount of starch grains and phytoliths. Corn was identified among the starch grains, and the phytoliths are forms assignable to grasses. This is a preliminary result, but its study may have great potential (Colobig, pers. comm., April 2022).

Ceramic Technology

Regarding the ceramic technology used in the area, I would like to highlight two manufacturing techniques, namely, the use of modeling and the coiling technique. The pottery-making technique of overlapping clay rolls was inferred from macroscopic indicators, such as horizontal grooves on the surfaces (seams) and thickening and thinning on potsherd profiles (Rye, 1994: 67–68; Balfet et al., 1992). As regards modeling, I highlight the presence of handles, a figurine foot, and the particular case of a vessel edge (Fig. 3). Changes in thickness are observed along the edge, where the modeling gesture begins, and then the ceramic object changes its morphology and takes a different direction that cannot be accurately defined due to the size of the potsherd. The modeling of the internal face accompanies the morphology of the external face. The curve shows the modeling, and there are traces of red paint on both sides. On the external face of the piece, a cut or groove appears to have been made with an instrument; it is wider at the bottom, varies in width towards the top, and looks worn. It could be inferred that the vessel, of which this fragment was a part, was used with a holding element, for a specific use linked to the shape of the vessel, possibly with a pouring spout. The potsherd profile, as well as the samples studied by our group, shows that the clay was fired at a low temperature, less than 800 °C (González, 2005; Rivas González et al., 2021).

Based on different archaeometric analyses, we advanced our understanding of how hunter-gatherer-fisher populations used pottery vessels in the lower Salado River basin during the Late Holocene. For over 15 years, we have conducted research on the greasy residue on ceramic pots through the chemical technique of gas chromatography, the analysis of the diet through stable isotopes in human bone remains, and studies of isotopic ecology from archaeological resources of the microregion. We integrated these analyses and added the first results of food

Fig. 3 Potsherd: (**a**) external face and (**b**) internal face

processing studies based on stable isotope studies (δ13C and δ15N) performed on potsherds (González, 2005; Frère et al., 2016a; González & Frère, 2017). The results of the stable isotope study showed that the vessels were used for preparing different recipes with a variety of ingredients from internal aquatic resources such as small- and/or medium-sized herbivores, fish, and vegetables. So far, the use of these independent lines of analysis has strengthened the inferences made about the resources prepared and consumed in the microregion. Isotope analyses in ceramic potsherds combined with studies already carried out on ceramic technology, greasy residue on pottery, archaeological and current fauna resources, as well as isotope studies on human remains, have conferred greater precision to generate consistent archaeological interpretations. The results of the stable isotope study on potsherds showed that the pots may have been used to prepare meals containing proteins from internal aquatic animals. Different recipes with a variety of ingredients of animal origin were probably used in the preparation of food, which shows the decisions and culinary choices made by the inhabitants of the Salado River Basin.

This is how in the isotopic results obtained, the δ13C values of freshwater fish, terrestrial animals consuming C3 plants, and aquatic birds tend to overlap. The isotopic values found for the pottery confirm that heterogeneous mixtures of these foods were prepared in the vessels. However, the isotopic values of the vessels studied do not clearly reveal coypu consumption, although the archaeofaunistic and paleodietary data, as well as the chemical studies of greasy residue on pottery, indicate a predominant consumption of this small herbivore (Frère et al., 2010, 2016a).

Lithic Technology

Various explanations of the material culture were proposed after the 1980s. The first interpretations were made from the technological organization framework (Flegenheimer et al., 1995). Also, conciliatory positions were adopted by combining theoretical and methodological trends.

I emphasize the influence of the works of Kopytoff and Appadurai, who argued that objects cannot be understood simply as fixed and static, but their meaning is transformed by their accumulated biographies (Kopytoff, 1991), and things can be viewed as active and socially alive through situations such as exchange (Appadurai, 1986).

Using this theoretical framework, we studied the weapon system for the Río Salado Depression. We have recovered 15 spearheads from different sites (La Guillerma 1, La Guillerma Ñandú, San Ramón 5, San Ramón 7, and Techo Colorado), all chronologically located in the Late Holocene (Vigna et al., 2014; González et al., 2021), except for a medium-sized/large piece, interpreted as a throwing spear or dart tip with an atlatl. The rest of the pieces surveyed are small and were possibly used with a bow and arrow. The spearheads were made from flakes as base forms, mostly from the Sierras Bayas Group (SBG), quartzite, and, in smaller proportions, ftanite and silicified dolomite.

We determined that the lithic resources unavailable in the microregion were transported from quarries located at different distances. Also, techniques were used for the exhaustive exploitation of the rock, such as bipolar cutting and intentional breakage of artifacts (González de Bonaveri & Horovitz, 1991; Flegenheimer et al., 1995; González, 2005; González de Bonaveri, 2002; Weitzel & González, 2012; Frère, 2015). Regarding these raw materials, it was observed that the quartzite used corresponds to the varieties from Tandilia, predominantly from the SBG. In addition, quartzite rocks from the Balcarce Formation and others from the Tandilia system were identified, such as ftanites (chert) and ocher. Also, raw materials from the Atlantic coast were found. These are coastal boulders that were brought to the sites by anthropic transport. Silicified limestone was also found, and its provenance was assigned to the territory of Uruguay and the Argentine province of Entre Ríos, as well as Meseta del Fresco (La Pampa). Petrographic sections were made in two specimens belonging to LG5 and one to LG2 (Fig. 1), and their study indicated the presence of diagnostic elements assimilable to the Queguay deposits in Uruguay (González, 2005; Zárate et al., 2002–2003). The "boleadoras" (rock balls) were made from diabase, granite, and tonalite, orthoquartzite from the Balcarce Formation, amphibolites, metaquartzite, muddy sandstone, and siltstone (Vecchi & González, 2018). Likewise, a chrysocolla bead (a copper mineral) was found at the LG1 site, and two discoidal fragments of sumptuary stones were found at the SR5 and LG1 sites. These pieces were not made at the site since no signs of the manufacturing steps were identified; they are necklace beads made from raw materials from a distant source (González et al., 2007).

Fauna: Small Prey and Fish

The human groups that inhabited the Salado River basin were mainly devoted to gathering, hunting, and fishing. They were hunter-gatherer-fisher populations with residential and logistical mobility, which may have conditioned the planned and effective time of occupation in each of the sites, and their use of specific sectors of the landscape may have been somewhat intensive (González, 2005). Considerable evidence of reduced mobility is shown by the type of raw materials and resources that appear at the sites. In all the cases analyzed, the highest percentage corresponds to resources easily available in the vicinity: game animals, gathered products, and raw materials for pottery-making. The animal remains identified at the sites correspond almost without exception to the local wildlife, especially small animals such as coypus, overo lizards, armadillos, birds, and fish. Part of the significance of these resources can be understood from models that predict the conditions under which hunter-gatherer-fisher societies may or may not have been able to intensify and the set of resources that may have triggered intensification (Binford, 2001; Johnson & Hard, 2008; Corbat et al., 2017). In the Salado wetland, the archaeofaunistic remains provided the opportunity to evaluate the incidence of minor species. We observed that the economic strategy emphasized the use of small- and medium-sized fauna associated with internal aquatic environments. This is supported by the discovery of numerous remains of birds of the families Anatidae, Tinamidae, Ardeidae, and Rallidae, the last one being the most widely represented, through *Fulica armillata* (red-gartered coot) and *Fulica leucoptera* (white-winged coot). Other species recorded were *Anas platalea* (red shoveler), *Coscoroba* sp. (coscoroba swan), *Chloephaga* sp. (bustard), and *Eudromia elegans* (elegant crested martineta). A very low proportion of skeletal remains exhibited thermal alterations and cut marks. Long bone diaphyses were used as instrument support, and thus bone technology was part of the integral use of the area's faunal resources (Escosteguy et al., 2017a, b). Among rodents, with the species *M. coypus* (coypu) being the most abundant, direct evidence of anthropic exploitation was recognized in cut marks associated with the different stages of exploitation of the prey, as well as fresh fractures and thermal alterations (González, 2005; Escosteguy, 2011; Escosteguy et al., 2012). Although different species of smaller rodents were identified, such as *Cavia aperea* (wild cavy), *Ctenomys* sp. (tuco-tuco), *Holochilus brasiliensis* (Brazilian marsh rat), and *Reithrodon auritus* (bunny rat), only occasional consumption of *C. aperea* was proposed due to the presence of heat-altered elements, anatomical abundance, the context of discovery, and the lack of evidence of damage caused by other predators or natural death in situ (González de Bonaveri, 2002).

When severe flooding events occur in the main rivers of the Río de la Plata basin, especially the Paraná, their influence reaches the Samborombón Bay. This phenomenon transforms the Salado River into a temporary corridor for the ichthyofauna and allows the movement of certain fish such as the mullet (*Mugil* sp.) (López et al., 2001: 13) and the southern black drum (*Pogonias courbina*). From a radiocarbon dating on fish bone (toad catfish) that yielded an age of 1340 ± 40 BP (GX-26477

AMS), we were able to initiate the discussion about the role of fishing in this wetland of the Salado River. It should be noted that this chronology is consistent with radiocarbon dates obtained for the same LG5 site (Fig. 1), for charcoal remains and, by the OSL technique, for pottery remains (González, 2005). A mixed ichthyological association has been recorded in these water bodies, for both freshwater and marine (euryhaline) taxa. Estuarine species such as *Mugil* sp. (mullet) and *Pogonias courbina* (southern black drum) were identified. Other taxa include *Rhamdia quelen* (South American catfish), *Hoplias argentinensis* (tararira), *Australoheros facetum* (chameleon cichlid), *Corydoras paleatus* (peppered corydoras), *Pimelodus maculatus* (spotted catfish), *Hypostomus commersoni* (vieja de agua), and *Oligosarcus jenynsii* (dientudo). Recently, *Synbranchus marmoratus* (swamp eel) was determined by Sergio Bogan.[1] The size and weight of some specimens were estimated from the study of their vertebrae and spines. The black drum was documented in numerous assemblages along the coast of the Río de la Plata estuary, in both Argentina and Uruguay (Politis & Leon, 2010; Beovide, 2011). This fish resource, which can weigh up to 30 or 40 kg, is a great source of protein. The swamp eel was also recognized in the lowlands of the Laguna Merín basin, Uruguay (Bica-Mendez, 2020), which shows the anthropogenic consumption of this resource at the macroregional level during the Late Holocene.

Closing Remarks

The model proposed for the human occupations of the Salado River, Buenos Aires Province, shows a pattern of partial sedentarism characterized by the occupation of residential bases, reduced mobility or periodic redundancy, and longer stays. This is articulated with the sporadic, less intensive use of places, whose contexts differ in the amount and variation of material culture (González, 2005). Based on these results, I will mention some of the topics and projects deriving from the studies conducted in the Salado River, Buenos Aires. This research has benefited from the incorporation of Dr. María Magdalena Frère. Her contribution has been very valuable, particularly in the research on ceramic technology and archaeological heritage and legislation. Equally important is her work as joint director of the archaeological project (2020-3-A-139-1) of the Provincial Directorate of Museums and Heritage Preservation of Buenos Aires Province, carried out since 2007. My opportunity to direct four UBACyT projects (SECyT UBA) and three PICT projects (ANPCyT) since the year 2003 made it possible to increase archaeological information at local and regional scales, thus consolidating a research group that studies the use of resources and technology by hunter-gatherers in the Buenos Aires Pampas, encompassing the Tandilia range, southwestern Buenos Aires, and the Salado River Depression. This PICT team, led by Nora Flegenheimer, Cristina Bayón, and I,

[1] Área Paleontología Fundación de Historia Natural Félix de Azara, Universidad Maimónides.

leads a large group of young researchers, which has given a major boost to the research lines and new problems that have been incorporated into the agenda. In these fields of study, new radiocarbon dates were made, which allowed us to specify the chronology of the occupations.

Several young researchers and members of these projects obtained post-graduate scholarships from national organizations (PICT and UBACyT), and some of them were also able to start their teaching careers at the university. Among those who used information about the Salado River Depression for their projects, Paula Escosteguy works on archaeofauna and taphonomy; Virginia Salerno focuses on Public Archaeology; Celeste Weitzel and Rodrigo Vecchi worked on lithic technology using data from the Salado.

New fellows and thesis students were incorporated in the 2010s. Doctoral fellow Miranda Rivas González works on the characteristics of pottery firing by conducting macroscopic and archaeometric studies of archaeological and experimental materials and clays (Rivas González et al., 2021). In addition, doctoral fellow and Professor Alejandro Fernández has begun actualistic taphonomy studies, carried out experimental observations with earthworms, and evaluated the damage caused by insects on bone materials. Thus, termite marks have been identified for the first time in the faunal assemblages of the microregion (Escosteguy & Fernández, 2017; Escosteguy et al., 2021).

Olivia Sokol is developing a 3D virtualization project of materials from excavations and private collections, and Micaela Grzegorczyk is studying the collecting practices of metal detector users. This line of inquiry also yielded results that have contributed to the archaeological study of the materials in the microregion. Fernando Balducci compares the ceramic technologies of hunter-gatherer-fisher societies from the Northern Salado with the studies on the Salado of Buenos Aires.

When we started our archaeological investigations, we engaged the local residents in heritage care activities. We included high school students, who became propagators of archaeological research, and ceramists from a local college. In this way, we created a line of work on Public Archaeology (Salerno, 2011–2012; Salerno & González, 2014).

We were able to perform impact studies when the Comprehensive Master Plan was carried out for the Salado River basin in Buenos Aires Province, which began in the late 1990s (see Sokol et al. in this publication). This project aims to expand and channel the river to gain land and reduce the risk of flooding in the territory of the province. Currently, section IV of the project is being implemented, and since the end of 2019, the works have focused on the stages that reach the Chivilcoy district, in Buenos Aires Province. In this framework, the archaeological work aimed to articulate with different professionals participating in these works. Sections I, II, and III were developed without archaeological impact studies. During these first years of project execution, we were able to record the transformations in the landscape and the impacts of the work in sectors with archaeological sensitivity. Thus, in 2004, alterations were observed at the La Postrera site, where we had worked 2 years earlier. Among the modifications, we recognized the cutting of logging forests, large removals of sediments, and widening of roads. In addition, on that

occasion, ceramic material was recorded on the surface, as a result of these soil alterations. In 2016, our team returned to the site and continued documenting the tremendous changes in the landscape, including the removal of a ravine, among other incidents (González et al., 2017).

As a corollary, an integral part of our work involves communication and social dissemination of archaeological activity and scientific knowledge, since the past is inherent to archaeologists, public officials, teachers, and neighbors. We are all responsible for the care and protection of the material remains. In this way, the directors and technical staff of the Museo Pampeano, along with the teachers of Escuela de Cerámica de Chascomús, reformulated the Archaeological room, where educational activities are carried out for children and adults to disseminate information about the life of the pre-Hispanic settlers. Also, the teachers of the Chascomús ceramics school have made a box, namely, a teaching tool, with replicas of pottery material that is used during guided tours. This didactic box sometimes leaves the walls of the museum and visits the schools.

Acknowledgments The author wishes to thank the organizers of the VI Congress of Archaeology of the Pampean Region, especially Dr. Diana Mazzanti, for her invitation to deliver this conference and to present it in this publication. I am grateful to Cristina Bayón, Alicia Francese, Lidia Nacuzzi, and Magdalena Frère for their comments on previous versions of this work and to Olivia Sokol and Mariana Sabbatella for their help with maps and figures. The results obtained from this research are the product of the financial support granted continuously over time by the Secretaría de Ciencia y Técnica de la Universidad de Buenos Aires and Agencia Nacional de Promoción de la Investigación, el Desarrollo Tecnológico y la Innovación del Ministerio de Ciencia y Tecnología.

References

Ameghino, F. (1943). *La antigüedad del Hombre en el Plata* (2 tomos). Intermundo.

Appadurai, A. (1986). *The social life of things*. Cambridge University Press.

Balfet, H., Fauvet-Berthelot, M. F., & Monzón, S. (1992). *Normas para la descripción de vasijas cerámicas*. México, Centre D'Études Mexicaines et Centroaméricaines (CEMCA).

Barnard, A. (2014). Defining hunter-gatherers: Enlightenment, romantic, and social evolutionary perspectives. In V. Cummings, P. Jordan, & M. Zvelebil (Eds.), *The Oxford handbook of the archaeology and anthropology of hunter- gatherers*. Oxford University Press.

Beovide, L. (2011). La presencia de cultígenos desde el quinto milenio en el registro del curso medio Platense: revisión y proyecciones. In *Avances y perspectivas en la Arqueología del Nordeste* (pp. 155–173). SantísimaTrinidad ST Ser Gráficos.

Berón, M. A. (1997). Mobility and subsistence in a semiarid environment: The Río Curacó Basin (La Pampa, Argentina). *Quaternary of South America and Antarctic Peninsula, 10*, 133–166.

Bica-Méndez, C. A. (2020). Peces y pesca en las tierras bajas de la Laguna Merín. In *Análisis de la ictiofauna recuperada en el sitio arqueológico CH2D01 (Rocha, Uruguay)*. Universidade Federal de Pelotas.

Binford, L. (2001). *Constructing frames of reference: An analytical method for archaeological theory building using ethnographic and environmental data sets*. University of California Press.

Brea, M., Zucol, A., & Mazzanti, D. (2003). Determinación de combustibles vegetales en Cueva El Abra, provincia de Buenos Aires. In Arqueología Argentina en los Inicios de un Nuevo Siglo, Actas del XIV Congreso Nacional de Arqueología Argentina, tomo 3 (pp. 693–700).XIV

Congreso Nacional de Arqueología Argentina, Rosario 2001 : 17 al 21 de septiembre de 2001. Facultad de Humanidades y Artes, Universidad de Rosario

Brea, M., Zucol, A., & Mazzanti, D. (2001). Determinación de combustibles vegetales en Cueva El Abra, provincia de Buenos Aires. In *Arqueología Argentina en los Inicios de un Nuevo Siglo, Actas del XIV Congreso Nacional de Arqueología Argentina, tomo 3* (pp. 693–700).

Ceruti, C., & González, M. I. (2007). Modos de vida vinculados con ambientes acuáticos del nordeste y pampa bonaerense de Argentina. *Relaciones de la Sociedad Argentina de Antropología, XXXII*, 101–140.

Cobb, H. (2014). Materials, biographies, identities, experiences: New approaches to materials in hunter-gatherer studies. In V. Cummings, P. Jordan, & M. Zvelebil (Eds.), *The Oxford handbook of the archaeology and anthropology of Hunter- Gatherers*. Oxford University Press.

Corbat, M., Zangrando, F. J., Gil, A. F., & Chiavazza, H. D. (2017). Explotación de peces e intensificación en ambientes áridos: comparando el registro en humedales del centro-occidente de Argentina. *Latin American Antiquity, 28*(2), 196–212.

Escosteguy, P. (2011). *Etnoarqueología de nutrieros. Una propuesta metodológica aplicada al registro arqueológico de la Depresión del Salado y del Noreste de la provincia de Buenos Aires* [Unpublished doctoral thesis] (p. 308). Facultad de Filosofía y Letras, Universidad de Buenos Aires.

Escosteguy, P. D. (2020). The experimental butchering of coypu (Myocastor coypus): Implications for the analysis of the archaeofaunal record. *Journal of Archaeological Science: Reports, 31*(102330). https://doi.org/10.1016/j.jasrep.2020.102330

Escosteguy, P., & Fernández, A. E. (2017). The effects of bioturbation by earthworms. Preliminary results of an actualistic taphonomy experiment. *Journal of Taphonomy, 15*, 11–17.

Escosteguy, P., Salemme, M., & González, M. I. (2012). Myocastor coypus ("coipo", Rodentia, Mammalia) como recurso en los humedales de la Pampa bonaerense: patrones de explotación. *Revista del Museo de Antropología, 5*, 13–30.

Escosteguy, P., González, M. I., & Frère, M. M. (2014). La fauna menor de la localidad arqueológica San Ramón (Pampa bonaerense, Argentina). *Archaeofauna, 24*, 295–313.

Escosteguy, P., Salemme, M., & González, M. I. (2017a). Tecnología ósea en la Depresión del Río Salado (provincia de Buenos Aires). *Arqueología, 23*(3), 65–90.

Escosteguy, P., Scabuzzo, C., & González, M. I. (2017b). Análisis bioarqueológico de los restos del Arroyo El Siasgo, (supuesto Homo caputinclinatus de Ameghino, 1910). *Revista Argentina de Antropología Biológica, 19*(2), 4. https://doi.org/10.17139/raab.2017.0019.02.04

Escosteguy, P. D., Fernández, A. E., & González, M. I. (2021). Insect marks on bones from La Guillerma Archaeological Locality (Salado River Depression, Buenos Aires, Argentina). *Quaternary, 4*(4), 45. https://doi.org/10.3390/quat4040045

Flegenheimer, N. (1986–1987). Excavaciones en el sitio 3 de la localidad arqueológica Cerro La China (provincia de Buenos Aires). Relaciones, *XVIII*(1), 7–27.

Flegenheimer, N., Bayón, C., & de Bonaveri, M. I. G. (1995). Técnica simple, comportamientos complejos: la talla bipolar en la Arqueología bonaerense. *Relaciones de la Sociedad Argentina de Antropología, XX*, 81–110.

Francese, A., Migueliz, G., & Sabbatella, M. (2011). *Cerámica y Arqueología. Producción interdisciplinaria del conocimiento*. Instituto Superior de Formación Artística, Escuela de cerámica de Chascomús, Buenos Aires Province.

Frère, M. M. (2015) *Tecnología cerámica de los cazadores-recolectores-pescadores de la microrregión del río Salado, provincia de Buenos Aires* [Unpublished doctoral thesis] (p. 322). Facultad de Filosofía y Letras, Universidad de Buenos Aires.

Frère, M. M., Constenla, D., Bayón, C., & González, M. I. (2010). Estudios actualísticos sobre recursos silvestres mediante el empleo de análisis químico. In M. Berón, L. Luna, M. Bonomo, C. Montalvo, C. Aranda, & M. Carrera Aizpitarte (Eds.), *Mamül Mapu pasado y presente desde la arqueología pampeana* (pp. 215–226). Libros del Espinillo.

Frère, M. M., González, M. I., Chan, D., & Flores, M. (2016a). Evidencias químicas de preparación de alimentos en la alfarería de la depresión del río Salado. *Anuario de Arqueología, 8*, 141–151.

Frère, M. M., González, M. I., & Greco, C. (2016b). Continuity in the use of shallow sites of the Salado River Basin in the Pampean Region, Argentina. *Radiocarbon, 58*(4), 921–933. https://doi.org/10.1017/RDC.2016.63

González, M. I. (2005). *Arqueología de alfareros, cazadores y pescadores pampeanos* (Colección). Sociedad Argentina de Antropología.

González de Bonaveri, M. I. (2002). *Los cazadores-recolectores-pescadores de la cuenca inferior del río Salado (Región Pampeana)*. Facultad de Filosofía y Letras, Universidad de Buenos Aires.

González de Bonaveri, M. I., & Horovitz, L. (1991). Desechos de talla del Sitio L.G.1, Partido de Chascomús, Provincia de Buenos Aires. *Shincal, 3*(2), 52–63.

González, M. I., & Frère, M. M. (2009). Talares y paisaje fluvial bonaerense: arqueología del río Salado. *Intersecciones en Antropología, 10*, 249–265.

González, M. I., & Frère, M. M. (2017). Análisis de isótopos estables en cerámica arqueológica del río Salado Bonaerense. *Comechingonia, 21*(2), 229–254. https://doi.org/10.37603/2250.7728.v21.n2.26782

González, M. I., Frère, M. M., & Fiore, D. (2007). Redes de interacción en la cuenca inferior y media del Salado. In C. Bayón, A. Pupio, M. I. González, N. Flegengheimer, & M. M. Frère (Eds.), *Arqueología en las Pampas* (1st ed., pp. 365–384). Sociedad Argentina de Antropología.

González, M. I., Escosteguy, P. D., & Frère, M. M. (2016). *Estudio ictioarqueológico y la presencia de corvina negra en La Guillerma 5 (Depresión del río Salado, provincia de Buenos Aires)*. III Encuentro Latinoamericano de Zooarqueología.

González, M. I., Frère, M. M., Salerno, V., & Escosteguy, P. D. (2017). Estimaciones del impacto arqueológico de las obras hidráulicas en la cuenca del río Salado. In *Simposio internacional de patrimonio cultural y comunidad local. Buenas prácticas de gestión*.

González, M. I., Escosteguy, P. D., Salemme, M. C., Frère, M. M., Weitzel, C., & Vecchi, R. (2021). Assessing strategies for coypu hunting and use in the Salado River Depression (Buenos Aires Province). In *Ancient hunting strategies in Southern South America* (pp. 59–81). Springer. https://doi.org/10.1007/978-3-030-61187-3_3

Greslebin, H. (1930). In R. Dorcasberro (Ed.), Algunos datos sobre la arqueología del partido de Chascomús (pp. 29–47), Compilado por R. Doscaberro. Edición única realizada en el 150 aniversario de la fundación de Chascomús Municipalidad de Chascomús.

Johnson, A., & Hard, R. (2008). Exploring Texas archaeology with a model of intensification. *Plains Anthropology, 53*, 137–153.

Kopytoff, I. (1991). La biografía cultural de las cosas: la mercantilización como proceso. In A. Appadurai (Ed.), *La vida social de las cosas. Perspectiva cultural de las mercancías* (pp. 89–122). Grijalbo, CONACULTA.

López Osornio, M. (1942). Paraderos Querandíes. Contribución al estudio de la historia de los indígenas del país. *Cuadernos Folclóricos, 5*, 2–45.

López, H. L., Baigún, C. R. M., Iwaszkiw, J. M., Delfino, R. L., & Padin, O. H. (2001). *La Cuenca del Salado: Uso y posibilidades de sus recursos pesqueros*. Editorial de la Universidad de La Plata.

Mazzanti, D. L. (1997). Archaeology of the eastern edge of the Tandilia Range (Buenos Aires, Argentina). *Quaternary of South America and Antarctic Peninsula, 10*(1994), 211–227.

Páez, M. M., Paleo, M. C., Meroni, M. P., & Pastorino, S. (1999). El bosque de tala como recurso potencial: interpretación arqueológica y palinológica. In *Resúmenes del XIII Congreso Nacional de Arqueología Argentina* (pp. 369–370). Córdoba.

Parodi, L. R. (1940a). La distribución geográfica de los talares en la provincia de Buenos Aires. *Darwin, 4*(1), 33–56.

Parodi, L. R. (1940b). Los bosques naturales de la provincia de Buenos Aires. *Anales de la Academia Nacional de Ciencias Exactas, Físicas y Naturales, 7*, 79–90.

Pérez Meroni, M., Paleo, M. C., Pochettino, M. L., & Lema, V. S. (2010). Procesamiento y consumo de vegetales por grupos cazadores-recolectores del Holoceno tardío, en los partidos de Magdalena y Punta Indio, provincia de Buenos Aires. In M. Berón, L. Luna, M. Bonomo, C. Montalvo, C. Aranda, & M. C. Aizpitarte (Eds.), *Mamül Mapu: Pasado y presente desde la arqueología pampeana* (pp. 215–226). Editorial Libros del Espinillo.

Politis, G. (1984). *Arqueología del Área Interserrana Bonaerense* [Unpublished doctoral thesis, 2 volumes]. Facultad de Ciencias Naturales y Museo, Universidad Nacional de La Plata.

Politis, G. (1988). Paradigmas, modelos y métodos en la arqueología de la Pampa bonaerense. In *Arqueología Contemporánea Argentina. Actualidad y perspectivas* (pp. 59–107). Búsqueda, editors.

Politis, G. (2000). Los cazadores de la llanura. In *Nueva Historia Argentina. Los pueblos originarios y la conquista* (pp. 61–103). Editorial Sudamericana.

Politis, G., & Barros, P. (2006). La región pampeana como unidad espacial de análisis en la arqueología contemporánea. *Folia Histórica del Nordeste, 16*, 51–73.

Politis, G., & Leon, D. C. (2010). Patrones adaptativos de los cazadores-recolectores pescadores de la margen occidental del Paraná inferior-Plata. In G. Cocco & M. R. Feuillet Terzaghi (Eds.), *Arqueología de cazadores recolectores en la cuenca del Plata* (pp. 63–86). Centro de Estudios.

Politis, G., & Madrid, P. (2001). Arqueología Pampeana: estado actual y perspectivas. In *Historia Argentina Prehispánica. Tomo II* (pp. 737–814). Brujas.

Prates, L., & Bonomo, M. (2017). Los ambientes acuáticos en arqueología. *Arqueología, Dossier, 23*(3), 11–33.

Rivas González, M., Conconi, S., Frere, M. M., & González, M. I. (2021). Estimación de temperaturas de cocción de alfarerías prehispánicas en el río Salado, Pampa Argentina. *Cuadernos del Instituto Nacional de Antropología y Pensamiento Latinoamericano, 30*(2), 25–40.

Rye, O. (1994). *Pottery technology. Principles and reconstruction* (Manuals on archaeology 4). Taraxacum.

Salemme, M. (1997). Progress in the archaeology of the Pampean Region (Argentina, South America). An up-to-date overview of recent investigations. *Quaternary of South America and Antarctic Peninsula, 10*, 121–132.

Salerno, V. (2011–2012). *Trabajo arqueológico y representaciones del pasado prehispánico en Chascomús* [Unpublished doctoral thesis]. Facultad de Filosofía y Letras, UBA.

Salerno, V., & González, M. I. (2014). Conocimiento en relación. Reflexiones sobre el trabajo de campo arqueológico en el curso medio e inferior del Río Salado bonaerense. *Revista del Museo de Antropología Universidad Nacional de Córdoba, 7*(1), 25–38.

Scabuzzo, C., & González, M. I. (2007). Un acercamiento a la dieta de las poblaciones prehispánicas de la depresión del Salado durante el Holoceno tardío. In C. Bayón, A. Pupio, M. I. González, N. Flegenheimer, & M. M. Frère (Eds.), *Arqueología de las Pampas, tomo I* (pp. 59–73). Sociedad Argentina de Antropología.

Vecchi, R., & González, M. I. (2018). Artefactos formatizados por picado, abrasión, pulido y modificados por uso en la Depresión del río Salado, provincia de Buenos Aires, Argentina. *Revista del Museo de Antropología, 11*(2), 85–100.

Vigna, M. (2010). Experimentación con artefactos de ortocuarcita: implicaciones para el análisis de los conjuntos líticos de la microregión del río Salado, Región Pampeana, Argentina. *Boletín de Arqueología Experimental, 8*, 37–49. UAM.

Vigna, M., González, M. I., & Weitzel, C. (2014). Los cabezales líticos de la microrregión del río Salado bonaerense, Argentina. Diseños e historias de vida. *Intersecciones en Antropología, 15*, 55–69.

Weitzel, C. (2012). Cuentan los fragmentos. Clasificación y causas de fractura de artefactos formatizados por talla. *Intersecciones en Antropología, 13*(1), 43–55.

Weitzel, C., & González, M. I. (2012). Técnicas y decisiones sociales para el empleo de rocas. Nuevos datos para La Guillerma 5 (área del río Salado, Región Pampeana). *Relaciones de la Sociedad Argentina de Antropología, 37*(2), 243–264.

Zárate, M. (2010). El paisaje pampeano a través del tiempo. In M. Berón, L. Luna, M. Bonomo, C. Montalvo, C. Aranda, & M. Carrera Aizpitarte (Eds.), *Mamül Mapu: pasado y presente desde la arqueología pampeana* (pp. 215–226). Libros del Espinillo.

Zárate, M., González de Bonaveri, M. I., Flegenheimer, N., & Bayón, C. (2002–2003). Sitios arqueológicos someros: el concepto de sitio en estratigrafía y sitio de superficie. *Cuadernos del INAPL, 19*, 635–653.

New Isotopic Analysis of the Los Tres Cerros 1 and Cerro de las Pajas Blancas 1 Archaeological Sites (Paraná River Delta, Argentina)

Clara Scabuzzo, Rodrigo Costa Angrizani, and Mariano Bonomo

Abstract In the last 15 years, our scientific team has characterized the pre-Hispanic subsistence in the Upper Delta of the Paraná River (Argentina) by performing different archaeological analyses. These included the study of artifacts (pottery, lithic, and bone tools), archaeobotanical and archaeofaunal remains, and stable isotopes. Integrating this information allowed us to confirm that the indigenous populations linked to the Goya-Malabrigo archaeological entity had a mixed economy that combined hunting, fishing, and gathering of wild resources with horticulture. The main objective of this chapter is to discuss the human diet in the pre-Hispanic settlements of the Upper Delta during the Late Holocene. To achieve this objective, the results of $\delta^{13}C_{col}$ and $\delta^{15}N$ analyses of the human bones of 11 individuals from two archaeological sites and 14 samples of currently cultivated crops are presented. Finally, using a multivariate diet reconstruction model, which incorporates $\delta^{13}C_{ap}$, $\delta^{13}C_{col}$, and $\delta^{15}N$ values, human paleodiets were characterized. The results suggest the consumption of continental resources within a food substrate based mainly on C_3 plant producers and, in some cases, the inclusion of C_4 food. This variation in diet also appears to be determined by the site of provenance of the samples. Finally, these results lead us to discuss the importance of wild and domesticated plants for the Goya-Malabrigo entity. They also allow us to overcome the dualistic visions that oppose hunter-gatherer and farmer societies, showing the importance of mixed economies as successful social and economic strategies for life in the wetlands of the Paraná River.

C. Scabuzzo (✉) · R. C. Angrizani
CICYTTP-CONICET, Provincia de Entre Ríos-UADER, División Arqueología, Facultad de Ciencias Naturales y Museo, Universidad Nacional de La Plata. Dr. Materi y España (3105), Diamante, Entre Ríos, Argentina

M. Bonomo
CONICET- División Arqueología, Facultad de Ciencias Naturales y Museo, Universidad Nacional de La Plata. Paseo del bosque s/n° (1900), La Plata, Buenos Aires, Argentina
e-mail: mbonomo@fcnym.unlp.edu.ar

G. F. Bonnat et al. (eds.), *Current Research in Archaeology of South American Pampas*, The Latin American Studies Book Series,
https://doi.org/10.1007/978-3-031-55194-9_2

Keywords La Plata basin · Pre-hispanic diets · Goya-Malabrigo · Mixed economies

Introduction

The reconstruction of subsistence strategies developed throughout human history has been one of the main objectives of archaeology (Bonomo et al., 2019a). Despite changes in the theoretical paradigms of anthropology and archaeology, the relevance of subsistence has been the key to classifying human groups for more than 250 years (Pluciennik, 2001). Subsistence studies have mostly focused on identifying antagonisms between hunter-gatherers and farmers. However, it is now recognized that these practices are not necessarily mutually exclusive (Smith, 2011). Evidence from different parts of the world shows that human societies have combined a wide range of strategies to obtain wild and domesticated resources, as reflected in the formation of mixed economies (Kent, 1989).

In the last 15 years, archaeological research carried out by our team has provided information to characterize pre-Hispanic subsistence in the Upper Delta of the Paraná River, from different lines of research. These include the study of plant microremains and macroremains present in sediments, pottery, and grinding artifacts (Bonomo et al., 2011; Sánchez et al., 2013; Colobig et al., 2015, 2018), the analysis of archaeofaunal remains (Bastourre, 2014, 2018; Bonomo et al., 2019b), the study of oral health evidence (Ramos van Raap, 2018, 2020), and carbon and nitrogen stable isotope analyses (Bonomo et al., 2017; Salazar-García et al., 2018). Together, this information has led us to rethink the traditional view on the subsistence of the populations that occupied southern North Eastern Argentina during the Late Holocene, traditionally classified as hunter-gatherer-fisher groups (Ceruti, 2003). In this sense, the new data indicate that the indigenous populations of the Upper Delta of the Paraná River had a mixed economy in which wild resources were obtained through hunting, fishing, and gathering (of palms [*Arecaceae*], carob [*Prosopis* spp.], and wild rice [*Oryza* spp.]), combined with food production through the practice of horticulture of maize (*Zea mays*), squash (*Cucurbita* spp.), and beans (*Phaseolus* spp.) (Bonomo et al., 2011; Politis & Bonomo, 2012, 2018).

Particularly, stable isotope analyses applied to paleodietary issues have contributed to discussing the changes in subsistence patterns, the incorporation of agricultural practices, and the exploitation of marine environments, among others (Panarello et al., 2021). In recent years, in different regions, isotopic studies incorporated the multivariate model of Froehle et al. (2012). This made it possible to discuss terrestrial lipids in hunter-gatherer subsistence along the southern coast of Tierra del Fuego during the Late Holocene (Alunni et al., 2021), the paleodiets and subsistence patterns in hunter-gatherer groups of the eastern Pampa-Patagonia transition during the Middle Holocene (Flensborg et al., 2018), and general trends and intragroup variability in human paleodiet from pre-Hispanic hunter-gatherers located in the southwest of La Pampa province (Lynch Ianniello et al., 2020). In Northeastern

Argentina, isotopic studies constitute an independent line of information for evaluating the differences in the dietary patterns of different pre-Hispanic groups and the role of C_3 and C_4 vegetables in the diet (Loponte & Acosta, 2007; Loponte & Kozameh, 2009; Loponte et al., 2016; Ottalagano & Loponte, 2017; Bonomo et al., 2017; Salazar-García et al., 2018; Costa Angrizani & Scabuzzo, 2019).

In the case of plants, two factors have biased their study: on the one hand, the humidity conditions of the river and lake environments of the Paraná River have not favored the preservation of plant remains, except when they are carbonized; on the other hand, the lack of application of adequate recovery techniques, such as flotation, and analysis of microremains, such as starches and phytoliths, which began to be applied only a few years ago. This situation has led to reconstructions of the isotopic ecology of the area based on the values obtained from the different animal resources, with no reference to the isotopic values of the plants. To overcome this bias in the reference frameworks, we include $\delta^{13}C_{col}$ and $\delta^{15}N$ values of current crops from the region.

The aim of this chapter is to improve the knowledge of the human diet in the pre-Hispanic settlements of the Upper Delta of the Paraná River during the Late Holocene. To this end, the isotopic results of 11 individuals from two archaeological sites and 14 samples of currently cultivated crops are presented. In addition, the human paleodiets are characterized using a multivariate diet reconstruction model incorporating $\delta^{13}C_{ap}$, $\delta^{13}C_{col}$, and $\delta^{15}N$ values.

Ecological Setting

The Paraná River is one of the most important rivers in South America, and, together with the Uruguay and Paraguay rivers, it forms the La Plata Basin, which extends over 3,100,000 km^2. From its headwaters in central-eastern Brazil (at the confluence of the Rio Grande and the Paranaíba), this long river flows through Brazil, Paraguay, and Northeastern Argentina until it reaches the Río de la Plata. In Argentina, the Paraná River runs through the provinces of Misiones, Corrientes, Chaco, Santa Fe, Entre Ríos, and Buenos Aires. Along its course, it encounters a great variety of environments with diverse climatic and geological characteristics (Iriondo et al., 2007). Traditionally, this river basin has been divided into different sections, namely, Upper Paraná, Middle Paraná, and Lower Paraná. In the third sector, the river forms a deltaic environment, 320 km in length by an average of 60 km in width, made up of numerous islands, channels, and water sources that are flooded in different seasons with varying frequency, intensity, and duration (Malvárez, 1997).

The forests that extend along the banks of the Paraná maintain their own ecological characteristics in the different environments along the river path. The direction of water flow explains the richness and specific diversity of the deltaic ecosystem (Burkart, 1957; Malvárez, 1997). The following species are found in these forests: ambay pumpwood (*Cecropia adenopus*), dragon's blood (*Croton urucurana*), earleaf nightshade (*Solanum mauritianum*), yellow poinciana (*Peltophorum dubium*),

aguay (*Pouteria gardneriana*), *canelón* (*Myrsine laetevirens*), *white timbó* (*Pithecellobium multiflorum*), pacara earpod tree (*Enterolobium contortisiliquum*), *yellow laurel* (*Nectandra falcifolia*), *yellow lanza* (*Terminalia triflora*), queen palm (*Syagrus romanzoffiana*), Humboldt's willow (*Salix humboldtiana*), and cockspur coral tree (*Erythrina crista-galli*), among others (Cabrera, 1953). The grasslands of the region consist grasses of the following botanical subfamilies: Arundinoideae (*Arundineae* and *Aristideae* tribes), Bambusoideae, Chloridoideae (*Chlorideae, Eragrostideae, Pappophoreae, and Sporoboleae* tribes), Danthonioideae, Panicoideae (*Andropogoneae* and *Paniceae* tribes), Phalarideae, Pooideae (*Agrostidinae, Aveneae, Meliceae, Stipeae,* and *Triticeae* tribes), and Stipoideae (Burkart, 1957).

The fauna includes different mammals (marsh deer [*Blastocerus dichotomus*], pampas deer [*Ozotoceros bezoarticus*], brocket [*Mazama gouazoubira*], capybara [*Hydrochoerus hydrochaeris*], coypu [*Myocastor coypus*], maned wolf [*Chrysocyon brachyurus*], jaguar [*Panthera onca*], Geoffroy's Cat [*Oncifelis geoffroyi*], among many others), birds (different families, e.g., Phalacrocoracidae, Ardeidae, Anatidae, Podicipedidae, Rallidae, and Psittacidae), reptiles (e.g., Argentine giant tegu [*Salvator merianae*] and broad-snouted caiman [*Caiman latirostris*]), mollusks (*Diplodon* sp., *Ampullariidae* sp.), and a great abundance of fish with around 300 species of the orders Characiformes and Siluriformes (López, 2001). The Paraná Delta shows an important biological diversity, which offers a wide range of potential resources for the consumption of human populations. Most of the vegetation corresponds to the photosynthetic group C_3. This is related to the intermediate temperate climate temperatures and the adaptation of the C_3 group species to the flooded lowlands of the Delta (Madanes et al., 2015). In this sense, some studies show a high contribution of C_3 plants in the diet of herbivorous animals in the middle Paraná River area (Marchese et al., 2014).

Archaeological Entities and Archaeological Sites

Archaeological research carried out in Northwestern Argentina (NEA) established the existence of two main archaeological entities, Goya-Malabrigo and Guaraní. Both entities have well-differentiated characteristics in terms of chronology, technology and material culture, settlement patterns, burial modalities, and diet, among others (Bonomo, 2012).

The Goya-Malabrigo archaeological entity (analogous to the *Ribereños Plásticos* culture defined by Serrano (1950, 1972)) encompasses the archaeological record of the populations that occupied the riparian zones of the Middle and Lower Paraná and the lower basin of the Uruguay River from 2400 years BP to at least the seventeenth century. Until recently, these populations were characterized as hunter-gatherers and fishermen whose subsistence was based on the exploitation of a great diversity of wild resources typical of these fluvial environments. Among the animals consumed were mammals (*Myocastor coypus*, *Hydrochoerus hydrochaeris*, *Cavia*

aperea, *Ozotoceros bezoarticus*, and *Blastocerus dichotomus*), freshwater fish of the orders Characiformes (e.g., *Salminus brasiliensis*, *Leporinus obtusidens*, *Prochilodus platensis*, and *Hoplias malabaricus*) and Siluriformes (e.g., *Pterodoras granulosus*, *Pimelodus clarias*, *Rhamdia sapo*, and *Pseudoplatystoma corucans*) birds, and freshwater mollusks (Ceruti, 2003). Likewise, within this characterization, it was suggested that the Goya-Malabrigo groups, close to the conquest, expanded their subsistence through the incorporation of cultivated vegetables (Ceruti, 2003). This innovation may have reached the area with the arrival of Amazonian groups identified with the Guaraní tradition (e.g., Kurc, 1995; Rodríguez, 2004). At present, from paleobotanical remains recovered in archaeological excavations, we know that plant cultivation in the sites assigned to the Goya-Malabrigo entity dates back to about 1000-1750 years BP (Sánchez et al., 2013; Colobig et al., 2015, 2019). Therefore, the practice of horticulture in the area predates the arrival of the Guaraní by at least 300 years and can reach up to about 1000 years before (Bonomo et al., 2017).

The Guaraní archaeological entity corresponds to horticultural groups that would have arrived in the Paraná Delta and lower Uruguay Basin about 700 years BP (Bonomo et al., 2015). The subsistence of this group has been characterized by a complex agroforestry management system, which combined the exploitation of wild resources associated with horticulture of species such as maize (*Zea mays*), cassava (*Manihot esculenta*), beans (*Phaseolus* spp.), and squash (*Cucurbita* spp.) (Noelli, 1993; Noelli et al., 2003). In the Paraná Delta, the settlements related to this entity are located in the lower sector (e.g., Outes, 1918; Lothrop, 1932; Loponte et al., 2011; Bonomo, 2013; Costa Angrizani et al., 2015). However, in the Upper Delta, pottery vessels and sherds associated with the Guaraní pottery tradition have been found in contexts assigned to the Goya-Malabrigo entity (Politis et al., 2018).

The samples of human skeletal remains on which the isotopic studies were carried out were collected from two sites located in the northeast of Argentina, both associated with the Goya-Malabrigo entity. The Los Tres Cerros 1 locality (LTC1) is located in the Victoria County (Entre Ríos Province, Argentina), and the Cerro de las Pajas Blancas 1 site (CDLPB1) is located in the San Jerónimo County (Santa Fe Province, Argentina), both locations adjacent to the main channel of the Paraná River (Fig. 1).

LTC1 is part of an archaeological locality that comprises three human-made earthen mounds (Politis et al., 2011). LTC1 is an anthropic mound with semi-permanent occupations that were linked to the Goya-Malabrigo archaeological entity (Politis & Bonomo, 2012, 2018). The site occupations were dated to the Late Holocene, between 560 and 1220 uncalibrated years 14C BP, according to 21 radiocarbon dates (Politis et al., 2011; Scabuzzo et al., 2015). The abundant archaeological material recovered at the site defines it as a settlement where domestic and productive activities were developed. Besides, the bioarchaeological record indicates that the mound was used as a burial ground. The characteristic pottery of the Goya-Malabrigo entity was recovered at the site: vessels with incised decorative designs and zoomorphic appendages, and truncated cone-shaped pottery pieces modeled with bird head appendages called "*campanas*" (bells) (Fig. 2).

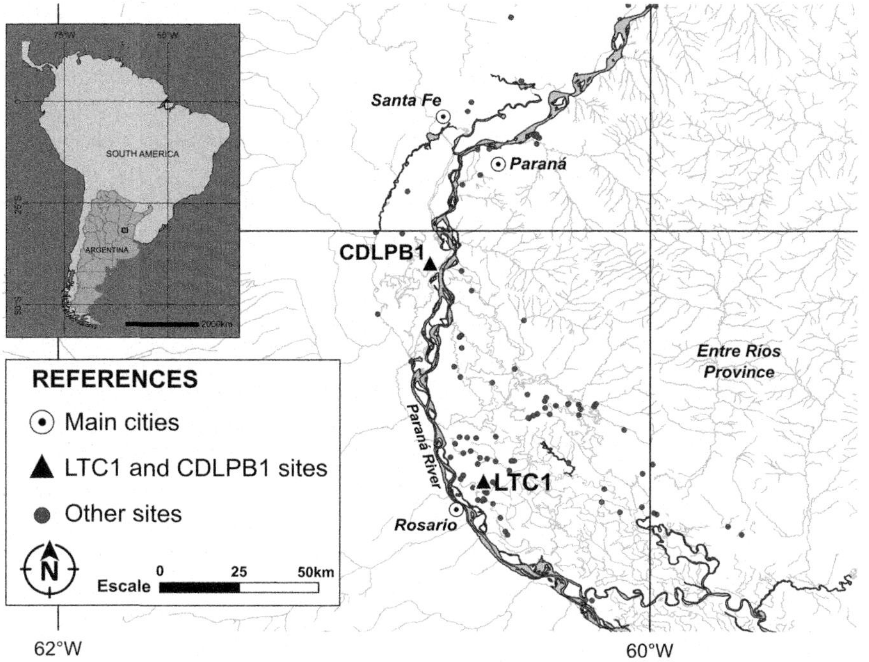

Fig. 1 Location of the Los Tres Cerros 1 and Cerro de Las Pajas Blancas 1 archaeological sites

The bioarchaeological analysis allowed us to establish the burial of at least 35 individuals, of different sex and age categories (Scabuzzo & Ramos van Raap, 2017; Ramos van Raap et al., 2019). The burial modalities were varied, including simple primary burials and simple and multiple funeral packages; bone accumulation and isolated bones were also recorded (Scabuzzo & Ramos van Raap, 2017). Paleopathological analyses of oral health and physiological fitness revealed that these populations had mixed diets, Fish was probably the most widely exploited resource, and the consumption of wild and domesticated vegetables had a complementary role in the diet. Likewise, different modes of food processing directly influenced oral health and dental wear (Ramos van Raap, 2018, 2020).

The CDLPB1 site corresponds to a site displaying occupations on a ridge. Although the geoform where the site is located originated from the natural cycles of fluvial deposition, anthropic interventions, such as the use of fire to consolidate the occupation floors, have been observed (Bonomo et al., 2019b). Based on the characteristics of the pottery styles found in recent excavations (Bonomo et al., 2019b), it is possible to infer that the site was initially occupied by Goya-Malabrigo groups. Likewise, the presence of two Guaraní vessels, one of which is a funerary urn, reported in the 1940s (Badano, 1940) and Guaraní pottery fragments (e.g., corrugated, unguiculate, brushed, and polychrome ceramics) recently recovered in the upper layers of the sedimentary sequence, could indicate that the site was occupied by these groups at later times (Fig. 3). So far, two radiocarbon dates point to

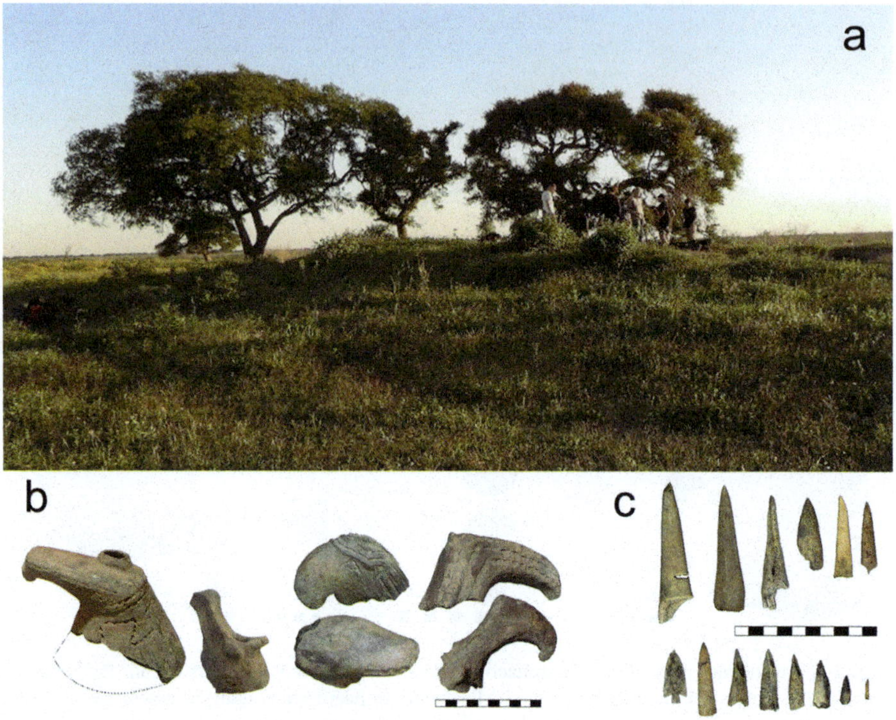

Fig. 2 Los Tres Cerros 1 archaeological site. References: (**a**) Los Tres Cerros 1 earth mound; (**b**) Pottery artifacts (*campanas* and zoomorphic appendages); and (**c**) Bone artifacts (harpoons, awls, and spearheads)

occupation of the site between 640 and 650 uncalibrated years 14C BP (Bonomo et al., 2019b). The bioarchaeological record is composed of four individuals with different states of preservation and integrity. One of them comes from a primary burial in an extended dorsal position with varied mortuary accompaniment (ear plugs, *tembetá* lip plug, and pendant made with feline tooth). Bioarchaeological analysis indicated that this skeleton corresponds to a young adult male individual. The other three individuals recovered are represented by mixed fragments and the burial modalities could not be inferred (Bonomo et al., 2019b). Based on the results of the analysis of different kinds of evidence recovered at this site (pottery, faunal and botanical remains, and human skeletons), it can be stated that various domestic activities related to food processing, the manufacture of artifacts made of different materials (pottery, bones, teeth, and shells), and ritual activities related to the burial of the dead were carried out at this site (Bonomo et al., 2019b).

From the zooarchaeological and paleobotanical studies conducted at both sites, some trends in terms of the resource exploitation and subsistence strategies adopted can be mentioned. Archaeofaunal analyses conducted at sites LCT1 (Bastourre, 2014, 2018) and CDLPB1 (Bonomo et al., 2019b) indicate a predominant

Fig. 3 Cerro de las Pajas Blancas 1 archaeological site. References: (**a**) Cerro de las Pajas Blanca 1 view; (**b**) Pottery associated with Goya-Malabrigo archaeological entity (vessels and zoomorphic appendages); (**c**) Body adornments (ear plugs, *tembetá* lip plug, and pendant made with feline tooth); (**d**) Pottery sherds associated with Guarani archaeological entity; and (**e**) Funerary urn reported by Badano (1940), and currently on display at the Museum of Natural and Anthropological Sciences "Prof. Antonio Serrano"

consumption of Characiformes and Siluriformes freshwater fishes. In addition, small (*Myocastor coypus*) and large rodents (*Hydrochoerus hydrochaeris*) are predominant, as well as cervids (*Blastocerus dichotomus*, *Ozotoceros bezoarticus,* and *Mazama* sp.) which are predominant among mammals, while reptilian, birds, and freshwater mollusks are also present.

Paleobotanical macro and microremain analyses of sediment samples and archaeological artifacts from both sites provide data on the management and consumption of vegetables in these settlements. Macroremains associated with the botanical families Solanaceae, Chenopodiaceae, Polygonaceae, Poaceae (Maideae and Oryzoideae), and Amaranthaceae (Colobig et al., 2015) and microremains related to *Zea mays*, Cucurbitaceae, and Oryzoideae (Sánchez et al., 2013) have been identified in the sedimentary samples from site LTC1. At the CDLPB1 site, microremain analyses have revealed the presence of *Phaseolus* sp. starch grains, *Zea mays* in ceramic sherds (Bonomo et al., 2011), and phytoliths related to *Zea mays* and Oryzoideae in sediments (Bonomo et al., 2019b; Colobig & Costa Angrizani, 2021). The macroremains detected at this site include charred grains of *Oryza* spp. and fragments of *Zea mays* corb, *Geoffroea decorticans,* and the families Amaranthaceae and Malvaceae (Bonomo et al., 2019b; Costa Angrizani et al., 2021).

The combined results of archaeofaunal and paleobotanical studies suggest the development of diversified subsistence strategies that formed mixed diets. In this sense, the capture of animals through fishing and hunting was complemented by the collection of wild plants (e.g., *Oryza* spp. and *Geoffroea decorticans*) and the cultivation of domesticated vegetables (*Zea mays*, *Phaseolus* spp., and Cucurbitaceae).

Samples and Methods

The human bone remains analyzed in this work are protected at Facultad de Ciencias Naturales y Museo (UNLP) and Laboratorio de Arqueología del CICYTTP (CONICET). The bioarchaeological studies were carried out following the Code of Ethics for the Study, Conservation, and Management of Human Remains of the Populations of the Past (Aranda et al., 2014) and National Law No. 25,743. Stable isotope analyses were performed on 12 samples, corresponding to a minimum number of 11 individuals, adults of both sexes (Table 1). Of the 12 samples, 10 were collected from Los Tres Cerros 1 and two from Cerro de las Pajas Blancas 1 (Table 1). The human remains analyzed correspond to Late Holocene inhumations.

A total of 32 isotopic data were provided by three different laboratories: The University of Cape Town (South Africa), the University of Georgia (USA), and the Laboratorio de Isótopos Estables en Ciencias Ambientales (Argentina). Some of the results have already been presented, while others are presented for the first time in this article. Isotope values in collagen and apatite were reported as the $^{13}C/^{12}C$ ratio in parts per thousand (‰) related to the Vienna-Pee Dee Belemnite standard (V-PDB) ($\delta^{13}C$) and $^{15}N/^{14}N$ in parts per thousand related to the AIR standard ($\delta^{15}N$) (Ambrose, 1990, 1993). In order to evaluate the quality of the collagen, the C:N ratio was calculated. Throughout the text, the guidelines for the publication of isotopic results of Samec et al. (2019) were followed. The samples presented for the first time in this article were analyzed and processed in the Laboratorio de Isótopos Estables en Ciencias Ambientales (IANIGLA-CONICET and UTN-FRSR), and a Flash 2000 Elemental Analyzer, coupled to Thermo Delta Advantage IRMS, was used. Hydroxyapatite analyses were carried out at the University of Wyoming Stable Isotope Facility (USA).

For eight samples of the human skeletal remains, $\delta^{13}C$ values of both fractions and $\delta^{15}N$ were available and, therefore, it was possible to calculate discriminant functions 1 and 2 proposed by Froehle et al. (2012) and apply a multivariate diet reconstruction model proposed by the same authors. The model is based on the holistic use of $\delta^{13}C$ isotope values of both fractions and $\delta^{15}N$ and on the calculation of discriminant functions to carry out the paleodietary reconstructions (Froehle et al., 2012: 353). This model proposes five clusters that differ in protein components and overall diet. In this sense, each of the defined clusters represents different types of diet according to the amount of C_3 and C_4 foods and marine proteins. Clusters 1 and 4 group diets based on C_3 food, although cluster 4 also includes 30% of C_4 food. On the other hand, clusters 2, 3, and 5 differ from the previous ones in

Table 1 Bioarchaeological and isotopic information on the individuals of Los Tres Cerros 1 (LTC1) and Cerro de las Pajas Blancas 1 (CDLPB1) sites

Site	Individual	Burial	Sex	Age at death	Age ^{14}C BP	$\delta^{13}C_{col}$. (‰ V-PDB)	$\delta^{13}C_{ap}$. (‰ V-PDB)	$\delta^{15}N$ (‰AIR)	References
LTC1	LTC1-P2	P	M	Indet. Adult		−19.5		11.2	Salazar-García et al. (2018)
	LTC1-P3	P	F	50–59	849 ± 45	−20.1	−14.7	12.4	Bonomo et al. (2017)
	LTC1-P4	P	F	35–44	801 ± 46	−19.9	−14.7	12.2	This paper
	LTC1-P4	P	F	35–44	801 ± 46	−19.6		12.0	Salazar-García et al. (2018)
	LTC1-P6	P	PM	19–20		−19.6	−13	12.1	This paper
	LTC1-S1	S		Indet. Adult	650 ± 70	−21	−16.7	11.1	Bonomo et al. (2017)
	LTC1-S3	S	M	Indet. Adult	802 ± 48	−19.7		12.0	Salazar-García et al. (2018)
	LTC1-S4	S	Indet.	Indet. Adult		−19.1		10.7	Salazar-García et al. (2018)
	LTC1-S5	S	Indet.	Indet. Adult		−19.8	−14.9	11.8	This paper
	LTC1-F	S	Indet.	Indet. Adult	775 ± 85	−20.6	−17.4	13.5	Bonomo et al. (2017)
CDLPB1	CDLPB1-ENT 1	P	PM	20–24	n/d	−16.1	−10.6	11.6	This paper
	CDLPB1-ENT 2-2	Indet.	Indet.	Indet. Adult	n/d	−19	−10.1	11.7	This paper

that they involve diets with higher consumption of C_4-type foods, and particularly cluster 3 is the only one that includes marine foods (Froehle et al., 2012). In this study, the main contribution of the multivariate model is that it allows us to quantitatively differentiate between C_3 and C_4 energy intake.

In order to draw stronger inferences about the diet of the analyzed individuals, the isotopic values of the most relevant animal and plant resources in the diet of these populations were surveyed according to the available archaeofaunal and paleobotanical information. Table 2 shows the isotopic values of the fauna, including those published by us and other authors (Scabuzzo & González, 2007; Loponte, 2008; Loponte et al., 2011; Loponte & Corriale, 2013; Corriale & Loponte, 2015; Salazar-García et al., 2018). The analyzed species correspond to cervids (*Blastocerus dichotomus* and *Ozotoceros bezoarticus*), rodents (*Myocastor coypus* and *Hydrochoeris hydrochaeris*), and freshwater fish (Siluriformes). Values of $\delta^{13}C_{col}$ and $\delta^{15}N$ were obtained for these species (Table 2).

Given the importance of vegetables for these populations and the difficulty of analyzing archaeological plant remains, this study included the isotopic values of 14

Table 2 Stable isotope values of faunal remains from the Paraná River Delta

Especies	$\delta^{13}C_{col}$. (‰ V-PDB)	$\delta^{15}N$ (‰AIR)	References
Ozotoceros bezoarticus			Salazar-García et al. (2018), Loponte (2008)
N	4	2	
Mean	−18.9	4.3	
Maximum value	−18.7	5.2	
Minimum value	−19.7	3.5	
Blastocerus dichotomus			Loponte and Corriale (2013)
N	11	0	
Mean	−20.6		
Maximum value	−17.1		
Minimum value	−22		
Myocastor coypus			Salazar-García et al. (2018), Loponte et al. (2011), Scabuzzo and González (2007)
N	5	2	
Mean	−19.1	4	
Maximum value	−15.9	4.5	
Minimum value	−21.2	3.5	
Hydrochaeris hydrochaeris			Loponte et al. (2011), Corriale and Loponte (2015)
N	19		
Mean	−15		
Maximum value	−11.7		
Minimum value	−20.2		
Freshwater fish			Salazar-García et al. (2018), Scabuzzo and González (2007), Loponte (2008)
N	3	3	
Mean	−19.2	6.4	
Maximum value	−21.7	7.9	
Minimum value	−18	5.2	

current crops (Table 3). The vegetables analyzed were as follows: maize (*Zea mays*), manioc (*Manihot esculenta*), sweet potato (*Ipomoea batatas*), pumpkin (*Cucurbita maxima*), squash (*Cucurbita moschata*), and beans (*Phaseolus vulgaris*), which were obtained from two street markets, Chacra de Posadas (Samples 1–7) and Franca of Puerto Iguazú, Misiones Province (Samples 8–14). In some cases, it was possible to find out the origin of the vegetables, as follows: the cities of Roca (black and white cassava), Andrade (Brazilian pumpkin, purple sweet potato), and the surroundings of Posadas (black and red beans). It is important to mention that one of the limitations of the results obtained here is the lack of information on crop management. We do not have information on the irrigation modalities and the use of fertilizers during the production of these vegetables that could have impacted the isotopic values obtained.

Table 3 Stable isotope values of the current plants

Sample	Species	Part	Dry weight (g.)	$\delta^{13}C$	$\delta^{15}N$	C:N	%C	%N
M1	*Zea mays*	Seed	9	−9.9	2.7	16.06	43.8	2.73
M2	*Manihot esculenta*	Pulp	7	−25.3	5	23.25	47.26	2.03
M3	*Manihot esculenta*	Pulp	10	−25.7	6.9	39.25	42.04	1.07
M4	*Cucurbita maxima*	Pulp	6	−24.5	2.5	35.97	42.07	1.17
M5	*Ipomoea batatas*	Pulp	7	−26.1	2.5	36.04	40.82	1.13
M6	*Phaseolus vulgaris*	Seed	12	−24.3	2.5	13.95	39.73	2.85
M7	*Phaseolus vulgaris*	Seed	12	−25.7	1.7	11.51	43.18	3.75
M8	*Zea mays*	Seed	15	−9.8	5.3	22.3	46.02	2.06
M9	*Manihot esculenta*	Pulp	7	−25.9	7.1	66.97	43.53	0.65
M10	*Cucurbita maxima*	Pulp	5	−24.9	4	31.64	44.37	1.4
M11	*Cucurbita moschata*	Pulp	4	−25.2	7.6	7.76	42.44	5.47
M12	*Ipomoea batatas*	Pulp	7	−25.9	6.8	84.94	45.4	0.53
M13	*Phaseolus vulgaris*	Seed	12	−26.4	3.4	13.23	41.21	3.11
M14	*Phaseolus vulgaris*	Seed	14	−27	4.2	17.83	39.55	2.22

The samples were first processed at Laboratorio de Radiocarbono del Centro de Investigaciones Geológicas (LATYR, CONICET-UNLP) and Laboratorio de Preparación de Muestras para Análisis Isotópicos (INCUAPA-CONICET-UNCPBA) and then analyzed at the Laboratorio de Isótopos Estables en Ciencias Ambientales of San Rafael (IANIGLA-CONICET and UTN-FRSR). The main parts of the corn cob (husk leaves, styles, seeds or kernels, and stalks) and of the squash fruit (stalk, peel, pulp, placenta, and seeds) were first separated for processing. They were initially dried in a SIMCIC oven for 48 h at 45–50 °C. The manioc and sweet potato tubers were dried for 4 days at 45 °C ± 3 °C in an oven and then freeze-dried to remove the remaining moisture. For the isotopic measurements of $\delta^{13}C_{col}$ and $\delta^{15}N$, a Flash 2000 Elemental Analyzer, coupled to Thermo Delta Advantage IRMS, was used. The $\delta^{13}C_{col}$ values of the present plants were corrected for the Suess effect with an increase of 1.5‰ (Tessone et al., 2010; Killian Galván et al., 2015; Dombrosky, 2020).

Results

Isotopic Values of the Resources

Table 2 presents the amount of isotopic data available, arithmetic means, and maximum and minimum values for each of the animal species. In the case of medium-to-large herbivores (*Ozotoceros bezoarticus* and *Blastocerus dichotomus*), the mean values of $\delta^{13}C_{col}$ are −18.9‰ and −20.6‰, respectively. This indicates a predominant consumption of C_3 photosynthetic pathway vegetables. The $\delta^{15}N$ values of *Ozotoceros bezoarticus* (mean = +4.3‰) are consistent with the expected values for

animals with herbivore habits. Rodents (*Myocastor coypus* and *Hydrochaeris hydrochaeris*) show mean values of $\delta^{13}C_{col}$ (−19.1‰ and −15‰, respectively) indicating, in the first case, a predominant consumption of C_3 plants, and, in the second case, the more highest values indicate C_4 plant intake (Corriale & Loponte, 2015). Likewise, the $\delta^{15}N$ values of *Myocastor coypus* are similar to those of the other herbivores considered here. Finally, the mean $\delta^{13}C_{col}$ values obtained from Siluriformes fish (−19.2‰) are similar to those obtained for terrestrial herbivores and *Myocastor coypus*. The $\delta^{15}N$ value is the highest in relation to the other animal resources (mean + 6.4‰).

To sum up, the isotopic values of the different resources are similar, which makes it difficult to differentiate the consumption of each resource from these analyses. In this sense, an overlap was observed in the means of $\delta^{13}C_{col}$ of cervids, *Myocastor coypus,* and fish, with *Hydrochaeris hydrochaeris* exhibiting more highest values, although with a wider range of variation (Bonomo et al., 2017). In connection with the nitrogen values, there was also an overlap between the different resources, although the fish stood out with higher values.

Table 3 shows the $\delta^{13}C_{col}$ and $\delta^{15}N$ data of the current plants, and the part analyzed is specified in each case. In relation to the ^{13}C, maize exhibited the most higher values (mean = −9.8‰), as expected for the C_4 photosynthetic pattern. On theother hand, squash, beans, sweet potato, pumpkin, and manioc showed lower $\delta^{13}C$ values (mean = −25.6‰), characteristic of C_3 photosynthetic patterns. The $\delta^{15}N$ exhibited heterogeneous values, with a mean of +4.4‰ (ranging from +1.7‰ to +7.6‰), and for some vegetables, such as manioc (mean = +6.3‰), the values were higher. Considering all the resources and despite the overlapping values, it is possible to establish different isotopic groups made up of terrestrial herbivores and coypu, capybara, fish, C_3, and C_4 plants.

Isotopic Values of Humans

Table 1 shows the isotopic data of the 12 human samples, and Table 4 shows the isotopic data and the %C, %N, and C:N ratio values of the human samples presented for the first time in this article. The samples exhibited a C:N ratio within the expected range for samples without diagenetic alterations (values between 2.9 and 3.6, DeNiro, 1985) and acceptable parameters of %C and %N. Table 5 presents the descriptive statistics for each of the isotopic parameters measured in the human skeletal remains. Considering the whole sample and without discriminating between sites, we found that the δ^{13} C_{col} arithmetic mean exhibited a mean of −19.5 ± 1.2‰ with minimum values of −21‰ and a maximum value of −16.1‰. On the other hand, the result obtained from the inorganic fraction also showed variations, with a mean of −14 ± 2.6‰ (ranging between −17.4 and −10.1‰). The $\delta^{15}N$ exhibited generally high values (mean 11.9 ± 0.2‰, minimum value +10.7‰, and maximum value +13.5‰). When we performed an analysis by discriminating the data according to the site of origin, we observed some differences between the different

Table 4 Isotopic data of the human samples presented for the first time in this article

Site	Individual	Lab. Code	Sample	$\delta^{13}C_{col}$. (‰ V-PDB)	$\delta^{13}C_{ap}$. (‰ V-PDB)	$\delta^{15}N$ (‰AIR)	C:N	%C	%N
LTC1	LTC1-P4	MSR-1447	Rib	−19.9	−14.7	12.2	3.3	40.8	14.6
	LTC1-P6	MSR-1443	Rib	−19.6	−13	12.1	3.2	41.9	14.8
	LTC1-S5	MSR-1449	Rib	−19.8	−14.9	11.8	3.2	41.3	14.8
CDLPB1	CDLPB1-ENT 1	MSR-1445	Rib	−16.1	−10.6	11.6	3.2	42.2	15.2
	CDLPB1-ENT 2-2	MSR-1446	Rib	−19	−10.1	11.7	3.3	41.5	14.6

Table 5 Mean, maximum, and minimum values of the human samples from the sites Los Tres Cerros 1 (LTC1) and Cerro de las Pajas Blancas 1 (CDLPB1)

Site	Values	$\delta^{13}C_{col}$. (‰ V-PDB)	$\delta^{13}C_{ap}$. (‰ V-PDB)	$\delta^{15}N$ (‰AIR)
LTC1	Mean	−19.9 ± 0.6	−15.2 ± 1.6	11.9 ± 0.8
	Max. value	−19.6	−13	13.5
	Min. value	−21	−17.4	10.7
CDLPB1	Mean	−17.6 ± 2.1	−10.3 ± 0.4	11.7 ± 0.1
	Max. value	−16.1	−10	11.7
	Min. value	−19	−10.6	11.6

parameters. In this sense, the samples from the LTC1 site showed more lower $\delta^{13}C_{col}$ values (mean = −19.9 ± 0.6‰) than those from the CDLPB1 site (mean = −17.6 ± 2.1‰). The same trend was observed for apatite values (LTC1 mean = −15.2 ± 1.6‰ and CDLPB1 mean = −10.3 ± 0.4‰). Finally, ^{15}N displayed similar values in human bone samples from both sites (LTC1 mean = +11.9 ± 0.8‰ and CDLPB1 mean = +11.7 ± 0.1‰) (Table 5).

Figure 4 presents the values of $\delta^{13}C_{col}$ and $\delta^{15}N$ of both human samples and animal and plant resources. For the interpretation of isotopic values, an average fractionation of 1‰ (ranging between 0‰ and 2‰) in the $\delta^{13}C_{col}$ and 4‰ (ranging between 3‰ and 5‰) in the $\delta^{15}N$ values between predator and prey was considered (Bocherens & Ducker, 2003; Tykot et al., 2009). On the X-axis ($\delta^{13}C_{col}$), it is observed that the values obtained from human remains are close to those of terrestrial mammals, fish, and C_3-type vegetables. However, some values, such as that of the individual CDLPB1-Ent 1, are higher, which would reflect the incorporation of C_4 food. On the Y-axis, as expected, humans present higher values of $\delta^{15}N$, which is compatible with protein consumption. Particularly, fish could explain the highest ^{15}N values, although, at this point, it is interesting to note that there are vegetables, such as squash and sweet potato, with high nitrogen signals. However, it is important to remember that there are certain agricultural practices such as the use of fertilizers that could contribute to an increase in plant nitrogen values.

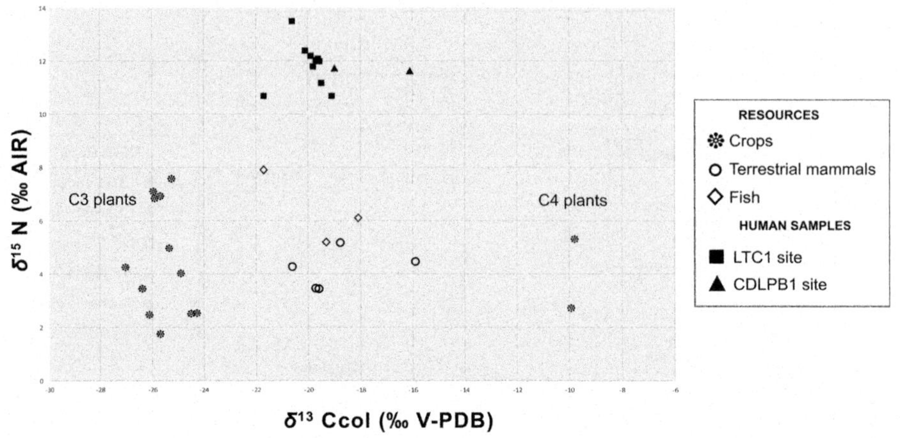

Fig. 4 Human and resources isotopic values

Table 6 Stable isotope values of the human samples and discriminant functions

Site/individual	$\delta^{13}C_{col}$. (‰ V-PDB)	$\delta^{13}C_{ap}$. (‰ V-PDB)	$\delta^{15}N$ (‰AIR)	F1	F2
LTC1-P3	−20.1	−14.7	12.4	−7.2765	2.1136
LTC1-P4	−19.9	−14.7	12.2	−7.18456	2.02759
LTC1-P6	−19.6	−13	12.1	−6.43452	1.32933
LTC1-S1	−21	−16.7	11.1	−8.8595	1.9713
LTC1-S5	−19.8	−14.9	11.8	−7.24776	1.85104
LTC1-F	−20.6	−17.4	13.5	−8.2685	3.7924
CD2PB1 ENT 1	−16.1	−10.6	11.6	−3.2235	0.5367
CDLPB1 ENT 2-2	−19	−10.1	11.7	−5.1328	−0.00295

According to the model proposed by Froehle et al. (2012), the values of human samples from both sites are mostly distributed within clusters 1 and 4 (Table 6). Cluster 1 involves diets based 100% on C_3 food (protein and energy), and includes five of the LTC1 individuals (LTC1-P3, LTC1-P4, LTC1-P6, LTC1-S5, and LTC1-F). Cluster 4 is characterized by $\geq$ 65% protein C_3, energy 70% C_3, and 30% C_4, and includes all individuals of the CDLPB1 site. Finally, one sample (LTC1-S1) falls outside the clusters and near cluster 1 (Fig. 5).

Discussion

The isotopic data presented in this work show that the human groups that occupied the Upper Paraná River Delta in the Late Holocene fed on diets that shared a common base. However, the application of the multivariate model allowed us to identify

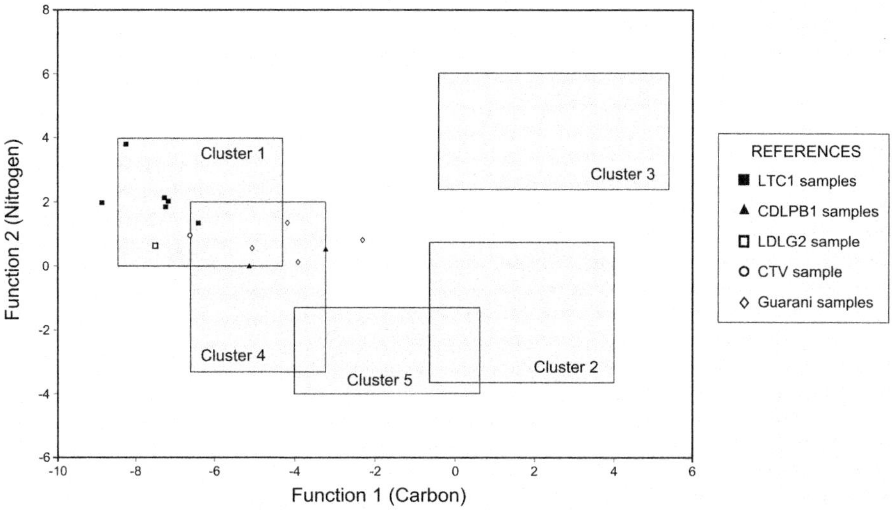

Fig. 5 Results obtained from the application of the model proposed by Froehle et al. (2012)

differences in the distribution of individuals within the clusters. The application of this model allowed us to note that the isotopic results obtained are due to a predominant intake of C_3-type foods (clusters 1 and 4). In this sense, the model refers to both clusters including between 100% and 65% C_3 protein and between 100% and 70% C_3 energy. Considering the archaeofaunal and paleobotanical analyses (seeds, cob, phytoliths, and starch grains), this result would reflect a dominant consumption of proteins from fish and herbivorous animals (coypus and deer) and energy provided for C_3-type vegetables (e.g., *Oryza* spp., *Geoffroea decorticans*, *Phaseolus* spp., and Cucurbitaceae spp.).

Generally speaking, this trend coincides with the results obtained in previous works where we presented isotopic data from four sites located in the Upper Paraná Delta (Los Tres Cerros 2 [LTC2], Laguna de los Gansos 2 [LDLG2], Cerro Tapera Vázquez [CTV], and Cerro Grande del Paraná Pavón [CGPP]) assigned to the Goya-Malabrigo archaeological entity (Bonomo et al., 2017; Salazar-García et al., 2018). The published data correspond to $\delta^{15}N$ and $\delta^{13}C_{col}$ performed on both bone fractions. For these samples, the $\delta^{13}C_{col}$ varies between −20.6‰ and −18.6‰ (mean −19.6‰), while the $\delta^{15}N$ varies between 10.2‰ and 12.4‰ (mean 11.5‰). Finally, the $\delta^{13}C_{ap}$ values vary between −13.2‰ and −14.4‰ (mean = −13.8‰) (Bonomo et al., 2017; Salazar-García et al., 2018). From a comparative point of view, the isotopic values of $\delta^{13}C_{col}$ and $\delta^{15}N$ of LTC2, LDLG2, CTV, and CGPP tend to overlap with those presented here from the LTC1 and CDLPB1 sites. Conversely, the $\delta^{13}C_{ap}$ values of these sites (LTC2, LDLG2, CTV, and CGPP) tend to overlap only with LTC1.

In two samples from the LDLG2 and CTV sites, it was possible to apply the multivariate model of Froehle et al. (2012). In both cases, the individuals fall within cluster 1, in the same way as individuals of the LTC1 site (Fig. 5). The same trend was observed by Ottalagano and Loponte (2017) by subjecting the isotopic results

of 23 individuals from six Goya-Malabrigo sites to the same multivariate test. The results indicate that the Goya-Malabrigo individuals analyzed fall into the box of diets composed of 100% protein and energy intake related to a C_3 photosynthetic pathway (Ottalagano & Loponte, 2017), coinciding with the main trend obtained in this work.

However, our analysis also showed cases (CDLPB1-ENT1 and CDLPB1-ENT2-2) in which the consumption of C_3 proteins and energy was accompanied by a 30% contribution of C_4 energy (cluster 4). Among the resources identified in the studied archaeological contexts, this result could be related to the incorporation of *Zea mays* in the diet, which was identified in the record through both macro and microremains. At site CDPB1, this cereal was detected in starch grains in ceramic sherds (Bonomo et al., 2011), phytoliths, and cob fragments in sediments extracted from the occupation layer (Bonomo et al., 2019b; Colobig & Costa Angrizani, 2021). In turn, charred seeds and phytoliths identified as *Zea mays* were recovered from sediments from site LTC1 (Sánchez et al., 2013; Colobig et al., 2015). The concentration of cruciform biosiliceous morphotypes specifically associated with maize leaf in certain sedimentary layers can be interpreted as a strong indication of local cultivation of this vegetable from ca. 1000 years BP (Colobig et al., 2019). A second source of C_4 food could come from the consumption of animals whose diets are based on C_4 vegetables, as is the case of capybara.

Likewise, in the cases addressed in this study, the isotopic variation determined by the presence of C_4 foods is related to the site of origin of the samples. In this sense, while most of the individuals buried at site LTC1 are grouped within cluster 1, those from site CDLPB1 are concentrated within cluster 4. These differences in diet can be linked to the internal variations in the dietary patterns of individuals linked to the Goya-Malabrigo entity who occupied different sectors within the upper Paraná River Delta. Although in both cases there is consistent evidence of the presence of *Zea mays* in artifacts and sediments, it is possible to infer that the intake of this food was more frequent among the inhabitants of CDLPB1 than those of LTC1.

Another hypothesis that may explain this variation is the presence or influence of populations associated with the Guaraní archaeological entity at site CDLPB1. The effective occupation of this site by Guaraní groups is still under discussion. The finding of a funerary urn and ceramic fragments stylistically associated with this entity allows us to infer at least the existence of contacts (circulation of goods or people) with these populations. Comparison with isotopic results from two Guaraní sites from the Uruguay River (Ensenada del Bellaco [EDB] and La Correntina [LC]) and one from the Lower Delta (Arroyo La Glorieta [ALG]) provides interesting parameters for this discussion (Bonomo et al., 2017; Costa Angrizani & Scabuzzo, 2019). By subjecting the data from these sites to the Froehle et al. (2012) model, we observed that two of the Guaraní samples fall into cluster 4 and only one falls into an intermediate position between clusters 4 and cluster 1. It is interesting to note that the samples from Guaraní sites are grouped within the same cluster in which those from site CDPB1 are positioned (Fig. 5), indicating the predominant consumption of C_3 protein/energy, with a 30% contribution of C_4 energy. Overall, the results of the isotopic analyses and the faunal and botanical evidence reaffirm the

idea that the populations of the Upper Paraná Delta deployed subsistence strategies that articulated the capture of animals through hunting and fishing, the collection of wild plants, and the production of cultivated vegetables. As observed from ethnographic data analysis on a global scale, fishing has been a relevant subsistence strategy among small-scale societies with mixed economies (Ahedo et al., 2021). In our case study, archaeofaunal evidence (Bastourre & Azpelicueta, 2020) and the results of oral health and physiological fitness analyses indicate the high importance of fish in the diet of the groups that occupied the Upper Paraná Delta (Ramos van Raap, 2018, 2020). Likewise, the isotopic data presented, particularly the high 15N values, are consistent with the majority consumption of freshwater fish. The predominance of C_3 protein and energy in our results was to be expected considering the prevalence in the environment of plants and animals associated with this photosynthetic pathway. Both wild plants with nutritional importance (palms [Arecaceae], carob [*Prosopis* spp.], and wild rice [*Oryza* spp.]) and most of those that were cultivated (e.g., squash [*Cucurbita* spp.] and beans [*Phaseolus* spp.]) correspond to the C_3 group. Wild rice deserves special consideration, as wild species of this vegetable were consumed by indigenous populations in the South American lowlands long before the introduction of Asian variants (Hilbert et al., 2017). Being adapted to flooded lands, such as the Paraná Delta, this plant resource may have acquired special importance during periods of river flooding and represent a significant source of C_3 energy.

Although the isotopic signal of C_4 plant consumption is not a majority, the paleobotanical evidence leaves no doubt about the consumption and cultivation of *Zea mays* in these contexts around 1000 years BP and probably before too. Moreover, the presence of microremains and macroremains of domesticated vegetables of the C_3 type reinforces the idea that crops did not enter the settlements only by exchange and that horticulture was not a sporadic or incipient practice.

Conclusions

In this work, new isotopic results were obtained for both current plant species and human skeletons from two sites linked to the Goya-Malabrigo archaeological entity. Obtaining isotopic values of present-day plants from the region contributes to a better characterization of the isotopic ecology of the area. In addition, the application of multivariate analysis to interpret the data obtained from the organic and inorganic fractions of human remains has allowed weighting of the diets of the analyzed individuals according to a model that evaluates protein and energy intake. The results obtained indicate the predominance of C_3-type foods complemented, in some cases, with C_4-type resources. The comparison of this information with the archaeofaunal and paleobotanical records of the sites addressed made it possible to identify the animal and plant resources that were exploited by the populations occupying the Upper Paraná Delta in the Late Holocene. The interpretations made in this work reinforce the idea that the subsistence patterns of these groups were based on the

exploitation of aquatic and terrestrial resources, together with the incorporation of a wide range of vegetables, some of which were cultivated by these populations (Bonomo et al., 2017).

The body of evidence presented in this paper reaffirms that the populations assigned to the Goya-Malabrigo archaeological entity developed and maintained for several centuries a mixed economy that articulated the exploitation of resources available in the aquatic environment (animals and plants) with the introduction of domesticated plants and the production of their own food. As noted for other regions of South America, the high productivity and richness of aquatic environments may have facilitated the persistence of mixed economies (Chanca et al., 2021). In the eastern Amazon, the adoption of polyculture agroforestry, which combined the cultivation of different vegetables with the exploitation of wild species and aquatic resources, was the subsistence basis for the development of complex societies 4,500 years ago (Maezumi et al., 2018). Around 1,000-1750 years BP, the indigenous populations that occupied our study area installed an analogous system that lasted until the European conquest and colonization in the seventeenth century.

In this framework, it can be proposed that the experiences of adoption and consolidation of horticultural practices in the Upper Paraná River Delta may have benefited from the abundance and predictability of the aquatic resources available in the surroundings of the settlements. As reported by de Souza and Riris (2021), the pulse of population growth that occurred in the La Plata Basin from 2,000 years BP onwards was driven by populations that introduced a diverse package of cultivated plants along with active forest management (Goya-Malabrigo, Jê, and Guaraní). Likewise, mixed economies reduced territorial mobility and favored population growth and the development of long-lasting cultural practices during most of the Late Holocene (Chancas et al., 2021).

Acknowledgments This work was supported by scientific projects of Agencia Nacional de Promoción Científica y Tecnológica PICT 2014-0813 and PICT 2016-1064 directed by Dr. C. Scabuzzo and Dr. M. Bonomo. We thank Norma Hilgert (Instituto de Biología Subtropical, CONICET-UNaM) and Lorena Salvateli (Museo Histórico y Arqueológico Andrés Guacurarí) for providing the plants for analysis. We would also like to thank María Lelia Pochettino, Victoria Roca, Eduardo Apolinaire, Florencia Mari, Ignacio Echeverría, Andrea Di Bastiano, Paula Vitale, and Adolfo Gil for their great help in different instances of the study. Finally, we thank the two anonymous reviewers who helped to improve this work.

References

Ahedo, V., Zurro, D., Caro, J., & Galán, J. M. (2021). Let's go fishing: A quantitative analysis of subsistence choices with a special focus on mixed economies among small-scale societies. *PLoS One, 16*(8), e0254539.

Alunni, D., Kochi, S., Tessone, A., & Zangrando, A. (2021). Consumo de grasas entre cazadores-recolectores de la costa sur de Tierra del Fuego durante el Holoceno tardío: una evaluación desde la zooarqueología y los isótopos estables. *Comechingonia, 25*(3), 195–228.

Ambrose, S. H. (1990). Preparation and characterization of bone and tooth collagen for stable carbon and nitrogen isotope analysis. *Journal of Archaeological Science, 17*, 430–451. https://doi.org/10.1016/0305-4403(90)90007-R

Ambrose, S. H. (1993). Isotopic analysis of paleodiets: Methodological and interpretive consideration. In M. K. Sandford (Ed.), *Investigations of ancient human tissue: Chemical analyses in Anthropology* (pp. 50–130). Gordon and Breach.

Aranda, C. M., Barrientos, G., & Del Papa, M. C. (2014). Código deontológico para el estudio, conservación y gestión de restos humanos de poblaciones del pasado. *Revista Argentina de Antropología Biológica, 16*(2), 111–113. https://doi.org/10.17139/raab.2014.0016.02.05

Badano, V. (1940). Piezas enteras de alfarería del litoral existentes en el museo de Entre Ríos. Notas arqueológicas II. *Memorias del Museo de Entre Ríos, 14*, 4–23.

Bastourre, L. (2014). Estudios arqueofaunísticos en el Delta Superior del Paraná: el sitio Los Tres Cerros 1 (Provincia de Entre Ríos, Argentina). *Revista Chilena de Antropología, 30*, 109–115. https://doi.org/10.5354/0719-1472.2014.36282

Bastourre, L. (2018). Del registro ictioarqueológico a las prácticas alimentarias: el caso de Los Tres Cerros 1 (Delta Superior del Paraná, Entre Ríos). *Revista del Museo de La Plata, suplemento resúmenes, 3*(1), 107.

Bastourre, M. L., & Azpelicueta, M. M. (2020). Del registro ictioarqueológico a las prácticas alimentarias: el caso de Los Tres Cerros 1 (Delta superior del Paraná, Entre Ríos). *Relaciones de la Sociedad Argentina de Antropología, XLV*(1), 13–57.

Bocherens, H., & Ducker, D. (2003). Trophic level isotopic enrichment of carbon and nitrogen in bone collagen: Case studies from recent and ancient terrestrial ecosystems. *International Journal of Osteoarchaeology, 13*, 46–53. https://doi.org/10.1002/oa.662

Bonomo, M. (2012). *Historia prehispánica de Entre Ríos*. Fundación de Historia Natural Félix de Azara.

Bonomo, M. (2013). Reanálisis de la colección de Samuel Lothrop procedente del Delta del Paraná. *Relaciones de la Sociedad Argentina de Antropología, 38*(1), 169–198.

Bonomo, M., Aceituno, F., Politis, G. G., & Pochettino, M. L. (2011). Pre-hispanic horticulture in the Paraná Delta (Argentina): Archaeological and historical evidence. *World Archaeology, 43*(4), 557–579. https://doi.org/10.1080/00438243.2011.624710

Bonomo, M., Costa Angrizani, R., Apolinaire, E., & Noelli, F. S. (2015). A model for the Guaraní expansion in the La Plata Basin and littoral zone of southern Brazil. *Quaternary International, 356*, 54–73. https://doi.org/10.1016/j.quaint.2014.10.050

Bonomo, M., Scabuzzo, C., Politis, G., & Zucol, A. (2017). Stable carbon and nitrogen isotope studies in the Paraná River Delta (Argentina): An approach to prehispanic diets. *Latin American Antiquity, 28*(1), 105–126. https://doi.org/10.1017/laq.2016.6

Bonomo, M., Skarbun, F., & Bastourre, L. (2019a). *Subsistencia y alimentación en arqueología*. Editorial de la Universidad Nacional de La Plata.

Bonomo, M., Costa Angrizani, R., Torino, R., Alvarez, M., & Moreira, G. (2019b). Nuevas investigaciones arqueológicas en el sitio Cerro de las Pajas Blancas 1 (Delta superior del río Paraná, Santa Fe). In *Libro de Resúmenes del XX Congreso Nacional de Arqueología Argentina* (pp. 93–95). Universidad Nacional de Córdoba.

Burkart, A. (1957). Ojeada sinóptica sobre: La vegetación del Delta del Río Paraná. *Darwin, 11*(3), 457–561.

Cabrera, A. L. (1953). Esquema fitogeográfico de la República Argentina. *Revista del Museo de la ciudad Eva Perón (NS), VIII*, 87–168. Sección Botánica.

Ceruti, C. N. (2003). Entidades culturales presentes en la cuenca del Paraná Medio (margen entrerriana). *Mundo de Antes, 3*, 111–135.

Chanca, I., Borges, C., Colonese, A. C., Macario, K., Toso, A., Fontanals-Coll, M., dos Anjos, R., Muniz, M., Pereira, R., Talamo, S., & Milheira, R. G. (2021). Food and diet of the pre-Columbian mound builders of the Patos Lagoon region in southern Brazil with stable isotope analysis. *Journal of Archaeological Science, 133*, 105439. https://doi.org/10.1016/j.jas.2021.105439

Colobig, M. M., & Costa Angrizani, R. (2021). El manejo de las plantas en el Noreste argentino a partir del análisis de microrrestos biosilíceos en sedimentos: Sitio Cerro de las Pajas Blancas 1 (Departamento de San Jerónimo, Provincia de Santa Fe). In *Livro de Resumos do XXI Congresso da Sociedade de Arqueologia Brasileira* (p. 101). Sociedade de Arqueologia Brasileira.

Colobig, M. M., Sánchez, J. O., & Zucol, A. (2015). Análisis de macrorrestos vegetales en el sitio arqueológico Los Tres Cerros 1 (isla Las Moras, Victoria, Entre Ríos). *Revista del Museo de Antropología, 8*(1), 115–124.

Colobig, M. M., Zucol, A., & Silva, C. B. (2018). Análisis de microrrestos de secuencias sedimentarias del sitio Laguna de los Gansos 1 y 2 (Delta Superior del río Paraná, Argentina). In G. Politis & M. Bonomo (Eds.), *Goya-Malabrigo: arqueología de una sociedad indígena del noreste argentino* (pp. 333–346). Editorial UNICEN.

Colobig, M. M., Iriarte, J., & Zucol, A. (2019). Evidencia de manejo de *Zea mays* durante el Holoceno tardío en el sitio Los Tres Cerros 1, Delta superior del río Paraná. In A. Laguens, M. Bonnin, & B. Marconetto (Eds.), *Libro de Resúmenes XX Congreso Nacional de Arqueología Argentina: 50 años de arqueologías* (pp. 110–111). Universidad Nacional de Córdoba. Facultad de Filosofía y Humanidades.

Corriale, M. J., & Loponte, D. (2015). Use of stable carbon isotope ratio for foraging behavior analysis of capybara (*Hydrochoerus hydrochaeris*) from Esteros del Iberá, Argentina. *Mammalian Biology, 80*(2), 73–80. https://doi.org/10.1016/j.mambio.2014.12.007

Costa Angrizani, R., & Scabuzzo, C. (2019). Múltiples enfoques para abordar la alimentación en los contextos de ocupación guaraní del Río Uruguay inferior. In L. Salvatelli, V. Roca, & A. Schmitz (Eds.), *Libro de Resúmenes del VIII Encuentro de Discusión de Arqueología del Nordeste Argentino* (p. 18). EdUNaM.

Costa Angrizani, R., Maravilla, M. L., Schwerdt, O., Alvarez, M., & Ramos Van Raap, M. A. (2015). Estudio de la colección arqueológica procedente del sitio guaraní Arroyo La Glorieta (provincia de Buenos Aires, Delta inferior del río Paraná). *Comechingonia. Revista de Arqueología, 19*(1), 191–201.

Costa Angrizani, R., Scabuzzo, C., Colobig, M. M., Bonomo, M., Brea, M., Zucol, A. F., & Loperfido, M. A. (2021). Múltiplas abordagens para o estudo dos vegetais coletados, cultivados e consumidos pelas populações indígenas Goya-Malabrigo e Guarani que ocuparam o setor sul da Bacia do Prata (Argentina). In *Livro de Resumos do XXI Congresso da Sociedade de Arqueologia Brasileira* (pp. 102–103). Sociedade de Arqueologia Brasileira.

de Souza, J. G., & Riris, P. (2021). Delayed demographic transition following the adoption of cultivated plants in the eastern La Plata Basin and Atlantic coast, South America. *Journal of Archaeological Science, 125*, 1–11. https://doi.org/10.1016/j.jas.2020.105293

DeNiro, M. J. (1985). Postmortem preservation and alteration of in vivo bone collagen isotope ratios in relation to paleodietary reconstruction. *Nature, 317*, 806–809. https://doi.org/10.1038/317806a0

Dombrosky, J. (2020). A ~1000-year 13C Suess correction model for the study of past ecosystems. *The Holocene, 30*(3), 474–478. https://doi.org/10.1177/0959683619887416

Flensborg, G., Martínez, G., & Tessone, A. (2018). First approach to the paleodiet of hunter-gatherers through stable isotopes (δ13C and δ15N) in the eastern Pampa-Patagonia transition during the Middle Holocene. *Journal of Archaeological Science: Reports, 17*, 571–580.

Froehle, A. W., Kellner, C. M., & Schoeninger, M. J. (2012). Multivariate carbon and nitrogen stable isotope model for the reconstruction of prehistoric human diet. *American Journal of Physical Anthropology, 147*, 352–369. https://doi.org/10.1002/ajpa.21651

Hilbert, L., Neves, E. G., Pugliese, F., Whitney, B. S., Shock, M., Veasey, E., Zimpel, C. A., & Iriarte, J. (2017). Evidence for mid-Holocene rice domestication in the Americas. *Nature Ecology & Evolution, 1*(11), 1693–1698. https://doi.org/10.1038/s41559-017-0322-4

Iriondo, M. H., Paggi, J. C., & Parma, M. J. (2007). Introduction. In M. H. Iriondo, J. C. Paggi, & M. J. Parma (Eds.), *The Middle Paraná River: Limnology of a subtropical wetland* (pp. 1–4). Springer.

Kent, S. (Ed.). (1989). *Farmers as hunters: The implications of sedentism*. Cambridge University Press.

Killian Galván, V., Sanmartino, G., Castellano, V., Seldes, V., & Marban, L. (2015). Estudios de isótopos estables en huertas familiares actuales de la Quebrada de Humahuaca (Jujuy, Argentina). Su potencial aporte a los estudios paleodietarios del Noroeste argentino. *Revista del Museo de Antropología, 8*(2), 107–118.

Kurc, A. N. (1995). La disponibilidad de los recursos y su relación con el uso del espacio en el Paraná Medio. *Revista del Museo de Historia Natural de San Rafael, 27*(1–4), 75–81.

López, H. (2001). Estudio y uso sustentable de la Biota Austral: Ictiofauna Continental Argentina. *Revista Cubana de Investigaciones Pesqueras (special issue, electronic version)*, 1–40.

Loponte, D. (2008). *Arqueología del Humedal del Paraná Inferior (Bajíos Ribereños Meridionales)*. Asociación Amigos del Instituto Nacional de Antropología y Pensamiento Latinoamericano.

Loponte, D., & Acosta, A. (2007). Horticultores amazónicos en el Humedal del Paraná Inferior: Los primeros datos isotópicos de la dieta. In C. Bayón, A. Pupio, I. González, N. Flegenheimer, & M. Frère (Eds.), *Arqueología en las pampas* (pp. 75–93). Sociedad Argentina de Antropología.

Loponte, D., & Corriale, M. (2013). Isotopic diet values of the diet of *Blastocerus dichotomus* (Marsh Deer) in Paraná Basin, South America. *Journal of Archaeological Science, 40*(2), 1382–1388. https://doi.org/10.1016/j.jas.2012.10.020

Loponte, D., & Kozameh, L. (2009). Nuevos datos para el conocimiento de las dietas prehispánicas del Delta Superior. *Comechingonia, 12*, 115–118.

Loponte, D., Acosta, A., Capparelli, I., & Pérez, M. (2011). La arqueología guaraní en el extremo meridional de la cuenca del Plata. In D. Loponte & A. Acosta (Eds.), *Arqueología Tupi guaraní* (pp. 111–157). Instituto Nacional de Antropología y Pensamiento Latinoamericano.

Loponte, D., Carbonera, M., Corriale, M., & Acosta, A. (2016). Horticulturists and oxygen ecozones in the tropical and subtropical forests of Southeast South America. *Environmental Archaeology, 22*(3), 1–21. https://doi.org/10.1080/14614103.2016.1211382

Lothrop, S. (1932). Indians of the Paraná Delta, Argentina. *Annals of the New York Academy of Science, 32*, 77–232.

Lynch Ianniello, I., Arrieta, M., Bernardi, L., & Mendonça, O. (2020). Paleodietary trends in Late Holocene hunter-gatherers from SW La Pampa, Argentina. First results of Médano Petroquímica site through stable isotopes analysis (δ13C and δ15N). *Quaternary International*. https://doi.org/10.1016/j.quaint.2019.11.037

Madanes, N., Quintana, R. D., Kandus, P., & Bó, R. F. (2015). Species richness and functional groups of angiosperms from the Paraná River Delta region (Argentina). *Check List, 11*(6), 1–12. https://doi.org/10.15560/11.6.1803

Maezumi, S. Y., Alves, D., Robinson, M., de Souza, J. G., Levis, C., Barnett, R. L., de Oliveira, E., Urrego, D., Schaan, D., & Iriarte, J. (2018). The legacy of 4,500 years of polyculture agroforestry in the eastern Amazon. *Nature plants, 4*(8), 540–547. https://doi.org/10.1038/s41477-018-0205-y

Malvárez, A. I. (1997). *Las comunidades vegetales del Delta del río Paraná. Su relación con factores ambientales y patrones del paisaje* [Unpublished doctoral thesis]. Universidad de Buenos Aires.

Marchese, M. R., Saigo, M., Zilli, F. L., Capello, S., Devercelli, M., Montalto, L., Papporello, G., & Wantzen, K. M. (2014). Food webs of the Paraná River floodplain: Assessing basal sources using stable carbon and nitrogen isotopes. *Limnologica, 46*, 22–30. https://doi.org/10.1016/j.limno.2013.11.004

Noelli, F. (1993). *Sem Tekohá não há Tekó (em busca de um modelo etnoarqueológico da subsistência e da aldeia Guarani aplicado a uma área de domínio no delta do Jacuí-RS)* [Unpublished Master dissertation]. Pontifícia Universidade Católica do Rio Grande do Sul.

Noelli, F., Mota, L., Nanni, M., Lavado, M., Oliveira, E., Panek, C., Jr., Simão, A., Marques, A., Silva, J., & Jerônimo, F. (2003). Agricultores guarani na pré-história da região noroeste do Paraná. *Universidade e Sociedade, 19*, 63–69.

Ottalagano, F., & Loponte, D. (2017). Stable isotopes and diet in complex hunter gatherers of Paraná River Basin, South America. *Archaeological and Anthropological Sciences*, 9, 865–877. https://doi.org/10.1007/s12520-015-0308-6

Outes, F. (1918). Nuevos rastros de la cultura guaraní en la cuenca del Paraná Inferior. *Anales de la Sociedad Científica Argentina, 85*, 153–182.

Panarello, H., Tessone, A., Killian Galván, V., Samec, C., Kochi, S., & Pirola, M. (2021). 35 años de análisis de isótopos estables en la arqueología argentina: conceptos, fundamentos, metodología y aplicaciones. *Cuadernos del Instituto Nacional de Antropología y Pensamiento Latinoamericano, 30*(1), 1–41.

Pluciennik, M. (2001). Archaeology, anthropology and subsistence. *Journal of the Royal Anthropological Institute, 7*(4), 741–758. https://doi.org/10.1111/1467-9655.00087

Politis, G., & Bonomo, M. (2012). La entidad arqueológica Goya-Malabrigo (Ríos Paraná y Uruguay) y su filiación Arawak. *Boletín de la Sociedade de Arqueología Brasileira, 25*(1), 10–46. https://doi.org/10.24885/sab.v25i1.338

Politis, G., & Bonomo, M. (Eds.). (2018). *Goya-Malabrigo: arqueología de una sociedad indígena del noreste argentino*. Editorial UNICEN.

Politis, G., Bonomo, M., Castiñeira, C., & Blasi, A. (2011). Archaeology of the upper delta of the Paraná River (Argentina): Mound construction and anthropic landscapes in the Los Tres Cerros locality. *Quaternary International, 245*, 74–88. https://doi.org/10.1016/j.quaint.2011.02.007

Politis, G., Domic, A. I., Bonomo, M., & Capriles, J. (2018). Modelamiento de los parámetros ambientales de la distribución de Goya-Malabrigo mediante Maxent. In G. Politis & M. Bonomo (Eds.), *Goya-Malabrigo arqueología de una sociedad indígena del noreste argentino* (pp. 289–305). Editorial UNICEN.

Ramos van Raap, M. A. (2018) Paleopatología en poblaciones prehispánicas del sur del Noreste Argentino [Unpublished doctoral thesis]. Universidad Nacional de La Plata, La Plata.

Ramos van Raap, M. A. (2020). Tendencias sobre la dieta de las poblaciones prehispánicas del sur del Noreste argentino: Análisis de indicadores bucodentales. *Latin American Antiquity, 31*(4), 800–816. https://doi.org/10.1017/laq.2020.62

Ramos van Raap, M. A., Di Lorenzo, B., Scabuzzo, C., & Batalla, Y. (2019). Determinación de sexo a partir de métodos cuantitativos en entierros humanos del sitio Los Tres Cerros 1 (Delta Superior del río Paraná, Entre Ríos). *Revista Arqueología, 25*(3), 167–182.

Rodríguez, J. A. (2004). En busca de la tierra sin mal. El poblamiento de la Cuenca del Plata por los guaraníes prehistóricos. *Ciencia Hoy, 14*(80).

Salazar-García, D., Bonomo, M., & Scabuzzo, C. (2018). Análisis de la alimentación a partir del estudio isotópico de Carbono y Nitrógeno en sitios de la entidad arqueológica Goya-Malabrigo del Delta del río Paraná. In G. Politis & M. Bonomo (Eds.), *Goya-Malabrigo: arqueología de una sociedad indígena del noreste argentino* (pp. 201–216). Editorial UNICEN.

Samec, C., Pirola, M., & Killian Galván, V. (2019). Lineamientos para la publicación de resultados isotópicos en Antropología Biológica y Arqueología. *Revista Argentina de Antropología Biológica, 21*(2), 1–9. https://doi.org/10.24215/18536387e007

Sánchez, J. O., Colobig, M. M., Zucol, A. F., Politis, G., Bonomo, M., & Castiñeira, C. (2013). Primeros resultados sobre el uso prehispánico de los vegetales en el sitio arqueológico Los Tres Cerros 1 (Victoria, Entre Ríos, Argentina): análisis del registro biosilíceo. *Darwiniana, nueva serie, 1*(2), 201–219. https://doi.org/10.14522/darwiniana.2013.12.541

Scabuzzo, C., & González, I. (2007). Un acercamiento a la dieta de las poblaciones prehispánicas de la Depresión del Salado durante el Holoceno tardío. In C. Bayón, A. Pupio, I. González, N. Flegenheimer, & M. Frère (Eds.), *Arqueología en las pampas* (pp. 59–74). Sociedad Argentina de Antropología.

Scabuzzo, C., & Ramos van Raap, M. A. (2017). Nuevos resultados de los estudios osteológicos del sitio Los Tres Cerros 1 (Delta Superior del río Paraná). *Comechingonia, Revista de Arqueología, 21*(2), 201–228.

Scabuzzo, C., Ramos van Raap, M. A., Bonomo, M., & Politis, G. (2015). Estudios bioarqueológicos en el sitio Los Tres Cerros 1 (Delta Superior del río Paraná, Entre Ríos, Argentina).

Boletim do Museu Paraense Emílio Goeldi. Ciências Humanas, 10(2), 509–535. https://doi.org/10.1590/1981-81222015000200015

Serrano, A. (1950). *Los primitivos habitantes de Entre Ríos*. Biblioteca Entrerriana General Perón, Ministerio de Educación, Provincia de Entre Ríos.

Serrano, A. (1972). Líneas fundamentales de la arqueología del litoral (una tentativa de periodización). *Revista del Instituto de Antropología de Córdoba, XXXII*, 3–75.

Smith, B. D. (2011). The cultural context of plant domestication in Eastern North America. *Current Anthropology, 52*, 471–484. https://doi.org/10.1086/659645

Tessone, A., Samec, C., Killian Galván, V., & Panarello, H. (2010). Restos vegetales de origen arqueológico e isótopos estables del carbono: su posibilidad de uso en las reconstrucciones paleodietarias y paleoclimatológicas. In S. Bertolino, R. Cattaneo, & A. Izeta (Eds.), *La arqueometría en Argentina y Latinoamérica* (pp. 251–256). Universidad Nacional de Córdoba. Facultad de Filosofía y Humanidades.

Tykot, R., Falabella, F., Planella, M. T., Aspillaga, E., Sanhueza, L., & Becker, C. (2009). Stable isotopes and archaeology in central Chile: Methodological insights and interpretative problems for dietary reconstruction. *International Journal of Osteoarchaeology, 19*, 156–170. https://doi.org/10.1002/oa.1065

Life at the Shallow Lakes: Bioarchaeological Record of the Northern Pampean Shallow Lake Environment

Carolina Valeria Píccoli, Juan David Avila, and María Carolina Barboza

Abstract Skeletal remains have provided key evidence of early human presence in the northern area of the Pampas (Argentina). This contribution reviews the mortality profile and paleopathological data from this region, focusing on the study of human remains found in the Central Pampean Dunefields (General López, province of Santa Fe, Argentina). This includes the Laguna El Doce collection, one of the few sites in Argentina with human remains dating back to the Middle Holocene. The assemblage comprises commingled human remains and uncovered individual burials. Private collections gathered by local collectors and amateur archaeologists were also assessed. An estimated minimum number of 24 individuals was determined. The human remains were analyzed in terms of representativeness and preservation, age of death, sex, oral health, paleopathology, and trauma. The results were interpreted considering previous proposals regarding the features of Holocene hunter-gatherer dynamics. This contributed to a better understanding of past biocultural processes and their social ramifications, despite the post-depositional processes that have altered part of the bioarchaeological record in the region.

Keywords Hunter-gatherers · Commingled human remains · Mortality profile · Paleopathology · Holocene

---

C. V. Píccoli (✉) · M. C. Barboza
Departamento de Bioantropología y Evolución (FHumyAr, UNR) – CONICET, Centro de Estudios Interdisciplinarios en Antropología (CEIA, FHumyAr-UNR), Santa Fe, Argentina
e-mail: carolinapiccoli@fhumyar.unr.edu.ar; mcbarboza@conicet.gov.ar

J. D. Avila
Departamento de Arqueología (FHumyAr, UNR), Centro de Estudios Interdisciplinarios en Antropología (CEIA, FHumyAr-UNR), Santa Fe, Argentina

G. F. Bonnat et al. (eds.), *Current Research in Archaeology of South American Pampas*, The Latin American Studies Book Series,
https://doi.org/10.1007/978-3-031-55194-9_3

Introduction

Archaeological research in microregions and less-known areas of the Pampas has significantly increased over the last 20 years. This allowed new interpretations of the bioarchaeological record based on stable isotope analysis in paleodiet reconstruction (Barrientos, 2001; Flegenheimer et al., 2002; Scabuzzo & González de Bonaveri, 2007), burial customs and mortuary practices studies (Barrientos, 2001; Mazzia et al., 2004; Aranda, 2007; Berón & Luna, 2007; Martínez et al., 2007), skeletal geometric morphometric analysis in population dynamics studies (Barrientos & Pérez, 2005), paleopathology studies (Barrientos, 2001; Luna, 2008), and post-depositional alteration studies on bone remains (Di Donato, 2007). New chronological data were added to these studies, which allowed investigating human settlement dates and dynamics, which, in turn, generated original models about various social aspects of these populations (Barrientos, 2001; Politis & Madrid, 2001; Barrientos & Pérez, 2005; Barrientos & Masse, 2014).

Archaeological research in the Central Pampean Dunefields (CPD, sensu Messineo et al., 2018) has shown that shallow lake environments have been attractive settlement areas for humans since early times (Castellanos, 1924; Aguerre, 1996; Salceda & Méndez, 1999; Avila, 2011; Politis et al., 2012; Avila & Ceruti, 2013; Mendonça et al., 2013; Oliva et al., 2015; Oliva & Panizza, 2017; Scheifler et al., 2017; Messineo et al., 2018, 2019). In many of the archaeological sites, the concurrence of burials and remains of residential occupations (stone and bone tools and potsherds, among others) is observed, which has been understood as the result of the repeated occupation by temporally discrete hunter-gatherers' groups. Skeletal remains are found in uncovered sites, next to lakes and hillocks, along with lithic artifacts and, to a lesser extent, pottery elements and standardized bone technology made with guanaco bones (Álvarez, 2014; Ceruti, 1993). In addition to this, records of faunal remains mainly associated with subsistence consumption are present (Álvarez, 2018; Messineo & Scheifler, 2016). This information is consistent with stable isotope data of human skeletal remains, which suggest a diet primarily based on the intake of C3 plant-eating animals (Mendonça et al., 2013; Messineo & Scheifler, 2016; Scheifler et al., 2017; Oliva & Solomita Banfi, 2017; Messineo et al., 2018). In general, the chronology obtained thus far dates human occupation in the area to the Late Holocene. Some of the radiometric analyses associated yield dates corresponding to the Early and Middle Holocene (Table 1).

In most of the cases, the CPD bioarchaeological record is found uncovered, dispersed, and isolated, featuring a predominantly low anatomical completeness. Nevertheless, partially exposed articulated anatomical units were identified (Laguna Los Pampas and Chadilauquen, among others), which would correspond to simple primary burials (Aguerre, 1996; Politis et al., 2012; Mendonça et al., 2013; Messineo et al., 2018). The number of buried individuals suggests that during the Holocene, these sites were occupied by hunter-gatherers organized in small groups. The minimum number of individuals does not exceed a maximum value of 5 (Mendonça et al., 2013; Salceda & Méndez, 1999; Oliva et al., 2015; Pereyra, 2019; Messineo

Table 1 Radiocarbon dating information from the Central Pampean Dunefields sites

Site	Lab number	Material/species	14C year BP	References
Laguna de los Pampas	AA-90127	*Tooth, Homo sapiens*	8971 ± 77	Politis et al. (2012)
Laguna de los Pampas	AA-93221	*Tooth, Homo sapiens*	8835 ± 83	Politis et al. (2012)
Laguna El Doce	AA-89915	*Tooth, Homo sapiens*	8274 ± 68	Ávila (2011)
Laguna de los Pampas	AA-110831	*Tooth, Homo sapiens*	7653 ± 39	Messineo et al. (2019)
La Susana 1	AA-105970	*Bone, Lama guanicoe*	7446 ± 52	Scheifler et al. (2017)
Laguna de los Pampas	AA-110832	*Bone, Homo sapiens*	7089 ± 37	Messineo et al. (2019)
Laguna El Doce	AA-89914	*Bone, Lama guanicoe*	7026 ± 58	Ávila (2011)
Laguna de los Pampas	AA-106728	*Bone, Lama guanicoe*	7024 ± 45	Messineo et al. (2018)
Laguna Chica	PSUAMS-6965	*Bone, Homo sapiens*	6870 ± 30	Roca-Rada et al. (2021)
Laguna Cabeza de Buey 2	AA-104106	Charcoal	6801 ± 41	Messineo and Scheifler (2016)
Laguna de los Pampas	AA-110829	*Tooth, Homo sapiens*	6543 ± 37	Messineo et al. (2019)
Laguna Chica	UCIAMS-185303	*Bone, Homo sapiens*	6080 ± 15	Posth et al. (2018)
Laguna Chica	UCIAMS-185302	*Bone, Homo sapiens*	5930 ± 15	Posth et al. (2018)
Laguna de los Pampas	AA-106730	*Bone, Homo sapiens*	5924 ± 40	Messineo et al. (2019)
Laguna Chica	PSUAMS-6966	*Bone, Homo sapiens*	5865 ± 30	Roca-Rada et al. (2021)
Laguna de los Pampas	MAMS-24770	*Bone, Homo sapiens*	5819 ± 24	Messineo et al. (2019)
Laguna de los Pampas	AA-106729	*Bone, Lama guanicoe*	5815 ± 40	Messineo et al. (2018)
Laguna de los Pampas	AA-108848	*Bone, Homo sapiens*	5688 ± 36	Messineo et al. (2019)
Laguna de los Pampas	AA-93220	*Bone, Lama guanicoe*	5684 ± 61	Politis et al. (2012)
Laguna Cabeza de Buey 2	AA-100947	*Bone, Homo sapiens*	4150 ± 42	Messineo and Scheifler (2016)
Laguna Chaudilauquen	AA-89807	*Tooth, Homo sapiens*	3714 ± 56	Mendonça et al. (2013)
Laguna Chaudilauquen	AA-90593	*Bone, Homo sapiens*	3657 ± 52	Tapia et al. (2020)

(continued)

Table 1 (continued)

Site	Lab number	Material/species	14C year BP	References
Laguna Chaudilauquen	AA-89808	*Tooth, Homo sapiens*	3629 ± 56	Mendonça et al. (2013)
Laguna El Doce	LP-3101	*Bone, Homo sapiens*	3490 ± 120	Gallego and Avila (2019)
Laguna Cabeza de Buey 2	AA110833	*Bone, Ozotoceros bezoarticus*	3120 ± 26	Scheifler et al. (2023)
Huencú Nazar	AA-105969	*Bone, Lama guanicoe*	3000 ± 43	Scheifler (2020)
Laguna La Pestaña	LP-2690	*Bone, Homo sapiens*	2960 ± 70	Oliva and Panizza (2017)
Huencú Nazar	PSUAMS#5738	*Bone, Lama guanicoe*	2925 ± 20	Scheifler (2020)
Laguna del Venado	AA-95516	Charcoal	2674 ± 38	Oliva and Solomita Banfi (2017)
Laguna Salalé	LP-1667	*Bone, Homo sapiens*	2650 + 70	Oliva et al. (2015)
Laguna El Doce	AA-89919	Pottery	2350 ± 180	Ávila (2011)
Las Marías	No data	*Bone, Homo sapiens*	2140 ± 80	Avila and Ceruti (2013)
Las Marías	No data	*Bone, Homo sapiens*	1880 ± 90	Avila and Ceruti (2013)
Laguna Chica	UCIAMS-185301	*Bone, Homo sapiens*	1750 ± 15	Posth et al. (2018)
Laguna Cuero de Zorro	LP-2679	*Bone, Homo sapiens*	1640 ± 60	Oliva and Panizza (2017)
Laguna Neicorehue	AA-107820	Pottery	1577 ± 20	Tapia et al. (2020)
Laguna El Doce	AA-89918	Pottery	1555 ± 85	Ávila (2011)
Laguna Las Lágrimas	LP-3670	*Bone, Homo sapiens*	1450 ± 50	Avila et al. (in press)
El Castillo	No data	*Homo sapiens*	1430 ± 60	Heider (2013)
Laguna Chaudilauquen	AA-107819	Pottery	1202 ± 20	Tapia et al. (2020)
Laguna Chaudilauquen	AA-111979	Pottery	657 ± 78	Tapia et al. (2020)

et al., 2019; Roca-Rada et al., 2021; Posth et al., 2018), although there are some exceptions (e.g. Laguna de los Pampas site, Politis et al., 2012) (Fig. 1).

This chapter summarizes and discusses the main results achieved over the investigations in the northern CPD shallow lake environment, in an attempt to contribute to bioarchaeological research of Holocene Pampean hunter-gatherers. The aim is to contribute to a better understanding of past biocultural processes and their social ramifications, despite the post-depositional processes that have altered part of the bioarchaeological record in the region. Human skeletal assemblages retrieved from the Laguna El Doce y Laguna Las Lágrimas sites are analyzed, in addition to osteological assemblages retrieved by local enthusiasts.

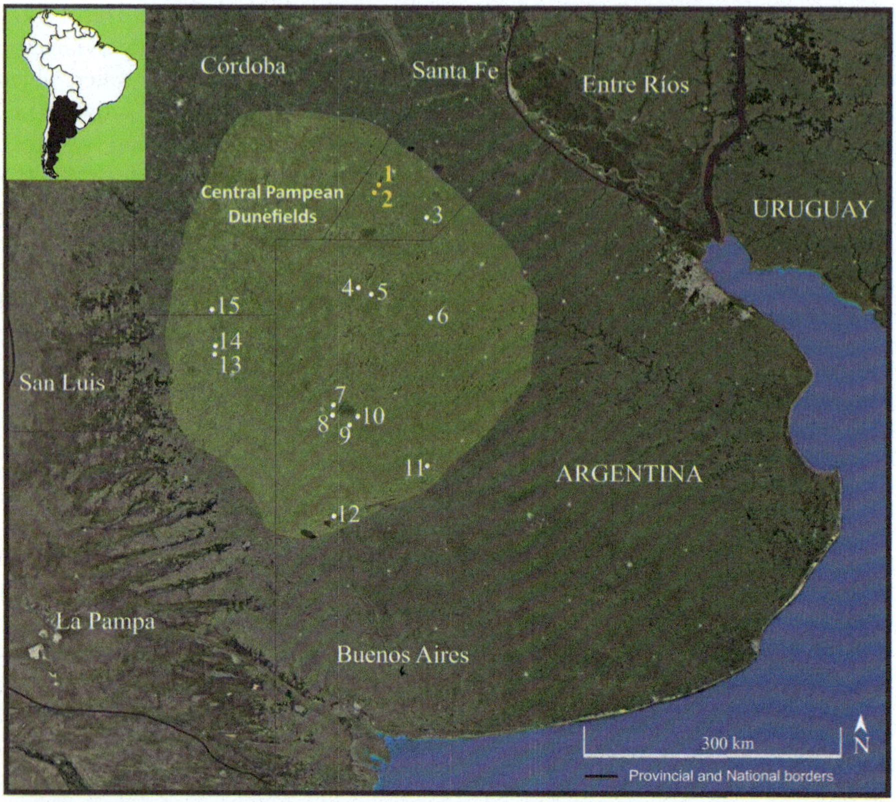

Fig. 1 Map showing the location of the Central Pampean Dunefields sites. References: (1) Laguna Las Lágrimas, (2) Laguna El Doce, (3) Laguna Las Marías, (4) Laguna Salalé, (5) Laguna La Pestaña, (6) Laguna de los Pampas, (7) Laguna Cuero de Zorro, (8) Huencú Nazar, (9) La Susana 1, (10) Laguna Chica, (11) Laguna Cabeza de Buey 2, (12) Laguna y arroyo Venado, (13) Laguna Chadilauquen, (14) Lagunas Neicorehue, and (15) Laguna El Castillo

Physical Setting and Archaeological Background

Human skeletal assemblages were found in different shallow lake beaches in the northern CPD (General López County, southwestern Santa Fe province, Argentina). The lakes of the northern CPD are the most prominent elements of the landscape. These dune fields (Zárate & Tripaldi, 2012) display a geological monotony that is characterized by the presence of a dunes and hillocks system. These dunes, a consequence of the action of southwestern winds, delimitate the depressed areas which promote shallow lake formation. Remains of these landforms are preserved on the edge of some of the shallow lakes, on which water erosion carved ravines (Iriondo & Kröhling, 2007; Zárate & Tripaldi, 2012).

The shallow lakes fulfill key environmental functions since they represent a unique habitat for flora and fauna, they are important enclaves of human activity,

and they serve as river regime moderators by storing excess water. The shallow lakes' water surface and volume fluctuate considerably during various pluviometric events, alternating between extensive periods of drought and excess water, which affects water availability. This process is intensified by anthropic activity (i.e., changes in land use, increased urban development, and regulation systems). Although the area of a shallow lake responds to dry and wet pluviometric events by reducing or expanding its surface, the shallow lake's morphometric behavior also depends on factors such as economic use, winds, runoff water, and algae proliferation (Aliaga et al., 2016).

These processes rearrange sediments and can cause the spatial movement of archaeological materials, which generates associations of materials on the banks. Archaeological materials display a surface distribution that is caused by the dynamics of taphonomic agents that produce their dispersion, deterioration, and loss, depending on their specific transportability predispositions (Ayuso et al., 2006; Avila, 2011; Avila et al., 2013; Cornaglia Fernández, 2013; Gabrielloni, 2012; Gabrielloni & Píccoli, 2011; Píccoli & Gabrielloni, 2012). Due to exposure, some artifacts and bones are recovered by collectors and amateur archaeologists.

This superficial exposure of lithic instruments, sherds, and bone remains allowed Ceruti (1987, 1992, 1993), during the 1980s and 1990s, to identify a series of archaeological sites in the northern CPD shallow lake environment, including the Laguna Las Lágrimas and Laguna El Doce sites (Fig. 1). In 1999, Avila and his team began systematic investigations in the area (Avila, 2011; Avila & Ceruti, 2013; Avila et al., 2013), which resulted in two of the area's most important osteological collections in the Laguna El Doce (LED)—*idem* El Doce (Ceruti, 1987)—(33°54′18″S, 62°08′45″W), and Laguna Las Lágrimas (LLL)—*idem* Las Lágrimas (Ceruti, 1987)—(33°51′58″S, 62°06′48″W) (Avila, 2011) (Fig. 2).

At the LED site, the associated material evidence comprises a variety of lithic instruments manufactured by knapping (e.g., scrapers, knives, and projectile points), and artifacts made by pecking, grinding, polishing, and/or abrading (e.g., *bola* balls and grinding artifacts). The raw materials used for tool production are scattered over a distance of more than 300 km away from the shallow lake, surely in outcrops located somewhere in the Tandilia Northern Hills, the Ventania Southern Hills, and the hills of Córdoba and San Luis province (Avila, 2011). The presence of these allochthonous lithic resources allows us to associate them with different groups settled in the provisioning areas. Likewise, the presence of allochthonous faunal material (sea snail, *Adelomenon brasilensis*) allows us to consider various mobility or resource exchange circuits (Cornaglia Fernández, 2013). The fauna recovered at this site is typical of semi-arid environments and wetlands. Archaeofaunal analyses revealed a high faunal usage, both for consumption and manufacturing instruments, mainly from guanaco bone. In addition, extinct mammalian species were identified. The presence of anthropic marks on these remains (Cornaglia Fernández, 2013; Cornaglia & Buc, 2013) accounts for the coexistence of this fauna with hunter-gatherer populations during various occupation events. Likewise, the detection of species from various environments accounts for an environment that, despite its

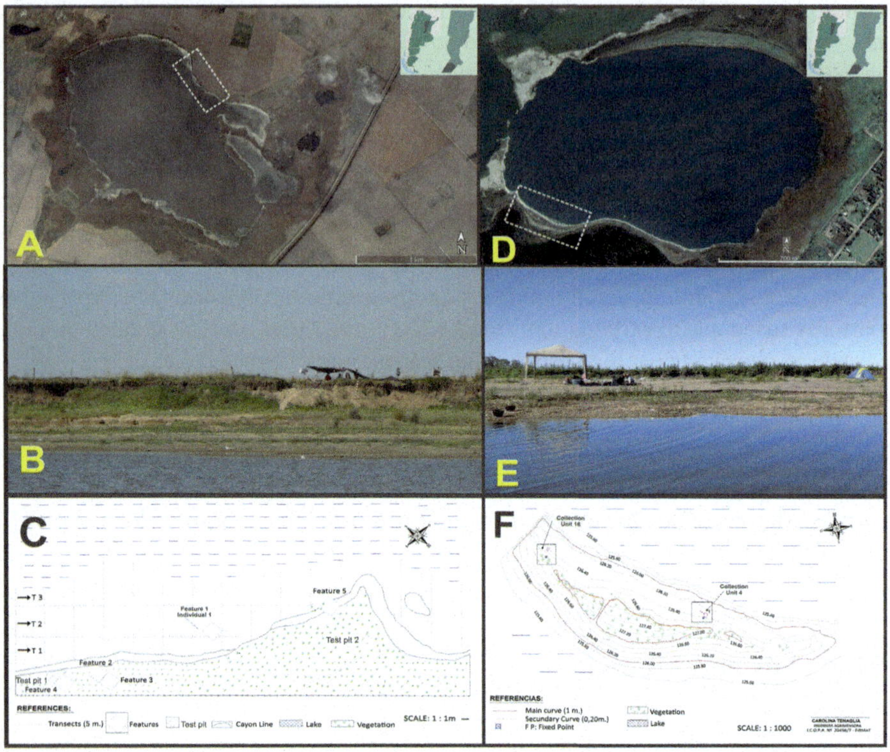

Fig. 2 Map showing the location of Laguna El Doce and Laguna Las Lágrimas sites: (**a**) Location of the Laguna El Doce site. (**b**) View of the NE sector of the El Doce Lake. (**c**) Planview of the El Doce site. (**d**) Location of the Laguna Las Lágrimas site. (**e**) View of the SW sector of the Laguna Las Lágrimas site. (**f**) Planview of the Laguna Las Lagrimas site

dynamism and the climate oscillations of the Latest Pleistocene and the Holocene, was attractive to human populations.

At the LLL site, the presence of species typical of semi-arid environments and wetlands is detected, as in LED. These are associated with lithic instruments made of raw materials suggesting various mobility or resource exchange circuits, as in the case of the LED site. The presence of pottery allows us to consider that some of the occupation periods were characterized by a more intensive collection and storage of plant food (Avila et al., 2022).

Evidence suggests that these shallow lake sites represented areas where multiple activities took place, such as the differential processing of plant and animal products, mineral pigments, and clays, among others (Avila, 2011; Cornaglia Fernández, 2013). Along with these, funerary practices were also carried out. The evidence of these burial events mostly comprises scattered and fragmented skeletal remains, uncovered and dispersed mainly due to shallow lake volume variations and wind (Gabrielloni, 2012). The level of fractures observed is closely related to the burial

conditions and the shape of each anatomical part (Píccoli, 2009). Along with the above, the LED sample is particularly degraded by floralturbation agents, especially in bone and teeth remains recovered near the water mark, and by the sediment's high carbonate content, which covered some of the samples completely with $CaCO_3$ (Gabrielloni, 2012).

At the LED site, separate depositional events occurred in the Early, Middle, and Late Holocene (Table 1). The human skeletal assemblage comprises mostly disarticulated and fragmented skeletal remains scattered on the shallow lake beach. Also, some skeletal elements were found in their anatomical relation. One consists of an incomplete subadult skeleton in the left lateral position, half-buried and with bones partially fused due to carbonation. Another comprises four adult articulated thoracic vertebrae retrieved near the shallow lake's water mark (Ayuso et al., 2006).

At the LLL site, skeletal assemblages half-buried were identified on the southwestern banks. Within archaeological collection unit 4 (Fig. 2), a partial axial skeleton and articulated upper limbs, in the left lateral position, and, at a distance of 110 m away, part of the lower limbs, also articulated. In addition, a cluster of commingled disarticulated bones was recovered within archaeological collection unit 16 (Fig. 2) (Avila et al., 2022).

Materials and Methods

All the human skeletal remains from LED and LLL sites were analyzed, as well as private collections (PC) whose origin could be estimated through unstructured interviews with the collectors. Most of this set was apparently retrieved just before the field work in the LED site began, with the exception of one element (skull), which was probably found in the Melincué shallow lake.

In the case of the isolated disarticulated elements, bones and teeth are considered the unit of analysis. Whenever possible, anatomical identification and laterality were determined. The size and composition of the sample were quantified using taxonomic (Number of Identified Specimens—NISP) and relative abundance (Minimum Number of Elements—MNE) measures (Lyman, 1994; Mengoni Goñalons, 1988). The refitted bone specimens were defined as units of analysis (Píccoli, 2009). Fragmentation intensity was assessed by determining the fragmentation index (FI) (Luna, 2008) and completeness index (Buikstra & Ubelaker, 1994).

Age estimation of subadult individuals was carried out through tooth formation and eruption sequence (Ubelaker, 1980), primary ossification of the coxal bones, epiphyseal fusion progression (Johnston, 1961, 1962), and various osteometric relationships of clavicle, humerus, radius, ulna, ilium, femur, tibia, and fibula (Scheuer & Black, 2000)—femur, humerus, radius, ulna, ilium, tibia, and fibula (Merchant & Ubelaker, 1977). Whenever possible, sex was considered when estimating age. In adult individuals, age was estimated based on cranial suture obliteration (Meindl & Lovejoy, 1985), tooth wear degree in 18 teeth (Lovejoy, 1985; Brothwell, 1993; L'Heureux, 1998; Barrientos & L'Heureux, 2001), and relief and contour

transformation of the pubic symphysis (Todd, 1920, 1921) and the auricular surface of the ilium (Lovejoy et al., 1985).

As sex estimation criteria differ depending on the subject's age, different analyses were performed in accordance with the two large age categories (subadults and adults). In the case of subadults, since no specific method was found for newborn (up to 0.5 years old) and juvenile (10–18 years old) individuals, sex determination was only applied in the remaining subcategory (child, 0.5–10 years old). Deciduous teeth were measured (sensu Hillson, 1986, 1996; Mayhall, 1992, 2000; Hillson et al., 2005). In the case of adult individuals, skull and pelvis traits were studied (Buikstra & Ubelaker, 1994; Barboza, 2015). Additionally, the following long bone dimensions were taken: the circumference of the femoral shaft middle third (Black, 1978), and the maximum and transverse diameter of the tibia at nutrient foramen height and its circumference (Slaus & Tomicic, 2005). The existence of sexual dimorphism was revealed by plotting the distribution of these dimensions (Píccoli, 2009). Tooth dimensions were also analyzed in order to sex adult individuals, observing dimorphic expression based on the cut-off point proposed by Luna (2008). Only those individuals whose skeletal parts could not be sexed with any technique were considered indeterminate (Píccoli, 2009).

In order to calculate the minimum number of individuals (MNI; Adams & Konigsberg, 2004) most of the elements comprising the sample were included, with the exception of ribs. Initially, anatomical part frequencies were established considering the observed completeness and fragmentation indexes (see above). Each anatomical element was numbered, so the same number in different sections represented uniqueness (Píccoli, 2009). Subsequently, the information of sex and age of death of each element was factored in; additionally, the number of right- and left-side elements was estimated. Next, using shared identifying features, pairs were defined, and their number was determined; the number of dissimilar elements was also added. To minimize overestimation, elements of indeterminate sex or that could not be assigned to any particular age category were considered to belong to one of the individuals whose sex/age had been determined. Individuals of indeterminate sex and/or age were reassigned, taking into account their relative contribution to the frequencies of the determined cases (*v.* Luna, 2008: 167). For this reason, the final results do not directly reflect the sum of the male, female, and indeterminate elements.

In order to assess metabolic and functional stress, we took into account different approaches that allow identifying indicators at the macroscopic level with the aid of a binocular magnifier. Tooth wear was visually evaluated using two ordinal and discrete techniques (Smith, 1984 for incisors, canines, and premolars, and Scott 1979 for molars; in Buikstra & Ubelaker, 1994). Compensatory physiological processes of tooth wear were analyzed. The active eruption process was assessed using the UEDCA ratio criterion (L'Heureux, 2014). Oral infectious and degenerative processes were surveyed. Abscesses were recorded depending on the type and size of the lesion, as well as indicating the location. Special attention was paid to their differentiation from periodontal disease. Teeth were recorded as lost antemortem if their alveolus had been partly or completely closed. The presence of caries was

determined using a dental probe (Buikstra & Ubelaker, 1994; Clarke & Hirsh, 1991). Caries were recorded according to Buikstra and Ubelaker (1994: 55). Rates of calculus (Buikstra & Ubelaker, 1994: 56) and location (for each tooth, indicating buccal or lingual surface) were also recorded. Recording of the hypoplasia and hypocalcified type considered the seven categories proposed by Buikstra and Ubelaker (1994: 56). Hypocalcification color was coded according to the standard presented by Buikstra and Ubelaker (1994: 57). Localization of the enamel defect was observed on two anatomical surfaces (facial and lingual); attrition rendered the occlusal observation impossible. All teeth with remnant enamel, and not broken due to post-mortem damage, were included. Cranial porotic lesions (cribra orbitalia and porotic hyperostosis) were recorded using the system recommended by Builkstra and Ubelaker (Buikstra & Ubelaker, 1994), which distinguishes degrees of expression, localization, and reaction. Osteoarthritis was identified by the presence of lipping, porosity, and eburnation in the articular surface of most commonly affected joints (shoulder, elbow, spine, hands, hips, and knees) (Buikstra & Ubelaker, 1994; Ortner, 2003).

Trauma and abnormal bone formation were examined macroscopically and recorded patterns based on Buikstra and Ubelaker (1994) standards. Regarding traumas, the locations, types, patterns, associated skeletal defects (abnormal bone loss; mature callus formation), and its healing processes were all considered. For abnormal bone formation, the observations included were type, proportion of the unit affected, presence of specific structures, and changes on periosteal/endostal surfaces. Changes specific to vertebral bodies and lipping in association with arthritis were recorded with the related pathologies.

Results

The LED human skeletal collection is mainly composed of disarticulated bones (Tables 2 and 3), in addition to the aforementioned articulated bones—an incomplete subadult skeleton (Fig. 3) and four adult thoracic vertebrae. On the other hand, in the case of the LLL sample, the percentage of disarticulated bones is lower (Table 2); all of these bones are from archaeological collection unit 16. In contrast, anatomical refitting of the adult bones from archaeological collection unit 4 was done. This adult skeleton presents many small elements, including carpals, phalanges, and the hyoid bone (Fig. 3).

The private collections, composed of isolated bones only, mainly comprise skull bones (Table 2), a fact explained by the greater visibility of these elements to the eyes of non-specialized collectors (Gabrielloni & Píccoli, 2011; Píccoli, 2009; Avila et al., 2022).

In the case of the LED site, 23% of the sample could not be assigned to any anatomical part. This is probably not closely related to bone mineral density, considering that, within the two large age categories, the percentage of unidentified adult

Table 2 Abundance measurement—adults

Adults									
Assemblage	LED			LLL			PC		
Element	NISP	MNE	FI	NISP	MNE	FI	NISP	MNE	FI
Calvaria	69	7	**0.10**	4	3	0.75	36	3	**0.08**
Face	5	4	0.80	2	2	1.00	10	2	0.20
Mandible	13	8	0.62	1	1	1.00	3	3	1.00
Cervical vertebra	9	7	0.78						
Thoracic vertebra	8	7	0.88	3	3	1.00	1	1	1.00
Lumbar vertebra	1	1	1.00	4	4	1.00			
Indet. Vertebra				1	1	1.00			
Clavicle	1	1	1.00	2	1	0.50			
Scapula	3	2	0.67	2	2	1.00			
Sternum	2	1	0.50						
Rib	37	31	0.84	9	9	1.00			
Humerus	20	13	0.65	4	4	1.00			
Ulna	7	7	1.00						
Radius	11	7	0.64	2	2	1.00			
Metacarpal	1	1	1.00						
Hand phalange	1	1	1.00						
Sacrum	1	1	1.00						
Coxa	15	8	0.53	4	3	0.75	1	1	1.00
Femur	40	19	0.48	5	4	0.80			
Patella	1	1	1.00						
Tibia	31	9	**0.29**	1	1	1.00			
Fibula	6	5	0.83						
Talus	2	2	1.00						
Metatarsal	3	3	1.00						
Foot phalange	1	1	1.00						
Fragment	77								
Total	365	147		44	40		51	10	

specimens (21%) is similar to that of subadults (24%). Likewise, age categories are represented by similar elements, although in different frequencies (Table 2, MNE).

Even though fragmentation hinders the observation of indicators which would allow further estimates (Tables 2 and 3, FI), the samples do display completeness profiles with significant observational possibilities. Most of the elements contain more than 75% of the anatomical part or element they belong to (Gabrielloni & Píccoli, 2011; Píccoli, 2009; Avila et al., 2022). However, in some cases, fragmentation hinders the assessment of features that would allow inferring human adaptive aspects, as in the calvaria and the facial skeleton which display the highest fragmentation of the LED assemblage (Tables 2 and 3). Post-depositional alterations in the alveolar region, as well as the presence of calcium carbonate on the surfaces of some of the teeth and alveolar cavities of the LED sample (Gabrielloni, 2012),

Table 3 Abundance measurement—subadults

Subadults						
Assemblage	LED			PC		
Element	NISP	MNE	FI	NISP	MNE	FI
Calvaria	**105**	4	**0.04**	**20**	1	**0.05**
Face	3	2	0.67	4	2	0.50
Mandible	6	2	0.33			
Cervical vertebra	2	2	1.00			
Thoracic vertebra	5	3	0.60			
Clavicle	1	1	1.00			
Rib	8	5	0.63			
Humerus	4	2	0.50			
Ulna	9	3	0.33			
Radius	4	4	1.00			
Carpal	1	1	1.00			
Metacarpal	5	4	0.80			
Hand phalange	13	12	0.92			
Ilium	3	3	1.00			
Ischium	5	2	0.40			
Pubis	1	1	1.00			
Sacrum	7	1	0.14			
Femur	12	5	0.42			
Tibia	5	4	0.80			
Fibula	5	2	0.40			
Calcaneus	1	1	1.00			
Talus	1	1	1.00			
Tarsal	4	4	1.00			
Metatarsal	4	4	1.00			
Foot phalange	7	7	1.00			
Fragment	71					
Total	292	80		24	3	

interfere with periodontal pathologies analyses, prevent measuring the dimensions required in the analysis of the enamel-alveolar crest distance, and impede the assessment of the occlusal surface's wear degree. As a result of the combined action of various post-depositional factors, 14% of the maxillae (n = 7) and 33% of the mandibles (n = 9) of adult individuals are fractured (LED). Regarding subadults, 20% of the maxillae (n = 5) are fractured (PC); in contrast, the mandibles (n = 2) are not fractured. In relation to these elements, 256 alveoli are observable. Finally, 3% of the 115 teeth belonging to adult individuals were found unattached (LED). In the case of subadults, loose teeth were not recovered; the 29 deciduous and 22 permanent teeth are in association with alveoli. In total, 38 postmortem tooth losses in adult individuals and 21 in subadults were observed. Incisors were the most frequently lost teeth (37% in adults; 81% in subadults), followed by premolars in adults (26%).

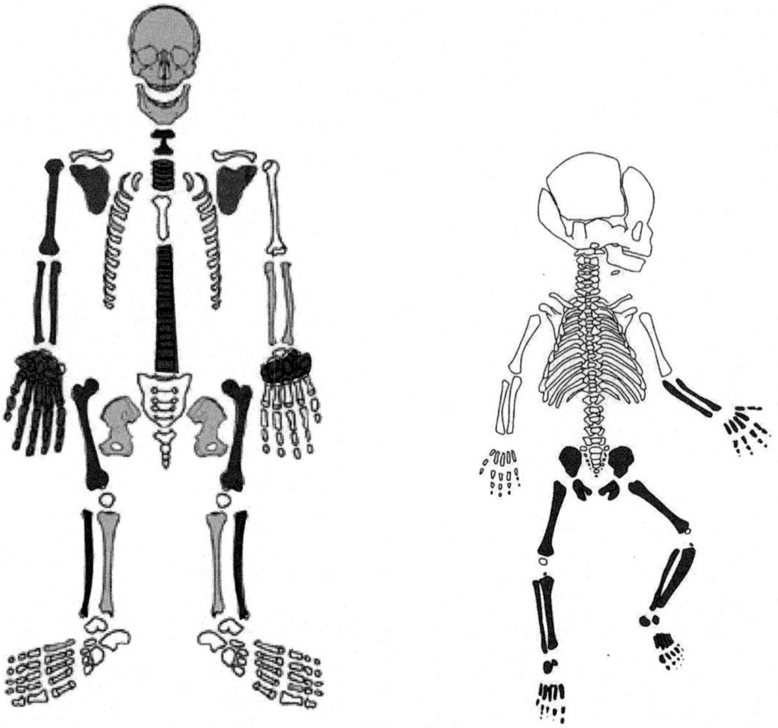

Fig. 3 Skeletal element representation. References: adult individual from Lagunas Las Lagrimas site (left) and child individual from Laguna El Doce site (right)

Age–Sex Structure

The probable age of death was estimated in 29 bone elements from LED, 2 elements from LLL, and 12 elements from PC. Within the subadult category, we were able to estimate the probable age of death in 16 elements from LED and 3 from PC. In the adult category, 13 elements were identified at the LED site, 2 at the LLL site, and 9 in the PC set (Table 4).

Only five elements assigned to the child subcategory could be sexed at the LED site. Two of the individuals were identified as female and three as male. Within the PC set, it was possible to identify only one subadult element as male.

At the LED site, it was possible to ascertain sex in 28 adult cases: 13 females and 15 males; at the LLL site, it was possible to sex two elements, both of them female; in the PC set, three elements were identified as female and one as male.

Table 4 Age at death determination

	Assemblage		
Age range	LED	LLL	CA
Newborn	1		
Child	12		2
Juvenile	3		
Indet. subadult			1
I	2		2
II	3		3
III	4	2	1
IV	4		
V			
Indet. adult			3

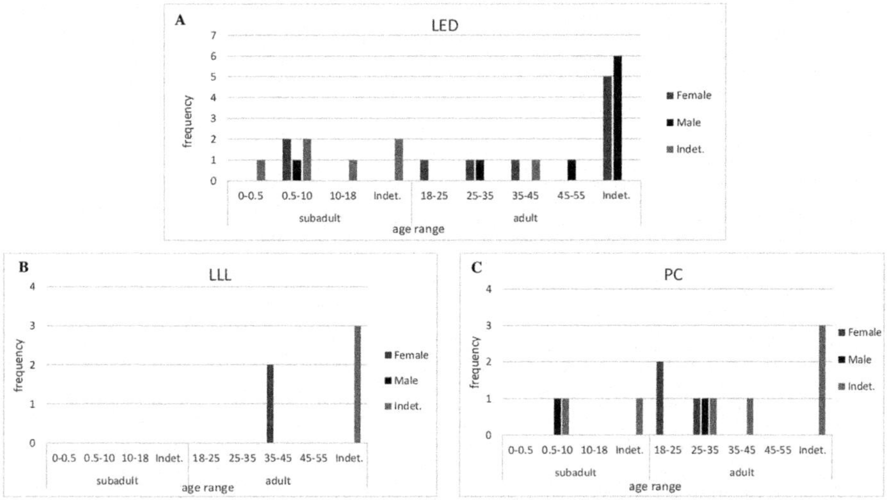

Fig. 4 Age and sex patterns (MNI) in Laguna El Doce assemblage (**a**), Laguna Las Lagrimas assemblage (**b**), and private collections (**c**)

Number of Individuals

In all cases, the age-at-death and sex could not be estimated in a high frequency (Fig. 4). Individuals of indeterminate sex and/or age were reassigned taking into account their relative contribution to the frequencies of the determined cases; consequently, the set of commingled and fragmented elements corresponds to an MNI of 16 in the LED site and an MNI of 3 in the LLL site. In both cases, the partially complete articulated skeletons must also be considered, so that the total MNI would add up to 17 and 4, respectively (Píccoli, 2009; Avila et al., 2022).

If the presumed origin of the elements of the PC set is sustained, its relative contribution should be included in the MNI calculation of the LED site; with the exception of a 35- to 45-year-old individual (i.e., the skull from Melincué shallow lake). The final MNI amounts to 19 individuals. Thus, two additional individuals should be included in the sample, a child individual of indeterminate sex and an 18-to 25-year-old female individual.

As seen in Fig. 4, all the individuals of different ages and sexes are represented. Taken as a whole, child individuals are the most represented among subadults. In adults, the most represented are the individuals in the 25- to 45-year-old range. Within the adult stage, a slight disparity regarding sex is also observed: female individuals are more represented than males (Fig. 5).

Oral Cavity Physiological Adaptation Indicators

In the subadult category (children), we identified a wear degree ranging from moderate cusp damage to considerable dentin exposure, but with a significant enamel presence in molars and in the circumference of the remaining teeth (i.e., mild = 56% to moderate = 44%). Considering that permanent molars erupt later in life, they

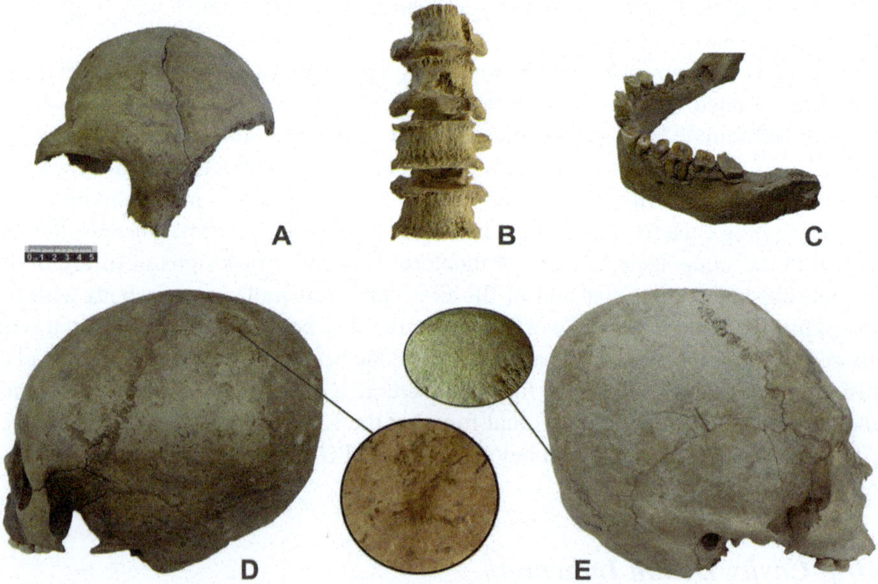

Fig. 5 (**a** and **e**) Porosities in the outer table of the cranial vault. (**b**) Marginal lipping on lumbar vertebrae. (**c**) Dental wear, lipping (second and third left molars), abscess (first left molar), and antemortem tooth loss (second left incisor; second and third right molar). (**d**) Depressed skull fracture

display a minimum degree of wear. Given the eruption sequence, the observed wear average patterns are not in contradiction with wear accumulation opportunities, with the exception of two male individuals from the LED site whose two second left molars display a higher wear degree than their first adjacent molars. One of these molars displays wear in an oblique, mesiodistal direction, contrary to the prevailing pattern and direction (flattening-horizontal). No compensatory physiological processes were determined in the subadult category because the wear degree present was not severe.

In the case of adult dentition, various wear degrees were observed. In the PC set, mild (9%), moderate (42%) and high (49%) wear degrees were observed; mild wear is associated with 18- to 25-year-old individuals. The remaining cases, in general, display a high wear degree, with the exception of a 25- to 35-year-old male individual from the LED site, with a mild-moderate wear degree. Anterior teeth were not found to be more worn than posterior teeth. Comparison of upper and lower wear by individuals was not possible because none of the elements was found with its antagonist. Regarding the wear of the first and second molars, an acceleration of the wear rate in the 25- to 45-year-old range is observed. Albeit there is a higher proportion of flattening in the horizontal direction, there is no prevailing wear pattern. However, at the LED site, convex (11%) and concave (30%) shapes are found in highly worn premolar, molar, and canine pieces, associated with 35- to 55-year-old individuals of both sexes (Fig. 5c). At the LLL site, albeit the flat shape prevails, flat and oblique directions are present in equal measure (31%). Concave shapes (38%) affect teeth indiscriminately and are always associated with high wear degrees.

Compensatory physiological processes as a response to wear in adult individuals are all associated with the highest wear degrees. In the LED assemblage, a 35- to 45-year-old female individual shows signs of active eruption in highly worn teeth, probably as a compensatory process given in response to the observed crown loss (no evidence of infections was observed). A 45- to 55-year-old male individual displays a lipping (inactive) near the third lower left molar (Fig. 5c), whereas an individual in the same age range but of indeterminate sex shows signs of tilting in the second upper left premolar and in the first upper left molar pieces, both with the maximum wear degree. It is worth mentioning that both cases are associated with abscesses that cause high wear and alveolar bone destruction which, in turn, lead to tilting (Píccoli, 2009). Inactive lippings were also detected near the two posterior molars of an adult female individual found at the LLL site and near the two posterior left molars of an individual belonging to the PC set.

Oral Cavity Health Indicators

Abscesses were only detected in adult individuals. The highest frequency (n = 22) is found in the LED assemblage, being in all cases of the complex and vertical angular shape type. These are located in the oral alveolar canal, with the exception of one

case located in the lingual region near the third lower right molar of a 45- to 55-year-old male individual. There is no clear pattern of abscess location in relation to tooth location. Most of the abscesses (63%, n = 13) were found in the mandibles. No abscesses were found in individuals within the first adult stages (18–35 years old) (Píccoli, 2009). At the LLL site, only three abscesses were detected, two associated with the first left premolars—upper and lower (periapical and complex types)—in a female individual, and a third involving the second upper right incisor (complex type) in a different female individual. In the PC set, an adult individual with high average wear presents two abscesses in the oral alveolar canals near the lower right canine and the first premolar.

Considering the degree of tooth wear and abscesses found in adult individuals, only for this age category did we find alveoli being resorbed (n = 28); 12 of these (43%) belong to the PC set. It is followed by the LLL site, as its assemblage represents 36% (n = 10) of the total. This process displays various frequencies for molars (54%, n = 15), premolars (29%, n = 8), incisors (14%, n = 4), and canines (4%, n = 1). The low prevalence of antemortem tooth loss in pieces with extreme wear and pulp cavity exposure was found to be influenced by the presence of physiological adaptation factors such as molar tilting.

A low prevalence of caries is found within the sample. Caries were identified in 5 of the 77 teeth in occlusion present within the LED assemblage, in the form of small cavities or fissures in three subadult individuals. The prevalence of caries was detected in the left molars. Two male child individuals show caries signs in the lower deciduous dentition. One of them presents lingual caries in the first molar and occlusal caries in the second molar; the other one presents a double cariogenic process (both occlusal and labial) in the second molar. In the only affected female child individual, the cariogenic process is confined to the first upper molar (Píccoli, 2009). In the PC set, an occlusal cavity was identified in the posterior dentition of an adult individual. Only a small amount of dental calculus was found on the upper dentition. The greatest prevalence was detected for the molars: on the buccal surface of a second molar belonging to a female child individual of the LED assemblage, and in the second and third right molars, as well as the second left incisor, in one of the middle-aged female individuals of the LLL assemblage.

Metabolic Stress Processes

A small proportion of tooth enamel development lesions was observed (n = 13%). The location of defects was recorded on the facial surface only. In the case of subadult individuals in the LED collection, enamel hypoplasia and dental hypocalcification signs were identified, in both deciduous and permanent teeth of child individuals of both sexes (two of each). Hypocalcification was recorded as a discrete boundary opacity. The individuals under study do not show signs of enamel hypoplasia on canines or incisors, the most susceptible teeth to hypoplasia (see Goodman et al., 1980). Only the molars show linear horizontal or vertical grooves

(with the exception of a second lower deciduous molars). All the individuals show linear grooves in permanent dentition; only two of them are also in the deciduous dentition (Píccoli, 2009). On the contrary, one child individual from the PC set shows linear grooves in the upper left deciduous canine (Gabrielloni & Píccoli, 2011).

Within the adult category in the LED site, only a 45- to 55-year-old male individual shows signs of linear hypoplasia in the lower right canine. No further information can be extracted from the remaining teeth due to their high wear degree (Píccoli, 2009), whereas in the case of the PC set, three female individuals in the first age ranges of the adult category (18–35 years old) are involved. As expected, not only incisors and canines are involved (a second left incisor and a right canine, both upper side) but also the first premolars (two uppers, both sides, and one lower right). They showed linear grooves, and also, one of them, linear horizontal pits. Even though molars are present, they showed no signs of hypoplasia. Furthermore, two of these individuals show evidence of hypocalcification in the form of white/cream lines, one in the second lower right incisor and the other one in the first lower left premolar. In the case of the LED site, hypocalcification is present as yellow, cream, or white lines and only in the left molars of child individuals: two males (second deciduous molar and first permanent molar, both lower) and one female (first permanent upper molar). Since enamel defects are more frequently observed in the middle and cervical thirds (Goodman & Armelagos, 1985; Larsen, 2000), high tooth wear degrees causing a partial, or even complete, crown removal probably hindered the possibility of detecting such indicators. As to the association with pathologies, enamel defects are associated with dental caries in two adult and three child individuals.

Cranial porotic lesion is another indicator of metabolic issues detected (Fig. 5a, e), only in the LED (n = 30%[1]) and PC assemblages (n = 20%). In the case of the LED site, two child individuals of indeterminate sex show porosity signs in the outer table of the diploë, adjacent to the sutures and symmetrically arranged in the parietal bone in one case and in the temporal bone in the other (Píccoli, 2009). The lesions were active in the first case (porosity shows that sharp edges and woven bone were recorded), and the other showed very indistinct porosity remodeled. It was also found in an adult male individual and a female individual 35–45 years old, showing indistinct porosity and true porosity in the outer table of the diploë, adjacent to the sutures of the parietal and frontal bones in the male individual and in the parietal and occipital bones in the female individual. The porosity was in the process of recovery at the time of death. In the PC set, only one case of porotic hyperostosis was found, in a probable female adult individual, whose features correspond to the distinctive morphology of recovered lesions adjacent to the lambdoid, sagittal, and coronal sutures. Whenever possible, no porosity signs were observed on the orbital surface.

[1] Percentages are calculated based on the elements present and observable for the pathology markers (i.e., not including element present but taphonomically modified).

Osteoarthritis

At the LED site, osteoarthritis was only found in seven elements (n = 14%). The most affected anatomical part belongs to the axial skeleton. A cervical vertebra presents porosity around the vertebral body as well as extensive osteophyte formation in the form of lipping affecting two-thirds of the body. The four articulated thoracic vertebrae and one lumbar vertebra show evidence of a mild manifestation comprising two-thirds of the vertebral surface (Fig. 5b). The other two anatomical parts with an osteoarthritic condition are the left femur and tibia of a male individual.

In turn, at the LLL site, osteoarthritis was detected in a 35- to 45-year-old female individual with generalized osteoarthritis in vertebrae (vertebral bodies and articular facets of neural arches), the right long and short bones (articular faces of the carpal bones), and the left ulna. The manifestation level ranges from mild to moderate, compromising no more than two-thirds of the observed area.

Trauma

In the LED sample, a total of six skull traumas were observed (n = 4%), three in an adult individual and the rest in two subadult individuals. One of the subadult individuals presents an oval-shaped depression (blunt with an almost imperceptible thickening of the edge), with a 15.5 mm maximum diameter. It is located on the external table of the right parietal bone near the sagittal suture. The other is a female child individual, who presents two elevations with a blunt edge (less than 18 mm) on the peribregmatic area of the external surface of both parietal bones. Moreover, on the internal surface, there are two depressions with a translucent thinning of the diploë. The adult case is a 35- to 45-year-old female individual with two oval-shaped depressions with maximum diameters of 5.2 and 4.5 mm and an almost imperceptible thickening of the edge, located on the external table of the right parietal bone near the sagittal suture. Both look clearly remodeled so they are antemortem trauma (Fig. 5D).

Within the PC set, only one case of trauma was identified (n = 8%), in an adult individual of indeterminate sex, as a series of sharp-edged linear depressions with a maximum length of 61.73 mm and a maximum width of 17.93 mm. It is located on the external table of a parietal bone, showing signs of bone loss (Gabrielloni & Píccoli, 2011).

Abnormal Bone Formation

Of the total number of analyzed elements (n = 227) and fragments (n = 148) in the LED site, a low percentage of them (2% of the elements and 1% of the fragments) show signs of abnormal bone formation. The cases in which it was observed belong exclusively to adult individuals, compromising femur, tibia, and fibula. Both femur and tibia belong to male individuals; the remaining samples (n = 2) correspond to diaphyseal sections of the fibula belonging to individuals of indeterminate sex. Normal morphology deviations are confined to a small area of the diaphysis (less than one-third, i.e., proximal, middle, or distal thirds), and in none of the cases, is the whole bone section involved. The affected surface tends to be irregular and of variable thickness, with the presence of hypervascularity in the form of pores. Osteoblastic production of new bone tissue is detected in the case of the tibia, fibula, and a fragment, forming a porous shell-like elevation associated with a trauma. In the case of the femur, a rather shallow depression is detected, whereas the fibula is characterized by a rough surface.

Life at the Shallow Lakes

Considering that Pampean hunter-gatherers are associated with low fertility rates, reduced population sizes, and high residential mobility (until the end of the late Holocene), the archaeological record of the area is usually characterized by the presence of highly dispersed burial areas and few burials per site, among others, which in turn negatively affect paleodemographic approaches (Bernardi et al., 2020). Additionally, in the CPD shallow lake environment, human skeletal remains are generally found in the surface, with a surface distribution caused by the dynamics of various agents that induce their dispersion, deterioration, and loss. The methodological decisions made here showed the importance of analyzing all of the retrieved remains, including those that were disarticulated, small, and with various fragmentation indexes. Integrating different methodologies for sex estimation and age-at-death determination allowed the inclusion of isolated bones, although the degree of reliability and accuracy compared with other methods is lower.

While the above-mentioned characteristics are typical of the CPD bioarchaeological record, previous studies have mainly focused on whole or unfragmented bones and teeth (Mendonça et al., 2013; Pereyra, 2019; Politis et al., 2012). The results obtained using the approach proposed here allowed reducing analytical biases, showing its potential for the area under consideration. The mortality profile and the paleopathological information presented here provide a global overview that allows establishing trends in relation to the general demographic features of the hunter-gatherers who occupied the northern CPD during the Holocene. This information can be compared with known paleodemographic profiles of similar societies.

The resulting mortality profile is similar to those of different CPD sites, which allows us to assume that the sites were occupied by small groups. In relation to the sites with late occupations, the MNI probably does not exceed a maximum value of 5 (e.g., Laguna Chadilauquen, Laguna Salalé, and La Pestaña), as in the LLL site, whereas sites with occupations dating to the Early and Middle Holocene—Laguna de los Pampas, Laguna Chica, and LED—present a higher MNI. The highest MNI corresponds to the LED site, which accounts for its recurrent occupation in various dimensions and hierarchies: residential, domestic, symbolic, and sacred.

Before adulthood, the hunter-gatherer profiles present a high mortality rate between birth and the first 5 years (Johnston, 1961, 1962; Luna, 2008; Ubelaker, 1980), as observed in the LED site. Among adults, the 20- to 35-year-old range comprises the most represented age group, decreasing in the later ranges (Bernal et al., 2004; Goñi et al., 2003; Luna, 2008), which is consistent with the LED mortality profile. This range coincides with the stages in which the individuals fully assume the social roles of reproduction and group subsistence. This exposes them to various stressors and their ensuing risks. Furthermore, 20- to 40-year-old female individuals are expected to be the most represented age group, which includes the determined range of the female adult individuals of the LED (including the PC set) and LLL sites (see Fig. 5). This period coincides with the full reproductive stage when the risks are greater due to the demands of pregnancy, childbirth, and lactation, which can lead to malnutrition and infection (Grauer, 1991; Hassan, 1973).

The low frequency of tooth enamel development lesions is expectable, since hunter-gatherer populations who lived in the Pampas during the Holocene have displayed low frequencies in both conditions, with variable prevalence (Aranda, 2007; Luna, 2008; Barrientos, 2001). This could be attributed to a diet mainly based on meat and animal derivatives (Cohen, 1997; Larsen, 1987, 2000), which is consistent with the faunal record retrieved from the LED and LLL sites. Even when the anterior dentition is most susceptible to hypoplasia (see Goodman et al., 1980), in the sample under study, the most commonly affected tooth is the first molar, in both deciduous and permanent teeth. Neither differences in frequency of enamel defects between anterior and posterior teeth nor high frequency in molars have been reported (e.g., Lovell & Dawson, 2003), which highlights the importance of record enamel defects on all teeth (Towle & Irish, 2020). Association with carious processes was recorded in five cases. Since hypoplasia is a reduction in the quantity of dental enamel and hypocalcification results in the enamel of lower quality (Lovell & Dawson, 2003), the enamel cannot protect the teeth effectively, allowing disease-promoting bacteria to cause dental diseases— caries being the most common (Lacruz et al., 2017).

Hypoplasia identified in this study, generally linear, was manifested on few teeth, which supports the presence of some acquired dysplasia form. It could be assumed that the Holocene hunter-gatherers under study had frequently suffered from physiological disorders during childhood (see Simalcsik et al., 2014; Towle & Irish, 2020). Additionally, in the case of child individuals from LED that show signs of both enamel development lesions observed (hypoplasia and hypocalcification), a genetic etiology should not be excluded (see Towle & Irish, 2020).

The location of defects gives insight into the timing of metabolic events. Taking into account the teeth affected, the age at which cusp enamel completion occurs (Schaefer et al., 2009: 87–95), the weaning age mean of New Word hunter-gatherers ($\underline{x}$ = 2.32 years old; Marlowe, 2005), and the cervical wall of enamel defects represents the period of recovery (Moes & Blatt, 2018), the developmental timing of most enamel defects in the sample has little to do with stress from weaning.

The association between dental enamel defects and cranial porotic lesions could be analyzed in four cases, where the calvaria and the splanchnocranium are present. Out of the two cases with hypoplasia and hypocalcification, porotic hyperostosis appears only in one of them; the rest either have dental enamel defects or cranial porotic lesions. Nevertheless, the presence of both enamel defects and porotic hyperostosis in the sample under study suggests that environmental conditions might have been causing metabolic stresses.

Since the skulls studied show either very indistinct porosity or true porosity (sensu Buikstra & Ubelaker, 1994) and the lesions are either recovered or in recovery, with the exception of one child individual, this also allows proposing that although the causal agents did affect them, the individuals were able to survive the initial episode of morbidity. It should be noticed that the presence of healing lesions in adult individuals suggests that the etiology of porotic lesions was also present in adult life. A relationship between porotic hyperostosis and anemia has been recognized, as a result of hemolytic anemias or iron-deficiency anemia (Rivera & Mirazon Lahr, 2017). As it was described for LED and LLL sites, diets were based on the consumption of semi-arid and wetland fauna, which provides a rich source of heme iron. Although dietary components could hinder iron absorption, iron-deficiency anemia is expected to cause a diploic expansion—subsequently unveiling its structure as visible porotic lesions—only in severe chronic cases (Rivera & Mirazon Lahr, 2017). For congenital forms, postcranial lesions and enlargement of the cranial bones (Suby, 2014) should have been present in the sample studied.

Gastrointestinal parasite infection (e.g., helminthes and schistosomiasis) may also cause or aggravate malnutrition and chronic blood loss, especially in polyparasitism cases (Suby, 2014). Some soil-transmitted parasites (e.g., *Eimeria* sp. and *Trichuris* sp.) were identified in modern wild guanaco populations in Argentina (González-Rivas et al., 2019). In shallow lake environments from Central Argentina (Cordoba province), helminthes (Heterakoidea) have been reported in archaeological sites where human remains were found (Ramírez et al., 2021). For nonparasitic infectious diseases, new bone formation should have been present in the sample studied. Since direct paleodietary and paleoparasitological evidence is not available, future studies should focus on nutritional and zoonotic factors (Suby, 2014).

An essential role in nutrition is played by teeth; they bite, tear, grind or crush, and chew food to facilitate their intake and digestion. In the sample studied, the activities which caused occlusal surface wear are mainly related to chewing for alimentary purposes. The absence of splinters, antemortem fractures, grooves, and indentations, among others, indicates that extra/para-masticatory uses were probably not a determining factor of the wear observed (see Turner II & Cadien, 1969).

Tooth wear is directly associated with diet, food preparation techniques, and age. Considering that it is likely that collected plant foods represented a sizeable percentage of what hunter-gatherers ate, especially during the Late Holocene, various plant organs (leaves, pods, seeds, etc.) must have been causes of wear (L'Heureux, 1998). It is possible to infer the intake of plants such as grasses or nuts given the presence of grinding tools, at both the LED and LLL sites. However, this does not preclude the use of these instruments in animal products (dried meat and bone, among others) and mineral (pigment) processing.

At the LED site, the presence of grinding tables, most of them flat and flat-concave, with straight parallel wear grooves, allows us to infer that these instruments were used with rectilinear sliding, vertical back and forth and circular sliding pressure movements, i.e., for the pulverization of intermediate substances such as vegetable, animal, and various mineral materials. These types of repetitive precision movements involving hands, arms, and neck cause muscle fatigue that, depending on the regularity and pauses in their execution, might lead to chronic muscle injuries and a greater compression of the intervertebral discs. This, in turn, results in degenerative alterations in the neck area, as observed in the LED and LLL assemblages, and in hands and arms, as observed in the LLL assemblage. The presence of osteoarthritis in the vertebral column and upper limb was also found in adult female individuals from the Arroyo Seco 2 site, chronologically assigned to the Early-Middle Holocene (Scabuzzo, 2012).

Among the abrasive agents that affected the masticatory apparatus, the region's edaphological features must also be considered, since sandy and silty sediments mixed with fine sand could contribute to the incorporation of abrasive particles during the processes of food handling on the ground or drying meat in the open (Puech, 1981; Puech et al., 1980) or during daily tasks due to the exposure of the mouth to suspended sediments (Jurmain, 1990). These observations match with phenomena that affected hunter-gatherer populations, especially in the Humid Pampas subregion. The only exception lies in the compensatory physiological adaptation processes as a response to wear, which show, in general, lower values than the averages found in pre-hispanic Pampean hunter-gatherers bioarchaeological samples (L'Heureux, 1998). Abscesses and antemortem tooth loss are pathologies with rather nonspecific causes and, hence, provide an uncertain polarity within the proposed tooth pathology profiles of different subsistence economies (L'Heureux, 2014). Within the sample under study, abscess formation and antemortem tooth loss are associated with individuals who display a high tooth wear degree and no signs of periodontal disease, dental calculus, or caries. We propose that wear caused the exposure and consequent pulp cavity infection and eventual tooth loss. This aspect had already been observed in individuals found at the Laguna El Pedernal site (Ceruti, 1992). Furthermore, Smith (1984) proposed that we should expect flat wear on the occlusal surface of hunter-gatherer dentition due to mastication of hard and fibrous foodstuff. However, L'Heureux (1998) found different wear shapes and directions in a bioarchaeological set associated with Pampean hunter-gatherer groups. This is understood in relation to the intake of plant food processed with lithic grinding instruments, and the presence of sand and stone within the game

meat that was cooked or dried in the open, aspects also proposed in the cases under study.

In agreement with Lukacs' model (Lukacs, 1989) of hunter-gatherer populations, the whole set shows a high prevalence of severe tooth wear, and a low prevalence of caries, dental calculus, and periodontal lesions. The masticatory processes and the high wear degree they entailed, mainly in adults, probably prevented carious lesions and dental calculus formation, provided that salivary pH changes and/or cariogenic diets existed. The existence of such a diet could be inferred from the relative caries prevalence, especially in subadult individuals, in relation to the results obtained by other researchers of hunter-gatherer societies (Larsen, 1987; L'Heureux, 1998). The presence of cariogenic diets among the Late Holocene Pampean hunter-gatherer groups has been proposed on the basis of the analysis of osteological collections from various sites in the southeastern Pampas (L'Heureux, 1998). Moreover, the recurrence of caries in subadults could be associated with the food types and preparation techniques used during their first years.

Just as a high functional demand of the masticatory apparatus was detected, a low level of various types of physical demands was also observed; this can be inferred from the results in relation to osteoarthritis. This coincides with the low incidence of degenerative pathologies observed in the osteological assemblage from Chenque I site—Pampean region (Luna, 2008). As in the affected male individual from the studied sample, osteoarthritis in the lower limbs was observed, at high frequency, in male individuals from the osteological collection of Arroyo Seco 2 site corresponding to Pampean hunter-gatherers from the middle Holocene. The cause could be related to travel on foot carrying belongings, carcasses of animals, and so on (Scabuzzo, 2012).

We observed traumas in the parietal bones of individuals from both age categories (i.e., subadult and adult); an evaluation based on sex could not be done, since only two of the affected individuals were sexed, both female. A similar frequency of oval-shaped traumas was observed for ancient pampean hunter-gatherers (Flensborg, 2011). The presence of healing reactions indicates that the injuries were not lethal, which implies the survival of individuals. The locations and types of the traumas in the studied sample suggest intentional and repeated blows to have been the cause. Depressed skull lesions and small discrete linear skull lesions are considered to be produced by intentional violent behaviors, rather than accidents (Nagaoka et al., 2017). The oval-shaped traumas (depression and elevation) are produced by the direct application of force (blunt force trauma), which in the studied sample could be caused by close-proximity assaults with slingstones or thrown *bola* balls or stones. Discrete linear lesions may be caused by sharp-edged objects (sharp force traumas) and can be caused by assaults with small lithic scrapers and knives. Nevertheless, the presence of depressed skull fractures and healing reactions suggest that traumas are related to violence that is not intended to kill opponents (Nagaoka et al., 2017). Considering that the presence of interpersonal violence is not the result of a single event, since there is a range of years that separates the death episodes of the represented individuals, and that shows a low-moderate frequency in the studied sample, social tension and conflicts between groups due to

socio-ecological pressures are not supported. As it was suggested for other Pampean bioarchaeological collections, it is related to common interpersonal violence practices, both in the domestic sphere and within the co-residential units (Flensborg, 2011).

Although a small sample size limits our interpretation, the information presented is consistent with the social dynamics of the hunter-gatherer populations who inhabited the area and contributes data reinforcing these dynamics from additional lines of research. The bioarchaeological record and artifacts analyzed at a regional scale suggest the existence of social interaction systems within this lake environment. This can be observed in the recurrence of burials throughout the Holocene and the circulation of various lithic raw materials originating mainly in the outcrops of the Tandilia hill ranges. Thus, the existence of a predominant mobility and exchange vector in a northwest-southeast-northeast orientation could be proposed, one which would enable access to resources and knowledge, and allow the interaction among different groups of the Pampean region.

Acknowledgments We would like to thank the community of San Eduardo, its communal presidents, Messrs. Luis Rodríguez and Leandro Rinaldi, and the Comisión de Cultura. We would also like to thank Sergio Denis, Alicia and Santiago Busso, Gladys Yaniselli, Laura and Andrés Brianza, Mauro Villegas, José María Bustos and his family, and Diego Palacios for their friendship, assistance, and logistical support during the various research stages; the CEIA members who also supported us during the various research stages. The anonymous reviewers are thanked for their careful reading of the manuscript and their insightful comments and suggestions. Any possible errors and/or omissions are the authors' responsibility.

References

Adams, B. J., & Konigsberg, L. W. (2004). Estimation of the most likely number of individuals from commingled human skeletal remains. *American Journal of Physical Anthropology, 125*, 138–151.

Aguerre, A. (1996). Arqueología de la Laguna Chadilauquen. El pasado en la Laguna Chadilauquen, Embajador Martini. In *Estudios Pampeanos* (1st ed., pp. 19–41). Instituto de Antropología Rural.

Aliaga, V. S., Ferrelli, F., Alberdi-Algarañáz, E. D., Bohn, V. Y., & Piccolo, M. C. (2016). Distribución y variabilidad de la precipitación en la Región Pampeana, Argentina. *Cuadernos de Investigación Geográfica, 42*(1), 261–280.

Álvarez, M. C. (2014). Tecnología ósea en el oeste de la Región Pampeana: identificación de las técnicas de manufactura a partir de evidencias arqueológicas y experimentales. *Chungara, Revista de Antropología Chilena, 46*(2), 193–210.

Álvarez, M. (2018). Análisis faunístico del sitio Laguna de los Pampas (partido de Lincoln, Región Pampeana). Aportes a los estudios de la subsistencia en el área Oeste de la Pampa Húmeda. *Arqueología, 24*(1), 127–146.

Aranda, C. (2007). Evaluación de la variabilidad de las formas de inhumación en el sitio Chenque I (Parque Nacional Lihué Calel, provincia de La Pampa). In C. Bayón, I. González, N. Flegenheimer, A. Pupio, & M. Frére (Eds.), *Arqueología en las Pampas* (pp. 109–128). Sociedad Argentina de Antropología.

Avila, J. D. (2011). Resultados de los fechados radiocarbónicos del sitio Laguna El Doce, Departamento General López, Provincia de Santa Fe. *Relaciones, XXXVI*, 337–343.

Avila, J. D., & Ceruti, C. N. (2013). El Holoceno temprano-medio y la ruta del poblamiento: Laguna El Doce, departamento General López, provincia de Santa Fe, Argentina. *Cuadernos del Instituto Nacional de Antropología y Pensamiento Latinoamericano, 4*(1), 22–33.

Avila, J. D., Gabrielloni, C., & Cornaglia Fernández, J. (2013). Nuevos avances vinculados a las ocupaciones tempranas en ambientes lagunares del sudoeste santafesino. Sitio Laguna El Doce como caso de estudio. *Cuadernos del Instituto Nacional de Antropología y Pensamiento Latinoamericano, 4*(1), 12–30.

Avila, J. D., Píccoli, C., Gallego, M., Barboza, C., Carvallo, M., Pusterla, S., Torri, M., & Avila, S. (2022). Nuevas evidencias de asentamientos durante el Holoceno tardío en el Campo de Dunas del Centro Pampeano: Laguna Las Lágrimas (Sudoeste de la provincia de Santa Fe). *Relaciones, 47*, 031.

Ayuso, M. P., Gabrielloni, J., Fernández, C., & Píccoli, C. (2006). Aproximación a los procesos tafonómicos sobre restos óseos humanos en ambientes lagunares ecotonales. Laguna El Doce sitio 1, San Eduardo, Provincia de Santa Fe. *Revista de la Escuela de Antropología, XII*, 163–172.

Barboza, M. C. (2015). *Variación esqueletal humana y contexto arqueológico* (BAR International Series) (p. 2703). Evaluación de Marcadores de Edad y Sexo.

Barrientos, G. (2001). Una aproximación bioarqueológica al estudio del poblamiento prehispánico tardío del sudeste de la Región Pampeana. *Intersecciones en Antropología, 2*, 3–18.

Barrientos, G., & L'Heureux, L. (2001). Determinación de la edad de muerte a través del análisis de la altura total de la corona dental en muestras del Holoceno temprano del Sudeste de la Región Pampeana. *Revista Argentina de Antropología Biológica, 3*(1), 7–21.

Barrientos, G., & Masse, W. B. (2014). The archaeology of cosmic impact: Lessons from two mid-Holocene Argentine case studies. *Journal of Archaeological Method and Theory, 21*(1, 3), 134–211.

Barrientos, G., & Pérez, I. (2005). Was there a population replacement during the late middle Holocene in the southeastern pampas of Argentina? Archaeological evidence and its paleoecological basis. *Quaternary International, 132*, 95–105.

Bernal, V., García Guraieb, S., & González, P. (2004). Perfiles de mortalidad de las muestras de restos óseos humanos procedentes del área del Lago Salitroso (provincia de Santa Cruz). In M. T. Civalero, P. Fernández, & S. Guraieb (Eds.), *Contra viento y marea. Arqueología de Patagonia* (pp. 362–373). Instituto Nacional de Antropología y Pensamiento Latinoamericano/ Sociedad Argentina de Antropología.

Bernardi, L., Arrieta, M. A., & Lynch Ianniello, I. (2020). Preliminary study of the age-at-death profile in a prehistoric skeletal sample from Médano Petroquímica Site, Argentina. *Journal of Archaeological Science: Reports, 34*, 102651. https://doi.org/10.1016/j.jasrep.2020.102651

Berón, M., & Luna, L. (2007). Modalidades de entierro en el sitio Chenque I. Diversidad y complejidad de los patrones mortuorios de los cazadores-recolectores pampeanos. In C. Bayón, I. González, N. Flegenheimer, A. Pupio, & M. Frére (Eds.), *Arqueología en las Pampas* (pp. 129–142). Sociedad Argentina de Antropología.

Black, T. (1978). Sexual dimorphism in the tooth-crown diameter of the deciduous teeth. *American Journal of Physical Anthropology, 48*, 77–82.

Brothwell, D. (1993). *Desenterrando huesos. La excavación, tratamiento y estudio de restos del esqueleto humano*. Fondo de Cultura Económico.

Buikstra, J., & Ubelaker, D. (1994). *Standards for data collection from human skeletal remains* (Arkansas archaeological survey research series no. 44). Faytteville.

Castellanos, A. (1924). Contribución al estudio de la Paleoantropología Argentina. Restos descubiertos en la Laguna Melincué (Provincia de Santa Fe). *Revista de la Universidad Nacional de Córdoba., 11*(N° 10–11–12), 279–310.

Ceruti, C. N. (1987). Investigaciones arqueológicas en el Departamento General López, provincia de Santa Fe: Informe preliminar. In *Informe de Carrera al CONICET período 1986–1988*. Manuscript.

Ceruti, C. N. (1992). Cambios climáticos y poblaciones prehistóricas en el Litoral Argentino. In M. Iriondo (Ed.), *El Holoceno en la Argentina* (Vol. 1, pp. 39–49). Cadinqua.

Ceruti, C. N. (1993). Arqueología. Para entender la Arqueología. *Nueva Enciclopedia de la Provincia de Santa Fe, IV*, 557–580. Ediciones Sudamérica Santa Fe.

Clarke, N. G., & Hirsh, R. G. (1991). Tooth dislocation: The relationship with tooth wear and dental abscesses. *American Journal of Physical Anthropology., 85*, 293–298.

Cohen, M. (1997). Does paleopathology measure community health? A rebuttal to "the osteological paradox" and its implications for world history. In R. Paine (Ed.), *Integrating archaeological demography. Mustidisciplinary approaches to prehistoric population* (242–259). Southern Illinois University.

Cornaglia Fernández, J., & Buc, N. (2013). Evidence of bone technology on the Santa Fe's Pampa lagoons. The LED site (Santa Fe Province, Argentina). In A. Choyke & S. O'Connor (Eds.), *From these bare bones: Raw materials and the study of worked osseous objects* (pp. 109–115). Oxbow books.

Cornaglia Fernández, J. (2013). Análisis taxonómico e inferencias paleoambientales en el sudoeste santafesino. El sitio arqueológico Laguna El Doce. Revista del Museo de La Plata Sección Antropología. 13 (87): 109–124. Universidad Nacional de La Plata.

Di Donato, R. M. (2007). Polvo al polvo... Variables tafonómicas humanas y no humanas. Su influencia en la alteración y preservación diferencial del registro bioarqueológico en un sitio funerario. In C. Bayón, I. González, N. Flegenheimer, A. Pupio, & M. Frére (Eds.), *Arqueología en las Pampas* (pp. 143–158). Sociedad Argentina de Antropología.

Flegenheimer, N., Guichón, R., & Scabuzzo, C. (2002). Restos óseos humanos en el sitio El Guanaco, Partido de San Cayetano. In D. Mazzanti, M. Berón, & F. Oliva (Eds.), *Del mar a los salitrales. 10000 años de historia pampeana en el umbral del Tercer milenio* (pp. 121–126). Universidad Nacional de Mar del Plata.

Flensborg, G. (2011). Lesiones traumáticas en cráneos del sitio Paso Alsina 1. Explorando indicadores de violencia interpersonal en la transición Pampeano-Patagónica Oriental (Argentina). *Intersecciones en Antropología, 12*, 45–59.

Gabrielloni, C. (2012). *Estudios tafonómicos sobre restos óseos humanos. Sitio Laguna El Doce, departamento General López, provincia de Santa Fe*. Lap Lambert Academic Publishing GmbH & KG.

Gabrielloni, C., & Píccoli, C. (2011). Análisis de una colección osteológica asistemática: departamento General López, provincia de Santa Fe. In *V Jornadas de Ciencia y Tecnología* (pp. 177–180). Secretaría de Ciencia y Tecnología, Universidad Nacional de Rosario.

Gallego, M., & Avila, J. D. (2019). Otras territorialidades. Visibilizando las ocupaciones de los pueblos originarios en el sur de Santa Fe desde la arqueología. In L. Bulacio, C. Pairoba & E. Orellano (comp.), *Ciencia y Tecnología 2018: divulgación de la producción científica y tecnológica de la UNR* (pp. 796–800). UNR Editora.

Goñi, R., Barrientos, G., & Cassiodoro, G. (2003). Condiciones previas a la extinción de las poblaciones humanas del sur de Patagonia: una discusión a partir del análisis del registro arqueológico de la cuenca del Lago Salitroso. *Cuadernos del INA, 19*(2000–2002), 249–266.

González-Rivas, C. J., Borghil, C. E., & De Lamo, D. A. (2019). Endoparásitos en guanaco (Lama guanicoe). Revisión de situación en Argentina y registros de la provincia de San Juan. *Revista de Investigaciones Veterinarias del Perú, 30*(1), 339–349. https://doi.org/10.15381/rivep.v30i1.14609

Goodman, A., & Armelagos, G. (1985). Factors affecting the distribution of enamel hypoplasias within the human permanent dentition. *American Journal of Physical Anthropology, 68*, 479–493.

Goodman, A., Armelagos, G., & Rose, J. (1980). Enamel high hypoplasias as indicators of stress in three prehistoric populations from Illinois. *Human Biology, 52*, 515–528.

Grauer, A. (1991). Patterns of life and death: The palaeodemography of medieval York. In H. Bush & M. Zvelevil (Eds.), *Health in past societies: Biocultural interpretations of human skeletal*

remains in archaeological context (BAR International Series 567) (pp. 67–80). Tempvs Reparatvm.

Hassan, F. (1973). On methods of population growth during the Neolithic. *Current Anthropology, 14*, 535–542.

Heider, G. (2013). Arqueología en el norte de la Pampa Seca: informe sobre compañas de prospecciones (provincias de Córdoba, San Luis, La Pampa). *Revista del Museo de la Plata. Sección Antropología, 13*(87), 95–124.

Hillson, S. (1986). *Teeth: Cambridge Manuals in Archaeology*. Cambridge University Press.

Hillson, S. (1996). *Dental anthropology*. Cambridge University Press.

Hillson, S., FitzGerald, C., & Flinn, H. (2005). Alternative dental measurements: Proposals and relationships with other measurements. *American Journal of Physical Anthropology., 126*, 413–426.

Iriondo, M., & Kröhling, D. M. (2007). Non-classical types of loess. *Sedimentary Geology, 202*, 352–368.

Johnston, F. (1961). Sequence of epiphyseal union in a prehistoric Kentucky population, Indian Knoll. *Human Biology, 33*, 66–81.

Johnston, F. (1962). Growth of the long bones of infants and young children at Indian knoll. *American Journal of Physical Anthropology., 20*, 249–254.

Jurmain, R. (1990). Paleoepidemiology of a Central California prehistoric population from CA-Ala-329: Dental disease. *American Journal of Physical Anthropology, 81*, 323–342.

L'Heureux, G. L. (1998). *Biología oral de las poblaciones prehispánicas del sudeste de la Región Pampeana. Provincia de Buenos Aires.* Unpublished Graduation thesis, Facultad de Humanidades y Artes, Universidad Nacional de Rosario.

L'Heureux, G. L. (2014). Indicadores fisiológicos y patológicos bucales en las poblaciones representadas en el sitio Arroyo Seco 2. In G. Politis, M. A. Gutiérrez, & C. Scabuzzo (Eds.), *Estado actual de las investigaciones en el sitio arqueológico Arroyo Seco 2, Partido de Tres Arroyos, Provincia de Buenos Aires, Argentina* (pp. 371–392). Universidad Nacional del Centro de la Provincia de Buenos Aires. Facultad Ciencias Sociales.

Lacruz, R. S., Habelitz, S., Wright, J. T., & Paine, M. L. (2017). Dental enamel formation and implications from oral health and diseade. *Physiology Reviews, 97*, 939–993. https://doi.org/10.1152/physrev.00030.2016

Larsen, C. (1987). Bioarchaeological interpretations of subsistence economy and behavior from human skeletal remains. In M. Schiffer (Ed.), *Advances in archaeological method and theory* (Vol. 10, pp. 339–445). Academic Press.

Larsen, C. (2000). Bioarchaeology. In *Interpreting behavior from the human skeleton*. Cambridge University Press.

Lovejoy, C. (1985). Dental wear in the Libben population. Its functional pattern and role in the determination of adult skeletal age at death. *American Journal of Physical Anthropology, 68*, 47–56.

Lovejoy, C. O., Meindl, R. S., Pryzbeck, T. R., & Mensforth, R. P. (1985). Chronological metamorphosis of the auricular surface of the ilium: A new method for the determination of adult skeletal age at death. *American Journal of Physical Anthropology, 68*, 15–28.

Lovell, N. C., & Dawson, L. (2003). Intra- and inter-tooth analysis of hypoplastic and hypocalcified enamel defects. *Journal of Human Ecology, 14*(4), 241–248.

Lukacs, J. (1989). Dental paleopathology: Methods for reconstructing dietary patterns. In M. Işçan & K. Kennedy (Eds.), *Reconstruction of life from the skeleton* (pp. 261–286). Alan R. Liss,.

Luna, L. (2008). *Estructura demográfica y estilo de vida de cazadores recolectores en un ambiente de desierto. Sitio Chenque I (parque nacional Lihué Calel, provincia de La Pampa).* B.A.R. International Series, 1886.

Lyman, R. (1994). *Vertebrate taphonomy*. Cambridge University Press.

Marlowe, F. W. (2005). Hunter-gatherers and human evolution. *Evolutionary Anthropology, 14*(2), 54–67. https://doi.org/10.1002/evan.20046

Martínez, G., Flensborg, G., Bayala, P., & López, R. (2007). Análisis de la composición anatómica, sexo y edad de dos entierros secundarios del sitio Paso Alsina 1 (Partido de Patagones, Provincia de Buenos Aires). In C. Bayón, I. González, N. Flegenheimer, A. Pupio, & M. Frére (Eds.), *Arqueología en las Pampas* (pp. 41–58). Sociedad Argentina de Antropología.

Mayhall, J. (1992). Techniques for the study of dental morphology. In S. Saunders & S. Katzenberg (Eds.), *Skeletal biology of past peoples: Research methods* (pp. 59–78). Wiley.

Mayhall, J. (2000). Dental morphology: Techniques and strategies. In M. Katzemberg & S. Saunders (Eds.), *Biological anthropology of the human skeleton* (pp. 103–134). Wiley.

Mazzia, N., Scabuzzo, C., & Guichón, R. (2004). Sobre cráneos, pelvis y otros huesos: Entierros humanos en el sitio El Guanaco. In G. Martínez, M. Gutiérrez, R. Curtoni, M. Berón, & P. Madrid (Eds.), *Aproximaciones contemporáneas a la arqueología pampeana. Perspectivas teóricas, metodológicas, analíticas y casos de estudio* (pp. 293–304). Facultad de Ciencias Sociales/Universidad del Centro de la Provincia de Buenos Aires.

Meindl, R., & Lovejoy, C. (1985). Ectocranial suture closure: A revised method for the determination of skeletal age at death based on the lateral anterior sutures. *American Journal of Physical Anthropology, 68*, 57–66.

Mendonça, O. J., Aguerre, A. M., Arrieta, M. A., & Pera, M. (2013). Investigaciones bioarqueológicas en la laguna Chadilauquen, Embajador Martini, departamento Realicó, provincia de La Pampa: Segunda etapa. *Revista del Museo de La Plata, 13*(87), 137–152.

Mengoni Goñalons, G. (1988). Análisis de materiales faunísticos arqueológicos. *Xama, 1*, 71–120.

Merchant, V., & Ubelaker, D. (1977). Skeletal growth of the protohistoric Arikara. *American Journal of Physical Anthropology, 46*, 61–72.

Messineo, P. G., & Scheifler, N. A. (2016). Investigaciones arqueológicas en el sitio Laguna Cabeza de Buey 2 (San Carlos de Bolívar, Buenos Aires). Cincuenta años después de las Industrias Culturales definidas por Bórmida. *Intersecciones en Antropología, 17*, 213–227.

Messineo, P. G., González, M. E., Álvarez, M. C., & Pal, N. (2018). Las ocupaciones humanas en la localidad arqueológica Laguna de los Pampas (Campo de Dunas del Centro Pampeano, Argentina) durante del Holoceno. *Latin American Antiquity, 29*(4), 736–753.

Messineo, P., Scheifler, N., Alvarez, M., González, M., Pal, N., Barros, M., & Politis, G. (2019). A model of human occupation in the central Pampean Dunefields of Argentina. *PaleoAmerica: A Journal of Early Human Migration and Dispersal. Londres, 5*(4), 378–391.

Moes, E., & Blatt, S. H. (2018). Linear Enamel hypoplasia in permanent dentition of children in the late Archaic and the late Prehistoric river valley. *Dental Anthropology, 31*(2), 3–17.

Nagaoka, T., Uzawa, K., Seki, Y., & Morales Chocano, D. (2017). Pacopampa: Early evidence of violence at a ceremonial site in the northern Peruvian highlands. *PLoS One, 12*(9), e0185421. https://doi.org/10.1371/journal.pone.0185421

Oliva, F., & Panizza, C. (2017). Análisis de la cerámica arqueológica del Área Ecotonal Húmedo Seca Pampeana. In A. Rocchietti, F. Ribero, & D. Reinoso (Eds.), *Investigaciones arqueométricas: técnicas y procesos* (pp. 105–120). Ciudad Autónoma de Buenos Aires.

Oliva, F., & Solomita Banfi, F. (2017). Distribución espacial de materiales termo-alterados en una estructura de combustión de la región pampeana: Análisis 3D. In A. Rocchietti, F. Ribero, & D. Reinoso (Eds.), *Investigaciones arqueométricas: técnicas y procesos* (pp. 189–204). Ciudad Autónoma de Buenos Aires.

Oliva, F., Panizza, M. C., Catella, L., Moirano, J., Morales, N., Algrain, M., Devoto, G., Iannelli, L., Oliva, C., Pereyra, B., & Sfeir, A. (2015). La construcción del pasado arqueológico en diferentes sectores del Área Ecotonal Húmedo-Seca Pampeana. Investigación y Extensión desde el Centro de Estudios Arqueológicos Regionales. *Revista de Antropología del Museo de Entre Ríos., 1*(2), 91–102.

Ortner, D. J. (2003). *Identification of pathological conditions in human skeletal remains* (2nd ed.). London Academic Press.

Pereyra, B. (2019). *Bioarqueología en norte y centro del Area Ecotonal Húmeda-Seca Pampeana (AEHSP) (Provincia de Buenos Aires): Estado de la cuestión, problemas y perspectivas.*

Unpublished Graduation thesis. Facultad de Humanidades y Artes. Universidad Nacional de Rosario.

Píccoli, C. (2009). *Estudio Bioarqueológico de la Colección Osteológica proveniente del Sitio Laguna El Doce (Departamento General López, Provincia de Santa Fe)*. Unpublished Graduation thesis. Facultad de Humanidades y Artes, Universidad Nacional de Rosario.

Píccoli, C., & Gabrielloni, C. (2012). Algunas piezas sueltas. Análisis de una colección de cerámica asistemática. In *VI Jornadas de Ciencia y Tecnología* (pp. 435–438). Secretaría de Ciencia y Tecnología/Universidad Nacional de Rosario.

Politis, G., & Madrid, P. (2001). Arqueología pampeana: Estado actual y perspectivas. In E. Berberián & A. Nielsen (Eds.), *Historia Argentina Prehispánica* (pp. 737–814). Editorial Brujas.

Politis, G., Messineo, P., González, M., Álvarez, M., & Favier Dubois, C. (2012). Primeros resultados de las investigaciones en el sitio Laguna de los Pampas (Partido de Lincoln, Provincia de Buenos Aires). *Relaciones de la Sociedad Argentina de Antropología, XXXVII*(2), 463–472.

Posth, C., Nakatsuka, N., Lazaridis, I., Skoglund, P., & Mallick, S. (2018). Reconstructing the deep population history of central and South America. *Cell, 175*(5), 1185–1197.e22. https://doi.org/10.1016/j.cell.2018.10.027

Puech, P. F. (1981). Tooth wear in La Ferrassie Man. *Current Anthropology, 22*(4), 424–425.

Puech, P. F., Albertini, H., & Mills, N. (1980). Dental destruction in Broken Hill man. *Journal Human Evolution, 9*, 33–39.

Ramírez, D. A., Vieira de Souza, M., Mayo Iñiguez, A., & Fabra, M. (2021). Primer estudio paleoparasitológico en restos humanos de la provincia de Córdoba (Holoceno tardío). *Revista Argentina de Antropología Biológica, 23*(1), 030. https://doi.org/10.24215/18536387e030

Rivera, F., & Mirazon Lahr, M. (2017). New evidence suggesting a dissociated etiology for cribra orbitalia and porotic hyperostosis. *American Journal of Physical Anthropology, 164*, 76–96. https://doi.org/10.1002/ajpa.23258

Roca-Rada, X., Politis, G., Messineo, P., Scheifler, N., Scabuzzo, C., González, M., Harkins, K. M., Reich, D., Souilmi, Y., Teixeira, J. C., & Llamas, B. (2021). Ancient mitochondrial genomes from the Argentinian Pampas inform the early peopling of the Southern Cone of South America. *iScience, 24*, 1–17. https://doi.org/10.1016/j.isci.2021.102553

Salceda, S., & Méndez, S. (1999). Nuevos hallazgos de restos óseos humanos en el noroeste bonaerense. In *III Jornadas Chivilcoyanas* (pp. 161–164). Province of Buenos Aires.

Scabuzzo, C. (2012). Estudios bioarqueológios de marcadores de estrés ocupacional en cazadores recolectores pampeanos del Holoceno temprano-medio. Análisis de la serie esqueletal de Arroyo Seco 2. *Revista Argentina de Antropología Biológica, 14*(1), 17–31.

Scabuzzo, C., & González de Bonaveri, M. I. (2007). Un acercamiento a la dieta de las poblaciones prehispánicas de la depresión del Salado durante el Holoceno tardío. In C. Bayón, I. González, N. Flegenheimer, A. Pupio, & M. Frére (Eds.), *Arqueología en las Pampas* (pp. 59–74). Sociedad Argentina de Antropología.

Scheifler, N. A. (2020). Fogones, combustible óseo y procesamiento faunístico en el sitio de Huencú Nazar (Sistema Lagunar Hinojo-Las Tunas, región pampeana, Argentina): implicancias para el uso del espacio por cazadores-recolectores. *Latin American Antiquity*, pp. 1–21. https://doi.org/10.1017/laq.2020.12

Schaefer, M., Black, S., & Scheuer, L. (2009). *Juvenile osteology: A laboratory and field manual*. Elsevier.

Scheifler, N., Messineo, P., & Antiñir, A. (2017). Cazadores-recolectores en el sistema lagunar Hinojo-Las Tunas (Región Pampeana – Área Oeste) durante el Holoceno temprano/medio y tardío. Primeros resultados de las investigaciones arqueológicas: Comechingonia. *Revista de Arqueología, 21*(1), 287–314.

Scheifler, N. A., Merino, M. L., Vitale, P., Kaufmann, C. A., Messino, P. G., Álvarez, M. C., & Bocherens, H. (2023). Isotopic ecology in Modern and holocene population of Pampas Deer (*Ozotoceros bezoarticus*) from Eastern Central Argentina. Implications for Conservation Biology and Ecological Models of Hunter-gatherer Subsistence. *Environmental Archaeology 28*:2, pp. 45–61. https://doi.org/10.1080/14614103.2020.1846451

Scheuer, L., & Black, S. (2000). *Developmental juvenile osteology*. Academic Press.

Simalcsik, R. D., Simalcsik, A., & Groza, V. M. (2014). Dental enamel hypoplasia: Investigations on the bones exhumed from the medieval necropole of Lozova (Republic of Moldova), XIVTH–XVTH centuries. *Memoirs of the Scientific Sections of the Romanian Academy Tome, XXXVII*, 85–96.

Slaus, M., & Tomicic, Z. (2005). Discriminant function sexing of fragmentary and complete tibiae from medieval Croatian sites. *Forensic Science International., 147*, 147–152.

Smith, B. H. (1984). Patterns of molar wear in hunter-gatherers and agriculturalists. *American Journal of Physical Anthropology, 63*, 39–56.

Suby, J. A. (2014). Porotic hyperostosis and cribra orbitalia in human remains from southern Patagonia. *Anthropological Science, 122*(2), 69–79.

Tapia, A., Salvino, A., & Aguerre, A. M. (2020). Chadilauquen y Neicorehue, dos asentamientos lagunares con cerámica en el Nordeste de la Pampa Seca. *Revista Relaciones de la Sociedad Argentina de Antropología, XLV*(2), 323–353.

Todd, T. W. (1920). Age changes in the pubic bone I: The male white pubis. *American Journal of Physical Anthropology, 3*, 285–334.

Todd, T. W. (1921). Age changes in the pubic bone. II: Pubis of the Male Negro – White hybrid. III: Pubis of White Female IV: Pubis of Female Negro—Wuite hybrid. *American Journal of Physical Anthropology, 4*, 1–70.

Towle, I., & Irish, J. D. (2020). Recording and interpreting enamel hypoplasia in samples from archaeological and palaeoanthropological contexts. *Journal of Archaeological Science, 114*, 25. https://doi.org/10.1016/j.jas.2020.105077

Turner, C. G., II, & Cadien, J. (1969). Dental chipping in Aleuts, Eskimos and Indians. *American Journal of Physical Anthropology, 31*, 303–310.

Ubelaker, D. H. (1980). *Human skeletal remains: Excavation, analysis, interpretation*. Smithsonian Institution.

Zárate, M., & Tripaldi, A. (2012). The Aeolian system of Central Argentina. *Aeolian Research, 3*(4), 401–417.

Paleodemography of Late Holocene Hunter-Gatherers in the Pampa-Patagonia Transition: Approaches from the Preliminary Analysis of the Médano Petroquímica Site, La Pampa Province, Argentina

Lila Bernardi, Ignacio Lynch Ianniello, and Mario Arrieta

Abstract Paleodemography refers to studying the population structure and dynamics of past societies. In the absence of written records, paleodemographic analyses offer a window for understating the demographics of populations in the past. This work addresses the analysis of human burials from the Médano Petroquímica site, a burial area located in the southwest of the Pampean region (Argentina) and dated to ca. 400 years BP. Radiocarbon dating suggests that hunter-gatherer societies may have used this site for a short period. The sex-age-at-death analysis of four multiple secondary burials and one single primary burial ($n = 75$) was performed. Subsequently, an abbreviated life table was prepared, and different paleodemographic parameters were calculated. The results showed that all age categories are represented in the sample. The age-at-death profile obtained shows a bimodal attritional mortality curve typical of hunter-gatherer societies. Juvenile index (JI), fertility index (FI), and population growth rates suggest that the analyzed sample represents a growing society. Furthermore, the juvenile index value suggests that it would not present bias between the different age categories. Finally, the importance of approaching this type of study is highlighted to be able to make inferences about the dynamics of prehistoric societies of the region at a time when important changes in the organization of these groups are proposed.

Keywords Population dynamics · Fertility index · Growth rates · Bioarchaeology

---

L. Bernardi (✉) · M. Arrieta
Laboratorio de Bioarqueología, Universidad Nacional de Río Cuarto, Córdoba, Argentina

CONICET, Córdoba, Argentina
e-mail: lbernardi@exa.unrc.edu.ar

I. L. Ianniello
Laboratorio de Bioarqueología, Universidad Nacional de Río Cuarto, Córdoba, Argentina

G. F. Bonnat et al. (eds.), *Current Research in Archaeology of South American Pampas*, The Latin American Studies Book Series,
https://doi.org/10.1007/978-3-031-55194-9_4

Introduction

Paleodemography is the discipline focused on the study of the population dynamics of past human societies (Frankenberg & Konigsberg, 2006; Hoppa, 2002). Paleodemographic studies are highly relevant for Bioarchaeology since they provide the tools to infer properties of ancient societies such as social structure and biological composition, space–time variations, health status, life expectancy, mortality, and fertility rates, as well as the influence of cultural, socioeconomic and environmental changes on population structure (Angel, 1969; Buikstra & Konigsberg, 1985; Frankenberg & Konigsberg, 2006; Hoppa, 2002; Meindl & Russell, 1998; Milner et al., 1989). However, paleodemographic studies present some difficulties because it is often challenging to apply the methodologies of demography to archaeological osteological samples, which often do not have written records. Currently, the use of assumptions (i.e., biological uniformitarianism, stable and/or stationary population, and representativeness of the sample, among others), the application of biomathematical models, and the improvement in excavation techniques and estimation of age and sex allow to address these difficulties (Eastbrook & Rissech, 2004; Hoppa & Vaupel, 2002; Murail et al., 2005; Roksadic & Armstrong, 2011). In any case, despite these solutions, researchers must be aware of the scope and limitations of the interpretations obtained through paleodemographic studies.

The Médano Petroquímica site is located in the southwest of the province of La Pampa, Argentina (Fig. 1a), and the radiocarbon dates obtained indicate that this cemetery would have been used during a limited period at the end of the late

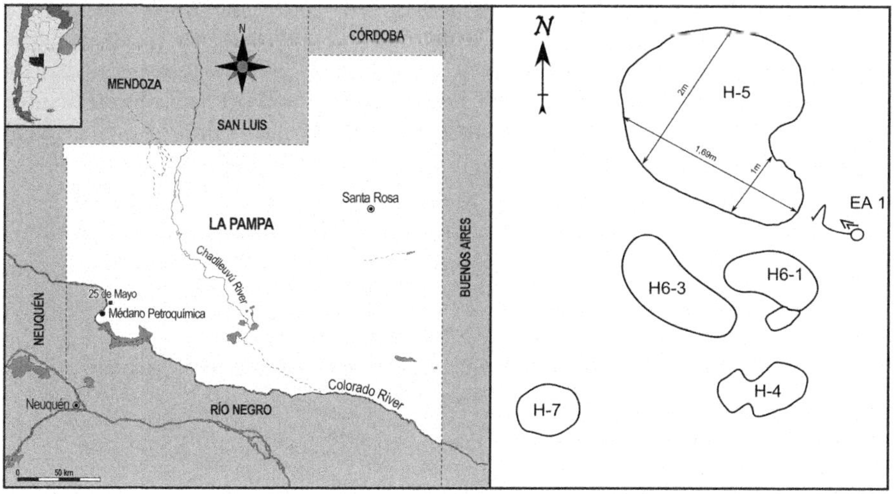

Fig. 1 (**a**) Location of the Médano Petroquímica site. (**b**) Spatial distribution of burials

Holocene (ca. 400 years BP) (Bottini et al., 2020a). For that period, several authors have proposed that the human groups in the Pampa-Patagonia transition region were experiencing a series of changes such as demographic increase, reduction in residential mobility, and changes in sociopolitical and economic organization (Berón, 2007, 2018; Berón & Politis, 1997; Madrid & Barrientos, 2000; Martínez, 2006, 2017; Mazzanti, 2006; Mazzanti & Quintana, 2001). Many of them associate these changes with the so-called Medieval Climate Anomaly (MCA), which would have been characterized by an increase in temperature and changes in rainfall patterns that led to the aridization of the region. This condition would have favored, on one hand, the concentration of human groups in those places with greater availability of critical resources such as water and, on the other, the subsequent decrease in residential mobility (Barrientos & Pérez, 2004; Berón, 2007, 2018). In addition, the appearance of formal burial areas close to these watercourses has been associated with low residential mobility and high population density (Barrientos, 2002). Another important point is that the time of use of the Médano Petroquímica site would also span the period of Hispanic-indigenous contact, which could also have had some impact on the societies that formed this cemetery. In this regard, Berón (2018) states that although the European conquerors were occupying territories far from the central Pampas region, their presence in nearby places could have exerted different types of pressure, which could translate, for example, into an increase in armed conflicts and the nucleation of hunter-gatherer groups.

The background information of paleodemographic studies in the Pampas, North Patagonia, and South Cuyo regions, although abundant, focuses mainly on the construction of mortality profiles (Flensborg & Bayala, 2017; Flensborg et al., 2015; Luna, 2018; Novellino et al., 2013; Peralta, 2019). However, the use of life tables and the several parameters currently proposed for paleodemographic analysis is still quite scarce. The low diffusion that the use of these tools has had would respond mainly to the strong criticism that this discipline has received since its inception. Among the works that have adopted this approach are Gordon et al. (2019) and Bernardi et al. (2020, 2022).

Hunter-gatherer societies would have had a small population size, high residential mobility, and different cultural practices in relation to mortuary behavior, which would result in limitations in terms of sample size that would influence their bioarchaeological record (García Guraieb et al., 2015). Considering the assumptions on which the paleodemographic interpretations are based, in this work, it is proposed to analyze the age-at-death profile together with different paleodemographic indices (juvenile index, annual fertility rate, fertility index, $D_{0\text{-}14}/D$ ratio, maternal mortality, and rates of population growth), to characterize the population dynamics and the changes in the societies that formed the Médano Petroquímica site. In addition, this analysis proposes to evaluate and discuss whether there was population growth in the hunter-gatherer societies of the Pampa-Patagonia transition during the end of the Late Holocene.

The Médano Petroquímica Site

The Médano Petroquímica site is located on a dune 2 km from the Colorado River, southwest of La Pampa province, Argentina (Fig. 1a). This is a burial area in which five multiple secondary burials (H4, H5, H6-1, H6-3, and H7) and a single primary burial (EA-1) were excavated (Fig. 1b). It has been estimated that about 120 skeletal individuals have been recovered from the site (Bernardi et al., 2022; Bottini et al., 2020b). The multiple burials contained numerous intentionally disarticulated individuals, whose skeletal remains would have been arranged in successively added "funerary packages" (Bottini et al., 2020a). The single primary burial consisted of an adult male individual (EA-1), in the right lateral decubitus position, with the skull resting on the right side and the orbits oriented toward the north (Fig. 1b). The pattern of funerary treatment in multiple burials would have consisted of accommodating the long bones, which would have been arranged in parallel bundles. The intentional dismemberment of the corpses would have been facilitated both by natural decomposition and by the use of cutting instruments (Bottini et al., 2020a). However, not all the corpses would have had the same treatment for the assembly in funerary packages (Bottini et al., 2020a). The general inclusion of individuals within the tomb seems to respond more to the aggregation of the different funerary packages than to their orderly arrangement.

The stages prior to the final burial of the deceased (primary deposit, fleshing and disarticulation of the corpse, assembly of packages, and move to the final deposit), added to the post-depositional processes, would have influenced the preservation of the bone remains (Bottini et al., 2020b). However, Bottini et al. (2020b) observed that the different bone elements presented high fragmentation rates (Fragmentation Index = 0.44), with the skulls being the most fragmented elements (Fragmentation Index = 0.03). In addition, all the bone elements had a very good representativeness (MAU % = 53.78%), presenting the right patellas (MAU%=16%) and the bones of the hands and feet (MAU% ≈ 30%) the lowest values (Bottini et al., 2020b). Regarding the sex and age-at-death estimation, it was observed that, despite the high fragmentation that the skulls and os coxae presented, the different anatomical and morphological features would not have been affected to a great extent (Bernardi, 2018, 2019). Sex was estimated in 98.4% of the ossa coxae and in 100% of the skulls, while age-at-death was estimated in 67% of the ossa coxae (Bernardi, 2019).

Grave goods were few, limited to a copper earring, a fragment of Venetian bead, numerous arrowheads and fragments thereof (several of them embedded in bone), and a large number of necklace beads made on mollusk shells. From the radiocarbon dates obtained (Bottini et al., 2020a), it was inferred that hunter-gatherer societies used the cemetery during a limited period of time (between 419 and 363 years BP) toward the end of the late Holocene, which would strengthen the population interpretations and/or inferences that could be carried out.

Materials and Methods

The human bone remains from five burials (EA-1, H4, H6-1, H6-3, and H7) from the Médano Petroquímica site were analyzed. The minimum number of individuals estimated using the anatomical correspondence technique (Todd & Frison, 1992) for these burials was 75. This estimate has been presented and analyzed in previous works (Bernardi et al., 2020, 2022).

Sex determination was performed following the criteria proposed by Buikstra and Ubelaker (1994). In adults, the morphology of os coxae, skull, and jaw was considered (Acsádi & Nemeskéri, 1970; Buikstra & Mielke, 1985; Loth & Henneberg, 1996, 2001; Milner, 1992; Phenice, 1969; Sutherland & Suchey, 1991; Walker, 2005), as well as the vertical diameters of the femoral and humeral heads. In juvenile skeletons, sex was estimated from the analysis of the morphological characters of the jaw and ilium (Loth & Henneberg, 2001; Luna et al., 2017; Molleson et al., 1998; Schutkowski, 1993).

Age-at-death estimation in adults was based on degenerative changes in the auricular surface (Buckberry & Chamberlain, 2002; Lovejoy et al., 1985; Meindl & Lovejoy, 1989) and the pubic symphysis (Brooks & Suchey, 1990; Gilbert & McKern, 1973; McKern & Stewart, 1957; Todd, 1921a, b). Tooth wear was also considered (Lovejoy, 1985; Mays, 2002). In juveniles, age was estimated based on sequence of formation and eruption of teeth (Ubelaker, 1999), epiphyses and ossification centers fusion (Johnston, 1962; Krogman & Isçan, 1986), and examination and determination of bone size (Fazekas & Kósa, 1978; Scheuer & Black, 2000). The size of the bone elements was obtained by using a digital caliper and an osteometric table. The results obtained, together with the field notes and the application of different techniques, were used to carry out the bilateral and intermembral assemblies (Byrd & LeGarde, 2014; Knüsel & Outram, 2004; Osterholtz, 2019; Schaefer, 2014; Todd & Frison, 1992). This allowed us to estimate the minimum number of individuals present in the studied sample (Todd & Frison, 1992).

Since the accuracy of age estimation techniques varies between adults and juveniles, the data were organized following the criteria proposed by Buikstra and Ubelaker (1994) for individuals older than 20 years (young, middle, and older adults). For subadults, five-year categories were used (except in the first category, which includes from perinatal to the first year of life). Therefore, the following age ranges were used: 0–0.9 years, 1–4.9 years, 5–9.9 years, 10–14.9 years, 15–19.9 years, 20–34.9 years (young adults), 35–49.9 years (middle adults), 50–64.9 years (older adults), and, finally, over 65 years. The use of the 0–0.9 years range responds mainly to the fact that infants under 1 year of age are generally more susceptible to diseases, nutritional restrictions, and environmental changes, among others (FitzGerald et al., 2006; Sánchez Romero, 2006). This situation has led to the postulate that the representativeness of a sample will also depend on the percentage of individuals less than 1 year of age that it presents (Volt & Atkinson, 2013).

Table 1 Description of the parameters used in the abbreviated life table (Ubelaker, 2007)

Parameters	Definition
D_x	Number of skeletons in age range x
d_x	Percentage of skeletons in age range x
l_x	Percentage of surviving individuals in age range x
q_x	Death probability in age range x (d_x/l_x)
T_x	Number of years remaining to live in age range x ($T_x = \Sigma L_x$)
L_x	Number of years lived by individuals in age range x ($L_x = c \times (l_x + l_{x+1})/2$)
$e^o{}_x$	Life expectancy in age range x ($e^o{}_x = T_x/l_x$)

The abbreviated life table was prepared by calculating, for each age range (x), the number of deaths (D_x), the percentage of deaths (d_x), the percentage of survivors (l_x), the probability of death (q_x), and the life expectancy ($e^o{}_x$). These parameters are detailed in Table 1.

The juvenile index (JI) was calculated (Bocquet-Appel & Masset, 1977). This is obtained by dividing the number of individuals whose estimated age falls in the 5–14 years range by the number of individuals whose age is equal to or greater than 20 years:

$$\mathrm{JI} = D_{5-14} / D_{20-\infty}$$

Based on the JI, Bocquet-Appel and Masset (1977) propose the annual fertility rate (${}_{35}F_{15}$) and the fertility index (FI). Both indices were calculated. ${}_{35}F_{15}$ considers the female population of reproductive age (15–50 years) and FI estimate the number of children per woman who survived reproductive age (35 years).

$${}_{35}F_{15} = 0.6601 \times \mathrm{IJ}^{0.5} + 1.2419 \times \mathrm{IJ}^{0.3} - 0.308 \pm 0.005$$

$$\mathrm{FI} = {}_{35}F_{15} \times 35$$

On the other hand, following the proposal of McFadden and Oxenham (2018a), the D_{0-14}/D ratio was calculated. This ratio was proposed by the authors as a paleodemographic alternative for commingled bone samples with a good representation of infants. This proportion has been shown to have a very good relationship with total fertility (McFadden & Oxenham, 2018a), so the authors have proposed it to estimate the fertility rate from the following equation:

$$\text{Fertility rate} = \left(7.734 \times D_{0-14} / D\right) + 2.224$$

McFadden and Oxenham (2018b) state that the D_{0-14}/D ratio and the fertility rate are very good predictors of the rate of natural increase of a population. Therefore,

through JI and the $D_{0\text{–}14}/D$ ratio, the population growth rate was calculated using the following equations:

$$\text{Population Growth Rate} = 1.484X + 1.485 \pm 0.0006$$

(Bocquet-Appel & Masset, 1977),

where $X = \log\left[200 \times \text{IJ} \times \left(D_{60-\infty}/D_{20-\infty}\right)\right]^{0.03}$

$$\text{Rate of Natural Increase}\ (\%) = \left(10.06 \times -D_{0-14}/D\right) - 1.61$$

(McFadden & Oxenham, 2018b).

Finally, considering the fertility rate and the rate of natural increase, maternal mortality rate was estimated (McFadden & Oxenham, 2019). This parameter represents women who died during pregnancy, childbirth, or within 42 days of termination of pregnancy, regardless of the duration and site of pregnancy, during a specific year, per 100,000 live births. The maternal mortality rate varies significantly over time and depends largely on the socioeconomic and health conditions of societies. For its calculation, the following formula is used:

$$\text{Maternal Mortality Rate} = \left(333.33 \times f_{F20-24/M20-24}\right) - 76.07$$

Results

The distribution by age and sex is presented in the histogram shown in Fig. 2. These same data were used to build the abbreviated life table (Table 2).

The JI value calculated for the sample was 0.29. This value indicates that, out of every 100 individuals, 29 would be subadults. The annual fertility rate was 0.193 (0.188–0.198), so the FI (i.e., the estimated number of children per woman who completed her reproductive cycle) would have been 6.75.

Regarding the estimators proposed by McFadden and Oxenham (2018a, b), the following results were obtained: the $D_{0\text{-}14}/D$ index was 0.453, which would indicate that about 45% of the sample would correspond to individuals under 14 years. This value was later used to calculate the fertility rate proposed by the authors (McFadden & Oxenham, 2018a), which was 5.73 children per woman who survived the reproductive period.

The population growth rates obtained using the formulas proposed by Bocquet-Appel and Masset (1977) and McFadden and Oxenham (2018b) were equal to 1.5 per 1000 and 2.95%, respectively. Finally, the maternal mortality rate was 254.26 per 100,000. The values of each of the calculated parameters are summarized in Table 3.

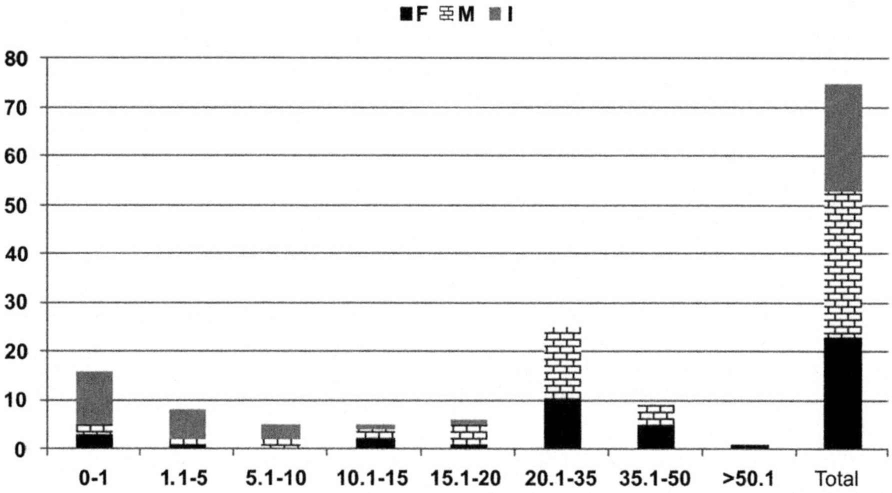

Fig. 2 Age and sex distribution for the burials analyzed (Bernardi et al., 2020)

Table 2 Abbreviated life table for the analyzed sample

Age range (years)	D_x	d_x	l_x	q_x	$e^0{}_x$
Perinate	16	21.33	100	0.21333	18.22
1–4.9	8	10.67	78.67	0.1356	22
5–9.9	5	6.67	68	0.0981	21.17
10–14.9	5	6.67	61.33	0.1088	18.2
15–19.9	6	8	54.66	0.1464	15.12
20–34.9	25	33.33	46.66	0.7143	12.3
35–49.9	9	12	13.33	0.9002	9.27
50–64.9	1	1.33	1.33	1	10
>65	0	0	0	0	0

Discussion

Previous studies (Bernardi et al., 2020, 2022) have shown that the Médano Petroquímica site presents an age-at-death distribution curve corresponding to an attritional mortality profile (according to Margerison & Knüsel, 2002), with modal peaks for the ranges that include infants under 1 year and young adults (20–34.9 years), which make the mortality curve take the form of a "U." In the Médano Petroquímica burials analyzed to date, individuals under 1 year of age comprise 21.33%, while individuals under 5 years of age at the time of death represent 32% of the total sample. In paleodemographic studies, the underrepresentation of infants has been a significant problem (McFadden & Oxenham, 2018a), since their representativeness is often affected by various factors, such as the taphonomic agents that operated at the site, cultural practices around the mortuary behavior, the difficulty in recognizing immature remains and their sex-age classification, and the use of inadequate excavation techniques, among others (Baker et al., 2005; García

Table 3 Values of the paleodemographic indicators calculated for the sample of the Médano Petroquímica site

Paleodemographic parameters	Value
JI (Bocquet-Appel & Masset, 1977)	0.29
${}_{35}F_{15}$ (Bocquet-Appel & Masset, 1977)	0.193
FI (${}_{35}F_{15} \times 35$) (Bocquet-Appel & Masset, 1977)	6.75
D_{0-14}/D (McFadden & Oxenham, 2017)	0.453
Fertility Rate (McFadden & Oxenham, 2017)	5.73
Maternal Mortality Rate (McFadden & Oxenham, 2019)	254.26
Population Growth Rate (Bocquet-Appel & Masset, 1977)	1.5‰
Rate of Natural Increase (McFadden & Oxenham, 2018a, b)	2.95%

Guraieb et al., 2015; Lewis, 2007). Because of this, children aged 0–4 years have often been excluded from paleodemographic studies (Milner et al., 2008; Seguy & Buchet, 2013). Therefore, a good representation of infants is key to supporting paleodemographic analysis that intends to obtain information that allows interpretations of population scope. This is why the use of some paleodemographic parameters is generally questioned and even left aside in bioarchaeological studies.

Some authors have suggested the use of certain percentage values of individuals younger than 1 or 5 years for hunter-gatherer samples to consider that the samples are representative (Volt & Atkinson, 2013). However, parameters such as the JI can be used for this purpose. Regarding this point, Jackes (2011) establishes a limit value for the JI equal to 0.38, above which it would be in the presence of a biased sample. Taking into account the value of JI obtained for the Médano Petroquímica (0.29), it could be suggested that the sample studied here lacks such a bias.

The peak that occurs in the age range that includes individuals aged under 1 year suggests a high infant mortality rate, which is expected for pre-industrial societies and is related to the high dependence of infants on adults for their subsistence, along with factors such as feeding patterns and care (FitzGerald et al., 2006). The immaturity of the immune system of infants, which makes them more prone to disease, should also be considered a highly influential factor in infant mortality (Lewis, 2007). Mazzeo (2007) divides infant mortality (0–0.9 years) into two large groups. The first is called the neonatal period and includes individuals who die during the first 4 weeks of life, where the main causes of death respond to endogenous factors. The second is called the post-neonatal period and includes the remaining individuals up to the year of age and is mainly related to exogenous factors. In this sense, one of the possible agents causing the death of the perinatal that make up this sample (16% of the total individuals analyzed were aged from 36 weeks of gestation to 3 months of extra-uterine life) may be due to endogenous causes, which would be influenced by the conditions that occur before and during childbirth, as well as exogenous causes, which depend on the environmental conditions immediately after birth (Lewis, 2007; Sánchez Romero, 2006). In this aspect, the response of the mother to different social, economic, and environmental factors indirectly influences the health status of the individual during pregnancy and/or the newborn.

After birth, the most critical moments in the life of infants would be given mainly by the practice of weaning and the beginning of the administration of substitute foods for breast milk (Buckley, 2000; FitzGerald et al., 2006; Katzenberg et al., 1996; Livi-Bacci, 2002; Stodder, 1997). At this time, when the infant no longer receives the nutrients and passive immunity provided by breastfeeding, the environmental and hygienic-sanitary conditions of societies are essential for survival (Herrada, 2002; Herring et al., 1998). This occurs mainly because the infant's digestive and immune systems are not yet fully formed, which leads them to suffer from diarrhea, parasitosis, and allergies to foods other than breast milk more frequently (Katzenberg et al., 1996). This is supported by previous results obtained from analyses of dental enamel hypoplasia carried out on individuals from the Médano Petroquímica (Bernardi, 2018), which suggest the presence of moderate to high incidence of metabolic and/or nutritional stress situations at an early age, between 2 and 6 years of age (81% of 47 individuals analyzed).

Finally, the high infant mortality can be related to life expectancy at birth and the fertility rate of the society that formed the site. From the analysis of the data provided by the abbreviated life table, it can be observed that life expectancy at birth was approximately 18 years and the probability of death was 0.21 (Table 1). In the immediately subsequent age range, life expectancy increases to 22 years, while the probability of death drops to 0.14. Therefore, after the first year of life, a critical period due to the immaturity of the immune and digestive systems (Stinson, 2000) and the need for extensive adult care, life expectancy increases, and the probability of death decreases. These values agree with those obtained by Gage (2000) for several prehistoric populations, in which life expectancy at birth varied between 18 and 25 years.

After 5 years of age, life expectancy begins to decrease gradually, while the probability of death decreases until the age of 20 years, the age from which the probability of death increases rapidly (Table 2).

Infant mortality is also associated with the fertility rate. McFadden and Oxenham (2018a), through their results, suggest that the 0–4 years category would be the age group most sensitive to changes in fertility. Furthermore, Wood et al. (1992) establish that life expectancy and the mean age at death are representative values of fertility in non-stationary populations (i.e., where birth and death rates are constant). In other words, the fertility rate impacts age distributions, since it directly influences the proportion of subadult individuals (Chamberlain, 2006). Furthermore, several authors (Konigsberg & Frankenberg, 1994; Milner et al., 1989; Sattenspiel & Harpending, 1983) propose that fertility has a greater impact on the age-at-death distribution than mortality. Given that the mortality peak for infants is directly related to fertility levels, Konigsberg and Frakenberg (2002) point out that a high presence of bone remains of individuals aged under 4 years in a burial site reveals a high level of fertility and, therefore, a possible trend toward a population increase.

In this regard, the annual fertility rate for women of reproductive age showed an average value of 0.193, while the female fertility index (i.e., the number of births per woman who survives her entire reproductive period) proposed by Bocquet-Appel and Masset (1977) was 6.75. This value would be consistent with those

reported for other hunter-gatherer populations (Bentley et al., 1993a, b; Campbell & Wood, 1988; Lanza & Valeggia, 2012; Pennington, 2001).

On the other hand, McFadden and Oxenham (2018a) propose the $D_{0–14}/D$ ratio as an alternative to the JI of Bocquet-Appel and Masset (1977). This ratio was proposed from data analysis on age at death, fertility rates, and population growth rates provided by the United Nations database (United Nations, 2017). When the $D_{0–14}/D$ ratio was calculated, a value of 0.453 was obtained, indicating that 45.3% of the analyzed sample comprises individuals under 14 years old. In another study, McFadden and Oxenham (2018b) propose the use of a formula that includes the $D_{0–14}/D$ ratio to calculate total fertility through a simple linear regression. Using this formula, a fertility rate of 5.73 children born per woman surviving to reproductive age was obtained. When the fertility rates obtained using the JI (Bocquet-Appel & Masset, 1977; 6.75) and the $D_{0–14}/D$ ratio (McFadden & Oxenham, 2018a; 5.73) were compared, it was observed that the latter was about one point lower. These values suggest that the fertility rate estimated for the Médano Petroquímica is high. The maximum biological fecundity for most populations is generally considered to be approximately 15 births per female if females start reproducing at the age of 15 years and do not breastfeed any of their children (Leridon, 2005). However, this would only occur at the individual level (Leridon, 2005). Therefore, it is very likely that the fertility rate values are always significantly lower than the maximum biological fecundity (McFadden & Oxenham, 2018a). Considering the proposal of Preston et al. (1989) (Table 4), in which the authors propose value ranges to classify the fertility of a population, both fertility rates obtained suggest that the Médano Petroquímica site would have presented a moderate to high fertility, an expected situation for societies with a certain degree of sedentary lifestyle, such as agropastoralists and complex hunter-gatherers (Bellwood & Oxenham, 2008; Bocquet-Appel, 2008; Oxenham et al., 2018).

Regarding the fertility rates used in this work, it is worth mentioning that the index proposed by McFadden and Oxenham (2018a) takes into account individuals aged between 0 and 5 years and uses linear regression in their calculation, unlike Bocquet-Appel and Masset (1977), who do not consider individuals under 5 years age at death and use a polynomial regression. As mentioned above, given that a high proportion of individuals under 1 year of age is mainly an indicator of fertility rather than mortality (Konigsberg & Frakenberg, 2002), that is, an increasing fertility rate will produce a greater number of deceased infants, the relationship proposed by McFadden and Oxenham (2018a) between the rates of age at death and the fertility rate would be linear. On the other hand, the omission, by Bocquet-Appel and Masset (1977), of the age group included in the range of 0–5 years, is due to the fact that the authors tried to minimize the error that can be generated by including an age group

Table 4 Fertility rates classification (*sense* Preston et al., 1989)

Fertility rates values	Classification
$X = 6.15$	High fertility
$X = 4.59$	Moderate fertility
$X = 1.64$	Low fertility

generally underrepresented in archaeological samples (i.e., as mentioned above, because immature bone remains are more fragile and, having poorer preservation, are often difficult to recover). Furthermore, in the D_{0-14}/D ratio of McFadden and Oxenham (2018a), this category of infants is considered, but the authors are careful to recommend its use for samples with a good representation of subadults. Regarding problems related to the accuracy of age estimation, which decreases with age, Bocquet-Appel and Masset (1977) group all adult individuals into a single category in the denominator, while McFadden and Oxenham (2018a), using the D_{0-14}/D ratio, reduce the potential impact of error in estimating the age at death by including all the individuals in the sample in the denominator. If this last point is considered, the ratio D_{0-14}/D would be less exposed to errors in age estimations, since these are more precise for subadults (0–14 years age range), while the denominator uses the total sample, without considering the age of the individuals that comprise it.

Concerning the modal peak recorded for young adults, it can be observed that this age range corresponds to 33.33% of the analyzed sample. This suggests that most individuals would have entered a situation of constant morbidity-mortality from very early ages. Probably, the tasks aimed at daily subsistence and protection and defense of the group, in addition to those involved in reproduction, demanded the participation of individuals before they completed their biological growth.

In the group of young adults, a greater presence of male individuals was recorded. Some authors suggest that this peculiarity would mean that young male adults from hunter-gatherer societies in this region would be most exposed to violent events (Berón, 2018; see below). As regards female individuals, although they are less represented than male individuals in the young adult category, it was in this range that the largest number of women was recorded (Fig. 2). This would be usual in this age range, since women are in full reproductive age and, therefore, are exposed to the complications or risks that pregnancy and childbirth entail, in addition to the nutritional requirements around lactation (Grauer, 1989; Powell, 1988). With respect to this last point, McFadden and Oxenham (2019) propose calculating the maternal mortality rate, that is, those deaths related to pregnancy, childbirth, and the days immediately after it. Given that maternal mortality can affect population dynamics and be reflected in the distribution of age at death, higher levels of maternal mortality are expected to generate a higher frequency of women in the fertile age ranges within the archaeological record (McFadden & Oxenham, 2019). The values of maternal mortality rates with which the authors worked ranged between 4.5 and 709 (McFadden & Oxenham, 2019). The value obtained for Médano Petroquímica was 254.26 deaths per 100,000. Although this value is in the first third of the range of variation found by the authors according to the populations studied, it is worth mentioning the limitations of applying this parameter. For example, in modern societies, women in labor receive medical attention and sanitary and septic conditions have improved considerably over time. Furthermore, finding information about these parameters for developing countries can be extremely difficult. Therefore, the values obtained by McFadden and Oxenham (2019) would not be suitable for direct comparisons with prehistoric societies. However, considering that the Médano

Petroquímica site would present a high fertility rate, it would also be expected to observe a high maternal mortality rate (McFadden & Oxenham, 2019).

Beyond the limitations in terms of establishing comparisons, it would be interesting to have data on this type of indicator for other samples in the region to draw interpopulation comparisons and analyze in more detail the relationships between maternal mortality and the prevalence of diseases and indicators of nutritional deficiencies, as well as to improve the understanding of the levels of maternal care in past societies. McFadden and Oxenham (2019: 145) express: "The estimated maternal mortality rate can be compared with those of other past populations to determine inter- and intrapopulation differences..." and "... may offer significant insights into the causes of maternal mortality, sophistication of maternal care, and female experience of pregnancy and childbirth in past populations." Nevertheless, considering that limited age ranges are used to calculate the maternal mortality rate, since age estimation methods in adult individuals are less precise and specific ages are difficult to establish, it would also be advisable to develop an indicator with a broader age-at-death range.

McFadden and Oxenham (2019) relate maternal mortality to fertility values and population growth, while Kelly (1995) suggests a greater relationship between interpregnancy intervals and fertility rates. The relationship between interpregnancy intervals and the final number of children per woman has already been discussed for Médano Petroquímica in a previous work (Bernardi et al., 2022). Additionally, different authors (Bellwood & Oxenham, 2008; Oxenham et al., 2018) suggest that the reduction in mobility or sedentary lifestyle reduces the period between births, leading to an increase in fertility and population growth. Given this idea, it could be proposed that the societies that formed the Médano Petroquímica site were experiencing an increase in fertility, which resulted in a population increase, as suggested by the calculated growth rates. This situation would strongly influence the maternal mortality rates experienced by these societies.

The values of the population growth rates obtained for the site were equal to 1.5 per 1000 (Bocquet-Appel & Masset, 1977) and 2.95% (McFadden & Oxenham, 2018b). Although these values are very dissimilar (the second value is almost 20 times greater than the first), both suggest that it would be a growing population. The differences in the values of these parameters respond to the fact that both estimates are obtained through different formulas, since one proposal considers individuals under 5 years of age at the time of death (McFadden & Oxenham, 2018b), while the other does not (Bocquet-Appel & Masset, 1977). Since the 0–4.9 years range was very well represented in the sample, it was expected that the McFadden and Oxenham (2018b) formula would provide a higher value. However, an annual growth rate equal to 29.5 per 1000 is even higher for current societies. According to the data provided for 52 countries by the United Nations Database (United Nations, 2017) for 1960, only 12 of the nations would present an annual growth rate higher than the value obtained for Médano Petroquímica. Therefore, the authors of this work prefer to be conservative and discuss the growth rate based on the formula proposed by Bocquet-Appel and Masset (1977).

The calculated growth rate equal to 1.5 per 1000 would correspond to values obtained for most pre-industrial societies (Cowgill, 1975; Ramírez, 2015). Cowgill (1975) states that until the late nineteenth century, the annual growth rates of most nations did not exceed 5–10 per 1000. In addition, Kohler and Reese (2014) state that demographic processes (variations in birth rates and growth rates) have always been slow processes. This is consistent with the results obtained from the JI and Mean Childhood Mortality (MCM) values (Jackes, 1992; Jackes et al., 1997), where Médano Petroquímica falls within the range for growing populations. The values suggested by these authors, below which a growing population can be considered, are JI equal to or less than 0.38 (for Médano Petroquímica the JI was 0.29) and MCM equal to or less than 0.135 (0.117 in the sample under study) (see Bernardi et al., 2022).

Concerning population dynamics, several authors (Berón, 2018; Berón & Politis, 1997; Madrid & Barrientos, 2000; Martínez, 2006, 2017; Mazzanti, 2006; Mazzanti & Quintana, 2001) have proposed that, toward the last 1000 years, the societies of the Pampa and Patagonia experienced important changes in its structures, such as increased demographic density, reduced mobility, and increased complexity in sociopolitical and economic organization. Some authors (Barrientos, 2002; Berón, 2007, 2018) have associated this series of organizational changes with the so-called Medieval Climate Anomaly (MCA). This condition would have favored the concentration of human groups in those places where critical resources such as water were stable and, consequently, the appearance of formal cemeteries in the vicinity of such watercourses. Therefore, the presence of areas with burials such as the Médano Petroquímica site, which is located approximately 2 km from the current course of the Colorado River, could be used as an argument in favor of this climatic event and the consequent decrease in residential mobility and increase in population density (Barrientos, 2002). In this regard, it is possible that the location of the site could have been intentionally chosen for its notorious visibility and landscape dominance, in addition to its proximity to natural watercourses. Then, probably, the Médano Petroquímica represented a point of the landscape with great symbolic value for the cohesion and survival of the group, which seasonally would return to the formal burial areas.

This complex socio-environmental scenario could have generated social tensions and situations of conflict between the groups and/or partialities that inhabited the Pampa-Patagonia transition. Berón (2018) recorded a notable increase in indicators of interpersonal violence at the Chenque I site (approximately 200 km to the east of the Médano Petroquímica site) for approximately 300–400 years BP. These situations would have occurred before the Hispanic-indigenous contact. It has been suggested that the reduction in mobility generated situations of competition for access to resources and control of exchange circuits and interaction nodes, which would have led to an increase in armed conflict scenarios (Berón, 2007, 2018; Salazar Siciliano & Berón, 2013). In a preliminary analysis of fractures of anthropogenic origin in EA1, H4, H6-1, and H7 burials of the Médano Petroquímica site, Arrieta et al. (2019) recorded the presence of penetrating fractures associated with violence in seven skeletal elements corresponding to five individuals, which represented

26.31% of the sample (19 adults individuals). This situation, added to the abundance of projectile points found in situ, both loose and impacting bone elements, would fit the proposal of Berón (2018). However, the statements of Barrientos and Gordón (2004), who affirm that there is no statistical support to directly relate the demographic increase in Northeast Patagonia during the MCA with the increase in interpersonal violence rates, should be considered. Although the authors do not state the absence of a relationship between population growth and competition for resources, they establish that the levels of violence developed to resolve these conflicts would not be higher than usual for these societies (Barrientos & Gordón, 2004).

Finally, it is necessary to highlight that although the changes that are evidenced from the late Holocene in the societies of Pampa-Patagonia have been studied from different lines of bioarchaeology, paleodemographic studies constitute a set of tools with great potential, since they can provide data from which inferences can be made regarding structural changes and population dynamics for the region. It is of great importance to always consider the existing limitations when attempting to make demographic inferences with samples that come from societies that do not have written records. However, the use of this type of indicator is a very valuable tool for understanding past societies, mainly regarding the health status of the population and its ability to respond to changes. In addition, the calculation of these parameters is easily reproducible and verifiable in archaeological samples that do not have an underrepresentation of subadults. Axtell et al. (2002) highlight that the lack of reproducibility of paleodemographic analysis has represented a great impediment in archaeology, which has only been overcome through the use of modeling. Therefore, from this work, the authors intend to encourage the use of these parameters in bioarchaeological investigations carried out in Argentina, so that researchers can have data that allow comparisons between the different late Holocene sites in the Pampa-Patagonia transition region. In this way, it is hoped to be able to reconstruct a broader and more precise panorama of the dynamics of the populations that inhabited the area.

Conclusions

Paleodemography provides a set of potentially valuable tools for analyzing and understanding the dynamics of prehistoric societies. The use and development of indicators, such as those analyzed here, help infer and compare different population patterns between archaeological sites that are close in space and time.

Environmental changes toward the end of the Late Holocene would have impacted the population dynamics of the societies that inhabited the Pampa-Patagonia transitional region. These changes would have caused a reduction in mobility and the nucleation of human groups in areas close to critical resources. The analysis of the burials from the Médano Petroquímica site supports this proposal. Parameters such as fertility rate, fertility index, juvenile index, and population growth rates suggest that this population was experiencing growth. Furthermore, the

obtained maternal mortality rate could also be closely associated with the increase in fertility.

Although the results obtained are in line with the scenario proposed for the Pampa-Patagonia societies for the late Holocene, it would enrich research in the region if these indices begin to be used in bioarchaeological studies. These parameters are easily reproducible and verifiable and allow comparisons between sites, which would help to have a general overview of the dynamics of the populations that inhabited the area.

Acknowledgments The authors would like to express their gratitude to the Secretaría de Cultura of La Pampa province and CONICET. They also thank Dr. Ana Margarita Aguerre and Lic. Lía Pera. The authors thank Germán Eberhardt for his collaborations in editing figures. In addition, they acknowledge the anonymous reviewers for providing helpful suggestions that greatly improved this paper. This research was funded by CyT UNRC (PPI GRF), FONCyT (PICT-2020-SERIEA-00581 GRF), and CONICET.

References

Acsádi, G., & Nemeskéri, J. (1970). *History of human life span and mortality*. Akadémiai Kiadó.

Angel, J. L. (1969). The bases of paleodemography. *American Journal of Physical Anthropology, 30*, 427–438.

Arrieta, M. A., Bottini, M. C., Bernardi, L., & Lynch Ianniello, I. (2019). Análisis preliminar de fracturas antrópicas en esqueletos del sitio Médano Petroquímica (La Pampa). In *Libro de Resúmenes del I Congreso Iberoamericano de estudios Sociales sobre Conflicto Armado* (pp. 59–60). Universidad Nacional de Luján.

Axtell, R. L., Epstein, J. M., Dean, J. S., Gumerman, G. J., Swedlund, A. C., Harburger, J., Chakravarty, S., Hammond, R., Parker, J., & Parker, M. (2002). Population growth and collapse in a multiagent model of the Kayenta Anasazi in Long House Valley. *Proceedings of the National Academy Sciences U.S.A., 99*(3), 7275–7279.

Baker, B., Dupras, T., & Tocheris, M. (2005). *The osteology of infants and children*. College Station, TX, Texas A&M University Press.

Barrientos, G. (2002). The archaeological analysis of death-related behaviors from an evolutionary perspective: Exploring the bioarchaeological record of early American hunter-gatherers. In G. Martínez & L. Lanata (Eds.), *Perspectivas Integradoras entre Arqueología y Evolución. Teoría, Métodos y Casos de Aplicación* (pp. 221–253). Serie teórica del INCUAPA N° 1, Facultad de Ciencias Sociales, UNCPBA.

Barrientos, G., & Gordón, F. (2004). Explorando la relación entre nucleamiento poblacional y violencia interpersonal durante el Holoceno tardío en el noroeste de Patagonia (República Argentina). *Magallania, 32*, 53–69.

Barrientos, G., & Pérez, I. (2004). La expansión y la dispersión de poblaciones del norte de Patagonia durante el Holoceno tardío: evidencia arqueológica y modelo explicativo. In T. Civalero, P. Fernández, & G. Guraieb (Eds.), *Contra Viento y Marea. Arqueología de la Patagonia* (pp. 179–195). INAPL y SAA.

Bellwood, P., & Oxenham, M. F. (2008). The expansions of farming societies and the role of the neolithic demographic transition. In J. P. Bocquet-Appel & O. Bar-Yosef (Eds.), *The neolithic demographic transition and its consequences* (pp. 13–24). Springer Science + Business Media. https://doi.org/10.1007/978-1-4020-8539-0_2

Bentley, G. R., Goldberg, T., & Jasieñska, G. (1993a). The fertility of agricultural and nonagricultural traditional societies. *Population Studies, 47*, 269–281.

Bentley, G. R., Jasieñska, G., & Goldberg, T. (1993b). Is the fertility of agriculturalists higher than that of non-agriculturalists? *Current Anthropology, 34*, 778–785.
Bernardi, L. (2018). *Los cazadores-recolectores del Médano Petroquímica (Colonia 25 de Mayo, Puelén, La Pampa). Caracterización de la estructura y composición del conjunto por medio de marcadores anatomofuncionales* [Unpublished doctoral thesis]. Universidad Nacional de Río Cuarto.
Bernardi, L. (2019). Evaluación de estimadores de edad y sexo en coxales adultos del sitio Médano Petroquímica. In M. Fabra, P. S. Novellino, M. A. Arrieta, & M. S. Salega (Eds.), *Metodologías para el estudio de restos esqueletales en Argentina: actualizaciones y nuevas perspectivas* (pp. 206–236). Universidad Nacional de Córdoba, Facultad de Filosofía y Humanidades.
Bernardi, L., Arrieta, M. A., & Lynch Ianniello, I. (2020). Preliminary study of age-at-death profile in a prehistoric sample from Médano Petroquímica site, Argentina. *Journal of Archaeological Science: Reports, 34*, 102651. https://doi.org/10.1016/j.jasrep.2020.102651
Bernardi, L., Arrieta, M. A., & Bottini, M. C. (2022). Estructura y dinámica poblacional de los cazadores-recolectores del sitio Médano Petroquímica (La Pampa, Argentina): inferencias a partir del uso de tablas de vida y otros indicadores paleodemográficos. *Chungara: Revista de Antropología Chilena, 54*(1), 55–70. https://doi.org/10.4067/S0717-73562021005002504
Berón, M. (2007). Circulación de bienes como indicadores de interacción entre poblaciones de la pampa occidental y sus vecinos. In C. Bayón, A. Pupio, M. I. González, N. Flegenheimer, & M. Frére (Eds.), *Arqueología en las Pampas* (pp. 345–364). Sociedad Argentina de Antropología.
Berón, M. (2018). Violencia, conflicto y guerra en las pampas: distintas aproximaciones desde el registro arqueológico del sitio Chenque I. In M. Berón (Ed.), *El sitio Chenque I: un cementerio prehispánico en la Pampa Occidental. Estilo de vida e interacciones culturales de cazadores-recolectores del Cono Sur Americano* (pp. 453–495). Sociedad Argentina de Antropología.
Berón, M., & Politis, G. (1997). Arqueología pampeana en la década de los 90', estado de las investigaciones y perspectivas. In M. Berón & G. Politis (Eds.), *Arqueología Pampeana en la Década de los 90'* (pp. 7–32). INCUAPA, Facultad de Ciencias Sociales de Olavarría, U.N.C.P.B.A. Museo de Historia Natural de San Rafael, province of Mendoza.
Bocquet-Appel, J. P. (2008). Explaining the neolithic demographic transition. In J. P. Bocquet-Appel & O. Bar-Yosef (Eds.), *The neolithic demographic transition and its consequences* (pp. 13–24). Springer.
Bocquet-Appel, J. P., & Masset, C. (1977). Estimateurs en paleodemographie. *L'Homme, 27*, 65–90.
Bottini, M. C., Arrieta, M. A., & Bernardi, L. (2020a). El comportamiento mortuorio en los cazadores-recolectores del sitio Médano Petroquímica (La Pampa, Argentina). Caracterización preliminar desde un abordaje tafonómico. *Chungara, Revista de Antropología Chilena, 52*(2), 241–259. https://doi.org/10.4067/S0717-73562020005000901
Bottini, M. C., Arrieta, M. A., & Bernardi, L. (2020b). Tafonomía en restos óseos humanos del SO de la región pampeana: el sitio Médano Petroquímica (Puelén, La Pampa, Argentina). *Revista del Museo de Antropología, 13*(3), 123–138. https://doi.org/10.31048/1852.4826.v13.n3.30056
Brook, S., & Suchey, J. M. (1990). Skeletal age determination based on the os pubis: A comparison of the Acsádi-Nemerkéri and Suchey-Brooks methods. *Human Evolution, 5*(3), 227–238.
Buckberry, J., & Chamberlain, A. (2002). Age estimation from the auricular surface of the ilium: A revised method. *American Journal of Physical Anthropology, 119*, 231–239.
Buckley, H. R. (2000). Subadult health and disease in prehistoric Tonga, Polynesia. *American Journal of Physical Anthropology, 113*(4), 481–505.
Buikstra, J. E., & Konigsberg, L. W. (1985). Paleodemography: Critiques and commentaries. *American Anthropologist, 87*, 316–333.
Buikstra, J. E., & Mielke, J. (1985). Demography, diet and health. In R. Gilbert & J. Mielke (Eds.), *The analysis of prehistoric diets* (pp. 360–422). Academic Press.
Buikstra, J. E., & Ubelaker, D. H. (Eds.). (1994). *Standards for data collection from human skeletal remains*. Arkansas Archeological Survey Research Series N° 44.

Byrd, J. E., & LeGarde, C. B. (2014). Osteometric sorting. In B. J. Adams & J. E. Byrd (Eds.), *Commingled human remains: Methods in recovery, analysis, and identification* (pp. 167–191). Academic Press.

Campbell, K. L., & Wood, J. W. (1988). Fertility in traditional societies. In P. Diggory, S. T'eper, & M. Potts (Eds.), *Natural human fertility: Social and biological mechanism* (pp. 39–69). Macmillan.

Chamberlain, A. (2006). *Demography in archaeology* (Cambridge manuals in archaeology). Cambridge University Press.

Cowgill, L. G. (1975). On causes and consequences of ancient and modern population changes. *American Anthropologist, 77*(3), 505–525.

Eastbrook, G. F., & Rissech, C. (2004). Program IDADE2 user's document.

Fazekas, I. G., & Kósa, F. (1978). *Forensic fetal osteology*. Akadémiai Kiadó.

FitzGerald, C., Saunders, S., Bondioli, L., & Macchiarelli, R. (2006). Health of infants in an Imperial Roman skeletal sample: Perspective from dental microstructure. *American Journal of Physical Anthropology, 130*(2), 179–189.

Flensborg, G., & Bayala, P. (2017). Bioarqueología. Sexo, edad, salud y patrones de mortalidad. In G. Martínez (Ed.), *Arqueología de cazadores-recolectores del curso inferior del Río Colorado* (Serie Monográfica N° 6) (pp. 146–174). INCUAPA, CONICET, UNICEN.

Flensborg, G., Martínez, G., & Bayala, P. (2015). Mortality profiles of hunter-gatherer societies: A case study from the eastern Pampa–Patagonia transition (Argentina) during the final late Holocene. *International Journal of Osteoarchaeology, 25*(6), 816–826.

Frankenberg, S. R., & Konigsberg, L. W. (2006). A brief history of paleodemography from Hooton to hazards analysis. In L. A. Beck & J. E. Buikstra (Eds.), *Bioarchaeology. The contextual analysis of human remains* (pp. 227–261). Elsevier Press.

Gage, T. B. (2000). Demography. In S. Stinson, B. Bogin, R. Huss-Ashmore, & D. O'Rourke (Eds.), *Human biology: An evolutionary and biocultural perspective* (pp. 507–551). Wiley-Liss.

García Guraieb, S., Goni, R., & Tessone, A. (2015). Paleodemography of Late Holocene hunter-gatherers from Patagonia (Santa Cruz, Argentina): An approach using multiple archaeological and bioarchaeological indicators. *Quaternary International, 356*, 147–158.

Gilbert, B., & McKern, T. (1973). A method for aging the female os pubis. *American Journal of Physical Anthropology, 38*, 31–38.

Gordón, F., Béguelin, M., Novellino, P., & Archuby, F. (2019). Inferencias paleodemográficas en el Noroeste de Patagonia a partir del sitio Aquihuecó, Provincia del Neuquén, Argentina. *Chungará (Arica), 51*(3), 363–380.

Grauer, A. L. (1989). *Health, disease and status in medieval york* [Unpublished Ph. Dissertation]. University of Massachusetts.

Herrada, C. R. (2002). *Dimensiones bioarqueológicas de los contextos funerarios. Estudios de los restos humanos de la necrópolis prehistórica de la cova des Càtarritx (Ciutadella, Menorca)* [Unpublished doctoral thesis]. Universidad Autónoma de Barcelona.

Herring, D. A., Saunders, S. R., & Katzenberg, M. A. (1998). Investigating the weaning process in past populations. *American Journal of Physical Anthropology, 105*(4), 425–439.

Hoppa, R. D. (2002). Paleodemography: Looking back and thinking ahead. In R. D. Hoppa & J. W. Vaupel (Eds.), *Paleodemography: Age distributions from skeletal samples* (pp. 9–28). Cambridge University Press.

Hoppa, R. D., & Vaupel, J. W. (2002). The rostock manifesto for paleodemography: The way from stage to age. In R. D. Hoppa & J. W. Vaupel (Eds.), *Paleodemography: Age distributions from skeletal samples* (pp. 1–8). Cambridge University Press.

Jackes, M. K. (1992). Paleodemography: Problems and techniques. In S. R. Saunders & M. A. Katzenberg (Eds.), *Skeletal biology of past peoples: Research methods* (pp. 189–224). Wiley-Liss.

Jackes, M. K. (2011). Representativeness and bias in archaeological skeletal samples. In S. Agarwal & B. Glencross (Eds.), *Social bioarchaeology* (pp. 107–146). Wiley-Blackwell.

Jackes, M. K., Lubell, D., & Meilejohn, C. (1997). Healthy but moral: Human biology and the first farmers of Western Europe. *American Antiquity, 71*, 639–658.

Johnston, F. E. (1962). Growth of the long bones of infants and young children at Indian Knoll. *American Journal of Physical Anthropology, 20*, 249–254.

Katzenberg, M. A., Herring, D. A., & Saunders, S. R. (1996). Weaning and infant mortality: Evaluating the skeletal evidence. *American Journal of Physical Anthropology, 101*(S23), 177–199.

Kelly, R. (1995). *The foraging spectrum: Diversity in hunter-gatherer lifeways*. Smithsonian Institution Press.

Knüsel, C. J., & Outran, A. K. (2004). Fragmentation: The zonation method applied to fragmented human remains from archaeological and forensic contexts. *Environmental Archaeology, 9*(1), 85–98.

Kohler, T. A., & Reese, K. M. (2014). Long and spatially variable Neolithic demographic transition in the North American southwest. *Proceedings of National Academy of Sciences U.S.A., 111*(28), 10101–10106.

Konigsberg, L. W., & Frankenberg, S. R. (1994). Paleodemography: "Not quite dead". *Evolutionary Anthropology: Issues, News, and Reviews, 3*(3), 92–105.

Konigsberg, L. W., & Frankenberg, S. R. (2002). Deconstructing death in paleodemography. *American Journal of Physical Anthropology, 117*, 297–309.

Krogman, W. M., & Isçan, M. Y. (1986). *The human skeleton in forensic medicine*. Thomas Publisher.

Lanza, N., & Valeggia, C. (2012). La modernización ¿Siempre disminuye la fecundidad? Análisis de la transición demográfica en una Población Toba. *Revista Argentina de Antropología Biológica, 14*(1), 33–43.

Leridon, H. (2005). Natural fertility and controlled fertility: Levels and models. In G. Caselli, J. Vallin, & G. Wunsch (Eds.), *Demography: Analysis and synthesis* (pp. 467–478). Academic Press.

Lewis, M. E. (2007). *The bioarchaeology of children: Perspectives from biological and forensic anthropology* (Cambridge studies in biological and evolutionary anthropology) (Vol. 50). Cambridge University Press.

Livi-Bacci, M. (2002). *Historia mínima de la población mundial*. Editorial Ariel.

Loth, S. R., & Henneberg, M. (1996). Mandibular ramus flexure: A new morphologic indicator of sexual dimorphism in the human skeleton. *American Journal of Physical Anthropology, 99*, 473–485.

Loth, S. R., & Henneberg, M. (2001). Sexually dimorphic mandibular morphology in the first few years of life. *American Journal of Physical Anthropology, 115*(2), 179–186.

Lovejoy, C. O. (1985). Dental wear in the Libben population: Its functional pattern and role in the determination of adult skeletal age at death. *American Journal of Physical Anthropology, 68*(10), 47–56.

Lovejoy, C., Meindl, R., Mensforth, R., & Barton, T. (1985). Multifactorial determination of skeletal age at death: A method and blind tests of its accuracy. *American Journal of Physical Anthropology, 68*, 1–14.

Luna, L. H. (2018). Primeras interpretaciones globales sobre el perfil de mortalidad del sitio Chenque I. In M. Berón (Ed.), *El sitio Chenque I: un cementerio prehispánico en la Pampa Occidental. Estilo de vida e interacciones culturales de cazadoresrecolectores del Cono Sur Americano* (pp. 101–136). Sociedad Argentina de Antropología.

Luna, L. H., Aranda, C., & Santos, A. L. (2017). New method for sex prediction using the human non-adult auricular surface of the ilium in the Collection of Identified Skeletons of the University of Coimbra. *International Journal of Osteoarchaeology, 27*, 898–911.

Madrid, P., & Barrientos, G. (2000). La estructura del registro arqueológico Laguna Tres Reyes 1 (provincia de Buenos Aires): nuevos datos para la interpretación del poblamiento humano del sudeste de la Región Pampeana a inicios del Holoceno Tardío. *Relaciones de la Sociedad Argentina de Antropología, 25*, 179–206.

Margerison, B. J., & Knüsel, C. J. (2002). Paleodemographic comparison of a catastrophic and an attritional death assemblage. *American Journal of Physical Anthropology, 119*(2), 134–143.

Martínez, G. (2006). Arqueología del curso medio del río Quequén Grande: estado actual y aportes a la arqueología de la región pampeana. *Relaciones de la Sociedad Argentina de Antropología, 31*, 249–275.

Martínez, G. (2017). Discusión. In G. Martínez (Ed.), *Arqueología de cazadores-recolectores del curso inferior del Río Colorado (provincia de Buenos Aires, Argentina). Aportes al conocimiento de las ocupaciones humanas Pampeano-Patagónicas* (Serie monográfica - número 6) (pp. 255–297). INCUAPA-CONICET-UNICEN.

Mays, S. A. (2002). Asymmetry in metacarpal cortical bone in a collection of British post-medieval human skeletons. *Journal of Archaeological Science, 29*, 435–441.

Mazzanti, D. L. (2006). La construcción de territorios sociales durante el Holoceno tardío. El caso de las sierras de Tandilia, Argentina. *Relaciones de la Sociedad Argentina de Antropología, XXXI*, 277–300.

Mazzanti, D., & Quintana, C. (2001). *Cueva Tixi: cazadores y recolectores de las sierras de Tandilia Oriental. Geología, Paleontología y Zooarqueología.* Laboratorio de Arqueología, Universidad Nacional de Mar del Plata, Publicación Especial 1.

Mazzeo, V. (2007). La mortalidad en la primera infancia en la ciudad de Buenos Aires en el período 1860-2002. *Papeles de población, 13*(53), 241–272.

McFadden, C., & Oxenham, M. F. (2018a). The D0-14/D ratio: A new paleodemographic index and equation for estimating total fertility rates. *American Journal of Physical Anthropology, 165*(3), 471–479.

McFadden, C., & Oxenham, M. F. (2018b). Rate of natural population increase as a paleodemographic measure of growth. *Journal of Archaeological Science: Reports, 19*, 352–356.

McFadden, C., & Oxenham, M. F. (2019). The paleodemographic measure of maternal mortality and a multifaceted approach to maternal health. *Current Anthropology, 60*(1), 141–146.

McKern, T., & Stewart, T. (1957). *Skeletal age changes in young American males: Analyzed from the standpoint of age identification.* Natick. Headquarters, Quartermaster Research and Development Command, Technical Report. EP 45.

Meindl, R., & Lovejoy, C. (1989). Age changes in the pelvis: Implications for paleodemography. In M. Isçan (Ed.), *Age markers in the human skeleton* (pp. 137–168). Thomas Publisher.

Meindl, R., & Russell, K. (1998). Recent advances in method and theory in paleodemography. *Annual Review of Anthropology, 27*, 375–399.

Milner, G. R. (1992). *Determination of skeletal age and sex: A manual prepared for the Dickson Mounds Reburial Team*. Dickson Mounds Museum.

Milner, G. R., Humpf, D. A., & Harpending, H. C. (1989). Pattern matching of age-at-death distributions in paleodemographic analysis. *American Journal of Physical Anthropology, 80*, 49–58.

Milner, G. R., Wood, J. W., & Boldsen, J. L. (2008). Advances in paleodemography. In M. A. Katzenberg & S. R. Saunders (Eds.), *Biological anthropology of the human skeleton* (pp. 561–600). Wiley.

Molleson, T., Cruse, K., & Mays, S. (1998). Some sexually dimorphic features of the human juvenile skull and their value in sex determination in immature skeletal remains. *Journal of Archaeological Science, 25*(8), 719–728.

Murail, P., Brukez, J., Houëti, F., & Cunha, E. (2005). A tool for probabilistic sex diagnosis using worldwide variability in hip-bone measurements. *Bulletins et Mémoires de la Société d'Anthropologie de Paris, 17*(3–4), 167–117.

Novellino, P. S., Aporta, N. B., D'Addona, L. A., Estrella, D., Bernal, V., Devincenzi, S., & Durán, V. (2013). Sitios de entierro de la localidad arqueológica Barrancas, Maipú (Mendoza, Argentina). *Intersecciones en Antropología, 14*(1), 271–277.

Osterholtz, A. J. (2019). Advances in documentation of commingled and fragmentary remains. *Advances in Archaeological Practice, 7*(1), 77–86.

Oxenham, M. F., Trinh, H. H., Willis, A., Jones, R. K., Domett, K., Castillo, C., & Buckley, H. (2018). Between foraging and farming: Strategic responses to the Holocene thermal maximum in Southeast Asia. *Antiquity, 92*(364), 940–957.

Pennington, R. (2001). Hunter-gatherer demography. In C. Panter-Brick, R. H. Layton, & P. Rowley-Conwy (Eds.), *Hunter-Gatherers: An interdisciplinary perspective* (pp. 170–204). Cambridge University Press.

Peralta, E. A. (2019). *Demografía humana, dieta y actividad en los límites de la dispersión agrícola: tendencias bioarqueológicas en el sur de Mendoza a finales del Holoceno tardío* [Unpublished doctoral thesis]. FFyL, Universidad de Buenos Aires.

Phenice, T. W. (1969). A newly development visual method of sexing the os pubis. *American Journal of Physical Anthropology, 30*, 297–301.

Powell, M. L. (1988). *Status and health in prehistory: A case of study of the Moundville Chiefdom*. Smithsonian Institution Press.

Preston, S. H., Himes, C., & Eggers, M. (1989). Demographic conditions responsible for population aging. *Demography, 26*, 691–704.

Ramírez, J. P. D. (2015). *Estudio Bioantropológico comparado de tres necrópolis históricas excavadas en el Término Municipal de Lucena (Córdoba)* [Unpublished doctoral thesis]. Universidad de Granada, Spain.

Roksandic, M., & Armstrong, S. D. (2011). Using the life history model to set stage(s) of growth and senescence in bioarchaeology and paleodemography. *American Journal of Physical Anthropology, 145*, 337–347.

Salazar Siciliano, G., & Berón, M. (2013). Diacríticos identitarios en las relaciones transcordilleranas. Circuitos económicos sociales y culturales entre el centro de Argentina, centro-oeste de Neuquén y la Araucanía chilena. In A. Nicoletti & P. Nuñez (Eds.), *Araucanía y Norpatagonia: territorialidad en debate. Perspectivas ambientales, culturales, sociales, políticas y económicas* (pp. 187–205). Instituto de Investigaciones en Diversidad Cultural y Procesos de Cambio. Universidad Nacional de Río Negro.

Sánchez Romero, M. (2006). Maternidad y Prehistoria: prácticas de reproducción, relación y socialización. In S. Mayor (Ed.), *Las mujeres en la prehistoria* (pp. 119–138). Museu de Prehistòria de València.

Sattenspiel, L., & Harpending, H. (1983). Stable populations and skeletal age. *American Antiquity, 48*, 489–498.

Schaefer, M. (2014). A practical method for detecting commingled remains using epiphyseal union. In B. J. Adams & J. E. Byrd (Eds.), *Commingled human remains: Methods in recovery, analysis, and identification* (pp. 123–144). Academic Press.

Scheuer, L., & Black, S. (2000). *Developmental juvenile osteology*. Academic Press.

Schutkowski, H. (1993). Sex determination of infant and juvenile skeletons: I. Morphognostic features. *American Journal of Physical Anthropology, 90*, 199–205.

Seguy, I., & Buchet, L. (2013). Finding the right models for pre-industrial populations. In I. Seguy & L. Buchet (Eds.), *Handbook of palaeodemography* (pp. 113–121). Springer International Publishing.

Stinson, S. (2000). Growth variation: Biological and cultural factors. In S. Stinson, P. Bogin, & H. D. O'Rourke (Eds.), *Human biology: An evolutionary and biocultural perspective* (pp. 425–463). Wiley-Liss.

Stodder, A. L. (1997). Subadult stress, morbidity, and longevity in Late Period populations on Guam, Mariana Islands. *American Journal of Physical Anthropology, 104*, 363–380.

Sutherland, L., & Suchey, J. (1991). Use of the ventral arc in pubic sex determination. *Journal of Forensic Sciences, 36*(2), 501–511.

Todd, T. (1921a). Age changes in the pubic bone. I: The male white pubis. *American Journal of Physical Anthropology, 3*, 285–334.

Todd, T. (1921b). Age changes in the pubic bone. III: The pubis of the white female. IV: The pubis of the female white-negro hybrid. *American Journal of Physical Anthropology, 4*, 1–70.

Todd, L., & Frison, G. (1992). Reassembly of bison skeletons from the Horner Site: A study in anatomical refitting. *BAR International Series, 578*, 63–63.

Ubelaker, D. H. (1999). *Human skeletal remains; excavation analysis, interpretation* (3rd ed.). Taraxacum.

Ubelaker, D. H. (2007). *Enterramientos Humanos. Excavación, Análisis, Interpretación*. Sociedad de Ciencias Aranzadi.

United Nations. (2017). *UNSD statistical database*. Available from https://unstats.un.org/unsd/databases.htm

Volk, A. A., & Atkinson, J. A. (2013). Infant and child death in the human environment of evolutionary adaptation. *Evolution and Human Behavior, 34*(3), 182–192.

Walker, P. (2005). Greater sciatic notch morphology: Sex, age, and population differences. *American Journal of Physical Anthropology, 127*, 385–391.

Wood, J. W., Milner, G. R., Harpending, H. C., & Weiss, K. M. (1992). The osteological paradox: Problems of inferring prehistoric health from skeletal samples [and comments and reply]. *Current Anthropology, 33*(4), 343–370.

20 Years Is Nothing. Revisiting Tapera Moreira Archaeological Locality, Province of La Pampa, Argentina

Mónica Berón, Manuel Carrera Aizpitarte, Florencia Paez, Eliana N. Lucero, and Sofía Bedrán

Abstract The Tapera Moreira archaeological locality is composed of five sites with different characteristics. It is located on the banks of the Curacó River, in the south-central sector of the province of La Pampa, Argentina. This set of sites represents the occupational redundancy of this landscape, ranging from the Middle Holocene to the final Late Holocene (ca. 4550 to 360 years BP), and constitutes the main chronological-cultural sequence of the Curacó area. This locality was investigated in two stages, the first between 1988 and 1995 and the second from 2015 to the present. Excavations were reopened at site 1 and a new burial was located at site 3. The objectives of this chapter are to present the results of the new excavations and then integrate these results with the previous ones (1988–1995) to assess patterns, trends, and differences. The data generated allow us to evaluate the trajectories and social interactions of hunter-gatherer populations on a macro-regional scale, as well as the mobility circuits and the interaction networks throughout a wide area that covers both oceans and from 36° to 40° LS. This information integrated with the

M. Berón (✉) · M. C. Aizpitarte
IDECU (National Council for Scientific and Technical Research – University of Buenos Aires), CONICET. Faculty of Philosophy and Letters, University of Buenos Aires. Faculty of Social Sciences, National University of the Center of the Province of Buenos Aires, Autonomus City of Buenos Aires, Argentina
e-mail: monberon@retina.ar

F. Paez
IDECU (National Council for Scientific and Technical Research – University of Buenos Aires), CONICET. Faculty of Philosophy and Letters, University of Buenos Aires, Autonomus City of Buenos Aires, Argentina

E. N. Lucero
IDECU (National Council for Scientific and Technical Research – University of Buenos Aires), National Agency for Scientific and Technological Promotion. Faculty of Philosophy and Letters, University of Buenos Aires, Autonomus City of Buenos Aires, Argentina

S. Bedrán
Faculty of Philosophy and Letters, University of Buenos Aires, Autonomus City of Buenos Aires, Argentina

G. F. Bonnat et al. (eds.), *Current Research in Archaeology of South American Pampas*, The Latin American Studies Book Series,
https://doi.org/10.1007/978-3-031-55194-9_5

knowledge generated in neighboring regions allows us to outline a panorama of the social and cultural dynamics of the last 5000 years.

Keywords Western pampa · Hunter-gatherers · Interaction networks · Social trajectories

Introduction

Prehispanic societies in the Southern Cone experienced numerous changes, of different kinds, throughout the Holocene, which were strongly emphasized in the latter part of this period, before the arrival of the European colonizers.

The Pampean region is a wide territory located in the center of Argentina. Different micro-regions with ecological, physiographic, and cultural differences are included (Politis & Barros, 2006). Basically, we can differentiate two major divisions, Eastern and Western Pampas, whose main distinctive feature is the rainfall regime, which decreases drastically toward the west. While the Eastern Pampas records a very deep and continuous tradition of archaeological investigations in time and space, the Western Pampas has had less emphasis on its knowledge of the past, which has only been recorded systematically during the last 40 years.

The radiocarbon data indicates that the earliest colonization of this large territory is around 14,000–14,500 calibrated years BP and has been defined by the site Arroyo Seco 2 in Eastern Pampa (Politis & Prates, 2019). Since then, there has been a growing population and colonization of the region, although with some hiatus, due to different factors. Some sectors, such as Western Pampas, record a gap of information between ca. 6000 and 5000 years BP, during the Middle Holocene, although this situation is changing through the availability of new data (Berón et al., 2022; Lucero et al., 2020; Martínez & Flensborg, 2018). A re-colonization of different landscapes with diverse environmental conditions has been recorded from ca. 5000 years BP onward. This situation turned out to be more complex during the Late Holocene when a demographic growth was recorded. During this stage, climatic conditions in the region stabilized, the exploitation of local resources was emphasized, formal burial structures appeared, and the presence of extra-regional elements such as allochthonous rocks, mineral raw materials, pottery, ornaments and foreign people increased, as a sign of large-scale interactions. These circumstances deepened notably during the last part of the Late Holocene.

The Western Pampa subregion, located in the southern sector of the province of La Pampa (Argentina), is an arid and semiarid environment that comprehends about 100,000 km^2 and is included between the 600- and the 200-mm isohyet (Fig. 1a). Most of the research area is comprise in the Pampean Sand Sea or Central Pampean Dunefield. The rainfall pattern is between 250 and 300 mm per year, determining a desert environment. It shows a highly variable geomorphology and some dominant

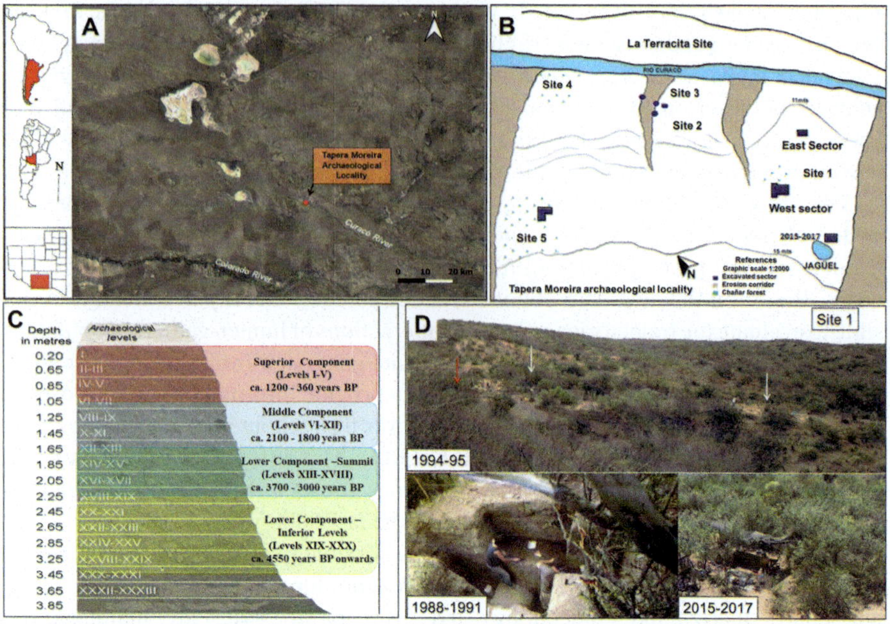

Fig. 1 Tapera Moreira archaeological locality. (**a**) Emplacement. (**b**) Freehand drawing, with the location of sites 1 to 5. (**c**) Stratigraphic section of site 1. (**d**) Panoramic photographs of different stages of excavation at site 1

physiographic features such as hills, eolian valleys, depressions, and plateaus, and secondary, but not least important, geomorphological features, such as dune fields, shallow lakes, and salpeter beds. Among the processes that permanently reshape this landscape, there are the high evapotranspiration records, intense sunlight, and dry riverbeds that rarely reach their base level (Berón, 2004, 2015, 2016).

The Tapera Moreira archaeological locality, the main topic of this chapter, witnessed these changes for at least the last 5000 years. It was investigated in the first stage during the years 1988 and 1995, obtaining a chronological-cultural sequence that has been the guiding structure of successive investigations. Since then, notorious progress has been made in the knowledge of this vast territory, based on the detection of numerous sites with diverse characteristics such as base camps, sites of specific activities, quarries-workshops, burial sites, and cemeteries.

Twenty years after the last interventions of the sites at the first stage, and since 2015, the excavations have been re-assumed, addressing new problems and questions based on the application of various analytical methodologies and the knowledge provided by research in other micro-regions and types of sites of Western Pampa that contribute to understand the social and cultural conformations of the diversity of landscapes of the region.

To evaluate the regional importance of the site, a portion of ground close to the *jagüel* was chosen, whose integrity allowed to establish six new excavation grids.

This new stage is not over yet. It was suspended for logistical reasons first and because of the pandemic later. Eighteen artificial levels (10 cm depth each) were excavated up to level XX. Much of the interest in retaking the excavations and analysis is based on the renewal of the work teams, which have deepened their knowledge of various aspects of the site based on the available collections, without having previously excavated it, all of which is the basis for its reopening (Fig. 1d).

The objectives of this chapter are to present the new results of excavations carried out between the years 2015 and 2017 in sites 1 and 3 and then integrate these results with the previous ones (1988–1995) to assess patterns, trends, and differences. Based on the new data obtained, added to those already available, it is possible to evaluate the trajectories and social interactions of hunter-gatherer populations on a micro- and macro-regional scale. The data generated from the analyses of the different materialities recovered in the locality provide information on the close and far mobility circuits and the interaction networks throughout a wide area that covers both oceans and from 36° to 40° South Latitude. This information integrated with the knowledge generated in neighboring regions allows us to outline a panorama of the social and cultural dynamics of, at least, the last 5000 years.

Comparing both, new and old records, it is possible to observe whether the interaction patterns that were previously identified from the lithic raw materials (obsidian, chert, quartz, and chalcedony) and the ceramic styles are maintained or strengthened. Additionally, we present the analysis of human remains from a new burial recovered in site 3.

Characterization of the Archaeological Locality Tapera Moreira

The Tapera Moreira archaeological locality is located on the right bank of the Curacó River, in the south-central sector of the province of La Pampa, Argentina (S 38°33′ and W 65°33′). It is composed of five sites with different characteristics. In this chapter, the characterization of sites 1 and 3, which are the ones that were recently re-excavated, is deepened. Sites 2 and 4 are superficial. Site 5, within the same chronology, stands out by the presence of trans-Andean ceramics of Chilean origin, corresponding to the red-on-white pottery tradition, alongside with local production containers.

Site 3 is a burial area, located on the margins of an extensive canyon of alluvial erosion. It concentrates on primary individual burials dated to the Late Holocene (Fig. 1b). In 1988, two human burial sites were found, 1 m apart from each other. In no case, elements associated with burials were observed, nor signs of the use of ocher. Individual 1 constitutes a primary burial, located in a genuflex posture, with the right dorsal decubitus position and the skull facing northwest. It is a mature-adult male with an estimated height between 169 and 177 mm. He has yielded a date of 2630 ± 60 years BP (Beta 82558; Table 1). Individual 2 is a female, and her skull is oriented toward the east. Both present intentional circular cranial deformation

Table 1 Summary of the chronology sequence obtained in sites 1 and 3 of the Tapera Moreira archaeological locality

Site/level	Laboratory acronym	Age ^{14}C	Method	Calibrated with 2s	Material
Site 1/II	UGAMS 7446	360 ± 25	AMS	cal AD 1491-1638	Organic adherence in pot
Site 1/III	Beta 81694	510 ± 67	Standard	cal AD 1301 (1421) 1485	Charcoal
Site 1/V	LP 265	1220 ± 67	Standard	cal AD 662 (779) 981	Charcoal
Site 1/VII	LP 343	1830 ± 86	Standard	cal BC 16 (cal AD 181, 188, 215) cal AD 409	Charcoal
Site 1/IX	Beta 81695	1900 ± 76	Standard	cal BC 48 (cal AD 86, 102, 121) cal AD 321	Charcoal
Site 1/X	LP 275	2140 ± 76	Standard	cal BC 389 (195, 195, 173) cal AD 23	Charcoal
Site 1/XI	LP 352	1860 ± 105	Standard	cal BC 87 (cal AD 131) cal AD 413	Charcoal
Site 1/XII	LP 358	1970 ± 95	Standard	cal BC 199 (cal AD 28, 41, 50) cal AD 243	Charcoal
Site 1/XIII	Beta 91936	3500 ± 86	Standard	cal BC 2107 (1876, 1842, 1812, 1798, 1778) 1618	Charcoal
Site 1/XIII	AA 35955	3685 ± 40	AMS	cal BC 2199 (2115, 2099, 2038) 1940	Charcoal
Site 1/XVII	LP 264	3040 ± 86	Standard	cal BC 1496 (1367, 1363, 1309) 1011	Charcoal
Site 1/XX	Beta 82557	2350 ± 76	Extended count	cal BC 762 (400) 205	Charcoal
Site 1/XXI	Beta 91937	4550 ± 60	AMS	cal BC 3502 (3348) 3027	Phalanx Guanaco
Site 1/Q XXV	UGAMS 7445	1750 ± 25	AMS	cal AD 256-304 314-417	Charcoal
Site 1/XXVI	Beta 91935	2200 ± 40	AMS	cal BC 388 (349, 318, 228, 221, 207) 119	Charcoal
Site 1/ XXVIII	Beta 82556	3900 ± 60	AMS	cal BC 2570 (2455, 2451, 2431, 2422, 2404, 2362, 2353) 2147	Charcoal
Site 1/ XXVIII	AA 35954	3995 ± 50	AMS	cal BC 2828 (2547, 2543, 2489, 2478, 2474) 2347	Charcoal
Site 3					
Site 3/ Individual 1	Beta 82558	2630 ± 60	AMS	cal BC 903 (802) 560	Femur fragment
Site 3/ Individual 3	UGAMS42899	1240 ± 20	AMS	cal AD 687-754 756-778	Third inferior right molar

(Berón, 2004). During the new excavations, Individual 3 was recovered and will be described in this chapter.

Site 1 is multi-component, with an occupational redundancy ranging from the Middle Holocene to the final Late Holocene, and constitutes the main chronological-cultural sequence of the Curacó area. The chronological range is between ca. 4550 and 360 radiocarbon years BP, during the last stage in coexistence with site 5 (Fig. 1b). Three cultural components were identified, each one with implications in the process of settlement and colonization of the subregion and not only at the intra-site level. These components comprise 32 artificial excavation levels and reach a depth of 3.50 m of archaeological fertility (Fig. 1c). The deepest levels excavated (XXXI and XXXII) are enriched with volcanic ashes of all sizes which are correlated with a total absence of evidence. Therefore, it represents the period when site 1 was first used consistently. Although there are no radiocarbon dates at these levels, it seems that its temporality precedes by several hundred years the process of human use of this space. This evidence suggests that some potent volcanic events occurred before the environment of site 1 was occupied (Berón, 2015).

Chronological-Cultural Sequence of Site 1

To differentiate analytical units in the stratigraphic sequence of the first stage of excavations at Tapera Moreira site 1, different methodological tools were used, including artifactual density quantification, contextual composition, radiocarbon dating, and sedimentological analysis. Based on the aforementioned variables, four units of analysis were delimited, made up of three components that were named, from the oldest to the most recent: Lower, Middle, and Upper Components. In turn, the Lower Component has been divided into two subcomponents: Summit and Inferior Levels.

The Inferior Levels range from level XXX, the deepest of archaeological fertility, up to level XIX. The sedimentary matrix presents a sand-loam texture, with the absence of fine silt, at the expense of a considerable increase in clay. In these levels, a low artifactual density is observed, with the absence of some typological groups that become frequent later (e.g., projectile points). In this new stage of excavations, the Inferior Levels remain not excavated yet. For this reason, we will refer to them, in more detail, in relation to the oldest excavations. This component shows signs of an initial occupation. Among the diagnostic elements of early occupation, a fishtail-type projectile point preform was recovered at 3.25 m depth. It was manufactured on a siliceous rock of excellent quality, of non-local origin. Also, active cores and flakes of this same silica, heat-treated, were found (Berón & Carrera Aizpitarte, 2023). With respect to the chronology, at the XXI level of the western sector, a radiocarbon dating of 4550 ± 60 years BP (Beta 91937; Table 1) was obtained. At the XXVIII level of the eastern sector, other two radiocarbon datings were obtained that were consistent with each other: 3900 ± 60 years BP and 3995 ± 50 years BP (Beta 82556 and AA35954, respectively; Table 1). It is considered that these are

minimum ages because the aforementioned evidence attests to a greater antiquity. Also, five discoidal stones were found in stratigraphic and superficial positions; these kinds of artifacts appear together with fishtail projectile points and raw materials with thermal treatment not available locally, in different locations of the South American cone, linked to the early moments of peopling (Capeletti, 2011; Castiñeira et al., 2011; Castro & Terranova, 2015; Cattáneo et al., 2016; Cornejo, 2008; Flegenheimer et al., 2003, 2013a, b; Hermo et al., 2013, 2015; Jackson & Méndez, 2007; Jackson et al., 1999, 2011; Laguens et al., 2007; López Romanelli, 2012; Magnin et al., 2020; Meneghin, 2011; Miotti et al., 2011, 2015; Miotti & Terranova, 2015; Nami, 2007; Rivero et al., 2015; Suárez, 2015; Terranova, 2021; among others).

The Summit of the Lower Component extends between levels XVIII and XIII of the stratigraphic sequence. The sedimentary matrix has a sandy-loam texture. A marked increase in artifacts is recorded, with greater variability of typological groups, which doubles the frequency recorded for the Inferior Levels. The radiocarbon chronology of this Component ranges between 3700 and 3000 years BP.

The Middle Component extends from excavation levels XII to VI. The textural composition differs from the Lower Component as there is an increase in the index of fine silt and a loamy-sandy texture. The number of artifacts and the variability of typological groups increase markedly. The chronology is consistent and represents a period of occupation between 2100 and 1800 years BP.

The Upper Component, levels V to I, represents the most recent unit of the site, with a chronology between 1200 and 360 years BP. The chronological range of this component includes other settlements in the Tapera Moreira locality, such as site 5. In both sites, the composition of the artifactual context shows great similarities, with smooth and decorated pottery being the most characteristic indicator of cultural change. This component presents the highest index of artifacts in general, as well as an intense use of chopped-polished artifacts, both grinding and bola stones. There are also ornaments and artistic representations (necklace beads, metal earrings, and engraved plates). An intense use of ocher is observed, represented by prepared crayons, and also recorded as adherences on grinding stones and lithic and malacological instruments. The variability of typological groups is similar to the previous unit, but the number of projectile points and scrapers predominates in the set. The sedimentary matrix has the same characteristics as that of the Middle Component.

Methodology

All the results presented in the tables and figures of this chapter correspond to the analysis of the evidence recovered in the excavations carried out between 2015 and 2017. In the final discussion, the trends of both stages of research are integrated.

Lithic materials were analyzed following the methodological guidelines of Aschero (1975, 1983) and Aschero and Hocsman (2004). Proposals from other

authors were also added. In the case of debris, the proposal of Bellelli et al. (1985–1987) was adopted, and for the cores, the guidelines of Paulides (2006) were chosen. Regarding the artifacts modified by chipping, polishing, and abrasion, the suggestions made by Berón (2004) were followed.

Determination of the different lithic raw materials was carried out mainly in a macroscopic and comparative way, because information was available from the Regional Base of Mineral Resources defined by Berón (2004, 2006) and research previously carried out in the subregion (Berón, 1996–1998, 2004, 2006; Berón et al., 1995; Berón & Curtoni, 2002; Curtoni et al., 2004; Carrera Aizpitarte, 2010, 2014; Carrera Aizpitarte et al., 2015; Velardez, 2018). In some cases, petrographic analyses were carried out.

Obsidian artifacts were analyzed using a portable X-ray fluorescence spectrometer (Bruker III–V) and a desktop spectrometer (QuantumX). Even though different instruments were used, the main elements (Rb, Sr, and Zr) were identified by both of them. The calibration for the Bruker III–V was verified by performing three replicate analyses on different international rock standards. These analyses were carried out by Martín Giesso, Michael Glascock, and Brandi McDonall, within the facilities of the Missouri University Research Reactor (MURR). For more details, see Berón et al. (2018).

For the qualitative analysis of the ceramic material, the classification methodology proposed by Berón (2004) in her PhD Thesis was followed. In it, several taxonomic units or ceramic groups were defined for the regional pottery, based on the systematization of certain technological and typological attributes (firing atmosphere, size of inclusions, surface treatment, and decoration techniques).

The faunal remains were anatomically and taxonomically identified from the comparative analysis with current reference samples. The same were subsequently quantified using traditional zooarchaeological measures and criteria (Mengoni Goñalons, 1999, 2006–2010). In cases where it was possible, the classification was made at the species level or in broader categories such as genus, family, or order. In other cases, more general names such as Bird and Mammal were used, also referring to the size of the animal (e.g., Large Mammal, Medium Bird, etc.).

For all the taxa determined, the following measurements were obtained: taxonomic and anatomical abundance, number of bone specimens identified by taxon (NISP), and minimum number of individuals (MNI) (Grayson, 1984; Klein & Cruz-Uribe, 1984; among others). To assess the integrity of the sample, the following taphonomic variables were considered: weathering (e.g., desiccation lines and cracks, exfoliation of cortical laminae and spalling), dissolution and chemical deposition (e.g., calcium carbonate and manganese oxide), presence of root marks, rodent marks, carnivore marks, trampling, type of fracture, and sedimentary abrasion (Behrensmeyer, 1978; Binford, 1981; Grayson, 1984; Haynes, 1980; Johnson, 1985; Lyman, 1994; Olsen & Shipman, 1988; Shipman, 1981). To identify processing indicators, the following variables were considered: thermal disturbance (Benett, 1999; David, 1990), type of fracture (Johnson, 1985), and cut marks (Binford, 1981; Shipman, 1981).

For the osteobiographical analysis of Individual 3 (site 3), the methods proposed by Buikstra and Ubelaker (1994) for sex determination were followed. A series of morphological variables defined for the coxal (e.g., ventral arch, subpubic concavity, ischiopubic ramus, and sciatic notch) and the characteristics of the mental eminence of the mandible were surveyed. To estimate the probable age of death, the characteristics of the pubic symphysis were analyzed according to the phases proposed by Todd (1921) and Brooks and Suchey (1990). Likewise, height was estimated according to the formulas for Mongoloids and Mexicans proposed by Genovés (1967), Steele and McKern (1969), and Trotter (1970).

In turn, radiocarbon and stable isotope analysis ($\delta^{13}C$, $\delta^{15}N$, and $\delta^{18}O$) were performed at the Center for Applied Isotope Studies (CAIS; University of Georgia, United States) and at the Laboratory of Stable Isotopes in Environmental Sciences – LIECA (Mendoza, Argentina). To reconstruct the diet and origin of the individual throughout his life, bone and dental elements were analyzed, which reflect the isotopic values of water and food consumed during their corresponding stages of formation.

The $\delta^{13}C$ analysis was performed on the mineral (apatite) and organic (collagen) fractions, $\delta^{15}N$ on collagen, and $\delta^{18}O$ on carbonates from dental apatite. Isotopic values were reported in parts per thousand (‰) according to different international standards: $\delta^{13}C$ according to the Pee Dee Belemnite (V-PDB) standard, $\delta^{15}N$ according to the AIR standard, and $\delta^{18}O$ according to the Vienna Standard Mean Ocean Water (VSMOW) (Ambrose, 1993).

Results

Lower Component Summit

The Lower Component has two portions: Summit and Inferior Levels. In this chapter, only the results of the Summit will be presented. This sub-component ranges from levels XVIII to XIII, whose chronology is ca. 3700 to 3000 years BP (Table 1).

Lithic

The lithic assemblage of the Lower Component Summit is made up of a total of 311 artifacts, of which 271 (87.1%) correspond to debris, 20 (6.4%) are retouched tools, 18 (5.8%) are cores, and 2 (0.6%) artifacts manufactured by chipping, abrasion, and/or polishing. In total, 10 raw materials were identified, plus a group of rocks classified as "undetermined." Basalt ($n = 137$; 44.1%) and siliceous rocks ($n = 118$; 37.9%) are the most represented stones, while the rest are recorded in percentages lower than 7% (Table 2).

Table 2 Raw materials identified by component, divided by artifactual class

Raw materials		Lower Component – Summit				Middle Component				Superior Component			
		Debris	F. tools	G. tools	Cores	Debris	F. tools	G. tools	Cores	Debris	F. tools	G. tools	Cores
Basalt	*n*	129	3	–	5	266	8	–	17	256	8	–	12
	%	44.1				37.7				13.7			
Chalcedony	*n*	–	–	–	–	1	1	–	1	5	2	–	1
	%	–				0.4				0.4			
Granite	*n*	–	1	2	–	1	–	3	–	–	–	3	–
	%	1.0				0.5				0.1			
Obsidian	*n*	–	–	–	–	1	–	–	–	1	–	–	–
	%	–				0.1				0.05			
Porphyry	*n*	1	–	–	–	2	1	–	–	9	–	–	1
	%	0.3				0.4				0.5			
Pumice	*n*	–	1	–	–	–	–	–	–	–	1	–	–
	%	0.3				–				0.05			
Quartz	*n*	11	–	–	–	14	1	–	1	14	1	–	–
	%	3.5				2.1				0.7			
Quartzite	*n*	2	–	–	–	1	–	–	–	9	–	–	–
	%	0.6				0.1				0.4			
Rhyolite	*n*	1	–	–	2	4	–	–	2	–	–	–	–
	%	1.0				0.8				–			
Sandstone	*n*	–	–	–	–	–	–	–	1	–	1	–	–
	%	–				0.1				0.05			
Siliceous rock	*n*	98	12	–	8	271	24	–	18	1398	60	–	44
	%	37.9				40.6				74.3			
Silicified limestone	*n*	2	–	–	–	11	1	–	2	39	5	–	–
	%	0.6				1.8				2.2			

Silicified wood	*n*	–	–	–	–	1	–	–	–	–	–	–	–
	%	–				0.1				–			
Silicified siltstone	*n*	8	2	–	2	13	3	–	2	15	1	–	2
	%	3.9				2.3				0.9			
Undetermined	*n*	19	1	–	1	91	4	–	4	132	1	–	–
	%	6.8				12.8				6.6			
Total	*n*	271	20	2	18	677	43	3	48	1878	80	3	60
	%	87.1	6.4	0.6	5.8	87.8	5.6	0.4	6.2	92.9	4.0	0.1	3.0

References: *F. Tools* Flake tools, *G. Tools* Ground tools

Fig. 2 Principal typological groups and cores, divided by components

The instruments recovered at the Summit of the Lower Component (*n* = 20) show a high rate of fragmentation (*n* = 13; 65%). Despite this, a total of 11 typological groups have been identified, with projectile points (*n* = 4; 18.2%) and side scrapers (*n* = 4; 18.2%) being the most represented (Fig. 2). In order of frequency, notches (*n* = 3; 13.6%), summary formatting artifacts (*n* = 2; 9.1%), and scrapers (*n* = 2; 9.1%) follow. Knives, hammers, drills, and gravers are the least represented groups (Table 3). Among the cores (*n* = 18), those of the amorphous type are frequent (*n* = 16; 88.9%), although isolated chipped ones were also recorded (*n* = 2; 11.1%) (Table 4). Most of them (*n* = 15; 83.3%) were found entire and are made of siliceous rocks (*n* = 8; 44.4%) or basalt (*n* = 5; 27.8%). To a lesser extent, silicified siltstone and rhyolite (*n* = 2; 11.1% each one) and an undetermined rock (5.5%) were recorded.

Fauna

In this unit, a bone set made up of 218 specimens was recovered, of which 86.7% could be identified at some taxonomic level. Within the group, guanacos, rodents, armadillos, small carnivores, birds of different sizes, reptiles, and anurans were identified. It was possible to determine, at the species level, the presence of guanaco (*Lama guanicoe*) and skunk (*Conepatus chinga*). The best-represented taxonomic group is rodents (34.4%), followed by guanaco (11%) and armadillos (10.1%). Among the resources with the highest economic yield, guanaco prevails, but the presence of two elements corresponding to *Rhea* sp. is relevant (Table 5).

Table 3 Typological groups represented by components

	Lower Component – Summit		Middle Component		Superior Component	
Flake and ground tools	*n*	%	*n*	%	*n*	%
Biface	–	–	1	2.2	5	6.0
Bola stone	–	–	3	6.5	1	1.2
Denticulates	–	–	1	2.2	3	3.6
Drill	1	4.5	1	2.2	5	6.0
FMR	2	9.1	5	10.9	2	2.4
Gaiman	–	–	–	–	1	1.2
Graver	–	–	3	6.5	–	–
Hammer	1	4.5	2	4.3	–	–
Knife	1	4.5	1	2.2	2	2.4
Mano	2	9.1	–	–	2	2.4
Net weight?	1	4.5	–	–	1	1.2
Notch	3	13.6	5	10.9	1	1.2
Projectile point	4	18.2	3	6.5	29	34.9
Rabot	–	–	1	2.2	–	–
RBO	–	–	3	6.5	4	4.8
Scraper	2	9.1	6	13.0	19	22.9
Side scraper	4	18.2	5	10.9	1	1.2
Spur between notches	1	4.5	3	6.5	1	1.2
Undetermined	–	–	3	6.5	4	4.8
Wedge	–	–	–	–	2	2.4
Total	22	100	46	100	83	100

References: *FMR* Flake with marginal retouch, *RBO* Oblique bevel retouch

Table 4 Cores and morphological types represented by components

	Lower Component – Summit		Middle Component		Superior Component	
Cores	*n*	%	*n*	%	*n*	%
Bipolar	–	–	13	27.1	23	38.3
Globular	–	–	2	4.2	1	1.7
Discoidal	–	–	1	2.1	–	–
Pyramidal	–	–	2	4.2	1	1.7
Unidirectional	–	–	–	–	1	1.7
Isolated flakes	2	11.1	8	16.7	–	–
Laminar	–	–	–	–	1	1.7
Undetermined	16	88.9	22	45.8	33	55.0
Total	18	100	48	100	60	100

In the case of guanacos, a minimum of two individuals were identified, corresponding to an adult and a subadult. Both age groups are represented by a small number of elements: MNE = 8 and MNE = 6, correspondingly. In the adult, the presence of

Table 5 Taxonomic representation by components

	Lower Component – Summit				Middle Component				Superior Component			
Taxa	NISP	NISP %	MNE	MNI	NISP	NISP %	MNE	MNI	NISP	NISP %	MNE	MNI
Artiodactyla	1	0.5	–	–	3	0.9	–	–	1	1.5	–	–
Lama guanicoe	24	11.0	19	2	72	21.6	53	3	26	39.4	22	3
Ozotoceros bezoarticus	–	–	–	–	2	0.6	2	1	3	4.5	3	1
Armadillos	22	10.1	18	2	35	10.5	30	2	11	16.7	10	2
Carnívora	–	–	–	–	2	0.6	–	–	–	–	–	–
Large Carnivore	–	–	–	–	1	0.3	–	–	–	–	–	–
Medium Felidae	–	–	–	–	2	0.6	2	–	1	1.5	1	–
Conepatus chinga	1	0.5	1	1	–	–	–	–	–	–	–	–
Rodentia	66	30.3	44	7	64	19.2	38	3	–	–	–	–
Cricetidae	1	0.5	–	–	–	–	–	–	–	–	–	–
Cavidae	6	2.8	4	2	6	1.8	4	3	–	–	–	–
Ctenomys sp.	2	0.9	2	1	–	–	–	–	–	–	–	–
Small Mammal	9	4.1	–	–	24	7.2	–	–	–	–	–	–
Medium Mammal	15	6.9	–	–	21	6.3	–	–	2	3.0	–	–
Large Mammal	7	3.2	–	–	20	6.0	–	–	5	7.6	–	–
Bird	4	1.8	–	–	1	0.3	–	–	–	–	–	–
Small Bird	4	1.8	–	–	4	1.2	–	–	–	–	–	–
Medium Bird	12	5.5	–	–	14	4.2	–	–	1	1.5	–	–

Tinamidae	6	2.8	6	3	7	2.1	7	3	–	–	–	–
Rhea sp.	2	0.9	2	1	3	0.9	3	1	3	4.5	3	1
Lacertilia	5	2.3	–	–	4	1.2	4	–	–	–	–	–
Anura	1	0.5	–	–	1	0.3	–	–	–	–	–	–
Chelonidae	1	0.5	–	–	1	0.3	–	–	–	–	–	–
Valve	–	–	–	–	–	–	–	–	3	4.5	–	–
Undetermined	29	13.3	–	–	46	13.8	–	–	10	15.2	–	–
Total	218	100	–	–	333	100	–	–	66	100	–	–

References: *NISP* number of identified specimens, *MNE* minimal number of elements, *MNI* minimal number of individuals

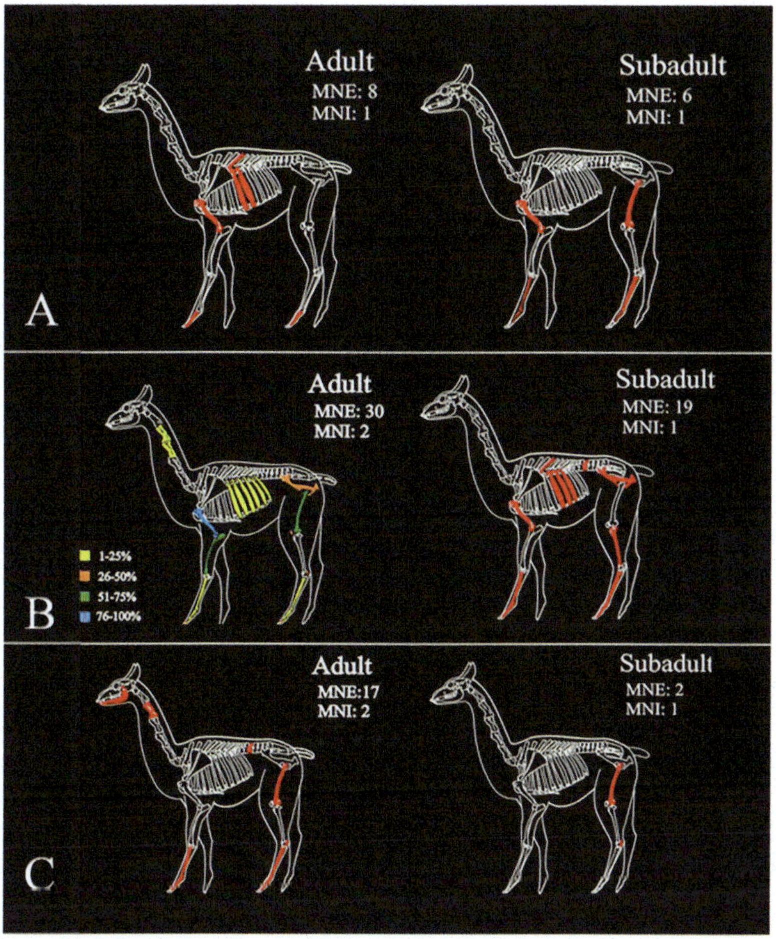

Fig. 3 Body part representation of *Lama guanicoe*. (**a**) Lower Component – Summit. (**b**) Middle Component. (**c**) Superior Component

the left humerus, a thoracic vertebra, two ribs, and phalanges 1 and 2 were observed. The second phalanges are the only whole elements. In the case of the subadult, fragments of the metapodium, first phalanx, vertebra, right femur, and left humerus were identified. In this way, between both individuals, no significant differences were observed in terms of the representation of skeletal parts (Fig. 3a).

Evidence of anthropogenic processing is limited to a fresh fracture present in a large mammalian long bone. It is considered that the absence of cut marks could be linked to the high percentage of calcium carbonate that the bones present in this component. In the case of guanaco, for example, it affects 79.2% of the group and may be hiding evidence related to the processing and consumption of this animal resource.

Middle Component

This component ranges from levels VI to XII and has five radiocarbon dates that place it in a chronological range of ca. 2100 to 1800 years BP (Table 1). In general terms, a significant increase in the density of the different evidence of the archaeological record is observed.

Lithic

The Middle Component lithic assemblage doubles the number of artifacts recovered at the Lower Component Summit and generally shows a similar distribution in the representation of the different artifact classes. Of the 771 artifacts, 87.8% (*n* = 677) correspond to debris, 6.2% (*n* = 48) to cores, 5.6% (*n* = 43) to flaked tools, and the remaining 0.4% (*n* = 3) to artifacts manufactured by chipping, abrasion, and/or polishing. Thirteen raw materials were identified, plus a group of "undetermined" rocks. Both basalt and siliceous continue to be the rocks with the greatest representation, although, unlike the previous levels, siliceous rock (*n* = 313; 40.6%) began to be the most abundant raw material, followed by basalt (*n* = 291; 37.7%) and the undetermined group (*n* = 99; 12.8%). The rest of the rocks are recorded in percentages lower than 3%. Among the latter, obsidian stands out, despite its minimal representation within the group, it allows us, based on its geochemical characteristics, to identify its source of origin (Table 2). This is a small debris recovered at level XI, whose source corresponds to Cerro Huenul (Berón et al., 2018, see Table 1, Fig. 2).

Unlike the previous component, the instruments (*n* = 46) recovered in the Middle Component are mainly found complete (*n* = 25; 54.3%). In total, 16 typological groups were identified, which indicates an increase in the variability of groups compared with those recorded in the Summit of the Lower Component. In this case, scrapers (*n* = 6; 13%), flakes with marginal retouch, notches, and side scrapers (*n* = 5; 10.9% each) are frequent (Fig. 2). Next in order of frequency are gravers, spurs between notches, projectile points, oblique bevel retouched (RBO), bola stone (*n* = 3; 6.5% each), and hammers (*n* = 2; 4.7%). Bifaces, knives, denticulates, drills, and rabots are the least represented groups of the Middle Component (Table 3). One of the projectile points is reactivated.

Among the cores (*n* = 48), whole pieces (*n* = 32; 66.7%) are more frequent than fractured ones (*n* = 16; 33.3%), and those made of siliceous rock and basalt predominate (*n* = 18; 37.5% and *n* = 17; 35.4%, respectively). Six morphological types were identified, with undetermined (*n* = 22; 45.8%), bipolar (*n* = 13; 27.1%), and isolated flakes (*n* = 8; 16.7%) being the most represented. Less frequently, they appear globular (*n* = 2; 4.2%), pyramidal (*n* = 2; 4.2%), and discoidal (*n* = 1; 2.1%) (Table 4). It is important to point out that among the siliceous rock cores, there are many small, bipolar-type ones (*n* = 6; 50%), and among the latter, most are exhausted (*n* = 4). On the contrary, the basalt cores (*n* = 16) were reduced using different techniques, and are active and medium-large in size. Although minor, two silicified

limestone cores that have been heat-treated were also recovered. A large rhyolite core stands out from the set (Fig. 2).

Fauna

In this component, 333 bone specimens were recovered, of which 86.2% (n = 287) could be classified taxonomically. At the species level, the presence of guanaco and pampas deer (*Ozotoceros bezoarticus*) was identified. The group is also made up of rodents, armadillos, carnivores, reptiles, anurans, and birds of different sizes. Among the latter, the genus *Rhea* sp. and the family Tinamidae were recorded. Regarding the frequency of specimens, guanaco predominates (21.6%), followed by rodents (21%) and armadillos (10.5%). Both *Rhea* sp. such as pampas deer, which constitute resources with a high energy return, have a very low frequency (0.9% and 0.6%, respectively). In this component, a greater taxonomic variability is highlighted than in the previous component, as well as the presence of a greater number of individuals. In this sense, a minimum of three individuals were quantified for guanaco, cavids, and tinamous, two for armadillos and one for pampas deer and *Rhea* sp. (Table 5).

The three individuals of guanaco correspond to two adults and one subadult. In the case of the first, the MNI could be calculated based on the presence of two right humeri and two left radiocubital. Regarding the representation of skeletal parts, in general terms, a predominance of the appendicular skeleton over the axial one is observed. In both limbs, the upper sector (humerus and femur) is the best represented. The mid sector also has a high frequency at the front end but is absent at the back end. All the elements of the lower sector are represented, but the second phalanges predominate. Other elements identified in the extremities correspond to cuneiform (carpal), patella, and calcaneus (tarsus). The axial skeleton is mainly represented by the pelvic girdle and, to a lesser extent, by the ribs and the cervical and caudal vertebrae.

In the subadult individual, the number of elements is less than in adults, but the anatomical representation is quite similar. The axial skeleton is represented by ribs, thoracic and lumbar vertebrae, and the pelvic girdle. In the front extremity, the absence of the middle sector (radius and ulna) was recorded, while in the rear extremity, the same thing happened with the upper sector (femur). In the case of the lower extremities, the presence of metapodia and the first and second phalanges are observed, but the third ones are absent (Fig. 3b).

Evidence of anthropogenic processing consists of cut marks and fractures made when the bones were fresh. Regarding the former, cut marks were identified in adult guanaco (humerus and three rib fragments), subadult guanaco (one rib fragment), pampas deer (radius), *Rhea* sp. (tarsus-metatarsus), medium bird (tarsus-metatarsus), medium mammal (long bone), and large mammal (four long bone fragments). Other possible cut marks were identified on a radioulnar and a second phalanx of an adult guanaco, the calcaneus of an armadillo, and the tarsus-metatarsus of a medium-sized bird. Additionally, fresh fractures were identified in guanaco (humerus, femur,

metapodium, and pelvis), pampas deer (radius), *Rhea* sp. (tarsus-metatarsus, tibia-tarsus), Tinamidae (tarsus-metatarsus), in a long bone fragment from a medium-sized mammal, and in six fragments from a large mammal. In the case of the guanaco, the cut marks could be linked to filleting and evisceration activities, while the fractures mentioned would indicate the consumption of bone marrow.

Superior Component

The Superior Component ranges from levels V to I and presents three radiocarbon dates that place it in a chronological range of ca. 1200 to 360 years BP (Table 1). It is the component with the highest artifactual density and technological diversity because ceramic technology is incorporated and bone technology increases.

Lithic

The Superior Component has the highest lithic artifactual density of the entire site 1 sequence, with an assemblage composed of 2021 artifacts. Of the total, 92.9% ($n = 1878$) corresponds to debris, 4% ($n = 80$) to tools, 3% ($n = 60$) to cores, and the remaining 0.1% ($n = 3$) to artifacts manufactured by chipping, abrasion and/or polishing.

The notable increase in the amount of debris recovered is noteworthy, which could be an indicator of the intensification of retouching activities carried out at the site during this chronological range. Twelve raw materials were identified, plus a group of "undetermined." Like the previous component, both siliceous rock ($n = 1502$; 74.3%) and basalt ($n = 276$; 13.7%) and the group of undetermined rocks ($n = 133$; 6.6%) continue to be the most represented. It is highlighted that the use of siliceous rocks increases markedly. The rest of the rocks have frequencies below 3% (Table 2). As in the Middle Component, the use of obsidian among the minority rocks is highlighted. Three pieces of debris were recovered in level III, coming also from Cerro Huenul (Berón et al., 2018, see Table 1, Fig. 2). It is important to mention that, although scarcely represented, the percentages of silicified limestone, chalcedony, and quartzite are the highest in the entire sequence.

Among the tools ($n = 83$), 17 different typological groups have been identified, showing stability in terms of the diversity of groups with respect to the previous component. However, there is a notable increase in the presence of projectile points ($n = 29$; 34.9%) and scrapers ($n = 19$; 22.9%) (Fig. 2). They are followed by bifaces and drills, although in a much lower frequency ($n = 5$; 6%). The rest of the groups are minimally represented, with percentages below 5% (Table 3). Among the instruments, fractured pieces are abundant ($n = 43$; 53.7%) and correspond mainly to the basal sections of projectile points.

Among the projectile points, it is possible to distinguish two modules, triangular of medium-small size and small, with a predominance of the latter. Similarly, among

the scrapers, a wide variability of modules can be observed, with a predominance of small, short, and wide ones. The presence of a lithic basalt retoucher or *gaiman* and a possible net weight made of pumice stone recovered in levels IV and V, respectively, stand out.

Among the cores ($n = 60$), whole pieces ($n = 47$; 78.3%) and those made of siliceous rock ($n = 40$; 66.7%) are frequent. Among the different morphological types identified, undetermined ($n = 33$; 55%) and bipolar ($n = 23$; 38.3%) are abundant. Less frequently, they appear as globular, laminar, pyramidal, and unidirectional ones ($n = 1$; 1.7%) (Table 4).

Fauna

Unlike the other components where the bone remains recovered in situ and from the sieve were analyzed, in this one, the analysis was limited to those recovered in situ, so a lower number of specimens are observed than that recorded in the previous components as well as the absence of previously recurrent taxonomic categories, like rodents. The bone sample is made up of 66 bone remains, of which 84.8% could be classified taxonomically. The presence of armadillos, medium and large birds (*Rhea* sp.), medium felines, large and medium mammals, and, at the species level, guanaco and pampas deer were identified. Three valve fragments were also recovered. Regarding frequency, guanacos (39.4%) and armadillos (19.7%) predominate, while other resources of possible economic importance such as pampas deer and *Rhea* sp. have little representation (4.5% each) (Table 5).

Regarding the guanaco, a minimum of three individuals could be identified, two adults and one subadult. In the case of the first, the MNI was obtained from the presence of two left cuneiforms. Both age groups have a low frequency of elements (MNE = 17 for adults and MNE = 2 for subadults). In adults, the presence of bone remains corresponding to the axial skeleton (hemimandible, cervical, and lumbar vertebrae) and appendicular skeleton (femur, carpus, tarsus, metapodium, and phalanges 1 and 2) is observed. The highest frequency is recorded in the lower extremities, where two metapodials, four first phalanges, and two second phalanges were identified. The skeleton of the subadult is represented by a left calcaneus and a fragment of the diaphysis of the right femur, the latter with fractures generated in the fresh state. The degree of fragmentation in the whole set of guanaco is high (91.7%). The only entire elements are two phalanges (first and second) and one of the cuneiforms (Fig. 3c).

Among the evidence of anthropic processing, cut marks, fractures in the fresh state, and impact negatives were detected. Traces related to the slaughter of the animals were observed on the femur, calcaneus, metapodium, and first phalanx of adult guanaco, also in the calcaneus of the subadult guanaco and in a long bone in which the taxonomic category could not be determined. Fresh fractures, meanwhile, were recorded in a long bone from a large mammal and in two femur fragments from adult and subadult guanaco. Due to the anatomical position of the cut marks,

they could be associated with disarticulation and filleting activities of the animals (Binford, 1984; Nilssen, 2000). Fractures and impact negatives may result from bone marrow consumption.

The recovery of two bone retouchers made on guanaco metapodium stands out, one of which presents thermal action, probably for the hardening of the active portion. Both come from excavation level IV.

Pottery

Tapera Moreira 1 is the site of the Western Pampa with the highest density of pottery in a single settlement locus, although its representation is based on fragments of variable dimensions. In this site, some fragments have also been recorded at levels below the Superior, but it is assumed that these are intrusive due to migrations resulting from the composition of the sedimentary matrix or the action of digging animals (Berón & Scarafoni, 1995).

During the last excavations, a total of 174 sherds were recovered, 94.2% ($n = 164$) of which came from the Superior Component, while the rest ($n = 10$; 5.8%) migrated through some of the aforementioned formation processes (Fig. 4a). Only 14 of the total (8.5%) are decorated using different techniques: burnished, slip, rhythmic incision, annular groove, and stamping like checkerboard pattern (Fig. 4b, c). Beyond its low representation in the analyzed archaeological record, the wide repertoire of decorative techniques denotes the great variability of techniques known by these societies. The presence of suspension or repair holes was also recorded (Fig. 4b).

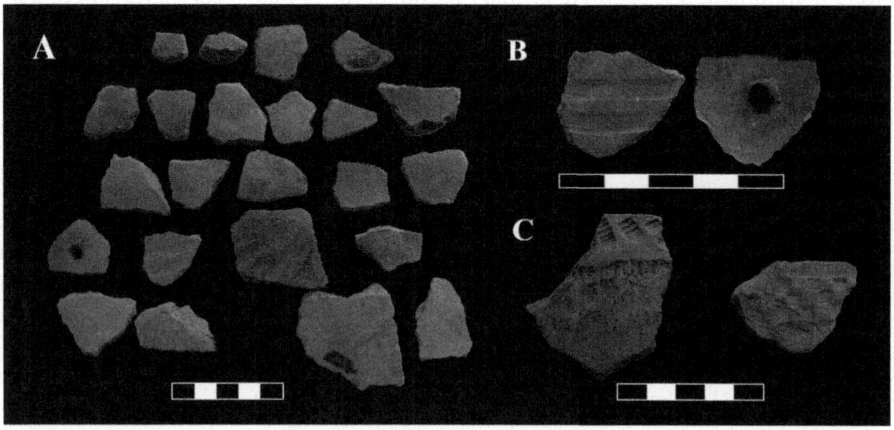

Fig. 4 Pottery. (**a**) Representative set of technological variability. (**b**) Sherds decorated by annular striations (*challa*) and suspension or repair hole, respectively. (**c**) Rhythmic incision, and checkerboard stamping sherds

Fig. 5 Site 3. (**a**) General view of the excavation of Individual 3. (**b**) Lower portion of the skeleton recovered in the 2015 excavation. (**c**) Superior portion of the skeleton recovered in 2017

Site 3, Individual 3

During the new excavations, Individual 3 was recovered (Fig. 5a). The lower portion of the skeleton was recovered in the 2015 excavation (Fig. 5b), while the upper portion was recovered in 2017, on a slope resulting from the collapse of the canyon wall, 1 m below the portion previously found (Fig. 5c). In both cases, the anatomical structure, vertebrae, articulated ribs, and even a southwest-northeast orientation were maintained. It is a simple primary burial, in the dorsal decubitus position, with flexed limbs, and without accompaniment.

Based on the osteobiographical analysis, it was determined that Individual 3 is a male, young adult (between 20 and 30 years old), with an estimated height between 159 and 167 cm (Lucero et al., 2017). It presents a radiocarbon date of 1240 ± 20 years BP (UGAMS42899; Table 1).

The isotopic analysis of the protein fraction of the diet ($\delta^{13}C_{col}$) shows a value of −17.17‰ in bone and −16.5‰ in tooth. The $\delta^{15}N$ shows a value of 12.3‰ in bone and 14.18‰ in tooth. In turn, the total diet ($\delta^{13}C_{apat}$) presents a value of −10.7‰ in bone and −10.3‰ in tooth. On the contrary, the analysis of $\delta^{18}O$ presents a value of −3.6‰ in bone and −0.3‰ in tooth. These data allow us to analyze part of the life history of the individual.

Discussion. Data Synthesis and Integration

In this section, a synthesis and interpretation of the stated data will be made, together with a comparison and integration with the previous ones (1988–1995), also looking for trends in the social trajectory of these populations at a macro-regional level.

The lithic record from the new excavations at site 1 comprises a total of 3103 artifacts. From a quantitative point of view, its distribution throughout the entire sequence presented here (from levels XVIII to I) shows changes in the artifactual density recovered by levels. A substantial increase in artifacts is observed in the Superior Component, with a maximum representation peak at level two. The same trend was recorded by Berón (2004) during the excavations carried out between the years 1988 and 1995.

Regarding the composition of the whole, between levels XII and I is where the greatest variability of typological groups represented has been recorded, as in previous excavations (Berón, 2004). The main typological groups throughout the sequence are cores, projectile points, scrapers, and side scrapers, although their representation varies between the different components.

Projectile points constitute the best-represented typological group in the entire stratigraphic sequence of site 1, although the best expression of this occurs in the Superior Component because 80.5% ($n = 29$) is concentrated in this unit. An index similar to this (70%) had been recorded in previous excavations (Berón, 2004). The side scrapers appear as one of the most abundant groups in the Summit of the Lower Component, and, together with the scrapers, they dominate the set of instruments in the Middle Component. Starting from the Superior Component, scrapers predominate alongside projectile points.

Cores are found in all three components, although they differ in size. Those recovered at the Summit of the Lower Component and the Middle Component are large and very large pieces, while those of the Superior Component are medium-small to small. This difference correlates with the type of raw material. Basalt or siltstone cores are large, and siliceous rock cores are small. The latter are usually depleted and associated with the bipolar reduction technique. The use of the bipolar technique increases its frequency in the Middle Component and very noticeably in the Superior Component. This is related to a decrease in logistics mobility and the development of other activities such as ceramic manufacturing.

This difference in the use and exploitation of the lithic resource responds mainly to the local availability of the three raw materials mentioned above: basalt, siliceous rock, and siltstone. Although all these rocks integrate the Manto Tehuelche (Calmels et al., 1996; Calmels & Casadío, 2004; Llambías, 1975), basalt and siltstone are predominant and with larger nodules than siliceous rock.

Projectile points recovered from the Summit of the Lower Component are mostly fractured. Those of the Middle Component are complete and are characterized by having different modules, which could respond to the use of different weapons systems. In the Superior Component, 62.0% are fractured and correspond mostly to the basal sections. The latter is possibly related to the recovery of shafts in the base camp.

It should be noted that artifacts manufactured by the chipping-polishing technique are recorded in all three components. Among the polished artifacts, four bola stones stand out, three of them from the Middle Component and one from the Superior.

In the entire sequence, basalt and siliceous rock are the majority stones; however, a change in frequencies between components is noted. Basalt begins as the predominant rock at the Summit of the Lower Component but is slightly surpassed by siliceous rock from the Middle Component, which widely increases in frequency in the Superior Component. This trend had already been observed in previous excavations (Berón, 2004). Carrera Aizpitarte and Berón (2021) propose that the rocks of the Manto Tehuelche were used differentially over time in the Pampa Occidental subregion. These authors note that, during the earliest times, basalt was the most exploited rock at the sites and that its frequency in the archaeological record decreased during the Late Holocene. Conversely, throughout this period, there was a constant increase in the use of siliceous rock. They propose that this could be because, as the occupation of space became effective, human groups improved and increased their knowledge about the lithic resources available in the environment, being able to recognize and select those of better quality for the making of their tools. This trend could explain the low frequency of basalt in all areas where late sites predominate (Carrera Aizpitarte & Berón, 2021).

On the contrary, from the Middle Component, an increase in the repertoire of raw materials used is observed. Non-local raw materials such as chalcedony, silicified limestone, quartzite, and obsidian, although underrepresented, increase notably in the Superior Component.

The provenances of some of these rocks are strong markers of macro-regional interaction vectors. The presence of fine-grained chalcedony and quartzite indicates a relationship to the east, with sources located in Eastern Pampa, mainly in the Tandilia and Ventania mountain systems (Bayón et al., 1999; Flegenheimer et al., 1996, 1999; among others).

In the case of the silicified limestone, it was determined by means of thin sections that they come from Meseta del Fresco, a highly available source located to the northwest, at ca. 150 km. Regarding the obsidian recovered in the latest excavations, the geochemical studies carried out indicate that they come from supply sources located in the province of Neuquén. These are added to the previous determinations, completing a panorama that indicates that the obsidians of the subregion come from six different sources located on both mountain slopes, between 34° and 40° South Latitude (Berón et al., 2018; Giesso et al., 2008; Páez et al., 2021).

The bone set recovered during the excavations carried out between 1988 and 1995 was analyzed by Salemme and Berón (2003). When comparing their results with those obtained in this new stage of reopening the site, differences and similarities between both sets are observed. The main differences are related to the frequency that some taxa present in each component. According to what was recorded by Salemme and Berón (2003), the frequency of guanacos is around 30% in the Lower and Superior Components, with a decrease (18%) in the Middle Component. In the excavations of 2015 and 2017, the trend is different, as a constant increase in the frequency of this species is observed over time: 15% in the Lower Component, 32% in the Middle, and 53% in the Superior. It must be taken into account, however, that in this last component, there may be an overrepresentation due to the fact that the analysis of sieve materials has not yet been incorporated. This clarification is valid for all taxa considered.

In the case of armadillos, quite similar frequencies are observed between the sets in the Lower and Superior Components. The main difference lies in the Middle

Component, in which Salemme and Berón (2003) register an increase with respect to the other two components, while in the set analyzed in this chapter, the increase in frequency is progressive from the Lower Component to the top.

On the other hand, in both sets, a similar trend is recorded with respect to frequency over time in rodents and pampas deer, with differences in terms of representation value. In both sets, the highest incidence of rodents occurs in the Lower Component and decreases steadily to the Superior. The pampas deer follow the opposite path, with a low frequency in the Lower Component (null in the case of the set recovered in 2015–2017) and a progressive increase until reaching its highest values in the Superior Component. However, Salemme and Berón (2003) indicate that in this last stage, the pampas deer reaches half of the representation of the guanaco, while in the new excavation stage, the pampas deer is scarcely represented.

The bone assemblages from both excavation stages of the site show similarities in terms of represented taxonomic variability and trends in the processing of animal resources. In both cases, the same taxonomic groups are present, which is expected given the short distance between the different excavated sectors. It stands out that in the two groups, a low frequency of *Rheids* and carnivores is recorded, as well as the recurrent presence of a wide variety of medium and small birds. In relation to subsistence, in both sets, it is observed that the guanaco was the main source of animal protein in the three components. Another remarkable similarity is that in both cases, these species are predominantly represented by the appendicular skeleton. Besides serving as food, the guanacos were a source of raw material for the manufacture of bone artifacts. Its relevance within the site is consistent with the importance it had for the hunter-gatherer societies that inhabited Pampa and Patagonia, not only for its role in subsistence but also for providing numerous by-products that were used in different spheres of these societies (Prates, 2009).

Pottery constitutes a technological innovation that is recorded as such in the Superior Component. Site 1 of Tapera Moreira is the one that concentrates on the largest sample of pottery in the entire Western Pampa. During the previous stage of work (1988–1995), a set of 703 fragments (617 in stratigraphy and 86 on the surface) of different sizes was recovered, representing a varied knowledge of manufacturing techniques, surface finishes (smoothed, polished, brushed, and rough), and decorative designs (burnished, slipped, linear incision, rhythmic incision, annular groove, and checkerboard stamping) that account not only for technological aspects but also for shared knowledge, social interactions, and information exchange at different spatial scales. With the assemblage recovered in the recent excavations ($n = 174$), a total sample of 877 fragments from site 1 of this locality has been completed. Most of them ($n = 164$) come from the Superior Component, while 10 of them (5.8%) are intrusive because they were recovered at excavation levels VI and VII (Middle Component), a situation also recorded in the first research stage, in similar percentages and that is fundamentally due to the action of digging animals or to the composition of the sedimentary matrix (Berón & Scarafoni, 1995; Berón, 2004).

In relation to the previous stages of research, the absence of the linear incised ceramic group, which is one of the best represented, is striking. Neither was the presence of shards corresponding to the Red-on-White Bichrome Tradition characteristic of Vergel and Valdivia groups from trans-Andean origin (the Chilean Araucanía) previously found in sites 1 and 5 of the Tapera Moreira locality (Adán et al., 2005). However, a fragment with ribbed decoration or annular striations (*challas*) was recorded, which is also diagnostic of the spatial extension of the social interaction of hunters-gatherers in this region (Berón et al., 2021). In turn, the presence of a new decorative design, stamped on the checkerboard, was recorded by means of the alternate imprint of Armadillo's plates (possibly Piche).

In reference to human remains recovered in this last stage, it is possible to identify continuity in the lifestyle among the individuals buried in the Tapera Moreira archaeological locality. Like Individuals 1 and 2, Individual 3 was partially exposed in the canyon profile, which confirms the use of site 3 as a burial area. All these cases are simple primary burials, in the dorsal decubitus position, with flexed limbs and without accompaniment.

The isotopic analysis made of the protein fraction of the diet ($\delta^{13}C_{col}$) for Individual 3 in bone and tooth indicates the presence of an important animal protein component in the diet. This is also supported by the values of $\delta^{15}N$. In turn, the total diet ($\delta^{13}C_{apat}$) is enriched with respect to the known values in the area (Barberena et al., 2018; Berón et al., 2009), which could indicate the possible direct or indirect consumption of some C_4-type complementary resources (Lucero et al., 2017). On the other hand, the analysis of $\delta^{18}O$ indicates that the geographic residence of the individual until the last years of life would be local, coinciding with the known ranges for the region (Barberena et al., 2018).

It should be noted that the radiocarbon date of 1240 ± 20 years BP (UGAMS42899) establishes a limit for the burial of primary and isolated individuals in areas close to residential sites. From this moment on, formal burial structures began to be used as referential burial spaces, as, for example, the Chenque I site. This is a hunter-gatherer cemetery used recurrently over several generations during the Late Holocene (ca. 1200 to 290 years BP). It constitutes one of the largest repositories of bioarchaeological remains in the province of La Pampa, with a preliminary MNI of ca. 250 (Barberena et al., 2018; Berón, 2018).

Final Remarks

The results of the new excavations reaffirm intrasite trends in the use of this landscape from the Middle Holocene to the final Late Holocene, with a sustained increase in artifactual density. At the same time, there are changes in the composition of the artifactual contexts and in the incorporation of new technologies, pointing toward a greater diversity of activities and a better knowledge of the surrounding environment, at different scales, as denoted by the use of diverse lithic raw materials.

Although the environmental conditions and the composition of the sediments do not favor the good conservation of the organic materials, the new excavations triple the number of bone instruments present in the Middle and Superior components, whose chronology corresponds to the final Late Holocene. However, we know that bone technology has been present in the Western Pampa since, at least, the Middle Holocene, according to recent findings of a burial with a complex trousseau whose components include three well-preserved bone artifacts (Berón et al., 2022).

Another technological innovation during the final Late Holocene is pottery. Its presence in a stratigraphic sequence of long continuity such as that of the Tapera Moreira site 1 accounts for the moment of its incorporation into the material repertoire of hunter-gatherers in the region, as well as the changes that it implies (Berón et al., 2021). Taking into account the chronology of the entire locality, the radiocarbon range of pottery in Western Pampa is ca. 1700 to 360 years BP. In addition, the use of ceramic containers brought about changes in food preparation, consumption behaviors, and the components of the diet, by being able to incorporate a greater amount of fat per intake and other kinds of foods as vegetal in the same cooking, as has been corroborated through archaeobotanical and isotopic studies, which are added to the faunal analysis outlined in this work (Musaubach & Berón, 2012, 2016).

Also, the presence of extra-regional decorative designs in pottery becomes a new support for symbolic representation and, therefore, for the transmission of information as well as the tangibility of the great extension of existing social networks. The allochthonous materials, although in low density, support the east–west interaction vectors throughout the macro-region. The use of distant resources, coming from different ecological spaces, allows us to establish complementary and reciprocal relations between these populations, as well as the existence of multiple networks for the circulation and exchange of goods, people, and information in the past.

It remains for future agenda to complete the excavations of the Lower Component and obtain new radiocarbon dates that allow to refine the chronological sequence of this component and confirm the age of the initial exploration signals.

Acknowledgements We want to thank the organizers of the IX CARPA for the invitation to publish this chapter. Also to two anonymous reviewers who, with their suggestions, helped to improve the final version of this work. The Secretary of Culture and the Heritage Department of the province of La Pampa and the authorities and community of Puelches provided logistical support to our investigations. We also thank the institutions that finance our field and laboratory research (UBA, CONICET, I+D+i Agency).

References

Adán, L., Mera, R., Uribe, M., & Alvarado, M. (2005). La tradición cerámica bícroma rojo sobre blanco en la región sur de Chile: los estilos decorativos Valdivia y Vergel. In *Sociedad Chilena de Arqueología, Actas del XVI Congreso Nacional de Arqueología Chilena* (pp. 399–410). Museo de Historia Natural.

Ambrose, S. (1993). Isotopic analysis of paleodiets: Methodological and interpretive considerations. In M. Sandford (Ed.), *Investigations of ancient human tissue* (pp. 59–129). Gordon and Breach Science.

Aschero, C. (1975). *Ensayo para una clasificación morfológica de artefactos líticos aplicada a estudios tecnológicos comparativos*. Unpublished report to CONICET CONICET.

Aschero, C. (1983). *Ensayo para una clasificación morfológica de artefactos líticos aplicada a estudios tecnológicos comparativos.* Apéndice A-C. Revisión. Unpublished report, Cátedra de Ergología y Tecnología. Facultad de Filosofía y Letras. Universidad de Buenos Aires. Ms.

Aschero, C., & Hocsman, S. (2004). Revisando cuestiones tipológicas en torno a la clasificación de artefactos bifaciales. In A. Acosta, D. Loponte, & M. Ramos (Eds.), *Temas de Arqueología. Análisis Lítico* (pp. 7–25). Universidad de Luján.

Barberena, R., Berón, M., & Luna, L. (2018). Isótopos estables en el sitio Chenque I: paleodieta y procedencia geográfica. In M. Berón (Ed.), *El sitio Chenque I. Un cementerio prehispánico en la Pampa Occidental. Estilo de vida e interacciones culturales de cazadores- recolectores del Cono Sur Americano* (pp. 367–395). Sociedad Argentina de Antropología.

Bayón, C., Flegenheimer, N., Valente, M., & Pupio, A. (1999). Dime cómo eres y te diré de dónde vienes: La procedencia de rocas cuarcíticas en la región pampeana. *Relaciones de la Sociedad Argentina de Antropología, XXIV*, 187–235.

Beherensmeyer, A. K. (1978). Taphonomic and ecological information from bone weathering. *Paleobiology, 4*, 150–162.

Bellelli, C., Guráieb, A., & García, J. (1985–1987). Propuesta para el análisis y procesamiento por computadora de desechos de talla lítica (DELCO- desechos líticos computarizados). *Arqueología Contemporánea, 2*, 36–53.

Benett, J. L. (1999). Thermal alteration of burned bone. *Journal of Archaeological Science, 26*, 1–8.

Berón, M. (1996–1998). Nuevos rumbos, viejos caminos. Cuenca Inferior del Chadileuvú (37°33′ a 38° 11′ Lat. Sur). Primer trabajo de campo 1996. *Palimpsesto, 5*, 106–118.

Berón, M. (2004). *Dinámica poblacional y estrategias de subsistencia de poblaciones prehispánicas de la cuenca Atuel-Salado-Chadileuvú-Curacó, provincia de La Pampa* [Unpublished Doctoral Thesis]. Facultad de Filosofía y Letras, Universidad de Buenos Aires.

Berón, M. (2006). Base regional de recursos minerales en el occidente pampeano. Procedencia y estrategias de aprovisionamiento. *Relaciones de la Sociedad Argentina de Antropología, XXXI*, 47–88.

Berón, M. (2015). Chronological distribution and disturbance factors to evaluate population dynamics in Western Pampas, Argentina. *Quaternary International, 356*, 74–88. https://doi.org/10.1016/j.quaint.2014.10.038

Berón, M. (2016). Dunes, hills, waterholes, and saltpeter beds: Attractors for human populations in western Pampa, Argentina. *Quaternary International, 422*, 163–173. https://doi.org/10.1016/j.quaint.2016.03.001

Berón, M. (Comp.). (2018). *El sitio Chenque. Un cementerio prehispánico en la Pampa occidental. Estilo de vida e interacciones culturales de cazadores-recolectores del Cono Sur americano.* Sociedad Argentina de Antropología.

Berón, M., & Curtoni, R. (2002). Propuestas metodológicas para la caracterización arqueológica de canteras y talleres de la Meseta del Fresco (La Pampa, Argentina). In D. Mazzanti, M. Berón, & F. Oliva (Eds.), *Del Mar a los Salitrales. Diez mil años de Historia Pampeana en el Umbral del Tercer Milenio* (pp. 171–184). Laboratorio de Arqueología, Facultad de Humanidades, Universidad Nacional de Mar del Plata.

Berón, M., & Scaraffoni, M. (1995). Acción de roedores en el sitio 1 de la Localidad Arqueológica Tapera Moreira, Provincia de La Pampa. Santa Rosa, La Pampa. In *Actas de las V Jornadas Pampeanas de Ciencias Naturales* (pp. 26–32).

Berón, M., Migale, L., & Curtoni, R. (1995). Hacia la definición de una base regional de recursos líticos en el área de Curacó. Una cantera taller: Puesto Córdoba (La Pampa, Argentina). *Relaciones de la Sociedad Argentina de Antropología, XX*, 111–128.

Berón, M., Luna, L., & Barberena, R. (2009). Isotopic archaeology in the Western Pampas (Argentina): Preliminary results and perspectives. *International Journal of Osteoarchaeology, 19*, 250–265.

Berón, M., Páez, F., Carrera Aizpitarte, M., Giesso, M., & Glascock, M. D. (2018). Procurement and circulation of obsidian in the province of La Pampa, Argentina. *Journal of Lithic Studies, 5*(2). https://doi.org/10.2218/jls.2993

Berón, M., Carrera Aizpitarte, M., & Páez, F. (2021). Social trajectories of Hunter-Gatherer societies in Central Argentina: Exploration and colonization of a desert landscape, La Pampa province, Argentina. In M. Bonomo & S. Archila (Eds.), *South American contributions to world archaeology. One world archaeology*. Springer. https://doi.org/10.1007/978-3-030-73998-0_4

Berón, M., Paez, F. N., Carrera Aizpitarte, M. P., & Lucero, E. N. (2022). Crónica de un enterratorio del Holoceno Medio en la Localidad Cerro de los Viejos (provincia de La Pampa, Argentina). Implicancias para el sector centro-meridional de la Diagonal Árida Sudamericana. *Chungara. Revista de Antropología Chilena, 54*(1), 5–31. https://doi.org/10.4067/S0717-73562021005001501

Berón, M., & Carrera Aizpitarte, M. (2023). Indicadores de poblamiento inicial en paisajes semidesérticos de la Pampa Occidental (Argentina). *Latin American Antiquity*: 1–28. https://doi.org/10.1017/laq.2023.47

Binford, L. (1981). *Bones: Ancient men and modern myths*. Academic Press.

Binford, L. (1984). *Faunal remains from Klasies River Mouth*. Academic Press.

Brooks, S., & Suchey, J. M. (1990). Skeletal age determination based on the os pubis: A comparison of the Acsadi-Nemeskeri and Suchey-Brooks methods. *Human Evolution, 5*, 227–238.

Buikstra, J., & Ubelaker, D. (1994). *Standards for data collection from human skeletal remains* (p. 44). Archaeological Survey Research Series.

Calmels, A., & Casadío, S. (2004). *Compilación Geológica de la Provincia de La Pampa*. Ediciones Amerindia.

Calmels, A., Aldacour, H., Fernández, E., & Carballo, O. (1996). Las unidades litoestratigráficas de Anzoátegui, provincia de La Pampa. In *Actas de la Sexta Reunión Argentina de Sedimentología y Primer Simposio de Arcillas* (pp. 97–104). Bahía Blanca.

Capeletti, L. (2011). Primer hallazgo de una punta de proyectil "cola de pescado" en la provincia de Entre Ríos. *Revista del Museo de La Plata. Arqueología, 12*, 86.

Carrera Aizpitarte, M. (2010). Maximización de rocas: el uso de los artefactos informales. In M. Berón, L. Luna, M. Bonomo, C. Montalvo, C. Aranda, & M. Carrera Aizpitarte (Eds.), *Mamül Mapu: pasado y presente desde la arqueología pampeana, Tomo II* (pp. 225–238). Libros del Espinillo.

Carrera Aizpitarte, M. (2014). *Estudios de las estrategias de aprovisionamiento lítico en las áreas Curacó, Bajos sin Salida, Valles Transversales y Centro-este (provincia de La Pampa, Argentina)* [Unpublished doctoral thesis] (p. 496). Facultad de Ciencias Sociales, Universidad Nacional del Centro de la Provincia de Buenos Aires.

Carrera Aizpitarte, M., & Berón, M. (2021). Explotación de recursos líticos en dos canteras prehispánicas de la provincia de La Pampa (Argentina): Meseta del Fresco y Manto Tehuelche. *Estudios Atacameños, 66*, 7–34.

Carrera Aizpitarte, M., Berón, M., & Blasi, A. (2015). Study of siliceous outcrops of Meseta del Fresco, La Pampa, Argentina. *Quaternary International, 375*, 27–43.

Castiñeira, C., Cardillo, M., Charlin, J., & Baeza, J. (2011). Análisis de morfometría geométrica a puntas cola de pescado. *Latin American Antiquity, 22*(33), 335–358.

Castro, J. C., & Terranova, E. (2015). Indicadores paleoindios en la provincia de Entre Ríos (Argentina). *Comechingonia. Revista de Arqueología, 19*(1), 183–190.

Cattáneo, R., Izeta, A., & Caminoa, J. (2016). A fishtail projectile point from the Southern Pampean Hills, Characato, Córdoba, Argentina. *PaleoAmerica, 2*(3), 274–276.

Cornejo, L. (2008). Huetelauquen: geometría de piedra. In J. Berenguer, C. Sinclaire, L. Cornejo, & M. Escobar (Eds.), *Pescadores de la niebla. Los changos y sus ancestros* (p. 67). Museo Chileno de Arte Precolombino.

Curtoni, R., Barros, P., & Berón, M. (2004). Meseta del Fresco: análisis de canteras y talleres. In C. Gradín & F. Oliva (Eds.), *La Región Pampeana. Su pasado arqueológico* (pp. 287–296). Venado Tuerto.

David, B. (1990). How was this bone burnt? In S. Solomon, I. Davidson, & D. Watson (Eds.), *Problem solving in taphonomy: Archaeological and palaeontological studies from Europe, Africa and Oceania* (Vol. 2, pp. 65–79). Tempus.

Flegenheimer, N., Kain, C., Zárate, M., & Barna, A. (1996). Aprovisionamiento de cuarcitas en Tandilia, las canteras del Arroyo Diamante. *Arqueología, 6*, 117–141.
Flegenheimer, N., Zárate, M., & Valente, M. (1999). El área de canteras Arroyo Diamante, Barker, Sierras de Tandil. In C. Diez Marín (Ed.), *Actas del XII Congreso Nacional de Arqueología Argentina* (pp. 134–148). Facultad de Ciencias Sociales y Museo, Universidad Nacional de La Plata.
Flegenheimer, N., Bayón, C., Valente, M., Beaza, J., & Femenías, F. (2003). Long distance tool stone in the Argentine Pampas. *Quaternary International, 109-110*, 49–64.
Flegenheimer, M., Mazzia, N., & Babot, M. (2013a). Estudios de detalle sobre una piedra discoidal pampeana. *Intersecciones en Antropología, 14*, 499–505.
Flegenheimer, N., Miotti, L., & Mazzia, N. (2013b). Rethinking early objects and landscape in the Southern Cone: Fishtail point concentrations in the Pampas and northern Patagonia. In K. Graf, C. Ketron, & M. R. Waters (Eds.), *Paleoamerican Odyssey* (pp. 359–376). Center for the Study of First Americans, A&M University.
Genovés, S. (1967). Proportionality of the long bones and their relation to stature among Mesoamericans. *American Journal of Physical Anthropology, 26*, 67–78.
Giesso, M., Berón, M., & Glascock, M. (2008). Obsidian in western Pampas, Argentina: Source characterization and provisioning strategies. *IIAOS, 38*, 15–18.
Grayson, D. (1984). *Quantitative zooarchaeology: Topics in the analysis of archaeological faunas*. Academic Press.
Haynes, G. (1980). Evidence of carnivore gnawing on Pleistocene and recent mammalian bones. *Paleobiology, 6*, 341–351.
Hermo, D., Terranova, E., Marchionni, L., Magnin, L., Mosquera, B., & Miotti, L. (2013). Piedras o litos discoidales en Norpatagonia: evidencias en la meseta de Somuncurá (Río Negro, Argentina). *Intersecciones en Antropología, 14*, 507–512.
Hermo, D., Terranova, E., & Miotti, L. (2015). Tecnología y uso de materias primas en puntas cola de pescado de la meseta de Somuncurá (provincia de Río Negro, Argentina). *Chungara. Revista de Antropología Chilena, 47*(1), 101–115.
Jackson, D., & Méndez, C. (2007). Litos discoidales tempranos en contextos paleoindios de Sudamérica. *Magallania, 5*(1), 75–84.
Jackson, D., Seguel, R., Báez, P., & Prieto, X. (1999). Asentamientos y evidencias culturales del Complejo Huentelauquén en la comuna de Los Vilos, Provincia del Choapa. *Anales del Museo de Historia Natural de Valparaíso, 24*, 5–28.
Jackson, D., Maldonado, A., Carré, M., & Seguel, R. (2011). Huentelauquén cultural complex: The early peopling of the Pacific Coast in the South-America Southern cone. In D. Vialou (Ed.), *Peuplements et préhistoire en Amériques* (pp. 221–232). Éditions du Comité des Travaux Historiques et Scientifiques.
Johnson, E. (1985). Current developments in bone technology. In M. B. Schiffer (Ed.), *Advances in archaeological method and theory* (pp. 157–235). Academic Press.
Klein, R. G., & Cruz-Uribe, K. (1984). *The analysis of animal bones from archeological sites*. University of Chicago Press.
Laguens, A., Pautassi, E., Sario, G., & Cattáneo, R. (2007). Fishtail projectile points from Central Argentina. *Current Research in the Pleistocene, 24*, 55–57.
Llambías, E. (1975). *Geología de la provincia de La Pampa y su aspecto minero*. Dirección de Minas de La Pampa. Manuscript.
López Romanelli, F. (2012). El yacimiento Los Ciervos: hallazgos paleontológicos y geológicos en sedimentos de la transición Pleistoceno-Holoceno (Departamento De Lavalleja, Uruguay). *Orígenes, 11*, 1–15.
Lucero, E., Gonzalez, G., & Diana, A. N. (agosto, 2017). *Análisis Bioarqueológico de un individuo procedente del Sitio 3 de la Localidad Arqueológica Tapera Moreira (Pcia. de La Pampa)*. Ponencia presentada en el VIII Congreso de Arqueología de la Región Pampeana Argentina, Luján, Argentina.

Lucero, E., Berón, M., González, G., & Diana, A. N. (2020). Osteobiografía de un individuo del Holoceno Medio, recuperado en la Localidad Arqueológica Cerro de los Viejos (Departamento Caleu Caleu, La Pampa). *Revista del Museo de Antropología, 13*(3), 193–206.

Lyman, R. L. (1994). *Vertebrate taphonomy*. Cambridge University Press.

Magnin, L., Terranova, E., & Lynch, V. (2020). Análisis espacial del sitio Amigo Oeste, Meseta de Samuncurá (Rio Negro, Argentina). *Revista del Museo de La Plata, 5*(1), 80–96. https://doi.org/10.24215/25456377e096

Martínez, G., & Flensborg, G. (2018). Nuevos contextos funerarios en la transición pampeano-patagónica oriental (Provincia de Buenos Aires, Argentina): aportes al Holoceno Medio y Tardío inicial. *Chungara. Revista de Antropología Chilena, 50*(3), 441–458.

Meneghin, U. (2011). Observaciones sobre algunos artefactos líticos discoidales registrados en el Uruguay. *Orígenes, 10*, 1–33.

Mengoni Goñalons, G. (1999). *El aprovechamiento del guanaco en el interior de Patagonia Meridional: transporte y procesamiento en la localidad arqueológica de Cerro de los Indios 1 (Santa Cruz)*. Sociedad Argentina de Antropología.

Mengoni Goñalons, G. (2006–2010). Zooarqueología en la práctica: algunos temas metodológicos. *Xama, 19-23*, 83–113.

Miotti, L., & Terranova, E. (2015). A hill full of points in Terra Incognita from Patagonia: Notes and reflections for discussing the way and tempo of initial peopling. *PaleoAmerica, 1*(2), 181–196.

Miotti, L., Hermo, D., Blanco, R. V., & Terranova, E. (2011). Puntas Cola de Pescado en el ecorrefugio de la meseta de Somuncurá (provincia de Río Negro, Patagonia argentina). In J. C. Jiménez López, C. Serrano Sánchez, A. González González, & F. J. Aguilar Arellano (Eds.), *Actas IV Simposio Internacional "El Hombre Temprano en América"* (pp. 149–172). Instituto de Investigaciones Antropológicas- INAH- UNAM- Museo del Desierto.

Miotti, L., Terranova, E., Hermo, D., & Blanco, R. (2015). Edenes en el Desierto. Señales de caminos y lugares en la historia de la colonización de la Patagonia argentina. *Antípoda. Revista de Antropología y Arqueología, 23*, 161–185.

Musaubach, M. G., & Berón, M. A. (2012). Cocinando en ollas en la Pampa Occidental. Datos desde la etnohistoria, el registro arqueológico y la arqueobotánica. In M. P. del Babot, M. Marschoff, & F. Pazzarelli (Eds.), *Las manos en la masa: arqueologías, antropologías e historias de la alimentación en Suramérica* (pp. 605–626). Museo de Antropología, Universidad Nacional de Córdoba, Instituto Superior de Estudios Sociales, Universidad Nacional de Tucumán.

Musaubach, M. G., & Berón, M. (2016). El uso de recursos vegetales entre los cazadores recolectores de la Pampa Occidental Argentina. *Latin American Antiquity, 27*(3), 397–413. https://doi.org/10.7183/1045-6635.27.3.397

Nami, H. (2007). Research in the Middle Negro River Basin (Uruguay) and the Paleoindian occupation of the Southern Cone. *Current Anthropology, 48*(1), 164–174.

Nilssen, P. J. (2000). An actualistic butchery study in South Africa and its implications for reconstructing hominid strategies of carcass acquisition and butchery in the Upper Pleistocene and Plio-Pleistocene [Unpublished PhD thesis]. Department of Archaeology, University of Cape Town.

Olsen, S. L., & Shipman, P. (1988). Surface modification on bone: Trampling versus butchery. *Journal of Archaeological Science, 15*, 535–553.

Paez, F., Berón, M., & Carrera Aizpitarte, M. (septiembre, 2021). *Aprovisionamiento y circulación de obsidianas en la Subregión Pampa Occidental, Argentina.* Puesta al día. Ponencia presentada en el VIII Congreso Nacional de Arqueometría.

Paulides, L. (2006). El núcleo de la cuestión: el análisis de los núcleos en los conjuntos líticos. In C. Pérez de Micou (Ed.), *El modo de hacer las cosas: artefactos y ecofactos en arqueología* (pp. 67–100). Facultad de Filosofía y Letras, Universidad de Buenos Aires.

Politis, G., & Barros, P. (2006). La región pampeana como unidad espacial de análisis en la arqueología contemporánea. *Folia Histórica del Nordeste, 16*, 51–72.

Politis, G. G., & Prates, L. (2019). Some opinions about Monte Verde: Response to Dillehay. *PaleoAmerica, 20*, 3–12.

Prates, L. (2009). El uso de recursos por los cazadores-recolectores posthispánicos de Patagonia continental y su importancia arqueológica. *Relaciones de la Sociedad Argentina de Antropología, XXXIV*, 201–229.

Rivero, D., Heider, G., & Pastor, S. (2015). La identificación de una punta cola de pescado en las Sierras Centrales, implicancias en un modelo de poblamiento del centro de Argentina. *Cuadernos del Instituto Nacional de Antropología y Pensamiento Latinoamericano, 24*(1), 151–155.

Salemme, M., & Berón, M. (2003). Análisis intrasitio del componente faunístico del sitio 1 de la Localidad Tapera Moreira. Diferencias y Tendencias. *Actas del XIII Congreso Nacional de Arqueología Argentina, III*, 325–345. Editorial Brujas.

Shipman, P. (1981). *Life history of a fossil*. Harvard University Press.

Steele, G., & McKern, T. (1969). A method for assessment of maximum long bone length and living stature from fragmentary long bones. *American Journal of Physical Anthropology, 31*, 215–227.

Suárez, R. (2015). Tecnología lítica y conjunto de artefactos utilizados durante el poblamiento temprano de Uruguay. *Chungara. Revista de antropología chilena, 47*(1), 67–84.

Terranova, E. D. (2021). Análisis de los desechos líticos del sitio Amigo Oeste (meseta de Somuncurá, provincia de Río Negro): aportes para el conocimiento de los procesos de producción lítica. *Revista de la Sociedad Argentina de Antropología, 46*(1), 267–290.

Todd, T. W. (1921). Age changes in the pubic bone. 5. Mammalian pubic metamorphosis. *American Journal of Physical Anthropology, 4*, 334–406.

Trotter, M. (1970). Estimation of stature from intact long limb bones. In T. D. Stewart (Ed.), *Personal identification in mass disasters*. National Museum of Natural History, Smithsonian Institution.

Velardez, S. (2018). Estrategias tecnológicas, secuencias de producción y distribución diferencial. El conjunto artefactual lítico del sitio Chenque I desde distintas perspectivas. In M. Berón (Ed.), *El sitio Chenque I. Un cementerio prehispánico en la Pampa Occidental. Estilo de vida e interacciones culturales de cazadores- recolectores del Cono Sur Americano* (pp. 177–223). Sociedad Argentina de Antropología.

Lithic Procurement, Mobility, and Social Interaction in Early Hunter-Gatherer Societies (~12,000 Cal. Years BP) in the Humid Pampas Sub-region, Buenos Aires, Argentina

Gustavo Federico Bonnat

Abstract The Pampas Region (Argentina) was inhabited by hunter-gatherer societies since ~14,000 cal. years BP, and around 32 archaeological sites have been identified associated with the early peopling of this area. Particularly, the archaeological records of 27 sites show evidence of human past activities, and the lithic artifacts are the most frequent. More than 15 types of rocks from different sectors of the humid Pampas Region have been identified and the use of these raw materials involved different procurement strategies. Using different spatial analysis scales, this chapter discusses the use of lithic resources, mobility ranges, procurement strategies, and social interaction patterns developed by early hunter-gatherer groups that occupied especially the Humid Pampas sub-region. Results show a direct procurement strategy (by special purpose and/or embedded strategies) for local resources located in the vicinity of the sites as well as at greater distances (~110 km) like orthoquartzite from the Sierras Bayas Groups (SBGOs) are a fine grain quartz rocks that show a restricted spatial availability but a wide regional distribution in their use. In addition, a possible indirect procurement strategy for resources located ~300 to 400 km from the sites was also recognized. Other rocks located at ~500 km from the study area may have entered the Pampas region with the first waves of the human population.

Keywords Hunter-gatherers · Lithic technology · Late pleistocene/early holocene · Pampa region · Argentina

G. F. Bonnat (✉)
INHUS, UNMDP-CONICET. Laboratorio de Arqueología Regional Bonaerense (LARBO), Facultad de Humanidades, Universidad Nacional de Mar del Plata (UNMDP), Mar del Plata, Buenos Aires, Argentina
e-mail: gbonnat@mdp.edu.ar

G. F. Bonnat et al. (eds.), *Current Research in Archaeology of South American Pampas*, The Latin American Studies Book Series,
https://doi.org/10.1007/978-3-031-55194-9_6

Introduction

The Pampas region (Argentina) is a large grassland located in southern South America, and it is divided into two sub-regions by the 600-mm isohyet, the dry Pampas to the west and the humid Pampas to the east (Fig. 1). Particularly, the humid Pampas sub-region was inhabited by hunter-gatherers since ca. 14,000 years cal BP (Politis et al., 2016), and around 32 archaeological sites attributed to the late Pleistocene and the early Holocene have been identified (Mazzanti, 2003; Flegenheimer et al., 2015; Martínez et al., 2015; Martínez & Mazzanti, 2017).

Archaeological research information suggests that the early hunter-gatherer occupation of the Pampas region had low population density and developed high residential mobility (Politis et al., 2004; Martínez, 2006). The highest concentration of archaeological sites is located in the Tandilia mountain and, to a lesser extent, in the inter-hills area (Fig. 1), probably due to the presence of critical subsistence resources (e.g., rocks, water, wood, and animals) and optimal locations for housing places in the ranges (Mazzanti, 1999; Martínez, 2006; Flegenheimer et al., 2015;

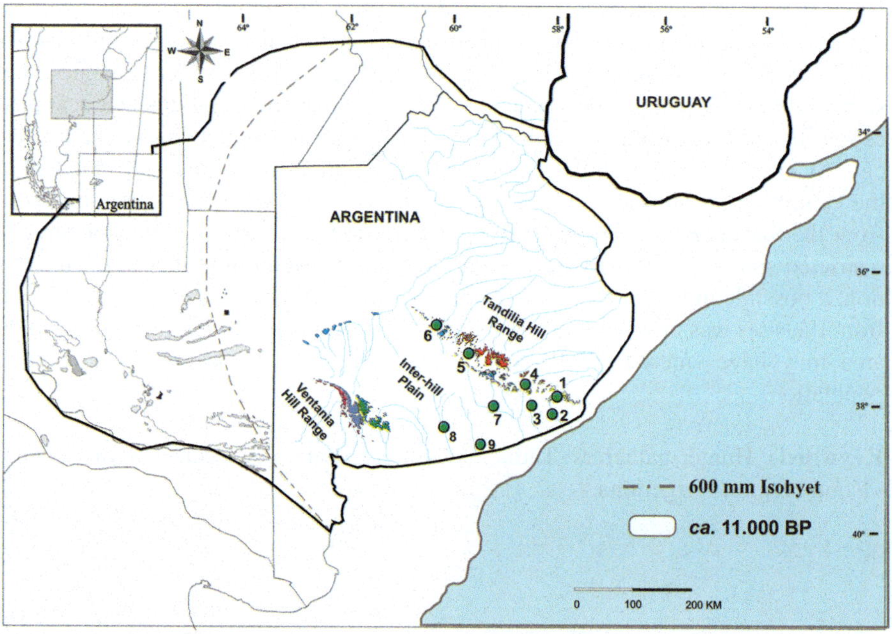

Fig. 1 Geographical location of the archaeological sites in the Pampa humid sub-region. *References:* (1) Cueva El Abra, Abrigo Los Pinos, Cueva Tixi, Cueva La Brava, Cueva Burucuyá, Alero El Mirador, Abrigo La Grieta, Cueva Ali Mustafa, and Alero Molina Site 1; (2) Amalia 2 Site 2; (3) Lobería 1 Site 1; (4) Cerro El Sombrero Cima (CoSC), Cerro El Sombrero Abrigo 1 (CoSA1), Cerro La China Sites 1, 2, and 3, Los Helechos, Cueva Zoro, and El Ajarafe; (5) Abrigo La Tinta; (6) La Moderna and Campo Laborde; (7) Paso Otero 4 and 5; (8) Arroyo Seco 2; and (9) El Guanaco Sites 1 and 2. The reconstruction of the Atlantic sea level during the late Pleistocene/early Holocene follows Ponce et al. (2011)

Politis et al., 2016; Messineo & Pal, 2020). For eastern Tandilia, an archaeological model for the use of the mountain-range landscape has been proposed (Mazzanti & Bonnat, 2013). On the one hand, some caves and rockshelters functioned as residential campsites with evidence of diverse socioeconomic activities (e.g., Cueva Tixi, Abrigo Los Pinos, and Cueva El Abra) (Fig. 1). On the other hand, there are sites with short-term occupation that were used for special tasks, possibly related to mobility circuits (e.g., daily foraging trips and logistic ranges) (Binford, 1980, 1983) and the use of landscape resources (e.g., Cueva Burucuyá, Abrigo La Grieta, Alero El Mirador, Cueva La Brava, Alero Molina 1, Amalia Site 2, and Lobería 1 Site 1) (Fig. 1; Mazzanti, 2003; Mazzanti & Bonnat, 2013). In this model, all of the archaeological sites were grouped in relation to the articulated use of three different environments: (1) valleys, creeks, and shallow lakes; (2) isolated outcrops located at the transition between the ranges and the plains; and (3) the Atlantic Ocean coast. In this sense, the use of these residential and ephemeral sites connects human occupations in caves and rockshelters in the landscape, allowing access to different resources (Mazzanti & Bonnat, 2013). Our current interpretation is that residential sites with diverse socioeconomic activities were those caves and rockshelters that were more habitable and accessible. On the other hand, smaller sites with open entrances and located at higher elevations were occupied for short terms for special tasks (Mazzanti & Bonnat, 2013).

The open-air sites in the inter-hills grassland have a wide space for use without spatial limitations as in the caves and rockshelters sites located in the Tandilia mountain system. Archaeological sites in this sector of the Pampas were characterized as base camps sites (e.g., Arroyo Seco 2; El Guanaco 1 and 2) and sites of specific activities (e.g., Paso Otero 5, La Moderna, and Campo Laborde) (Fig. 1). In all of those sites, research studies consider that short-time human occupations developed different activities, such as hunting, scavenging, and processing of modern and extinct fauna, as well as the last stages of stone tool production (Politis & Gutiérrez, 1998; Martínez, 2006; Armentano et al., 2007; Messineo, 2015; Politis et al., 2016, 2019, among others). However, several authors suggest that both Tandilia Mountain and Pampa's grassland were visited and occupied by the same hunter-gatherer groups (Mazzanti, 1999; Politis et al., 2004; Martínez, 2006; among others).

Zooarchaeological studies suggest a generalized subsistence strategy, which would include both the consumption of extinct taxa (e.g., *Eutatusseguini, Macrauchenia* sp., *Megatherium americanum,* and *Doedicurus clavicaudatus*), as well as smaller modern species such as the guanaco (*Lama guanicoe*), the Pampas deer (*Ozotoceros bezoarticus*), the Greater rhea (*Rhea americana*), the armadillos (*Dasypodidae* and *Chlamyphoridae*), and the plains vizcacha (*Lagostomus maximus*) (Mazzanti & Quintana, 2001; Martínez & Gutiérrez, 2004; Politis et al., 2004, 2019). Regarding the role played by megamammals in the subsistence of human groups during the initial settlement of the Pampa region, some authors propose that even within a generalized regional economy they would have had an important role (Politis et al., 2004), while others suggest that these animals would have been marginally consumed through hunting and/or scavenging (Politis & Gutiérrez, 1998;

Martínez & Gutiérrez, 2004; Gutiérrez & Martínez, 2008). However, there is a consensus that subsistence would have been based on a generalized strategy of resource capture, the guanaco (*Lama guanicoe*) was the main prey hunted by indigenous societies (Gutiérrez & Martínez, 2008) and other animals, such as some birds, rodents, and carnivores were occasionally exploited, although not necessarily as food (Kaufmann et al., 2021).

Particularly for the Pampas Humid sub-region, the lithic technology of early hunter-gatherers incorporates wide artifactual variability, among which the "fish-tail" projectile points (or Fell 1) are interpreted as the result of technological knowledge shared among the early hunter-gatherers of the Southern Cone (Flegenheimer et al., 2003; Politis et al., 2004). Numerous projectile points have been recovered from the Cerro El Sombrero Cima site (Flegenheimer et al., 2015: 172), although they have also been recorded, with less frequency, in other sites of the region (e.g., Paso Otero 5, Abrigo Los Pinos, Amalia Site 2, Cerro El Sombrero Abrigo, and Cerro La China Sites 1 and 2) (Martínez, 2006; Valverde, 2006; Bonnat et al., 2015; Flegenheimer et al., 2015; Bonnat, 2020a). The most frequently used rocks for stone tools production were fine-grained orthoquartzites from the Sierras Bayas Group (SBGO) and medium-grained orthoquartzites from the Balcarce Formation (BFO), although a wide diversity of rocks was also identified. Lithic technology studies allowed us to discuss various topics, such as the availability and location of raw materials, color selection in SBGO, procurement strategies, stone tools use and production, the use of different knapping techniques, artifact fractures, archaeometry studies, etc. (Flegenheimer et al., 2003; Bayón et al., 2006; Martínez, 2006; Armentano et al., 2007; Weitzel, 2010; Colombo, 2012; Colombo & Flegenheimer, 2013; Messineo & Barros, 2015; Bonnat, 2019, 2020a; Bonnat et al., 2021; among others).

In this scenario, the main goal of this chapter is to discuss the use of lithic resources, mobility ranges, procurement strategies, and social interaction patterns developed by early hunter-gatherer groups that occupied the Humid Pampas sub-region during the late Pleistocene and the early Holocene. In order to address this objective, frequencies of the raw materials, stages of lithic production sequence, and stone tools diversity of 27 archaeological sites were mainly considerate.

Study Area and Rocks Sources in the Pampa Humid Sub-region

The Pampa region is an extensive plain grassland located in the eastern portion of Argentina (between 30° and 39° south latitude and 57° and 65° west longitude) and is divided into two sub-regions by the 600-mm isohyet, the dry Pampas to the west and the humid Pampas to the east (Fig. 1). Paleoenvironmental studies developed on different proxy records (geoarchaeological, stratigraphic, zooarchaeological, and paleobotanical) in the Humid Pampas sub-region suggest that cold and arid conditions were present at the end of the Pleistocene and changed quickly to warmer and wetter conditions during the early Holocene (ca. 10,500 14C years BP) (Tonni et al., 1999; Zárate et al., 2000; Gutiérrez et al., 2011; Martínez et al., 2013, 2015; among others).

The south east area of the Pampas comprises a wide flat to slightly undulating surface landscape. This area is only interrupted by two major orographic systems: the Tandilia Ranges, located in the central and eastern Buenos Aires province, and the Ventania Ranges, located in the southwest. These two mountain ranges provide a diversity of lithic resources that are abundant and highly localized in the landscape.

Lithic resources in the Pampas Humid sub-region are abundant and highly localized. Thus, most of the rock sources used by early hunter-gatherers have been identified and are well known. Four main areas have been recognized as having potential sources: the Tandilia and the Ventania mountain ranges, isolated outcrops in the inter-hill plains, and isolated outcrops along the Atlantic coast (Fig. 1). In the Tandilia Ranges, several rock types used as raw material have been detected. Two different sedimentary rocks were the most used: the orthoquartzites from the Balcarce Formation (BFO) and the orthoquartzites from the Sierras Bayas Group (SBGO). The BFOs are located mainly in the eastern sector of Tandilia and decrease their abundance in the central and western sectors (Fig. 1). This rock is presented as subhorizontal strata in the hills and is characterized by medium to coarse granulometry, a product of the low cementation of quartz grains (Bayón et al., 1999). Also, there are very small quartz pebbles into veins in the Balcarce Formation (Mazzanti, 1997). The SBGO has restricted availability in the landscape and is located in the central and western parts of the Tandilia Ranges (Fig. 2). This rock appears as primary deposits in the hills (e.g., blocks and veins) and as secondary deposits in the creeks that originate in the hills (e.g., Arroyo Diamante). This SBGO has a finer grain size than BFO and also a wider range of colors. In archaeological terms, these rocks are better for tool knapping than BFO and have been widely used since the beginning of the settlement of the region (Flegenheimer et al., 1996; Bayón et al., 1999; Colombo, 2012). Also, in the central sector of Tandilia, other small blocks and pebbles of different rocks have also been identified, such as microcrystalline flint, diabase, and silicified dolomite (Bayón et al., 2006; Vecchi, 2011). In addition, in western Tandilia mountain, a primary source of a wide variety of rocks has been identified, such as quartz, chert (chert), and silicified dolomite (Messineo & Barros, 2015).

In the Ventania Ranges (Fig. 2), there are primary and secondary sources of metaquartzites, rhyolites, and orthoquartzites (Catella et al., 2013). On the Atlantic coast, secondary deposits of Patagonian pebbles and cobbles of different lithological compositions have been identified, among which basalt, andesites, flint, and rhyolites are mainly recognized (Bonomo & Prates, 2014). Finally, silicified limestone is another rock that appears in the early contexts of the Pampean region. This rock is found 400–500 km away from the sites (Fig. 2), at the province of Entre Ríos (Argentina) or in the Republic of Uruguay (Flegenheimer et al., 2003; Bonomo & Blasi, 2010; Loponte et al., 2010; Nami, 2017).

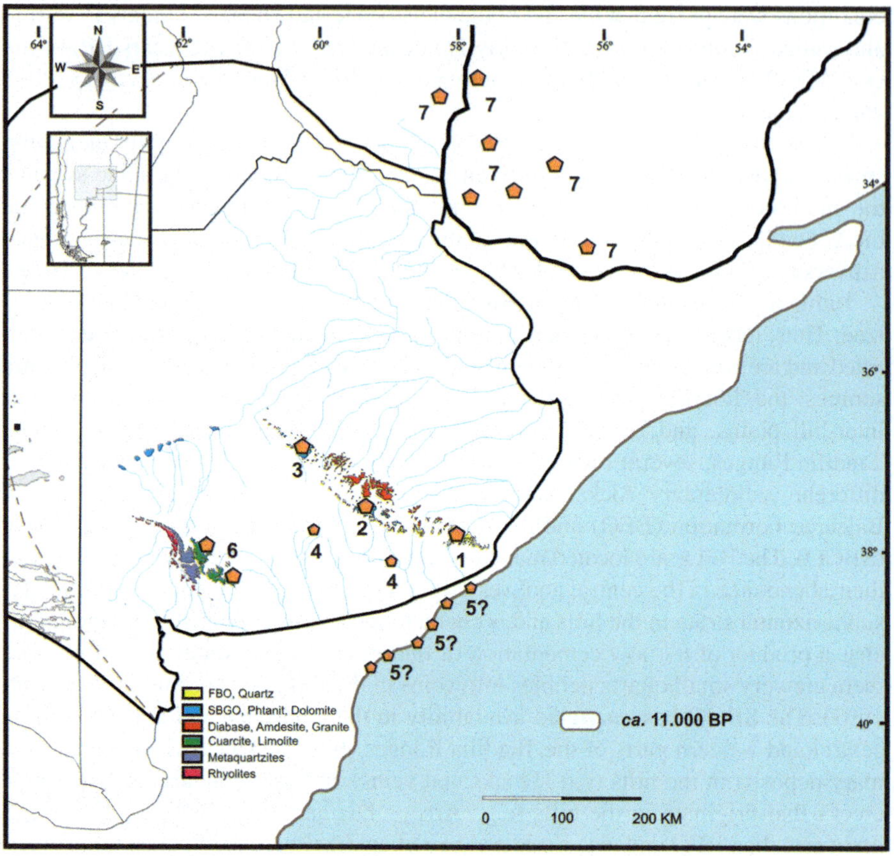

Fig. 2 Raw material sources in the Pampa humid sub-region. *References:* (1) eastern Tandilia ranges (FBO, and quartz); (2) east-central Tandilia ranges sources (SBGO, diabase, granite, chert, silicified dolomite, quartz, and flint rock); (3) western Tandilia ranges sources (SBGO, granite, silicified dolomite, phtanite, and quartz); (4) inter-hill plain sources (silicified tuff, quartzite sandstones, and coarse-grained quartzites); (5) Atlantic coast rocks (pebbles and cobbles of basalt, flint, quartz, and rhyolite); (6) Ventania sources (rhyolite, quartzite, andesite, metaquartzite, and orthoquartzite rocks); and (7) Entre Ríos and Uruguay sources (silicified limestone rocks)

Based on the availability of lithic resources and their location in the landscape, different micro-regional scale procurement ranges and models have been proposed in the Pampas humid sub-region (e.g., Bayón et al., 2006; Martínez, 2006; Barros et al., 2014; Messineo, 2015; Bonnat, 2019, 2020a). Following Melzer's (1989) and Gould and Saggers's (1985) proposal adapted for the Pampas humid sub-region (Bayón et al., 2006; Bonnat, 2020a), the lithic resources were considered in four categories of provenance: (1) immediately available are rocks within a 10-km radius away from the sites; (2) local resources involve rocks located between 10 and 60 km radius away from the sites; (3) non-local resource comprises rocks located between 60 and 120 radius away from the sites; and (4) long distance includes raw materials located more than 120 km away from the sites. Thus, following the models of those authors, this chapter assumed that in a landscape with plains and low mountain ranges (mainly the Tandialia ranges with a maximum of 300 m above sea level), movements and access to the lithic sources would not involve high costs of time and energy as in others high-altitude environments such the Andean mountain (Lucero et al., 2021).

Theoretical and Methodology Concepts

The information recovered from ethnoarchaeology and ethnography studies has provided a frame of reference for the mobility and procurement of lithic resources in hunter-gatherer societies (Binford, 1979, 1983; Binford & O'Connell, 1984; Erickson, 1984; Gould & Saggers, 1985; Bamforth, 1986, 1990; Meltzer, 1989; Torrence, 1989; Nelson, 1991; Brantingham, 2003; Andrefsky, 2009; Ditchfield, 2016; among others). However, the procurement of lithic resources and the way they are used depend on several factors, such as the landscape, the raw material distribution, the abundance and quality of the rocks, the distances to the sources, and a number of social factors that could include restrictions, accessibility, preferences, etc. Under this scenery, hunter-gatherer groups apply basically two different strategies in order to access the raw materials: direct procurement (either special purpose and/or embedded acquisition) and indirect procurement (Binford, 1979, 1983; Gould & Saggers, 1985; Meltzer, 1989; Nelson, 1991). Thus, if a hunter-gatherer group opts for a special-purpose procurement strategy, the group or part of them has to move through the landscape with the exclusive purpose of obtaining raw materials. On the other hand, opting for an embedded strategy implies that the costs of procurement are relatively low since these would be included in other activities linked basically to the subsistence of the hunter-gatherer groups (Binford, 1979). Finally, the third strategy is related to an indirect acquisition of the raw material in the source by human groups or part of them and then transferred to another population by one of several possible mechanisms like, among others, exchange, human mobility, social interaction, etc. (Meltzer, 1989).

Consequently, the uses of one or another strategy result in different material evidence. Therefore, following the proposal applied by Franco (2014) in Argentine Patagonia and including some modifications, this chapter explores mobility strategies developed by human groups in the past based on a series of expectations generated from the lithic record of the archaeological sites. For direct raw material acquisition and circulation in those sites located near the supply sources, it is expected that lithic assemblages show different artifact classes (e.g., cores, different flake types, and tools) (Franco, 2014). In addition, for raw material near archaeological sites, I expect a number of archaeological indicators: an increase in their frequency, as well as artifact size and thickness, more cortical index, discarding of large artifacts without resharpening, lower number of blow extractions in cores, increased volume of discarded cores, and more expedited stone tool knapping. On the other hand, when the distance between archaeological sites and sources supply increases, I expect a decrease in artifact size and number of artifacts, the presence of exhausted cores or tools, more evidence of tool and core maintenance, an increase in the number of core extractions, low cortex rates, evidence of a raw material economy, etc. Finally, indirect raw material acquisition is related to the procurement of three general groups of artifact classes: the acquisition of some kinds of stone tools, the acquisition of nodules or cores in cases of valuable raw materials or without internal impurities (Franco, 2014), and the acquisition of preforms and/or base forms (e.g., bifacials). Thus, the lithic assemblages should evidence a lower artifact frequency, smaller artifact size, more undifferentiated type forms (e.g., debris), an increase in the angles of discarded tools, a higher degree of exploitation of cores, as well as evidence of maintenance and resharpening.

Following previous lithic artifact classifications (Aschero, 1975, 1983; Inizan et al., 1995; Andrefsky, 1998, 2009), lithic assemblages were divided into general classes of artifacts: hammerstones, cores, debitage products (flakes, debris, chips, and fragments), and tools. For each of these classes, some techno-morphological variables (e.g., raw material, artifact size, cortex presence, and tool type) were considered in all cases when it was possible.

Results

In the Pampa region, around 32 archaeological sites with early evidence of hunter-gatherer population have been identified. However, this chapter only considers those sites located mainly in the humid Pampa sub-region that have evidence of lithic assemblages ($N = 27$) (Fig. 1). An important characteristic of the sites in this sub-region is that they are mainly located in the Tandilia mountain and in the inter-hills area. Because both archaeological research and publications have increased considerably in the region scenario in the last 30 years, this chapter considers those general and synthesis papers that different researchers have made. Thus, the reader can obtain more information from the references cited in these publications.

Eastern Portion of the Tandilia Ranges System

In the eastern portion of the Tandilia system, 11 archaeological sites dating to the late Pleistocene and early Holocene have been discovered in caves and rockshelters (Fig. 1, Table 1). Based on the archaeological record and its location on the landscape, these sites have been characterized as residential campsites or sites with ephemeral occupations (Mazzanti, 2003; Mazzanti & Bonnat, 2013).

Table 1 Archaeological sites with lithic assemblage evidence for the Pampa humid sub-region

Area	Sites	^{14}C years BP	Laboratory sample	Site type	References
Eastern Tandilia	Cueva El Abra	10,270 ± 200 9834 ± 65	AA-94641 AA-38098	Multiple activities (base campsite)	Mazzanti (2003), Bonnat (2020a)
	Abrigo Los Pinos	10,465 ± 65 10,415 ± 70 9570 ± 150	AA-24045 AA-24046 LP-630	Multiple activities (base campsite)	Mazzanti (1999), Bonnat (2020a)
	Cueva Tixi	10,375 ± 90 10,045 ± 95	AA-12130 AA-12131	Multiple activities (base campsite)	Mazzanti (1999), Bonnat (2020a)
	Cueva La Brava	10,178 ± 54 9670 ± 120 9264 ± 55	AA-94639 LP-550 AA-10345	Ephemeral occupation (specific activities)	Mazzanti (1999), Bonnat and Mazzanti (2015)
	Cueva Burucuyá	10,672 ± 56 10,000 ± 120	AA-94640 LP-863	Ephemeral occupation (specific activities)	Mazzanti (1999), Bonnat and Mazzanti (2021)
	Alero El Mirador	8920 ± 51	AA-98683	Ephemeral occupation (specific activities)	Bonnat (2020a)
	Abrigo La Grieta	Stratigraphy and cultural association	–	Ephemeral occupation (Specific activities)	Bonnat (2020a)
	Cueva Ali Mustafa	10,450 ± 38 10,125 ± 31 9032 ± 30 9060 ± 31	AA-110084 AA-108366 AA110082 AA110081	Ephemeral occupation (specific activities)	Mazzanti et al. (2019), Bonnat (2020b)
	Alero Molina 1	10,628 + 32	AA-111628	Ephemeral occupation (specific activities)	Mazzanti and Bonnat (2021)
	Amalia (Site 2)	10,705 ± 60 10,425 ± 75 10,340 ± 60	AA-103453 AA-35499 AA-103452	Ephemeral occupation (specific activities)	Bonnat et al. (2015)
	Lobería 1 (Site 1)	9787 ± 81	AA-81063	Ephemeral occupation (specific activities)	Bonnat (2020a)

(continued)

Table 1 (continued)

Area	Sites	^{14}C years BP	Laboratory sample	Site type	References
East Central Tandilia	Cerro El Sombrero Cima (CoSC)	Cultural association	–	Specific activities(replacement and re-equipping of projectile points)	Flegenheimer et al. (2015)
	Cerro El Sombrero Abrigo 1(CoSA1)	10,725 ± 90 10,675 ± 110 10,480 ± 70 10,270 ± 85 8060 ± 140	AA-4765 AA-4767 AA-5220 AA-4766 AA-5221	Domestic specific activities (processing of fresh skins)	Flegenheimer and Leipus (2007), Flegenheimer et al. (2015)
	Cerro La China 1	10,804 ± 75 10,790 ± 120 10,745 ± 75 10,730 ± 150 10,525 ± 75	AA-8953 AA-1327 AA-8952 I-1274 AA-8954	Multiple activities (several domestic activities)	Flegenheimer et al. (2015)
	Cerro La China 2	11,150 ± 135 10,560 ± 75	AA-8955 AA-8956	Ephemeral occupation (hunting activities)	Flegenheimer et al. (2015)
	Cerro La China 3	10,610 ± 180	AA-1328	Multiple activities (several domestic activities)	Flegenheimer et al. (2015)
	Los Helechos	9640 ± 40	Be-137747	Ephemeral occupation	Flegenheimer and Bayón (2000)
	Cueva Zoro	10,153 ± 61 10,094 ± 62 8859 ± 64	AA-82707 AA-82706 AA-85687	Ephemeral occupation	Mazzia (2013)
	El Aljarafe	8787 ± 41 8574 ± 42	AA-84039 AA-84037	Ephemeral occupation	Mazzia (2011)
Western central Tandilia	Abrigo La Tinta	10,414 ± 32	AA-112043	Ephemeral occupation	Rafuse and Masiggoge (2020)
Western Tandilia	La Moderna	7510 ± 370 7460 ± 80 7010 ± 100	TO-1507-2 TO-2610 TO-1507-1	Specific activities (Occasional megafauna processing event)	Politis and Gutiérrez (1998), Politis et al. (2019)
	Campo Laborde	10,690 ± 380 10,570 ± 170 10,655 ± 35 9730 ± 290 8720 ± 190 8080 ± 200	CAMS-171861 CAMS-171851 CAMS-171852 AA-71665 AA-55119 AA-55118	Specific activities (Hunting and primary processing)	Messineo (2015), Politis et al. (2019)

(continued)

Table 1 (continued)

Area	Sites	^{14}C years BP	Laboratory sample	Site type	References
Inter-hills plain	Paso Otero 5	10,440 ± 100 10,190 ± 120 9560 ± 50	AA-39363 AA-19291 GX-29795	Specific activities (lookout, secondary processing, and consumption)	Martinez (2006)
	Paso Otero 4	8556 ± 65	AA-87934	General activities (wooden artifacts manufacture and processing of animal skins and carcasses)	Barros et al. (2014)
	Arroyo Seco 2	12,200 ± 170 12,240 ± 110 12,170 ± 55 12,170 ± 45 12,155 ± 70 12,070 ± 140 11,770 ± 120 11,750 ± 70 11,730 ± 70 11,590 ± 90 11,320 ± 110 11,250 ± 105 11,190 ± 110 11,000 ± 100 10,500 ± 90	CAMS-58182 OXA-4591 OXA-15871 UCIAMS-142842 OXA-10387 OXA-9243 AA-62514 CAMS-16389 OXA-9242 AA-7964 AA-39365 AA-7965 AA-90118 OXA-4590 AA-9049	Multiple activities (short duration base camps, processing of animal carcasses)	Politis et al. (2014, 2016)
	El Guanaco 1	9250 ± 40	SR-6381	Multiple activities (multi-component sites)	Bayón et al. (2004)
	El Guanaco 2	9480 ± 69 9140 ± 120 8507 ± 84 8433 ± 84 8411 ± 80 8123 ± 82	AA-82713 AA-82713 AA-82712 AA-82705 AA-71658 AA-82710	Multiple activities (multi-component sites)	Flegenheimer et al. (2010)

Cueva El Abra

This site is located at the entrance of the La Vigilancia valley (Fig. 1) and was characterized as a residential campsite for multiple activities (Table 1) (Mazzanti & Bonnat, 2013). The lithic assemblage is integrated by 7864 artifacts, and this chapter only considers and discusses a random sample of around 25% of the lithic assemblage (see Bonnat, 2020a). Therefore, 3284 artifacts were analyzed and grouped into three general classes of artifacts: cores and cores fragment ($N = 47$), debitage products ($N = 3111$), and tools ($N = 90$). A great diversity of raw materials types was recognized ($N = 14$; Table 2) where the BFO, a local rock, was the most frequently used resource (50.15%) and is present in all stages of lithic artifact production (cortex flakes, internal flakes, cores, and different tool types). SBGO

Table 2 Raw materials absolute and relative frequencies from Pampa humid sub-region archaeological sites

Area	Sites	Tandilia rocks					Ventania rocks		Atlantic	Silicified			
		SBGO	BFO	Chert	S. Dolomite	Quartz	Riolite	Metaquartzite	Coast	limestone	Others	Total	References
Eastern Tandilia	Cueva El Abra	1352 (41.62)	1629 (50.15)	1 (0.03)	3 (0.09)	97 (2.98)	5 (0.15)	31 (0.95)	26 (0.83)	13 (0.4)	91 (2.8)	3248	Bonnat (2020a)
	Abrigo Los Pinos	612 (21.7)	1825 (64.72)	0	1 (0.04)	63 (2.23)	34 (1.2)	66 (2.34)	23 (0.85)	1 (0.04)	195 (6.91)	2820	Bonnat (2020a)
	Cueva Tixi	1400 (47.91)	1265 (43.29)	0	1 (0.03)	120 (4.1)	4 (0.14)	1 (0.03)	41 (1.41)	19 (0.65)	70 (3.4)	2921	Bonnat (2020a)
	Cueva La Brava	107 (75.89)	23 (16.31)	0	0	2 (1.42)	0	0	3 (2.13)	0	6 (4.25)	141	Bonnat (2020a)
	Cueva Burucuyá	15 (14.42)	86 (82.69)	0	0	0	1 (0.96)	0	0	2 (1.92)	0	104	Bonnat (2020a)
	Alero El Mirador	102 (45.33)	76 (33.78)	0	0	14 (6.22)	0	10 (4.44)	0	4 (1.78)	19 (8.45)	225	Bonnat (2020a)
	Abrigo La Grieta	58 (46.4)	39 (31.2)	0	0	10 (8)	0	2 (1.6)	0	11 (8.8)	5 (4)	125	Bonnat (2020a)
	Cueva Alí Mustafá	217 (53.85)	146 (36.23)	0	1 (0.25)	18 (4.47)	0	1 (0.25)	0	9 (2.23)	11 (2.72)	403	Bonnat (2020b)
	Alero Molina 1	45 (38.18)	53 (44.91)	0	0	8 (6.77)	0	2 (1.69)	2 (1.69)	3 (2.54)	5 (4.23)	118	Mazzanti and Bonnat (2021)
	Amalia (Sitio 2)	10 (33.33)	16 (52.33)	0	0	0	0	0	3 (9.99)	1 (3.33)	0	30	Bonnat (2020a)
	Lobería 1 (Sitio 1)	6 (66.66)	3 (33.33)	0	0	0	0	0	0	0	0	9	Bonnat (2020a)

Area	Sites	Tandilia rocks					Ventania rocks		Atlantic Coast	Silicified limestone	Others	Total	References
		SBGO	BFO	Chert	S. Dclomite	Quartz	Riolite	Metaquartzite					
East Central Tandilia	Cerro El Sombrero Cima (CoSC)	5928 (91.8)	0	127 (1.9)	0	0	0	0	0	77 (1.2)	312 (4.7)	6453	Flegenheimer et al. (2015)
	Cerro El Sombrero Abrigo 1 (CoSA1)	222 (55.5)	10 (2.5)	4 (1)	0	124 (31)	0	0	0	3 (0.7)	26 (6.5)	400	Bayón et al. (2006)
	Cerro La China 1	428 (67.61)	175 (27.64)	0	0	0	0	0	0	3 (0,47)	26 (4.1)	633	Bayón et al. (2006)
	Cerro La China 2	55 (65.47)	14 (16.66)	2 (2.38)	0	2 (2.38)	0	0	0	0	11 (13.1)	84	Bayón et al. (2006)
	Cerro La China 3	679 (62.35)	307 (28.19)	0	0	36 (3.3)	0	0	0	14 (1.3)	43 (3.9)	1089	Bayón et al. (2006)
	Los Helechos	8 (88.8)	0	0	0	0	0	1 (11.1)	0	0	0	9	Flegenheimer et al. (2015)
	Cueva Zoro	26 (54.16)	16 (33.33)	0	0	1 (2.08)	0	3 (6.25)	0	0	2 (4.16)	48	Mazzia (2013)
	El Aljarafe	20 (74)	5 (18.5)	0	1 (3.7)	0	0	0	0	0	1 (3.7)	27	Mazzia (2011)
Western central Tandilia	Alero La Tinta	14 (93.3)	0	1 (6.6)	0	0	0	0	0	0	0	15	Rafuse and Masiggoge (2020)
Western Tandilia	La Moderna	2 (0.1)	0	0	1 (0.05)	2016 (99.85)	0	0	0	0	0	2019	Messineo (2015)
	Campo Laborde	65 (49.6)	0	34 (27.8)	23 (17.6)	5 (3.8)	0	0	0	0	4 (3.1)	131	Messineo (2015)

(continued)

Table 2 (continued)

Area	Sites	Tandilia rocks					Ventania rocks		Atlantic	Silicified			
		SBGO	BFO	Chert	S. Dolomite	Quartz	Riolite	Metaquartzite	Coast	limestone	Others	Total	References
Inter-hills plain	Paso Otero 4	176 (93.6)	0	1 (0.5)	0	0	0	1 (0.5)	1 (0.5)	0	8 (4.7)	187	Barros et al. (2014)
	Paso Otero 5	48 (55.8)	0	1 (1.2)	1 (1.2)	4 (4.7)	0	0	4 (4.7)	1 (1.2)	27 (31.2)	86	Armentano et al. (2007)
	Arroyo Seco 2	2264 (93.1)	5 (0.2)	52 (2.1)	1 (0.05)	0	0	0	39 (1.6)	0	71 (2.9)	2432	Leipus and Landini (2014)
	El Guanaco 1	83 (90.2)	0	0	0	0	0	1 (1.08)	4 (4.34)	0	4 (4.34)	92	Bayón et al. (2004)
	El Guanaco 2	253 (76)	0	5 (1.5)	1 (0.3)	0	0	6 (1.8)	28 (8)	1 (0.3)	37 (10.9)	331	Vecchi (2011)

References: SBGO: orthoquartzite from the Sierras Bayas Group; BFO: orthoquartzite from the Balcarce Formation; S. Dolomite: silicified dolomite; Others: diabase, andesite, dacite, granite, quartzite, volcanic rocks, and indeterminables

(41.62%), a non-local rock located at ca. 110 km away from the sites, was the second-most important resource identified. Only the final stages of lithic artifact production using this stone have been identified (non-cortex, small and exhausted cores, small internal flakes, and evidence of tool resharpening), as well as a greater diversity of retouched artifacts (e.g., side scrapers, notched tools, end scrapers, knives, and multiple tools). Moreover, different knapping techniques such as bipolar flaking and unifacial and bifacial retouching have been identified in this resource (Bonnat, 2020a). In addition, flakes and tools made of a wide diversity of rocks were identified at lower frequencies (Table 2). Moreover, some supply sources of these artifacts are located at different distances from the site, like flint from the Atlantic coastal area at ca. 50 km, metaquartzite and rhyolites from the Ventania Range at ca. 400 km, and silicified limestone from the Entre Ríos province (Argentina) and/or in the Republic of Uruguay at ca. 500 km from the site.

Abrigo Los Pinos

Abrigo Los Pinos is located in the middle of the La Vigilancia valley (Fig. 1) and was characterized as residential campsites for multiple activities (Table 1). The lithic assemblage is composed of 2820 artifacts and was grouped into three general classes of artifacts: cores and cores fragment ($N = 34$), debitage products ($N = 2701$), and tools ($N = 85$) (Bonnat, 2020a). A great diversity of raw materials was recognized ($n = 12$; Table 2), where the BFO (64.72%), a local rock, was the most frequently used resource and all stages of lithic production (cortex flakes, cores, internal flakes, tool manufacture, maintenance, and different tool types) are represented. SBGO (21.7%), a rock located at ca. 110 km away from the sites, was the second-most important resource identified in the archaeological site (Table 2). This rock was mostly noted in artifacts at the last stages of production (non-cortex, small and exhausted cores, small internal flakes, and evidence of tool maintenance) and was used in a great diversity of retouched artifacts (e.g., side and end scrapers, notched tools, combined tools, etc.) made by different knapping and retouching techniques (unifacial, bifacial, and bipolar flaking). Also, a recycled SBGO fishtail projectile point and a preform of the same rock type were recovered (Valverde, 2006; Bonnat, 2020a). In addition, other rocks from a variety of sources located at different distances from the archaeological site were also identified in low frequencies (Table 2). Some flakes and tools were made in basalt and flint from Atlantic coastal rocks located at ca. 50 km, metaquartzite and rhyolites from the Ventania Range, sources located at ca. 400 km, as well as silicified limestone from the Entre Ríos province (Argentina) and/or in the Republic of Uruguay at ca. 500 km from the site.

Cueva Tixi

This site is located in San Pedro Valley (Fig. 1) and was characterized as a residential campsite for multiple activities (Table 2) (Mazzanti & Bonnat, 2013). The lithic assemblage is composed of 2921 artifacts and was grouped into three general classes of artifacts: cores and cores fragment ($N = 64$), debitage products ($N = 2790$), and tools ($N = 67$) (Bonnat, 2020a). A great diversity of raw materials was recognized ($n = 13$; Table 2), where the SBGO (47.91%), a rock located at ca. 110 km away from the sites, was the most frequently used resource and different stages of artifact production were detected in this rock (exhausted cores, internal flakes, debris, and several tool types). The BFO (43.29%), a local rock, is the second-most important resource identified on the site (Table 2) and is found in all stages of lithic production (e.g., cortex flakes, cores, internal flakes, tool manufacture, maintenance, and different tool types) (Bonnat, 2020a). Other rocks from a variety of sources located at different distances from the archaeological site were also identified in low frequencies (Table 2). Some flakes and tools were made in basalt and flint from Atlantic coastal rocks located at ca. 50 km, metaquartzite and rhyolites from the Ventania Range, sources located at ca. 400 km, as well as silicified limestone from Entre Ríos province (Argentina) and/or in the Republic of Uruguay at ca. 500 km from the site.

Cueva Burucuyá

This site is located in the middle sector of the La Vigilancia valley (Fig. 1) and was characterized as an ephemeral occupation for specific activities (Table 1) (Bonnat & Mazzanti, 2021). The lithic assemblage is composed of 104 artifacts and was grouped in two general classes of artifacts: debitage products ($N = 89$), and tools ($N = 15$). Four different rock types were recognized (Table 2), where the local BFO is the most frequently used rock type (82.9%), and many stages of stone-tool production were identified (e.g., cortex flakes, manufacture of cores, and production of flakes and tools like side scrapers and notch). In contrast, SBGO (14.42%), a rock located at ca. 110 km away from the sites, was identified in small internal flakes and tools, as well as low frequencies of other non-local rocks (Table 2). In addition, two small silicified limestone internal flakes were identified, and one of them has evidence of resharpening and maintenance. A multiple-use tool (scraper and notch) made of rhyolite was recovered (Bonnat & Mazzanti, 2021). The supply areas of these two rocks are located in the Ventania Range (e.g., rhyolites) at ca. 400 km from the sites and in Entre Ríos province (Argentina) and/or in the Republic of Uruguay (e.g., silicified limestone) at ca. 500 km from the site.

Abrigo La Grieta

Abrigo La Grieta is a small rockshelter located in the middle sector of the La Vigilancia valley (Fig. 1) and was characterized as an ephemeral occupation for specific activities (Table 1) (Bonnat, 2020a). The lithic assemblage is composed of 125 artifacts and was grouped into three general classes of artifacts: cores and cores fragment (*N* = 7), debitage products (*N* = 109), and tools (*N* = 9) (Bonnat, 2020a). Eight different rock types were recognized (Table 2), where the SBGO (46.4%), a rock located at ca. 110 km away from the sites, was the most frequently used resource and different stages of artifact production were detected in this rock (e.g., bipolar cores, internal flakes, debris, and side scrapers tools). The BFO (31.2%), a local rock, is the second-most important resource identified on the site (Table 2), and all stages of lithic production were identified (e.g., cortex flakes, cores, internal flakes, and tools like side scrapers) (Bonnat, 2020a). Moreover, in the lithic assemblage, 2 metaquartzite flakes were recognized from the Ventania Range, rock sources located at ca. 400 km, as well as 11 silicified limestone debitage products from the Entre Ríos province (Argentina) and/or in the Republic of Uruguay, source located at ca. 500 km from the site.

Alero El Mirador

This site is a small rockshelter located in the middle sector of the La Vigilancia valley (Fig. 1) and was characterized as an ephemeral occupation for specific activities (Table 1) (Bonnat, 2020a). The lithic assemblage is composed of 225 artifacts and was grouped into three general classes of artifacts: cores and cores fragment (*N* = 12), debitage products (*N* = 201), and tools (*N* = 12) (Bonnat, 2020a). Nine different rock types were recognized (Table 2), where the SBGO (45.33%), a rock located at ca. 110 km away from the sites, was the most frequently used resource and different stages of artifact production were detected in this rock (e.g., cores without cortex, bipolar cores, internal flakes, bipolar flakes, debris, and side scrapers tools). The BFO (33.78%), a local rock, is the second-most important resource identified on the site (Table 2), and all stages of lithic production were identified (e.g., cortex flakes, cores, internal flakes, and tools like side scrapers) (Bonnat, 2020a). In addition, other rocks from a variety of sources located at different distances from the archaeological site were also identified in low frequencies (Table 2). Three debitage products and one tool were made of silicified limestone from the Entre Ríos province (Argentina) and/or in the Republic of Uruguay a rock source located at ca. 500 km from the site. Moreover, ten metaquartzite flakes from the Ventania Range, sources located at ca. 400 km, were also recognized (Bonnat, 2020a).

Cueva La Brava

This site is located near La Brava lagoon (Fig. 1) and was characterized as an ephemeral occupation for specific activities (Table 1) (Bonnat & Mazzanti, 2015). The lithic assemblage is composed of 141 artifacts and was grouped into three general classes of artifacts: cores ($N = 2$), debitage products ($N = 131$), and tools ($N = 8$) (Bonnat, 2020a). Six different rock types were recognized (Table 2), where the SBGO (75.89%), a rock located at ca. 120 km away from the sites, was the most frequently used resource and different stages of artifact production were detected in this rock (e.g., internal flakes, flakes with marginal retouch, and undetermined tools). The BFO (16.31%), a local rock, was the second-most important resource identified on the site (Table 2), and most of the stages of artifact production were observed (e.g., cortex flakes, cores, internal flakes, and a side scraper) (Bonnat & Mazzanti, 2015). Other less-frequent lithic resources (e.g., local quartz and flint probably from the Atlantic coast at 60 km away) were identified as flakes and a flake with marginal retouch.

Amalia Site 2

This site is located on an isolated hill between the Tandilia mountain and the inter-hill plains area (Fig. 1) and was characterized as an ephemeral occupation for specific activities (Table 1) (Bonnat et al., 2015). The lithic assemblage consists of 30 artifacts grouped into two general classes of artifacts: debitage products ($N = 21$) and tools ($N = 9$) (Bonnat, 2020a). Five different rock types were recognized (Table 2), where the BFO (52.33%), a local rock, was the most frequent raw material used, and only some stages of artifact production were identified (internal flakes and tools like side scraper and notch). SBGO (33.33%), a rock located at ca. 110 km away from the sites, was the most frequently used and a few stages of artifact production were detected (e.g., internal flakes, bipolar flakes, side scrapers tools, and a fishtail projectile stem). In addition, other lees frequencies raw materials from a variety of sources located at different distances from the archaeological site were also identified (Table 2). Some cortex flakes and a tool were identified on basalt and flint from the Atlantic coast located at ca. 50 km of distance, as well as one tool on silicified limestone from Entre Ríos (Argentina) and/or in the Republic of Uruguay at ca. 500 km away from the sites (Bonnat et al., 2015).

Lobería 1, Site 1

This site is located on an isolated hill between the Tandilia mountain and the inter-hill plains area (Fig. 1), and it was used as an ephemeral campsite during the early Holocene (Table 1; Martínez & Mazzanti, 2017). The lithic assemblage is composed of nine artifacts and was grouped into three general classes of artifacts: hammerstone ($N = 1$), cores ($N = 2$), and debitage products ($N = 6$) (Bonnat, 2020a).

Two different rock types were recognized (Table 2), where the SBGO (66.66%), a rock located at ca. 80 km away from the sites, was the most frequently used resource and a few stages of artifact production were detected in this rock (e.g., cores without cortex and five internal flakes). The BFO (33.33%), a local rock, was the second-most important resource identified on the site (Table 2), and one hammerstone, a core, and an internal flake were detected (Bonnat, 2020a).

Cueva Alí Mustafá

Cueva Alí Mustafá is located in a tributary valley of the San Pedro hills (Fig. 1), and it was used as an ephemeral camp for specific activities (Table 1; Mazzanti et al., 2019).

The lithic assemblage is composed of 403 artifacts and was grouped into five general classes of artifacts: hammerstone ($N = 1$), cores ($N = 3$), debitage products ($N = 394$), biface ($N = 1$), and tools ($N = 4$) (Bonnat, 2020b). Nine different rock types were recognized (Table 2), where the SBGO (53.85%), a rock located at ca. 110 km away from the sites, was the most frequently used resource and different stages of artifact production were detected in this rock (e.g., internal flakes, debris, a biface, and side scrapers tools). The BFO (36.23%), a local rock, was the second-most important resource identified on the site (Table 2), and all stages of lithic production were identified (e.g., one hammerstone, three cores, cortex, and internal) (Bonnat, 2020b). In addition, other rocks from a variety of sources located at different distances from the archaeological site were also identified in low frequencies (Table 2). Some internal flakes and debris were recognized in metaquartzite from the Ventania Range, sources located at ca. 400 km, as well as silicified limestone from the Entre Ríos province (Argentina) and/or in the Republic of Uruguay at ca. 500 km from the site.

Alero Molina 1

This archaeological site is a small rockshelter located in the southeastern of the La Vigilancia hills (Fig. 1) and has evidence of an ephemeral human occupation dated at the Pleistocene/Holocene transition (Table 1; Mazzanti & Bonnat, 2021). The lithic assemblage is composed of 118 artifacts and was grouped into three general classes of artifacts: cores ($N = 5$), debitage products ($N = 109$), and tools ($N = 4$) (Mazzanti & Bonnat, 2021). Nine different rock types were recognized (Table 2), where the BFO (44.91%), a local rock, was the most frequently used resource and some stages of lithic production were identified (e.g., cortex flakes, cores, and internal flakes). The SBGO (38.18%), a rock located at ca. 110 km away from the sites, was the second-most important resource used on the site and different stages of artifact production were detected in this rock (e.g., cores without cortex, internal flakes, debris, debris, and composite tools) (Mazzanti & Bonnat, 2021). In addition, other rocks from a variety of sources located at different distances from the archaeological site were also identified in low frequencies (Table 2). Some flakes were made in basalt and flint from the Atlantic coastal rocks located at ca. 50 km; in metaquartzite from the Ventania Range, sources located at ca. 400 km, as well as silicified limestone from the Entre Ríos province (Argentina) and/or in the Republic of Uruguay at ca. 500 km from the site.

Central of Tandilia Ranges System

In the central eastern portion of the Tandilia Ranges, eight archaeological sites have been reported and one in the central western (Fig. 1). All of these sites are located on rockshelters and in open-air situations and present evidence of human occupation since the Pleistocene/Holocene transition (Table 1).

Cerro El Sombrero: Cima y Abrigo 1

Two sites have been identified on Cerro El Sombrero (Fig. 1), one of them is an extensive open-air site located on the flat top of the hill and the other is a rockshelter on the western slope (Flegenheimer et al., 2015). Cerro El Sombrero Cima (CoSC) site has been interpreted as a prominent place in the landscape that had a special significance for early human societies in the region. In this space, also used as a sighting place, different activities of daily life were carried out, such as the replacement of fishtail projectile points, the manufacture of the last stages of the operative chain of the same points, the maintenance of tools, and their use (Weitzel, 2010; Flegenheimer et al., 2015). The lithic stratigraphic assemblage is composed of 6453 artifacts and was grouped into two general classes of artifacts: debitage products ($N = 6195$) and tools ($N = 258$) (Flegenheimer et al., 2015). Eight different rock types were recognized (Table 2), where the SBGO (91.8%), a rock located at ca. 65–70 km away from the sites, was the most frequently used resource and different stages of artifact production were detected in this rock (e.g., internal flakes, bipolar flakes, debris, and different tool types like side scrapers, fishtail projectile point and performs, burin, notched, and knife). In addition, other rocks from a variety of sources located at different distances from the archaeological site were also identified in low frequencies (Table 2). Some internal flakes and fishtail points were recovered in silicified limestone from Entre Ríos province (Argentina) and/or from the Republic of Uruguay at ca. 500 km from the sites (Flegenheimer et al., 2003, 2015).

Cerro El Sombrero Abrigo 1 is a small rockshelter characterized as a camp for specific domestic activities (Flegenheimer & Leipus, 2007). The lithic assemblage is composed of 400 artifacts and was grouped into three general classes of artifacts: cores ($N = 8$), debitage products ($N = 366$), and tools ($N = 26$) (Bayón et al., 2006). Seven different rock types were recognized (Table 2), where the SBGO (55.5%), a rock located at ca. 65–70 km away from the sites, was the most frequently used resource and different stages of artifact production were detected in this rock (e.g., bipolar cores without cortex, internal flakes, bipolar flakes, debris, and different tools like side scrapers, end scrapers, knife, and denticulate). Quartz (31%), a local mineral, was the second-most important resource identified on the site (Table 2), and some stages of lithic production were identified (e.g., cores, cortex flakes, internal flakes, and tools) (Bayón et al., 2006). In addition, other rocks from a variety of sources located at different distances from the archaeological site were also identified in low frequencies (Table 2). Some internal flakes were recovered in silicified limestone from the Entre Ríos province (Argentina) and/or from the Republic of Uruguay at ca. 500 km from the sites (Bayón et al., 2006; Flegenheimer et al., 2003).

Cerro La China (Sites 1, 2, and 3)

The Cerro La China is located in the southern sector of the Sierra La Larga (Fig. 1). Three archaeological sites with evidence of early human occupation (Table 1) were identified in this mountain range (Flegenheimer et al., 2015). Cerro La China Site 1 (LCH1) is a small rockshelter characterized as a camp for domestic activities (Flegenheimer et al., 2015). The lithic assemblage is composed of 633 artifacts and was grouped into three general classes of artifacts: cores ($N = 15$), debitage products ($N = 561$), and tools ($N = 57$) (Bayón et al., 2006). Four different rock types were recognized (Table 2), where the SBGO (67.61%), a rock located at ca. 60–75 km away from the sites, was the most frequently used resource and some stages of artifact production were detected in this rock (e.g., cores without cortex, internal flakes, bipolar flakes, and side scrapers tools). The BFO (27.64%), a local rock, was the second-most important resource identified on the site (Table 2), and different stages of lithic production were identified (e.g., cortex flakes, cores, internal flakes, and flake with marginal retouch) (Bayón et al., 2006). In addition, other rocks from a variety of sources located at different distances from the archaeological site were also identified in low frequencies (Table 2). Some internal flakes were recovered in silicified limestone from the Entre Ríos province (Argentina) and/or from the Republic of Uruguay at ca. 500 km from the sites (Flegenheimer et al., 2003, 2015).

The Cerro La China Site 2 (LCH2) is an open-air site located in the hills, and the site was interpreted as a hunting place (Flegenheimer et al., 2015: 169). The lithic assemblage is composed of 84 artifacts and was grouped into three general classes of artifacts: cores ($N = 1$), debitage products ($N = 73$), and tools ($N = 10$) (Bayón et al., 2006). Five different rock types were recognized (Table 2), where the SBGO (65.47%), a rock located at ca. 65–70 km away from the sites, was the most frequently used resource and different stages of artifact production were detected in this rock (e.g., internal flakes, bipolar flakes, debris, and tools). The BFO (16.66%), a local rock, was the second-most important resource identified on the site (Table 2), and little evidence of lithic production was identified (e.g., flakes) (Bayón et al., 2006).

Finally, the Cerro La China Site 3 (LCH3) is also an open-air archaeological site (Fig. 1, Table 1), and according to the varied assemblage, the site has been considered a multipurpose place where several domestic activities were carried out (Flegenheimer et al., 2015: 171). The lithic assemblage is composed of 1089 artifacts and was grouped into three general classes of artifacts: cores (*N* = 36), debitage products (*N* = 979), and tools (*N* = 74) (Bayón et al., 2006). Seven different rock types were recognized (Table 2), where the SBGO (62.35%), a rock located at ca. 65–70 km away from the sites, was the most frequently used resource and different stages of artifact production were detected in this rock (e.g., cores, bipolar cores, internal flakes, bipolar flakes, bifacial flakes, debris, side different tool types, and flakes with blades modified by use). The BFO (28.19%), a local rock, was the second-most important resource identified on the site (Table 2) and different stages of lithic production were identified (e.g., cores, internal flakes, and tools) (Bayón et al., 2006). In addition, other rocks from a variety of sources located at different distances from the archaeological site were also identified in low frequencies (Table 2). Some internal flakes were recovered in silicified limestone from the Entre Ríos province (Argentina) and/or from the Republic of Uruguay at ca. 500 km from the sites (Flegenheimer et al., 2003, 2015).

Los Helechos

This site is a small rockshelter located on the west side of Cerro Chato (Fig. 1) and was used by an ephemeral human occupation during the early Holocene (Table 1) (Flegenheimer & Bayón, 2000). The lithic assemblage is composed of a few artifacts (*N* = 9) that include a side scraper and at least eight flakes. The main raw material is SBGO (99%) located around ca. 45–70 km away from the site, and one flake has been identified as a possible rock from the Ventania ranges located at ca. 300 km of distances (Flegenheimer et al., 2015).

El Aljarafe

El Aljarafe site is located at the hilltop in the south of Sierra Larga (Fig. 1) and was defined as a site of ephemeral human occupations possibly used as part of logistical trips within the Tandilia Ranges system (Table 1) (Mazzia, 2011). The lithic assemblage is composed of 27 artifacts and was grouped into three general classes of artifacts: debitage products (*N* = 22), an abrasive stone (*N* = 1), and tools (*N* = 4) (Mazzia, 2011; Flegenheimer et al., 2015). Four different rock types were recognized (Table 2), where the SBGO (74%), a rock located at ca. 30–60 km away from the sites, was the most frequently used resource, and the latest stages of artifact production were detected in this rock (e.g., internal flakes, bipolar flakes, bifacial reduction flakes, and flakes with marginal retouch tools). The BFO (18.5%), a local rock, was the second-most important resource identified on the site (Table 2), and a few stages of lithic production were identified (e.g., internal flakes and a tool flake with marginal retouch) (Mazzia, 2011).

Cueva Zoro

Cueva Zoro is an archaeological site located in the northern end of Sierra Larga (Fig. 1) defined as an ephemeral campsite occupied during the Pleistocene/Holocene transition (Table 1) (Mazzia, 2013). The lithic assemblage is composed of 48 artifacts and was grouped into two general classes of artifacts: debitage products ($N = 43$) and tools ($N = 5$). Six different rock types were recognized (Table 2), where the SBGO (44.16%), a rock located at ca. 30–60 km away from the sites, was the most frequently used resource, and the latest stages of artifact production were detected in this rock (e.g., internal flakes, bifacial reduction flakes, side scrapers, and others and flakes with marginal retouch tools). The BFO (33.33%), a local rock, was the second-most important resource identified on the site (Table 2), and some stages of lithic production were identified (e.g., cortex flakes and internal flakes) (Mazzia, 2013). In addition, other rocks from a variety of sources located at different distances from the archaeological site were also identified in low frequencies (Table 2). Some flakes made in metaquartzite from the Ventania Range, sources located at ca. 300 km, were identified, as well as a flint stone tool from undetermined sources.

Abrigo La Tinta

Abrigo La Tinta archaeological site is located in the central western region of the Tandilia mountain (Fig. 1) and is the first site in this sector of the landscape with human occupations dating back to the Pleistocene/Holocene transition (Table 1). The lithic assemblage is totally composed of internal flakes ($N = 15$) made on SBGO (92%) and chert (8%), both rocks located at ca. 50 km of distance from the site. The site was interpreted as an ephemeral camp where the last stages of lithic tool production were carried out (Rafuse & Massigoge, 2020).

Western Slope of the Tandilia Ranges System

The northwestern sector of the Tandilia Ranges system is the end of this mountain system, as well as the transition with a sector to the Pampean inter-hills area (Fig. 1). Two open-air archaeological sites with evidence of extinct fauna processing have been identified in this sector (Politis et al., 2019).

La Moderna

La Moderna archaeological site is located on the banks of the Azul Creek (Fig. 1) and is interpreted as a faunal procurement site during the early Holocene (Table 1). Thus, during a single event, an extinct *Doedicurus clavicaudatus* was butchered at the edge of an ancient swamp (Politis et al., 2019: 1). The lithic assemblage is composed of ca. 2000 quartz flakes (external and internal flakes) and 16 quartz tools (rock available in the area) that were manufactured by unifacial retouch, used and discarded on site (Politis & Gutiérrez, 1998). In addition, three tools made on SBOG and dolomite, rocks located at ca. 20 km from the site, were also recovered (Messineo, 2015).

Campo Laborde

The Campo Laborde site is located at the headwaters of the Arroyo Tapalqué, in the Olavarría district (Fig. 1), and the site was characterized as a hunting and/or procurement site for *Megatherium* on the banks of a swamp (Politis et al., 2019). This archaeological site presents a similar pattern to La Moderna since the lithic materials and bone remains were recovered in a paleo-swamp. Zooarchaeological studies carried out identified several modern fauna (*Rehidae, Lagostomus maximus, Dolichotis patagonus, Tayassu* sp., and *Artiodactyla*) as well as extinct mammals (*Megatherium americanum, Neosclerocalyptus* sp., and *Doedicurus* sp.), but only two of them (*M. americanum* and *Dolichotis patagonum*) show cultural cut marks and fractures evidence (Messineo, 2015; Politis et al., 2019). The lithic assemblage is composed of 131 artifacts and was grouped into two general classes of artifacts: debitage products ($N = 129$) and tools ($N = 2$) (Messineo, 2015: 350). Five different rock types were recognized (Table 2), where the SBGO (49.6%), a rock located at ca. 20 km away from the sites, was the most frequently used resource, and the last stages of artifact production were detected in this rock (e.g., internal flakes, debris, a side scraper, and a bifacial fractured tool). Chert (27.80%), a rock located at ca. 25 km away from the sites, was the second-most important resource identified on the site (Table 2), and all stages of lithic production were identified (e.g., cortex flakes and internal flakes) (Messineo, 2015). In addition, on silicified dolomite (17.6%), another rock available at ca. 25 km away from the sites, some external and internal flakes also were recovered, as well as in quartz (Table 2).

Inter-Hills Area

Arroyo Seco 2

Arroyo Seco 2 (Fig. 1) has the earliest evidence of human occupation in this region of Argentina (Table 1) and was characterized as a multi-activity settlement where extinct fauna has evidence of human consumption (Politis et al., 2014, 2016). The lithic assemblage is composed of 2432 artifacts and was grouped into three general classes of artifacts: cores ($N = 17$), debitage products ($N = 2329$), and tools ($N = 87$) (Leipus & Landini, 2014). Eight different rock types were recognized (Table 2), where the SBGO (93.1%), a rock located at ca. 110 km away from the sites, was the most frequently used resource and different stages of artifact production were detected in this rock (e.g., cores without cortex, bipolar cores, internal flakes, bipolar flakes, debris, and different tools types like side scrapers and other two informal tool flakes with marginal retouch and undetermined tools). In addition, other rocks from a variety of sources located at different distances from the archaeological site were also identified in low frequencies (Table 2). A flake and a bipolar core were made in basalt from coastal Atlantic rocks located at ca. 55 km from the site, as well as flakes on chert from the Tandilia mountain ranges located at ca. 150 km away from the sites (Politis et al., 2016).

Paso Otero Archaeological Locality

Paso Otero archaeological locality (Necochea district) is composed of several open-air sites with different chronologies, where sites 4 (PO4) and 5 (PO5) show evidence of early human occupation near the Quequén Grande River (Fig. 1; Table 1; Martínez, 2006). PO5 is located in the middle basin of the river, and the site was characterized as a brief human occupation camp linked to hunting and/or scavenging processing activities of different species of mammals (Martínez, 2006). A large accumulation of bones was identified, many of which showed a high degree of fragmentation and thermal alteration, which led to the proposal that these bone remains are not only evidence of fauna consumed by human groups but could also have been used as fuel for the hearths (Martínez & Gutiérrez, 2011). The lithic assemblage is composed of 86 artifacts and was grouped into two general classes of artifacts: debitage products ($N = 80$) and tools ($N = 6$) (Armentano et al., 2007). Nine different rock types were recognized (Table 2), where the SBGO (55.8%), a rock located at ca. 55 km away from the sites, was the most frequently used resource, and the latest stages of artifact production were detected in this rock (e.g., internal flakes and different stone tools like a peduncle basal fragment, a composite artifact, and a bifacial fragment artifact). In addition, other rocks from a variety of sources located at different distances from the archaeological site were also identified in low frequencies (Table 2). Some internal flakes made in basalt from Atlantic coastal rocks located at ca. 80 km and a fishtail projectile point in silicified limestone from the Entre Ríos province (Argentina) and/or in the Republic of Uruguay at ca. 500 km from the sites also were identified (Armentano et al., 2007).

On the other hand, the human occupation of PO4 is dated to the early Holocene (Table 1). The lithic assemblage is composed of 187 artifacts and was grouped into three general classes of artifacts: cores ($N = 1$), debitage products ($N = 157$), and tools ($N = 30$) (Barros et al., 2014). Seven different rock types were recognized (Table 2), where the SBGO (93.6%), a rock located at ca. 55 km away from the sites, was the most frequently used resource and different stages of artifact production were detected in this rock (e.g., a core without cortex, bipolar cores, internal flakes, debris, and different tools like end scrapers, notches, flake with marginal retouch, undetermined, and fragments tools). In addition, other rocks from a variety of sources located at different distances from the archaeological site were also identified in low frequencies (Table 2). Finally, some internal flakes made in basalt from Atlantic coastal rocks located at ca. 80 km and in metaquartzite from the Ventania Range, sources located at ca. 300 km, also were identified (Barros et al., 2014).

El Guanaco

The El Guanaco locality (San Cayetano district) is composed of two open-air archaeological sites (Fig. 1) located next to the El Lucero shallow lake, about 13 km from the Atlantic coast (Bayón et al., 2004; Flegenheimer et al., 2010). El Guanaco Site 1 is a multicomponent site where different domestic activities were carried out, and the lithic assemblage recovered was associated with remains of current and extinct fauna (*Macrauchenia patachonica* and *Equus* sp.) and rheid eggshells. The lithic assemblage is composed of 92 artifacts and was grouped into three general classes of artifacts: cores ($N = 4$), debitage products ($N = 83$), and tools ($N = 5$) (Bayón et al., 2004: 254). Four different rock types were recognized (Table 2), where the SBGO (90.2%), a rock located at ca. 100 km away from the sites, was the most frequently used resource and different stages of artifact production were detected in this rock (e.g., cores, internal flakes, debris, and tools like flake with marginal retouch and undetermined tools). Also, some external flakes made in rocks from the Atlantic coast located at ca. 15 km away, as well as in metaquartzite from the Ventania Range, sources located at ca. 300 km away, were identified as well (Bayón et al., 2004).

El Guanaco 2 was interpreted as a domestic tasks site in which lithic artifacts, mineral pigments, bone remains of current fauna, and rheid eggshells were recovered (Flegenheimer et al., 2010). The lithic assemblage is composed of 331 artifacts and was grouped into three general classes of artifacts: cores ($N = 1$), debitage products ($N = 315$), tools ($N = 5$), and unused clasts ($N = 10$) (Vecchi, 2011). Nine different rock types were recognized (Table 2), where the SBGO (76%), a rock located at ca. 100 km away from the sites, was the most frequently used resource and different stages of artifact production were detected in this rock (e.g., cores without cortex, internal flakes, bipolar flakes, debris, and tools). In addition, other rocks from a variety of sources located at different distances from the archaeological site were also identified in low frequencies (Table 2). Some flakes were made in basalt from coastal Atlantic rocks located at ca. 15 km, metaquartzite from the Ventania Range, sources located at ca. 300 km, as well as silicified limestone from the Entre Ríos province (Argentina) and/or in the Republic of Uruguay at ca. 500 km from the site.

Discussion

This study of 27 lithic assemblages from the Pampa humid sub-region provides evidence that a high diversity of raw materials was used by hunter-gatherers population indifferent frequencies and according to settlements patterns of mobility and socioeconomic requirements (Bayón et al., 2006; Martínez, 2006; Armentano et al., 2007; Flegenheimer et al., 2015; Messineo & Barros, 2015; Bonnat, 2019, 2020a; among others). On the one hand, some archaeological sites considered base camp or diverse activity sites (e.g., Arroyo Seco2, LCH 1 and 3, Cueva El Abra, Cueva Tixi, Abrigo Los Pinos, and El Guanaco 1 and 2) show a high diversity of raw materials as well as artifactual classes. On the other hand, some sites considered for specific activities (e.g., fauna and megafauna procurement in PO5, Campo Laborde, and La Moderna) or ephemeral occupations campsite (e.g., Cueva Burucuyá, Cueva La Brava, Lobería 1 Site 1, LCH 2, El Aljarafe, and Abrigo La Tinta) show a lower diversity rocks and artifact types than those sites. However, some of these sites with both ephemeral and more prolonged occupations (e.g., Amalia Site 2, Paso Otero 5, El Guanaco 2, and Arroyo Seco 2) have small lithic assemblages (Table 2) that evidence a wide diversity of materials, as well as the distances to the sources that are located in different places of the landscape and articulate the territories of the Atlantic coast with the mountain system (Armentano et al., 2007; Vecchi, 2011; Bonnat et al., 2015; Politis et al., 2014). In this scenery, the raw material acquisitions could involve different strategies, as a special-purpose, embedded, and indirect procurement.

Orthoquartzite from Tandilia Mountain Ranges

Rocks from the Tandilia mountain system were the most used by the early hunter-gatherers in the region, particularly the SBGO that has better qualities for knapping, and were the most selected and transported. In this sense, above the functional differences between archaeological sites and their location in the landscape, these rocks are present in all of them; thus, they were the most raw materials that circulated among the first human groups and their supply can be considered of regional importance (sensu Bayón et al., 2006). In the eastern Tandilia mountain, 11 archaeological sites were identified and the SBOG was used between 76% and 14% (Table 2). At the east-central and western-central portions of the Tandilia mountain (Fig. 1), nine archaeological sites were identified and SBGO was used between 93% and 54% (Table 2). In the inter-hills plain, this rock was also widely used in sites such as Paso Otero 4 (94%), Arroyo Seco 2 (93%), and El Guanaco 1 (90%) and less intensity in El Guanaco 2 (76%). Finally, in the western Tandilia mountain, SBOG was identified with less frequency in Campo Laborde (50%) and was practically not used in La Moderna, and the lithic management included other immediately available resources (e.g., quartz and silicified dolomite).

As noted above, the frequencies in the use of this rock are diverse according to the type of site and its landscape location. In addition, archaeological data also allow us to affirm that the procurement strategies, plans, and technological decisions involved in the management of lithic resources were also different (Armentano et al., 2007; Bayón et al., 2006; Bonnat, 2020a; Colombo, 2012; Flegenheimer et al., 2015; Messineo & Barros, 2015; Messineo & Pal, 2020; among others). In eastern Tandilia mountain, the use of this rock was different depending on the functionality and type of archaeological sites. In sites characterized as base camps (e.g., Cueva Tixi, Abrigo Los Pinos, and Cueva El Abra) almost all stages of lithic artifact production were identified, and some evidence for curated strategies like the optimization of raw material by bipolar knapping (Valverde, 2006) and a greater use of unifacial retouch for knapping different tool types were detected (Bonnat, 2020a). At the east-central Tandilia sites, there is evidence of more stages of lithic artifact production, bifacial technique involved in the manufacture of different stone tools (e.g., fishtail projectile points), and a great color selection of the SBOG (Colombo & Flegenheimer, 2013; Flegenheimer et al., 2015, among others). In the inter-hills area, lithic assemblages evidenced final stages of tool production and the use of retouched long edge and natural edge for specific activities in Paso Otero 4 and Arroyo Seco 2 (Leipus & Landini, 2014; Pal et al., 2019).

Based on this evidence and source areas identified in the Pampas humid subregion (Flegenheimer et al., 1996; Bayón et al., 1999; Colombo, 2012; Barros et al., 2021), the SBGOs were transported throughout different spatial scales and traveled distances between 20 and 120 km. In addition, the high frequency of this raw material in many archaeological sites suggests a fluid access of this rock, where the distances to the sources seem not to have been a problem, and probably, the better quality for tool knapping could be the main reason for their selection. Thus, in sites located at central Tandilia mountain, where hunter-gatherer groups inhabited neighbor source area (ca. 30 km), access to quarries could be frequent and without limitations. Therefore, human groups were able to develop a special-purpose procurement strategy (Binford, 1979; Gould & Saggers, 1985) by logistical groups during daily foraging trips. On the other hand, some lithic assemblages in sites located between 100 and 120 km away from the source supply (e.g., eastern of Tandilia mountain and inter-hill area) suggest a higher intensity in the use of the SBGO, such as low cortex indices, smaller size of artifacts, exhausted cores, and tools, more evidence of maintenance in tools and cores, as well as flakes extractions in cores. In this scenario, hunter-gatherer societies were able to develop a special-purpose procurement strategy (Binford, 1979; Gould & Saggers, 1985) with the main purpose of acquiring the SBGOs which were transported by logistical groups during specific trips to the sources and/or by residential bases movements during their annual logistic ranges (Binford, 1982, 1983; Kelly, 1995).

The BFO, a rock with lower quality and aptitude for knapping than SBGO, was mainly used as a local resource in sites located in the eastern of the Tandilia mountain, in a technological system characterized as expeditive strategies (Flegenheimer et al., 2015; Bonnat, 2019, 2020a). In this sector of the landscape, the percentages of the use of this rock vary according to the functionality of the sites, but in general terms, it was used between 83% and 16% (Table 2). In some sites (e.g., Cueva Burucuyá), it was the most used raw material involved in the tool production, while in other sites of restricted activities (e.g., Cueva La Brava), it was used in low frequencies (Bonnat & Mazzanti, 2015, 2021). In the central-eastern sector of Tandilia, this rock is also locally available and easy to obtain near archaeological sites; however, the lithic technological systems were not oriented to this lower-quality resource for carving. Thus, in some sites, the BFO was not used (e.g., Cerro El Sombrero Cima) or was used in lower frequencies (between 32% and 5%) also within an expeditive technological system (Mazzia, 2011; Flegenheimer et al., 2015). Consequently, in this sector of the Tandilia mountain, hunter-gatherer groups were able to develop a special-purpose procurement and/or embedded strategy (sensu Binford, 1979; Gould & Saggers, 1985) in procurement of this immediately available raw materials during daily foraging trips or during the exploitation territory where there are no costs and efforts involved. On the other hand, in the westernmost sites of Tandilia (e.g., Campo Laborde, La Moderna, and Abrigo La Tinta) and in the Inter hill area (e.g., Paso Otero 4 and 5, Arroyo Seco 2, and El Guanaco 1 and 2), this rock was practically not selected for manufacture of stone tools (Table 2) (Politis & Gutiérrez, 1998; Armentano et al., 2007; Flegenheimer et al., 2010, 2015; Vecchi, 2011; Barros et al., 2014; Leipus & Landini, 2014; Politis et al., 2019; Rafuse & Massigoge, 2020). Possibly, activities developed at these sites and the costs involved in the acquisition of raw materials from the Tandilia mountain were oriented to better quality rocks (e.g., SBGO, chert, and sea coastal rocks).

Other Raw Materials Used in the Pampa Humid Sub-region

Lithic Resources from the Tandilia Mountain Ranges

With the exception of some cases (e.g., La Moderna, Campo Laborde, Abrigo La Tinta, and Loberia 1 Site 1), the earliest archaeological sites evidence a wide diversity of raw materials used in less frequency (Table 2), which involve different procurement strategies and spatial mobility scales. Thus, some rocks from Tandilia mountain (e.g., chert, diabase, silicified dolomite, and quartz) were more used in those sites located near the source area (Table 3). In western Tandilia, the lithic assemblages of La Moderna were mainly manufactured on quartz (99.75%), and in Campo Laborde, half of the lithic assemblages (52%) were made on chert, silicified dolomite, and quartz (Messineo, 2015). In both cases, the raw materials used are on a local scale and very close to the sites (Fig. 1); thus, hunter-gatherer groups were able to develop a special purpose and/or embedded procurement strategy (Binford, 1979; Gould & Saggers, 1985) during daily foraging trips and logistic ranges (Binford, 1983) without much effort and costs involved.

Archaeological sites from eastern and east-central Tandilia mountain ranges provide evidence of a lesser use of these raw materials (Table 2); in both cases, only some stages of the lithic production sequence are represented in these sites. For the

Table 3 Area site location and estimated distances to source area

Area site location	Tandilia rocks							
	SBGO	BFO	Chert	Silicified Dolomite	Quartz	Ventania rocks	Atlantic rocks	Silicified limestone
Eastern Tandilia	100–120 km	1 km	100–120 km	100–120 km	1 km	350–400 km	50 km	400–500 km
East Central Tandilia	30–70 km	1 km	30–70 km	30–70 km	1 km	300–350 km	70 km	400–500 km
Western central Tandilia	50 km	1 km	50 km	50 km	1 km	200–250 km	150–200 km	400–500 km
Western Tandilia	20 km	20 km	20 km	20 km	20 km	200–250 km	200–250 km	400–500 km
Inter-hills plain	60–100 km	60–100 km	60–100 km	60–100 km	60–100 km	150–200 km	10–80 km	500–600 km

east-central area, chert, silicified dolomite, and diabase are locally available resources, but in the east-central area, these rocks are not available and they must have traveled about 100 km away from the quarry places (Bonnat, 2020a). However, all of these rocks are considered a secondary used resource and its acquisition could have been included in the development of other activities (sensu Binford, 1979), like the procurement of other better quality rocks in the hills during the logistic range (Binford, 1983). Finally, in the inter-hill plain area, these rocks were not widely used (less than 2%; Table 2). Thus, their use was sporadic or occasional and this raw material was not a main component of the technological systems of hunter-gatherer groups.

Rocks from the Ventania Range, Atlantic Coast Pebbles, and Silicified Limestone

Rocks from the Ventania mountain system (e.g., metaquartzites and rhyolites) entered into the technological systems as secondary or minority-use resources (sensu Bayón et al., 2006). These rocks are registered in low frequencies and were not recovered in all archaeological sites of the region (Table 2). Thus, rocks from Ventania traveled distances between 150 and 200 km to sites located in the inter-hill area (e.g., Arroyo Sesco 2, Paso Otero 5, and El Guanaco 2) and distances between 300 and 400 km for sites in central-eastern Tandilia (e.g., Cueva Zoro) and eastern Tandilia (e.g., Cueva Tixi, Abrigo Los Pinos, Cueva El Abra, Cueva Burucuyá, Alero El Mirador, Abrigo La Grieta, and Alero Molina 1) (Table 3). An interesting feature is that there is a higher frequency of these rocks in the sites farthest from the source area; however, there are mostly medium/small debris sizes and tools without elaborated retouching that shows little investment of time and energy in their manufacture (Bonnat, 2020a; Mazzanti & Bonnat, 2021). These rocks were probably obtained through exchange nets (indirect strategies) or during the development of other activities, like logistic and/or annual range movement (Binford, 1983) between the Ranges and the Atlantic coast, where the creeks have been used as vectors connecting these different landscapes (Bonnat et al., 2015; Bonnat, 2020a).

On the other hand, coastal sea lithic resources, including basalt, silex, and secondary rock sources from the Ventania Range, entered into the sites in low frequencies, and there is evidence mainly of debris and bipolar pieces from the pebbles knapping. Probably due to the distance to the sources (Table 3), these rocks were more used in the sites near the coast (e.g., El Guanaco 2) and then in less frequency in the inter-hill plain (e.g., Paso Otero 5) and eastern Tandialia Range (e.g., Cueva La Brava, Alero Molina 1, and Amalia Site 2) (Table 2). In those sites near the coast, where there are no immediate vicinity raw materials, a special purpose and/or embedded procurement of the coastal pebbles rocks could be developed; however, the production of lithic artifacts was oriented to the SBGOs of Tandilia mountain. Nevertheless, for these two rock sources, the river banks, the Atlantic sea coast and the paleo-shallow lakes near the coastline, may have been the most optimal routes for the circulation of these resources.

Finally, silicified limestone is one of the most interesting rocks in the archaeological record of the Pampa humid sub-region due to the location of the supply sources, distant ca. 500 km away. This rock is present in the form of tools (e.g., fishtail projectile point in Paso Otero 5), bifacial artifacts (e.g., Cerro La Cina Sites 1 and 3), and mostly small, indeterminate, and bifacial reduction debris (e.g., Amalia Site 2, Abrigo La Grieta, Cueva El Abra, Alero Molina 1, and Cueva Alí Mustafa, among others) (Table 2) that evidence the knapping intensity of this rock. These silicified limestone could have traveled along the Atlantic coast and entered the Buenos Aires territory through rivers and creeks that functioned as vectors contacting the different basins of human occupation (Mazzanti & Bonnat, 2013). However, it is hard to answer whether these rocks are the result of an indirect provisioning related to large-scale social interaction or the result of the first human migration that entered the Pampa region, brought them as a tool kit, and were replaced by local raw materials.

In summary, several archaeological sites show some evidence (sensu Franco, 2014) of an indirect procurement strategy for raw materials such as metaquartzite, rhyolite, silicified, and coastline pebbles: lower artifact frequency, smaller size, undifferentiated forms (debris) due to the high knapping, exhausted cores, maintenance evidence, etc. Probably, some of these evidences identified in silicified limestone are the result of an indirect acquisition of some instruments, base forms or performs in archaeological sites like Paso Otero 5, El Guanaco 2, Amalia Site 2, and Cerro La China Sites 1 and 3. However, the uses of secondary or minority resources would be evidence of an indirect supply, possibly as a result of annual ranges mobility (sensu Binford, 1983), in which there were moments of inter-group contact, circulation of different resources, and greater fluidity of knowledge, information, experiences, etc. Probably, the procurement of the main rocks used in all archaeological sites (SBGO) plays a central role in the circulation of these secondary rocks in different sectors of the landscape. Finally, it is interesting to note that within the mobility circuits that connected the hills, plains, and coast, some archaeological sites with the oldest chronologies located in the inter-hills area and near the ranges (e.g., Arroyo Seco 2, El Guanaco 2, Amalia Site 2, and Paso Otero 5) have lithic assemblages that show the articulation of these different environments.

Final Remarks

This chapter evaluated the use of lithic resources, procurement strategies, mobility ranges, and the use of landscape developed by early hunter-gatherer groups that occupied especially the Humid Pampas sub-region during the late Pleistocene and the early Holocene. Lithic raw materials can be acquired in a number of different ways, including special purpose, embedded, and indirect procurement. In general terms, with some exceptions, the new knowledge generated in the last two decades is consistent with the pattern of resource procurement proposed by Bayón et al. (2006), where SBGO is a main resource found in all archaeological sites and other raw materials are positioned in a second, third, and fourth order. Thus, a direct procurement without limitations seems to have been the scenario in which the SBGOs circulated over the whole Pampas landscape. Moreover, other less-used raw materials were probably obtained through exchange nets (indirect strategies) or during the development of other activities, such as SGBO procurement (embedded strategies).

Acknowledgements Financial support for this research was provided by the Agencia Nacional de PromociónCientífica y Tecnológica, Argentina (projects PICT 2008 N° 1390, PICT 2013 N° 1979 and PICT 2017 N° 1027), and the Universidad Nacional de Mar del Plata.

References

Andrefsky, W., Jr. (1998). *Lithics: Macroscopic approaches to analysis*. Cambridge University Press.

Andrefsky, W., Jr. (2009). The analysis of stone tool procurement, production and maintenance. *Journal of Archaeological Research, 17*, 65–103.

Armentano, G., Martínez, G., & Gutiérrez, M. A. (2007). Revisión del sitio Paso Otero 5: aspectos tecnológicos y fuentes de aprovisionamiento. In C. Bayón, A. Pupio, M. I. González, N. Flegenheimer, & M. Frere (Eds.), *Arqueología en las Pampas* (pp. 535–548). Sociedad Argentina de Antropología.

Aschero, C. (1975). *Ensayo para una clasificación morfológica de artefactos líticos aplicada a estudios tipológicos comparativos*. Unpublished manuscript, CONICET.

Aschero, C. (1983). *Ensayo para una clasificación morfológica de artefactos líticos aplicada a estudios tipológicos comparativos*. Apéndices A-C. Revisión. Unpublished manuscript, FFyL-UBA.

Bamforth, D. B. (1986). Technological efficiency and tool curation. *American Antiquity, 51*, 38–50. https://doi.org/10.2307/280392

Bamforth, D. B. (1990). Settlement, raw material, and lithic procurement in the central Mojave Desert. *Journal of Anthropological Archaeology, 9*, 70–104. https://doi.org/10.1016/0278-4165(90)90006-Y

Barros, M. P., Martínez, G., & Gutiérrez, M. A. (2014). Análisis de los materiales líticos del sitio Paso Otero 4 (partido de Necochea, provincia de Buenos Aires). Avances en el conocimiento de las estrategias tecnológicas en el curso medio del río Quequén Grande durante el Pleistoceno tardío-Holoceno. *Relaciones de la Sociedad Argentina de Antropología, XXXIX*, 119–144.

Barros, M. P., Álvarez, M. C., Kaufmann C., Gómez Peral, L., Rafuse, D., & Politis, G. G. (2021). *Una nueva cantera de cuarcita en el extremo occidental de Tandilia* (p. 139). Proceedings of the IX Congreso de Arqueología de la Región Pampeana Argentina.

Bayón, C., Flegenheimer, N., Valente, M., & Pupio, A. (1999). Dime cómo eres y te diré de dónde vienes: La procedencia de rocas cuarcíticas en la región pampeana. *Relaciones de la Sociedad Argentina de Antropología, XXIV*, 187–235.

Bayón, C., Flegenheimer, N., Zárate, M. A., & Deschamps, C. (2004). Y vendrán los arqueólogos en busca de un hueso, Sitio El Guanaco, partido de San Cayetano. In G. A. Martínez, M. A. Gutiérrez, R. P. Curtoni, M. A. Berón, & P. Madrid (Eds.), *Aproximaciones Contemporáneas a la Arqueología Pampeana. Perspectivas Teóricas, Metodológicas, Analíticas y Casos de Estudio* (pp. 247–258). FACSO-UNICEN.

Bayón, C., Flegenheimer, N., & Pupio, A. (2006). Planes sociales en el abastecimiento y traslado de roca en la Pampa Bonaerense en el Holoceno Temprano y Tardío. *Relaciones de la Sociedad Argentina de Antropología, XXXI*, 19–45.

Binford, L. R. (1979). Organization and formation processes: Looking at curated tecnologies. *Journal of Anthropological Research, 35*, 255–273.

Binford, L. R. (1980). Willow smoke and dog's tails: Hunter-gatherer settlement system and archaeological site formation. *American Antiquity, 45*, 4–20.

Binford, L. R. (1982). The archaeology of place. *Journal of Anthropological Archaeology, 1*(1), 5–31.

Binford, L. R. (1983). Long term land use patterns: Some implications for archaeology. In R. Dunnell, D. Grayson, & W. Koelz (Eds.), *Lulu linear punctated: Essays in honor of George Irving Quimby* (Vol. 72, pp. 27–53). Museum of Anthropology. University of Michigan.

Binford, L. R., & O'Connell, J. F. (1984). An Alyawara day: The stone quarry. *Journal of Anthropological Research, 40*(3), 406–432.

Bonnat, G. F. (2019). Raw material procurement and landscape use during the Pleistocene/Holocene transition in the eastern Tandilia Range (Buenos Aires, Argentina). *PaleoAmerica: A Journal of Early Human Migration and Dispersal, 5*(1), 62–72. https://doi.org/10.1080/20555563.2018.1564962

Bonnat, G. F. (2020a). *Análisis de la organización de la Tecnología Lítica de Grupos Cazadores-recolectores tempranos del área de Tandilia Oriental, Buenos Aires*. Sociedad Argentina de Antropología. Edición Tesis doctoral. 978-987-1280-51-3.

Bonnat, G. F. (2020b). Lithic technological analysis of a new archaeological site (Cueva Alí Mustafá, 12,000 cal yr BP) of the Pleistocene/Holocene transition in Eastern Tandilia, Buenos Aires. Argentina. *PaleoAmerica: A Journal of Early Human Migration and Dispersal, 6*(3), 223–233. https://doi.org/10.1080/20555563.2019.1710934

Bonnat, G. F., & Mazzanti, D. L. (2015). Análisis de la tecnología lítica de las ocupaciones humanas efímeras durante la transición Pleistoceno-Holoceno: el caso de Cueva La Brava (Buenos Aires, Argentina). *Intersecciones en Antropología, 16*, 287–300.

Bonnat, G. F., & Mazzanti, D. L. (2021). Estrategia tecnológica expeditiva en la gestión de recursos líticos durante el postglacial. El caso de Cueva Burucuyá (Tandilia oriental, Argentina). *Comechingonia. Revista de Arqueología, 25*(2), 151–191.

Bonnat, G. F., Mazzanti, D. L., & Martínez, G. A. (2015). Tecnología lítica y contexto geoarqueológico de la ocupación temprana del Sitio 2 de la Localidad Arqueológica Amalia, provincia de Buenos Aires (Argentina). *Revista del Museo de Antropología, 8*(2), 21–32.

Bonnat, G. F., Desimone, P. M., Martínez, G. A., & Mazzanti, D. L. (2021). Raman spectroscopy study of ca 12.000 years cal BP lithic artifacts adhesions from archaeological sites in the Pampas region, Argentina. *Journal of Archaeological Science: Reports, 24*, 1–11. https://doi.org/10.1016/j.jasrep.2021.103101

Bonomo, M., & Blasi, A. (2010). Base regional de recursos líticos del Delta del Paraná. Estudio petrográfico de artefactos y afloramientos en el sur de Entre Ríos. *Cazadores-Recolectores del Cono Sur. Revista de Arqueología, 4*, 17–41.

Bonomo, M., & Prates, L. (2014). La explotación de depósitos secundarios de rodados en el curso medio del Río Negro y el litoral Marítimo Pampeano. In P. Escola & S. Hocsman (Eds.), *Artefactos líticos, movilidad y funcionalidad de Sitios en Sudamérica. Problemas y Perspectivas* (pp. 77–92). BAR International Series 2628.

Brantingham, P. J. (2003). A neutral model of stone raw material procurement. *American Antiquity, 68*(3), 487–510.

Catella, L., Manassero, M. J., Moirano, J., & Oliva, F. W. (2013). Nuevos aportes al estudio del aprovisionamiento de cuarcita en la Región Pampeana, Argentina. *Cuadernos del Instituto Nacional de Antropología y Pensamiento Latinoamericano-Series Especiales, 1*(2), 200–215.

Colombo, M. (2012). *Los cazadores recolectores pampeanos y sus rocas. La obtención de materias primas líticas vista desde las canteras arqueológicas del centro de Tandilia* [Unpublished doctoral Thesis]. Facultad de Ciencias Naturales y Museo, Universidad Nacional de La Plata.

Colombo, M., & Flegenheimer, N. (2013). La elección de rocas de colores por los pobladores tempranos de la región Pampeana (Buenos Aires, Argentina): Nuevas consideraciones desde las canteras. *Boletín del Museo Chileno de Arte Precolombino, 18*(1), 125–137.

Ditchfield, K. (2016). An experimental approach to distinguishing different stone artifact transport patterns from debitage assemblages. *Journal of Archaeological Science, 65*, 44–56. https://doi.org/10.1016/j.jas.2015.10.012

Ericson, J. (1984). Toward the analysis of lithic production systems. In J. Ericson & B. Purdy (Eds.), *Prehistoric quarries and lithic production* (pp. 1–9). Cambridge University Press.

Flegenheimer, N., & Bayón, C. (2000). New evidence for early occupations in the Argentine Pampas, Los Helechos site. *Current Research in the Pleistocene, 17*, 24–26.

Flegenheimer, N., & Leipus, M. (2007). *Trabajar en un espacio reducido, Cerro El Sombrero, Abrigo 1. Proceedings of the XVI Congreso Nacional de Arqueología Argentina, Jujuy 2007.* Tras las huellas de la materialidad. Tomo 1: 441–444. Jujuy, Facultad de Humanidades y Ciencias Sociales, Universidad Nacional de Jujuy.

Flegenheimer, N., Kain, S., Zárate, M., & Barna, A. (1996). Aprovisionamiento de cuarcitas en Tandilia, las canteras del Arroyo Diamante. *Arqueología, 6*, 117–141.

Flegenheimer, N., Bayón, C., Valente, M., Baeza, J., & Femenías, J. (2003). Long distance tool stone transport in the Argentine Pampas. *Quaternary International, 109–110*, 49–64. https://doi.org/10.1016/S1040-6182(02)00202-1

Flegenheimer, N., Bayón, C., Scabuzzo, C., Mazzia, N., Vecchi, R., Weitzel, C., Frontini, R., & Colombo, M. (2010). Early-Holocene human skeletal remains from the Argentinean pampas. *Current Research in the Pleistocene, 27*, 10–12.

Flegenheimer, N., Mazzia, N., & Weitzel, C. (2015). Landscape and rocks in the East-Central portion of the Tandilia range (Buenos Aires Province, Argentina). *Paleoamerica: A Journal of Early Human Migration and Dispersal, 1*(2), 163–180. https://doi.org/10.1179/2055556315Z.00000000017

Franco, N. V. (2014). Lithic artifacts and the information about human utilization of large areas. In P. Escola & S. Hocsman (Eds.), *Lithic artifacts, mobility and site functionality problems and perspectives* (BAR International Series S2628) (pp. 116–127). British Archaeological Reports.

Gould, R. A., & Saggers, S. (1985). Lithic procurement in Central Australia: A closer look at Binford's idea of embeddedness in archaeology. *American Antiquity, 50*(1), 117–136. https://doi.org/10.2307/280637

Gutiérrez, M. A., & Martínez, G. (2008). Trends in the faunal human exploitation during the Late Pleistocene and Early Holocene in the Pampean region (Argentina). *Quaternary International, 191*, 53–68.

Gutiérrez, M. A., Martínez, G. A., Luchsinger, H., Grill, S., Zucol, A. F., Hassan, G., Barros, M. P., Kaufmann, C. A., & Álvarez, M. C. (2011). Paleoenvironments in the Paso Otero locality during Late Pleistocene-Holocene (Pampean region, Argentina): An interdisciplinary approach. *Quaternary International, 245*, 37–47.

Inizan, M. L., Reduron-Ballinger, M., Roche, H., & Tixier, J. (1995). Technologie de la pierretaillée. *Préhistoire de la Pierre Taillée*, 4. Cercle de Recherches et d'Etudes Préhistoriques, France.

Kaufmann, C. A., Álvarez, M. C., Messineo, P. G., Barros, M. P., Bonomo, M., & Heider, G. (2021). Guanaco hunting strategies in the southeastern Pampas during the late Holocene. In J. B. Belardi, D. Bozzuto, P. M. Fernández, E. A. Moreno, & G. A. Neme (Eds.), *Ancient hunting strategies in southern South America* (The Latin American Studies Book Series) (pp. 83–112). Springer. https://doi.org/10.1007/978-3-030-61187-3_4

Kelly, R. (1995). *The foraging spectrum: Diversity in Hunter-Gatherer lifeways*. Smithsonian Institution Press.

Leipus, M., & Landini, C. (2014). Materias primas y tecnología: un estudio comparativo del material lítico. In G. G. Politis, M. A. Gutiérrez, & C. Scabuzzo (Eds.), *Estado Actual de las Investigaciones en el sitio arqueológico Arroyo Seco 2* (Serie Monográfica 5) (pp. 179–227). INCUAPA-CONICET UNICEN.

Loponte, D., Tchilinguirian, P., & Sacur, S. R. (2010). Caracterización de afloramientos de calizas silicificadas de la provincia de Entre Ríos (Argentina) y su vinculación con los circuitos de abastecimiento prehispánico. In M. Feuillet, M. Terzaghi, B. Colasurdo, J. Sartori, &

S. Escudero (Eds.), *Avances y Perspectivas en la Arqueología del Nordeste* (pp. 125–139). ST Servicios Gráficos.

Lucero, G. F., Castro, S. C., & Cortegoso, V. (2021). GIS modeling of lithic procurement in highlands: Archaeological and actualistic approach in the Andes. *Journal of Archaeological Science: Reports, 38*, 1–11.

Martínez, G. (2006). Arqueología del curso medio del río Quequén Grande: estado actual y aportes a la arqueología de la región pampeana. *Relaciones de la Sociedad Argentina de Antropología, XXXI*, 249–275.

Martínez, G., & Gutiérrez, M. A. (2004). Tendencias en la explotación humana de la fauna durante el Pleistoceno final y Holoceno en la Región Pampeana (Argentina). In G. G. Mengoni (Ed.), *Zooarchaeology of South America* (pp. 81–98). BAR International Series 1298.

Martínez, G., & Gutiérrez, M. A. (2011). Paso Otero 5: A summary of the interdisciplinary lines of evidence for reconstructing early human occupation and paleoenvironment in the Pampean region, Argentina. In D. Vialou (Ed.), *Peuplements et Préhistoire de l'Amérique* (pp. 271–286). Muséum National d'Histoire Naturelle, Département de Préhistoire.

Martínez, G. A., & Mazzanti, D. L. (2017). Evidencia geoarqueológica de la transición Pleistoceno-Holoceno en reparos rocosos de Tandilia oriental (provincia de Buenos Aires). *Relaciones de la Sociedad Argentina de Antropología, XLII*(1), 83–106.

Martínez, G. A., Mazzanti, D. L., Quintana, C. A., Zucol, A. F., Colobig, M. M., Hassan, G., Brea, M., & Passeggi, E. (2013). Geoarchaeological and paleoenvironmental context of the human settlement in the Eastern Tandilia Range, Argentina. *Quaternary International, 299*, 23–37.

Martínez, G. A., Mazzanti, D. L., & Quintana, C. A. (2015). Asentamientos del Holoceno medio en Tandilia oriental. Aportes para el conocimiento de la dinámica poblacional de la región pampeana, Argentina. *Relaciones de la Sociedad Argentina de Antropología, XL*(1), 209–231.

Mazzanti, D. L. (1997). Excavaciones arqueológicas en el sitio Cueva Tixi, Buenos Aires, Argentina. *Latin American Antiquity, 8*, 55–62.

Mazzanti, D. L. (1999). Ocupaciones humanas tempranas en Sierra La Vigilancia y Laguna La Brava, Tandilia Oriental, provincia de Buenos Aires. In D. Marin (Ed.), *Actas del XII Congreso Nacional de Arqueología Argentina, Tomo III* (pp. 149–155). Universidad Nacional de La Plata.

Mazzanti, D. L. (2003). Human settlements in caves and rockshelters during the Pleistocene Holocene transition in the eastern Tandilia range, Pampean Region, Argentina. In M. Salemme, N. Flegenheimer, & L. Miotti (Eds.), *Where the South Winds Blow* (pp. 57–61). Texas A&M University Press.

Mazzanti, D. L., & Bonnat, G. F. (2013). Paisajes arqueológicos y cazadores-recolectores de la transición Pleistoceno-Holoceno. Análisis de las cuencas de ocupación en Tandilia oriental, provincia de Buenos Aires, Argentina. *Relaciones de la Sociedad Argentina de Antropología, XXXVIII*(2), 521–541.

Mazzanti, D. L., & Bonnat, G. F. (2021). Alero Lolina I. Funcionalidad y cronología de un reparo pequeño durante el Pleistoceno final y el Holoceno tardío, Tandilia oriental (Argentina). *Relaciones de la Sociedad Argentina de Antropología, XXXII*(1), 363–372.

Mazzanti, D. L., & Quintana, C. A. (2001). *Cueva Tixi: Cazadores y Recolectores de las Sierras de Tandilia. I. Geología, Paleontología y Zooarqueología*. Publicación Especial 1. LARBO UNMDP.

Mazzanti, D. L., Bonnat, G. F., Donadei, C. J. P., & Martínez, G. A. (2019). Cueva Alí Mustafá: un nuevo sitio arqueológico con ocupación temprana en las Sierras Orientales de Tandilia (Argentina). *Arqueología, 25*(2), 103–116.

Mazzia, N. (2011). El Aljarafe: Un espacio serrano ocupado efímeramente en diferentes momentos del Holoceno (Tandilia, Provincia de Buenos Aires). *Revista del Museo de Antropología, 4*, 33–46.

Mazzia, N. (2013). Cueva Zoro: nuevas evidencias sobre pobladores tempranos en el sector centro oriental de Tandilla. *Intersecciones en Antropología, 14*, 93–106.

Meltzer, D. J. (1989). Was stone exchanged among Eastern North American Paleoindians? In C. Ellis & J. Lothrop (Eds.), *Eastern Paleoindian lithic resource use* (pp. 11–39). Westview Press.

Messineo, P. G. (2015). Technological organization in hunting/scavenging and butchering sites of megamammals in the Pampa grassland (Argentina). *PaleoAmerica: A Journal of Early Human Migration and Dispersal, 1*, 343–359.

Messineo, P. G., & Barros, M. P. (2015). Lithic raw materials and modes of exploitation in quarries and workshops from the center of the Pampean grassland of Argentina. *Lithic Technology, 40*(1), 3–20. https://doi.org/10.1179/2051618514Y.0000000007

Messineo, P. G., & Pal, N. (2020). Cazadores-recolectores tempranos en los pastizales pampeanos de Argentina: una síntesis del poblamiento. *Boletín Americanista, LXX*(2), 9–38. ISSN: 0520-4100. https://doi.org/10.1344/BA2020.81.1001

Nami, H. (2017). Silcrete as a valuable resource for stone tool manufacture and its use by Paleo-American Hunter–Gatherers in Southeastern South America. *Journal of Archaeological Science: Reports, 15*, 539–560.

Nelson, M. (1991). The study of technological organization. *Archaeological Method and Theory, 3*, 57–100.

Pal, N., Barros, M. P., Gutiérrez, M. A., & Martínez, G. (2019). Modos de uso de la tecnología lítica durante el Holoceno temprano y medio: el caso del sitio Paso Otero 4 (Región Pampeana, Argentina). *Chungara Revista de Antropología Chilena, 51*(3), 325–337.

Politis, G. G., & Gutiérrez, M. A. (1998). Gliptodontes y cazadores-recolectores de la Región Pampeana (Argentina). *Latin American Antiquity, 9*(2), 111–134.

Politis, G. G., Messineo, P. G., & Kaufmann, C. A. (2004). El poblamiento temprano de las llanuras pampeanas de Argentina y Uruguay. *Complutum, 15*, 207–224.

Politis, G. G., Gutiérrez, M. A., & Scabuzzo, C. (2014). Introducción. In G. G. Politis, M. A. Gutiérrez, & C. Scabuzzo (Eds.), *Estado Actual de las Investigaciones en el sitio arqueológico Arroyo Seco 2* (Serie Monográfica 5) (pp. 11–28). INCUAPA-CONICET UNICEN.

Politis, G. G., Gutiérrez, M. A., Rafuse, D. J., & Blasi, A. (2016). The arrival of *Homo sapiens* into the Southern Cone at 14.000 years ago. *PLoS One, 28*, 1–28. https://doi.org/10.1371/journal.pone.0162870

Politis, G. G., Messineo, P. G., Stafford, T., Jr., & Lindsey, E. (2019). Campo Laborde: A giant ground sloth kill and butchering site in the Argentine Pampas. *Science Advances, 5*(3), 1–9.

Ponce, F., Rabassa, J., Coronato, A., & Borromei, A. (2011). Palaeogeographical evolution of the Atlantic coast of Pampa and Patagonia from the last glacial maximum to the Middle Holocene. *Biological Journal of the Linnean Society, 103*(2), 363–379.

Rafuse, D. J., & Massigoge, A. (2020). Abrigo La Tinta: Preliminary results of a New Late Pleistocene-Early Holocene rockshelter from the Pampas Region (Tandilia Range), Argentina. *PaleoAmerica: A Journal of Early Human Migration and Dispersal, 6*(3), 299–301. https://doi.org/10.1080/20555563.2019.1684047

Tonni, E. P., Cione, A. L., & Figini, A. J. (1999). Predominance of arid climates indicated by mammals in the pampas of Argentina during the Late Pleistocene and Holocene. *Palaeogeography, Palaeoloclimatology, Palaeoloecology, 147*, 257–281. https://doi.org/10.1016/S0031-0182(98)00140-0

Torrence, R. (1989). Tools as optimal solutions. In R. Torrence (Ed.), *Time, energy and stone tools* (pp. 1–6). Cambridge University Press.

Valverde, F. (2006). Estrategias de conservación y economía de materias primas líticas en contextos tempranos de las Sierras de Tandilia oriental. *Cazadores-Recolectores del Cono Sur. Revista de Antropologia, 1*, 179–187.

Vecchi, R. (2011). *Bolas de boleadora en los grupos cazadores-recolectores de la pampa bonaerense* [Unpublished doctoral Thesis]. Facultad de Filosofía y Letras, Universidad de Buenos Aires.

Weitzel, C. (2010). *El estudio de los artefactos formatizados fracturados. Contribución a la comprensión del registro arqueológico y las actividades humanas* [Unpublished doctoral Thesis]. Facultad de Filosofía y Letras, Universidad de Buenos Aires.

Zárate, M., Kemp, R., Espinosa, M., & Ferrero, L. (2000). Pedosedimentary and palaeoenvironmental significance of a Holocene alluvial sequence in the southern Pampas, Argentina. *The Holocene, 10*(4), 481–488.

Lithic Technology in the Semi-arid Center of Argentina: Use of Internodal Spaces Among Hunter-Gatherers in the Late Holocene

Guillermo Heider and María Laura Salgán

Abstract In arid and semi-arid environments of central Argentina, archaeological research agendas are implemented to understand the ways in which human groups explored and inhabited these spaces. In this chapter, we seek to address the use of space and mobility of hunter-gatherer populations that occupied a sector of the arid and semi-arid center of Argentina, known as the *Travesía Puntana* (Puntana Traverse). This name is probably motivated by the difficulties faced by the Hispanic-Creole populations to generate routes and enclaves in its interior, mainly due to the scarcity of surface water. However, such a European notion is contested by the archaeological evidence. The sites are located in dunes with little or no vegetation, which functioned as ecorefuges thanks to the presence of high-quality water in the nearby groundwater. It was recently proposed that the *Travesía Puntana* was part of an internodal space where human activities were infrequent, ephemeral, or non-existent in relation to neighboring regions. To investigate the role of this internode we carried out the technological study of 66 sets of surface lithic artifacts and the geochemical analysis of obsidian artifacts. These two lines of research are adequate to answer questions about social interaction and circulation of information or objects. The results show the input of rocks and tools with different vectors and distances from non-local sources. We believe that in these landscapes, the dunes favored dynamics of "population convergence," as places of potential shared use.

Keywords *Travesía puntana* · Technological management · Geochemistry · Ecorefuge · Population convergence

G. Heider (✉)
CONICET-CCT San Luis. Departamento de Geología, Universidad Nacional de San Luis, San Luis, Argentina

M. L. Salgán
Instituto de Evolución, Ecología Histórica y Ambiente – CONICET y Universidad Tecnológica Nacional San Rafael, Mendoza, Argentina
e-mail: lsalgan@mendoza-conicet.gob.ar

G. F. Bonnat et al. (eds.), *Current Research in Archaeology of South American Pampas*, The Latin American Studies Book Series,
https://doi.org/10.1007/978-3-031-55194-9_7

Introduction

Several landscapes in central and western Argentina are characterized by an arid and semi-arid climate, with a scarce supply of surface water, deep and/or salty groundwater, episodic rainfall limited to summer periods, and a marked annual water deficit. In the sixteenth century, these spaces were considered geographical barriers capable of isolating large regions (Pastor & Boixados, 2016). Moreover, they were called *Travesía*, as a synonym for a passageway or alleyway exposed to risks linked to climatic conditions or unfamiliarity with the terrain. This denomination was imposed by the Spaniards and Creoles who tried to cross them to join the mountain oases of the west (Mendoza and San Juan) with the highland settlements of central Argentina (San Luis and Córdoba). This chapter concentrates on one of these spaces, called Travesía Puntana (hereafter TP). Although this space has been relegated in archaeological research agendas, in favor of those where prehistoric climatic conditions have been judged to be more conducive to human occupation (Greslebin, 1924, 1928; Heider & Curtoni, 2016), the archaeological record from the TP demonstrates human presence since at least the middle Holocene (Heider & Rivero, 2018; among others).

Our project assumes that the ecological contexts in which social groups make their decisions must be taken into account for the construction of models of occupation and use of space. Thus, plausible social strategies of use by local groups would be restricted to a limited stock of responses linked to the characteristics of the environment (Butzer, 1989; Dearing, 2006; Dincauze, 2000; Jones et al., 1999; Leroy, 2006; Morales et al., 2009, among many others). In this sense, the TP has been characterized as an Internodal Space due to its biogeographic and archaeological characteristics (Heider et al., 2019), internodal spaces being understood as intermediate spaces with respect to relatively permanent settlements or regions with high densities of stable population (Nielsen, 2006: 34). Following Nielsen (2017), we understand internodes as areas where human activities and interactions are less diverse, infrequent, ephemeral, or nonexistent in relation to relatively permanent settlements. However, it is pertinent to mention that nodes and internodes are reflections of the spatiality of social practices, not of essential characteristics of geography. They are always socially constituted, even if their configuration necessarily relates to local biogeography, especially to the distribution of resources such as water and biomass (Nielsen, 2017; Yacobaccio, 1994). In this framework, the aim of this chapter is to analyze the role that the TP, understood as an internode, could have played in the dynamics of regional circulation and interaction (Fig. 1). To this end, and within the framework of local biogeography, we present the results of lithic resource management from 66 surface sites. Second, we wonder about the motives that led human groups to occupy this hostile place. In this sense, we approach the question from the proposal of Nielsen (2006, 2017), who mentions two types of occupation in the internodes: transit and extractive. Complementarily, we integrate the results of geochemical studies of obsidian samples recovered from the TP sites and their surrounding areas (nodal in our scale) into the TP record. The excellent

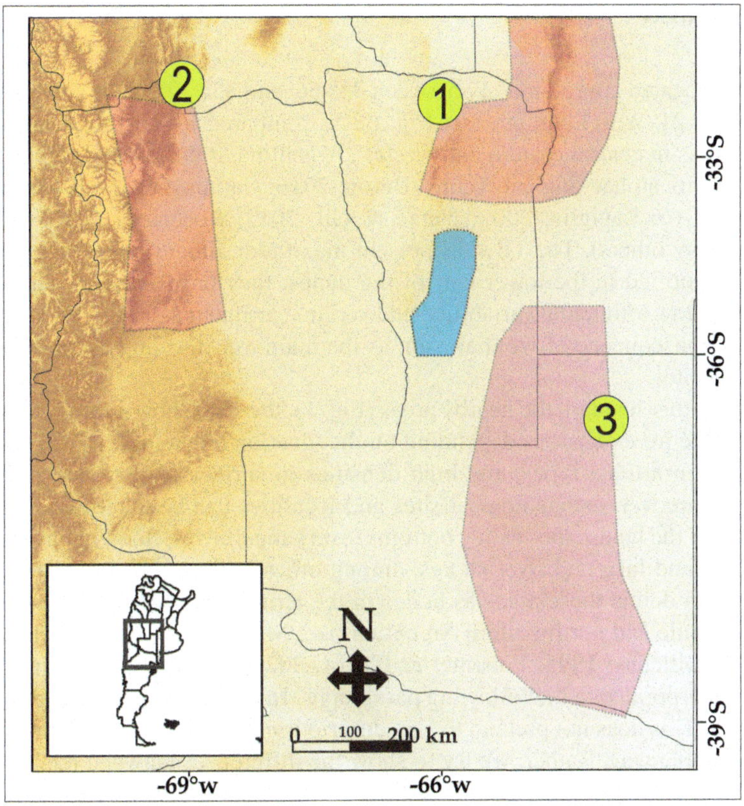

Fig. 1 The area in blue is the internal node of the Travesía Puntana. The areas in violet are nodes in our scale of analysis: (1) Central Mountains; (2) Andean Piedemonte of Mendoza; and (3) Northern Pampa Seca and Southern Pampa Seca-Transversal Valley

traceability of this raw material allows the possibility of discussing circulation vectors. Finally, we discuss the role of the TP in the circulation of human groups that occupied the arid to semi-arid environments of central Argentina.

At this point, it is pertinent to make a clarification. As mentioned, the chapter analyzes the TP record from the perspective of Internodal Archaeology, whose origin and development are strongly concentrated in agro-pastoral societies of the Central-Southern Andes. In contrast, the human groups that inhabited central Argentina were immersed in processes of cultural incorporation, sedentarization, and increased social complexity (in very general terms). Even more so, if this process is analyzed from the long trajectory of the occupation of space. For this reason, and as mentioned, we use concepts from biogeography as a complement to address our questions. This view lacks originality, although it has few antecedents, the closest in time and space being the one proposed by Barberena and collaborators (2017).

Background

The TP is located approximately between 33° 66′ and 35° 20′ South latitude and 65° 44′ and 66° 21′ West longitude (see Fig. 1). As compared with neighboring archaeological areas in central eastern and western Argentina, it presents particular characteristics in its archaeological record (Berón, 2016; Cattáneo et al., 2013; Heider & Curtoni, 2016; Lagiglia, 2002; Neme & Gil, 2012; Rocchietti & Ribero, 2015; among many others). The TP sites present maximum dimensions of 50 m^2 and are usually identified in the lower part of the dunes; they have very low stratigraphic resolution, low artifactual variability, and variable artifactual density (Heider, 2015). This is a consequence of wind erosion as the main transforming agent of the landscape and sites.

On the other hand, in the nearby areas (Fig. 1), the sites show stratigraphic potential, variable m^2 extensions depending on the location (some over 10,000 m^2), high variability in artifact types, and high densities in surface and/or stratigraphy (see Fig. 1). Moreover, archaeological sites and localities can be identified in different elements of the landscape (valley bottoms, overhangs, caves, flood plains, structural highlands, and lake and river shores, among others). These characteristics are not sufficient to define these areas as nodes in the terms proposed by colleagues from northern Chile and northwestern Argentina (i.e., Martel et al., 2017; Nielsen, 2017; Núñez & Dillehay, 1995; Pimentel & Barros, 2020). For this reason, information will be incorporated in the following paragraphs. In all cases concentrated in the late Holocene. This does not pretend to be a description that allows us to define nodes in an indubitable way. Rather, we try to show the differences between the TP and the eastern and western spaces. All this will be taken up in the discussion, trying to validate the application of the archaeology of internodal spaces in other social configurations.

A review for the late Holocene of the Central Mountains (see Fig. 1) shows evidence of maize cultivation, complemented by beans, squash, potatoes, and quinoa during the late Holocene (Lopez et al., 2020). The information is consistent with plot management framed in the "weed-crop-domesticated" complex. The development of a diversified economy, consequently, was accompanied by high residential mobility, co-residential group fission–fusion mechanisms, and the abandonment of cultivated fields for crop cultivation in search of food. Evidence allows us to propose that the social and political systems of the Sierras de Córdoba and San Luis were based on flexible patterns of subsistence and mobility (i.e., Lalinde et al., 2018; Medina et al., 2016, 2019).

The archaeological record of central western Argentina constitutes an empirical reference to understand the characteristics and causes of the agricultural dispersion process in South America (Iniesta et al., 2020). Research conducted so far presents this region as one of the southern limits in the record of pre-Hispanic cultigens in the eastern Andean slope of America (i.e., Lagiglia, 1982; Gil, 1997–1998; Gil et al., 2018; Iniesta et al., 2020). Specifically, the Atuel River valley has been proposed as the southern limit to where the first agricultural settlements reached

(Lagiglia, 1982). Without delving into an intense discussion, it can be noted that various colleagues propose that the first farmers settled in northern Mendoza in ca. 2000 years BP (Chiavazza, 2013; Iniesta et al., 2020). In the south, on the contrary, a forager economy continued to predominate (Gil et al., 2014, 2018). A non-strict boundary with buffer area conditions was proposed for the Atuel River (Gil & Neme, 2010). Agricultural processes are related to significant regional changes in terms of population growth, colonization of new areas, intensification, and environmental changes. In light of new research, researchers believe that the landscape and occupational dynamics would have been complex and varied, with connection networks with the plains and plains of the eastern slope and links with the NOA (Argentine North-west) and the Bolivian altiplano (Barberena et al., 2020; Iniesta et al., 2020).

The Pampas Region of Argentina (see Fig. 1) has heterogeneous environmental conditions. This complex environmental panorama, together with its uses and social transformations, gave rise to a system of interaction, exchange, social complementarity, and population dynamics (sensu Berón, 2006). Several lines of evidence (i.e., bioarchaeological, archaeobotanical, ceramic analysis, isotopic, lithic resource provenance, and documentary sources) reflect a notorious process of interaction and regional integration. The strongest evidence of that process is concentrated in the late Holocene (i.e., Berón, 2013, 2016; Heider, 2016a, b; Tapia, 2008). Recently Berón et al. (2017) rethought local social dynamics with a view of internodal archaeology. Both proposals explore the value of this theoretical approach for societies that were hunter-gatherers for most of their historical trajectory. This chapter is based on these proposals for the region and takes them up in the discussions.

As argued by Nielsen (2006), the study of the archaeological record of Internodal Spaces offers independent and complementary information to that generated by the evidence of the more permanent or stable nodes. Thus, following Nielsen (2017), we understand internodal spaces as areas where human activities and interactions are less diverse, infrequent, ephemeral, or non-existent in relation to relatively permanent settlements (on an intra-regional scale). Internodal spaces also include regions characterized by high densities of stable populations (on an interregional scale). Barberena and collaborators (2017) point out that biogeography is a theoretical instance capable of articulating diverse questions and lines of evidence. To this end, these authors highlight the need to operationalize the concepts of nodes and internodes to confront them with archaeological records/their archeological records and thus make it possible to evaluate the relationship between the ecological configuration of the landscape and the spatial distribution of population nodes and internodes. Our research project uses biogeographic models (Borrero 1994–1995) to interpret the archaeological record from the moments of initial exploration of southern San Luis and Córdoba (Heider, 2015; Heider & Rivero, 2018; among others). In this framework, the proposal of Sturm and collaborators (2016), who define marginal environments as continental-terrestrial landscapes with strong seasonality and marked variations in temperature, precipitation, and biomass between years, is appropriate. This perspective is thus taken to describe the biogeographic characteristics of the space studied in this work.

From the climatic point of view, the TP is included in a continental temperate zone, with semi-arid to arid conditions. Annual precipitation averages around 450 mm in the east and 350 mm in the west (Demaría, 2008). Rainfall occurs mainly during the summer season and is associated with the Atlantic domain and moisture inputs from the Amazon region (*Chaco Jet*) (Garreaud et al., 2009). The average annual maximum temperatures are 23/24 °C during January, with extreme temperatures of around 44 °C. The average annual minimum temperatures are $^7/_8$ °C during the month of July, with historical records of around −10 °C. The frost regime is variable, and the number of frosts is decreasing. In the west of the TP, up to 125 frost days can be reached between autumn, winter, and spring (Servicio Meteorológico Nacional, 2021).

The TP is located within the phytogeographic Pampean province, which constitutes the most extensive grassland ecosystem in South America, covering 700,000 km^2 (Cabrera, 1976; Oyarzabal et al., 2018; Soriano, 1992). It shows a strong dominance of a few species of C4 summer-cycle perennial grasses (Demaria, 2008), including cow grass (*Sorghastrum pellitum*), bitter straw (*Elyonurus muticus*), and olivillo (*Hyalis argentea*). Additionally, there are small islands of chañar (*Geoffroea decorticans*). The most outstanding aspect of zoogeography is its marked transformation process, which is a consequence of local extinctions and/or replacement of several species in their ecological niches (Demaría, 2008). Exotic fauna (wild boar – *Sus scrofa*, *Cervus elaphus*, and hare – *Lepus europaeus* – among others) have successfully occupied the niches. Moreover, it is not currently possible to observe specimens of guanaco (*Lama guanicoe*), even though their skeletal remains are present in the archaeological record (Heider, 2015). The ñandu and the pampas deer still survive and have begun a slow recovery process in marginal sectors, limited to western grassland systems (Álvarez & Heider, 2019; Dellafiore et al., 2001) (Fig. 2).

From the geomorphological point of view, the PT is included in the Pampean Sand Sea. This is a sandy deposit generated by eolian transport (Iriondo, 1990, 1999; Iriondo & Kröhling, 1995). The broad and diverse cover of the Pampean Sand Sea was systematized by Zárate and Tripaldi (2012), who defined eight aeolian units based on the types of deposits, geoforms, and geological-structural context. The TP is specifically located in the unit called Western Pampean dunefields (Zárate & Tripaldi, 2012), where a variety of dunes can be identified. Large subcircular deflation units stand out, up to 2 km in diameter and 20 m of slope between the peripheral depositional zones and the base of the inner basin (Tripaldi & Forman, 2007). Stratigraphic sections indicate eolian deposition between ca. 12,000 and 1000 *cal.* years BP. In this same area, lacustrine records show incipient water bodies in the interdunes ca. 1500 to 900 *cal.* years BP, which evolved into shallow lagoons with wet and dry phases (Vilanova et al., 2015, 2017, 2022). A pervasive eolian sand sheet appears to have accumulated during most of the Holocene, possibly reflecting a weakened influence of the SAMS, probably under a semi-arid climate, at least 200 mm drier than today (Forman et al., 2014; Villanova et al., 2022). Currently, water tables reach the surface in many parts of the landscape. This phenomenon began to generate strongly from the changes in the rainfall regime previously

Fig. 2 Typical landscape of the Puntana Traverse

mentioned. The shallow lake system is composed of more than 200 permanent or temporary lakes and ponds of variable size (Tripaldi et al., 2013). The surface area and depth of these water bodies vary between 20 and 100 ha and between 2 and 12 m, respectively. They are endorheic basins with shallow lakes being fed by atmospheric precipitation or groundwater. In summary, in addition to the aforementioned water scarcity, it is possible to establish that the TP resources are heterogeneously distributed in both time and space. Undoubtedly, resource abundance is affected by topography, evapotranspiration, and effective temperature, among other factors. The dunes were a hierarchical space in the general marginal context. The biogeographic characteristics described above provide a valuable element for understanding the internode role that we propose for the TP.

Nielsen (2006) argues that the study of the archaeological record of internodal spaces offers independent and complementary information to that generated by the evidence of the more permanent or stable nodes. Following this line, we believe that the study of internodes on an archaeological scale is a suitable agenda for integrating complementary theoretical domains, which seek to understand long-term human historical trajectories in that particular biogeographic and archaeological framework (Barberena et al., 2017).

In the geomorphological context of the TP, it is complex to infer chronologies and temporalities of occupation. Site formation processes, basically wind erosion, generate palimpsests of low chronological resolution. The earliest occupations are typical of the middle Holocene (Heider & Rivero, 2018). The evidence on this period is very scarce. They are limited exclusively to projectile points identified in archaeological sites and collections. Similar pieces have chronologies of ca. 8000 to 6000 years BP typical of that period in sites located in the Central Mountains located to the north (Heider & Rivero, 2018; Heider et al., 2022b). Moreover, a spatial analysis shows a distribution of points limited only to nearby areas of the Quinto River basin (less than 50 km), at the very limits of the TP (Heider, 2015). In this sense, an initial stage of exploration/colonization of the area linked to the use of the basin as a low-cost route was proposed (Heider & Ribero, 2018). However, the focus of this discussion is on an instance of settlement comparable to "effective occupation" (sensu Borrero, 1994–1995), typical of the late Holocene. Several previous articles discuss, in which sites were equipped for periodic return and domestic plant products from neighboring areas were consumed (Heider, 2016a, b; Heider & López, 2016, 2018; López et al., 2020).

From the studies of lithic technology organizations, the construction of a Regional Lithic Resource Base (sensu Ericson, 1984) is an unavoidable research step. Previous studies have shown that the TP lacks rocks suitable for knapping (Heider, 2015, 2016a). However, on a broader scale, it is possible to identify an environment that offers a rich structure of lithic resources (Fig. 3). The selection of rocks for petrographic study was limited to those that could eventually yield diagnostic results and was limited to those that could eventually yield diagnostic results. In this sense, it was decided not to carry out studies on quartz (of important presence in the central and northern center and north of the area), because this type of analysis does not provide significant results to differentiate varieties and/or origin. On the contrary, geological cuts were made in quartzite, chalcedonies, opal, and rhyolites (Heider, 2015, 2016a, b).

To the southeast of the TP, the Loma de los Pedernales quarry area is identified, around parallel 35°S and meridian 65°28′W (3.1), where a chalcedony of very good quality for knapping has been previously characterized (Heider & Demichelis, 2015). Its tabular or kidney-shaped forms are a consequence of its genesis as a liquid with a high siliceous charge, which precipitated filling cavities in a carbonate of the Miocene age. In this work, this chalcedony will be referred to as "Chalcedony A" (Fig. 3.1). Subsequent studies allowed the identification of the Alta de El Lechuzo chalcedony source, with macroscopic and lying characteristics similar to the former, although with larger relative sizes. It was defined as a secondary source of very good to excellent quality for knapping, distributed in a morphology located at 33°34′S and 66°47′W (Heider et al., 2017). This will be referred to as "Chalcedony B" (Fig. 3.2).

The Cerro Varela and Cerrillo de las Salinas rhyolite sources (Fig. 3.3) are located at a longitude of 62°32′W and latitudes of 34°01′S and 34°28′S, respectively. Both were characterized as primary sources (seams and outcrops) and secondary sources (dispersed rocks in dry riverbeds). Studies, still in progress, show good to very good quality raw material for knapping and a color range from brown to violet (Heider, 2015; Heider et al., 2017; Martínez et al., 2014).

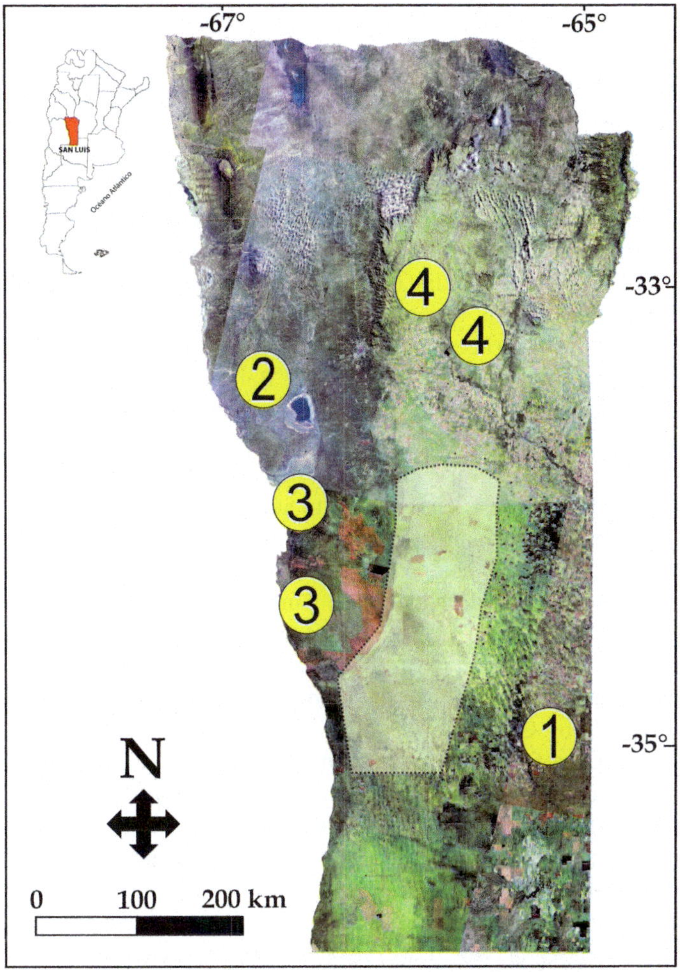

Fig. 3 Raw material sources near the Puntana Traverse. (1) Loma de los Pedernales; (2) Alto del Lechuzo; (3) Cerro Varela and Cerrillo de las Salinas; and (4) La Falla and El Chiquero. The discolored area is the TP

To the north, there is a wide variability of siliceous rocks (Fig. 3.4) in a quarry area called La Falla, where Borgo (2020) identified more than 40 quarry-workshop sites of a chalcedony of very good quality, generated in a pegmatite of Ordovician age. The assemblage includes crystalline and hyaline quartz, opals, and chalcedony (Borgo et al., 2019a, b; Heider et al., 2015; Heider et al., 2020).

Methodology

The survey and collection of the archaeological record involved a field sampling design based on postulates from Distributional Archaeology (sensu Ebert, 1992). Satellite images of geoforms were surveyed, in particular compound parabolic dunes and isolated dunes (Heider, 2015). Of the total identified ($n = 102$), 50% were prospected. Plots of cultivated areas or natural grasslands (10,000 m^2), both near and distant to the dunes, were delimited, covering the entire area. Due to the absence of previous field information, a first classical or opportunistic approach (Aldenderfer, 1998) was carried out. This was oriented to the recognition and evaluation of the characteristics of the superficial archaeological record. The second stage was based on stratified probability sampling (Renfrew & Bahn, 1998). In this sense, transects defined in the sites and a review of random units generated through the application of GIS were carried out. The surface distribution of artifacts was recorded from transects in which all the observed material was collected.

Lithic raw materials are a key element to study the settlement, subsistence, and mobility patterns of hunter-gatherer groups in different parts of the planet. This fact is a direct consequence of their excellent preservation, informative potential, and significance for people (Andrefsky, 2008; Binford, 1979, 1980; Borie et al., 2017; Gould & Saggers, 1985; Kelly, 2012; Torrence, 1984; among many others). The TP constitutes a very good example of this because lithic resources constitute 99% of the recovered material. To evaluate the modes of supply and rock transfer, the spatial limit proposed by Gamble (1993) for local resources was adopted. The author proposes a distance of 80 km as the limiting range for local resources in hunter-gatherer groups in arid areas and/or with low population density. This proposal, expressed in terms of linear distance, is articulated with that put forward by Bayón and Flegenheimer (2004), who consider raw materials located beyond 100 km from the supply areas as long distance.

The functionality of the sites was from the techno-typological study of the flaked artifacts, on the basis of which we also analyzed the technological trajectories. Specifically, the typology of Aschero (1975, 1983) and its reformulation, carried out at the beginning of the present century (Aschero & Hocsman, 2004), were applied. The results of the typological study were analyzed in light of the Organization of Technology postulates (Nelson, 1991). Given the characteristics of the TP (scarcity of previous research, extension of the area, and type of raw materials present, among others), we selected the typological variables that we considered of greater diagnostic value (i.e., raw material, cortex reserve, type of flakes, size, technical series, type of artifact, and type of core).

For the analysis of obsidian provenance, the X-ray fluorescence (XRF) geochemical method was used, which is a surface analysis technique, non-destructive and quick to apply (180 s per sample). The instrument used was the portable spectrometer Bruker Corporation (Tracer III-V, serial number K0557), which allows quantifying the presence of 13 elements (Rubidium (Rb), Strontium (Sr), Yttrium (Y), Zirconium (Zr), Niobium (Nb), Potassium (K), Titanium (Ti), Manganese

(Mn), Iron (Fe), Zinc (Zn), Gallium (Ga), Lead (Pb), and Thorium (Th)). The equipment was calibrated based on information from previously characterized obsidian samples from the reference collection of the Missouri University Research Reactor. Element concentrations were calculated in parts per million, and their results were directly compared with those of obsidian sources known for the region (i.e., Cortegoso et al., 2012; Giesso et al., 2011). Provenance determination was performed through tables of concentration measurements and bivariate element scatter plots (Fig. 2). Artifacts were assigned to sources whose diagnostic elements (Rb–Sr, Rb–Zr, and Sr–Zr) fell within the confidence ellipses with 90% probability and within the limits of the chemical variability recorded for each source.

Technological Strategies in the Punta a Traverse

Results of the Techno-Typological Analysis

Fieldwork in the TP identified a total of 1748 artifacts, recovered from 66 archaeological sites. As mentioned, in all cases, the sites were located within dunes without vegetation or, to a lesser extent, partially vegetated. Almost no other type of material was recorded in the record. A reduced set of 10 small ceramic sherds was classified as typical of the Central Mountains. These, together with some remains of glass-made artifacts, allowed anchoring temporalities to the late Holocene in a relative way (Heider, 2015, 2016a). Laboratory studies allowed us to divide the total number of artifacts into 285 tools, 63 cores, 69 non-modified flakes with traces of use, 393 knapping debris (this category includes both debris and debitage, to reduce data presentation categories), 879 flakes, and 59 ecofacts (see Table 1). At this point, it is pertinent to clarify that we include within the term "ecofacts" all the rocks present.

Figure 4 shows the raw materials identified in the TP sites. Rocks of good and excellent quality for knapping are notoriously similar: Chalcedony A is the most represented ($n = 342$). Quartz ($n = 335$) and Chalcedony B ($n = 331$) have similar proportions. Silicas (opals) numbered 311. Finally, 301 pieces of rhyolite were identified. The category "others" ($n = 128$) includes grinding artifacts made on igneous and metamorphic rocks (granite, schist, and andesite, among others). Obsidian Flakes ($n = 4$) were also recorded and were the subject of geochemical studies due to the diagnostic value of this volcanic rock based on its traceability (Heider et al., 2022b).

The presence and percentage of cortex reserve are directly related to the management of the raw materials identified, as well as the regional availability and entry into the archaeological record. The low presence of cortex in the rocks suitable for knapping is notorious. Only 6.2% of the total recovered ($n = 109$) have 50% or more cortex (Table 2). Only Chalcedony B and rhyolite have pieces with cortex. In the latter, the presence of very large flakes ($n = 9$) without evidence of shaping was observed. The highest percentage of cortex is represented by instruments shaped by

Table 1 Materials recovered in the Puntana Traverse

Sites TP	Nf.	Af.	Co.	Fnu.	Kd.	Fl.	Ec.	Total
Alto Pelado 1	1	3	–	2	4	5	–	14
Alto Pelado 2	1	3	2	2	3	8	–	18
Alto Pelado 3	1	2	4	1	3	1	–	11
Alto Pelado 4	1	3	1	1	–	5	–	10
Alto Pelado 5	1	–	–	–	7	14	–	21
Alto Pelado 6	1	3	1	–	5	8	–	17
Alto Pelado 7	1	2	–	1	6	5	–	14
Capelen 1	2	2	–	–	2	4	–	8
Capelen 2	2	–	–	1	2	4	–	7
Capelen 3	2	4	–	2	2	2	1	11
El 30	3	2	–	–	–	3	–	5
El Durazno	4	3	–	1	–	4	–	8
La Alborada A1	4	3	1	1	7	17	9	38
La Alborada A2	4	2	–	–	15	16	3	36
La Alborada A3	4	8	–	–	6	7	1	22
La Alborada B1	4	–	–	–	12	15	–	27
La Alborada B2	4	7	–	–	8	12	–	27
La Alborada C1	4	1	–	–	14	17	–	32
La Alborada C2	4	1	1	–	12	15	4	33
La Alborada C3	4	–	–	–	14	17	–	31
La Felisa 1	5	6	–	–	2	5	1	14
Laiseca 1	6	6	–	–	4	4	1	15
Laiseca 2	6	1	–	2	–	1	–	4
Laiseca 3	6	2	–	–	–	5	–	7
Loma de los Indios 1	7	9	5	–	8	19	–	41
Loma de los Indios 2	7	6	–	–	1	13	–	20
Loma de los Indios 3	7	12	5	–	6	19	–	42
Loma de los Indios 4	7	36	5	14	86	210	–	351
Loma de los Indios 5	7	12	1	5	2	26	–	46
Loma de los Indios 6	7	8	2	–	9	18	–	37
Loma de los Indios 7	7	11	4	1	18	94	–	128
Loma de los Indios 8	7	5	–	–	2	6	–	13
Loma de los Indios 9	7	7	1	3	8	16	–	35
Loma de los Indios 10	7	13	4	2	8	34	–	61
Loma de los Indios 11	7	8	–	1	4	10	–	23
Loma de los Indios 12	7	6	–	1	7	20	–	34
Loma de los Indios 13	7	10	4	1	–	9	–	24
Loma de los Indios 14	7	4	3	–	1	6	–	14
Loma de los Indios 15	7	4	6	–	2	2	1	15
Loma de los Indios 16	7	8	1	2	6	10	1	28
Loma de los Indios 17	7	7	–	–	–	6	1	14
Las Viscacheras	7	1	–	–	4	1	–	6

(continued)

Table 1 (continued)

Sites TP	Nf.	Af.	Co.	Fnu.	Kd.	Fl.	Ec.	Total
El Colorado	7	2	–	–	1	1	2	6
Los Toldos	7	–	–	–	5	8	–	13
El Venado	7	7	–	–	1	20	1	29
Los Pocitos	4	3	–	–	2	7	1	13
La Maroma 1	8	5	–	3	2	1	6	17
La Maroma 2	8	2	–	2	4	1	8	17
Médano de Bagüal 1	9	5	1	5	10	18	–	39
Médano de Bagüal 2	9	–	2	4	14	42	–	62
Médano del Cementerio	9	1	2	2	2	3	–	10
Médano del Norte	3	1	–	–	1	3	2	7
Médano de la Primavera	3	4	–	1	–	4	–	9
Médano del Vecino	4	–	–	–	1	4	3	8
El Hugo	4	4	1	–	2	2	2	11
Laguna del Bagual 1	10	4	2	1	10	11	1	29
Laguna del Bagual 2	10	–	–	–	4	–	1	5
Médano de los Loros	11	1	2	–	–	1	1	5
Médano de Marcelo	9	1	1	–	6	3	–	11
Médano de Pacheco	12	1	–	–	16	9	–	26
Médano de El Porvenir	12	3	–	2	4	8	–	17
Laguna de Batavia	12	1	–	–	–	2	1	4
Pueblo Ranquel Norte	5	2	–	1	–	5	–	8
Pueblo Ranquel Sur	5	–	–	–	–	4	5	9
San José de Escala 1	5	6	–	2	7	4	–	19
San José de Escala 2	5	1	1	2	1	5	2	12
Total	–	285	63	69	393	879	59	1748

Glossary: *Nf.* Site number to be observed in its position in Fig. 5, *Af* Artifacts, *Co.* Cores, *Fnu.* Natural filters with traces of use, *Kd* knapping debris, *Fl.* Flakes, *Ec* Ecofacts

picking and abrasion ($n = 52$). Only cores ($n = 18$) or primary flakes ($n = 39$) of chalcedony (A and B) and rhyolite show high percentages of cortex.

The raw materials identified in the knapping debris present a relatively homogeneous distribution: Chalcedony B ($n = 205$), Chalcedony A ($n = 182$), silica ($n = 153$), quartz ($n = 120$), rhyolite ($n = 109$), and others ($n = 110$). This last group includes 85 flakes from rocks of poor quality for knapping (igneous and metamorphic). They may be the result of unplanned actions such as the breakage of a firing pin or the fracture of a grinding artifact. They may even be the result of the formatization or refinement of grinding artifacts. In the remaining group, there are unidentified rocks of very good to excellent quality for knapping and two obsidian flakes (Heider, 2015; Heider et al., 2022a). The external flakes, corresponding to the first stages of artifact production, are a group of 233 elements: 116 primary and 117 secondary flakes, almost entirely of rhyolite and chalcedony A and B. The internal flakes, representative of intermediate and final production stages, are the majority: edge ($n = 286$), angular (172), and flat ($n = 116$). Finally, bifacial thinning flakes

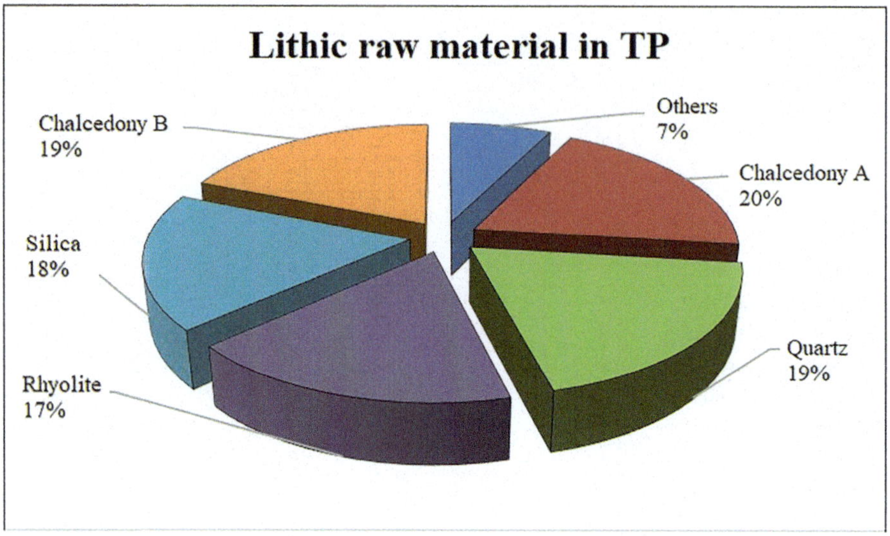

Fig. 4 Main raw materials registered in the TP

Table 2 Percentage of cortex by raw material

Raw material	0%	1–25%	26–50%	51–75%	76–100%	Total
Chalcedony A	143	117	67	15	0	342
Chalcedony B	135	127	56	13	0	331
Quartz	258	44	33	0	0	335
Rhyolite	172	63	37	20	9	301
Silica	257	45	9	0	0	311
Obsidian	3	1	0	0	0	4
Others	11	23	37	32	20	123
Total	979	420	239	80	29	1747

($n = 28$), reactivation flakes ($n = 5$), and bipolar flakes ($n = 42$) were also identified. Regardless of this point, its presence is related to the lack of local rocks (Heider, 2015).

The overall size of the recovered materials shows a predominance of the small category, with 51.5% ($n = 901$). This is followed by medium small (20.1%; $n = 362$), very small (16%; $n = 282$), medium large (5.4%; $n = 95$), large (3.6%; $n = 61$), and very large (3.3%; $n = 47$). A more detailed analysis, where fractured elements are discarded, shows that the three smaller categories (very small, small, and medium small) dominate in the knapping products (debitage and flakes), while 70% of the flaked artifacts, ecofacts, grinding artifacts, and cores are in medium large to very large categories. The larger flaked formatized artifacts correspond to rhyolite pieces. Among the medium-small or smaller elements, the presence of silica tools of good quality for knapping is notorious.

Typological studies allowed us to derive archaeological expectations about the technological management or mode of supply and transport of raw materials. The results show an important variability in the identified technical series (Aschero, 1975, 1983). Retouching and re-knapping (partially extended and extended) occurred mostly in bifacial pieces, projectile points, and composite instruments. In almost all cases, this technique was applied on high-quality siliceous or chalcedonic rocks. In this group of instruments, the greatest amount of bifacial flaking was also recorded. In the remaining cases, the artifacts were made or reactivated by direct or alternating unifacial knapping. The bipolar technique for obtaining base forms was recorded in a small number of flakes ($n = 42$). However, the active grinding artifacts have hollows on the dorsal face, which were identified as anvils. These were proposed as indirect evidence for the use of the bipolar technique (Heider, 2016a, b).

Nineteen typological groups were identified. Although this variability is broad, it is less than that identified regionally (Heider, 2015, 2016a). The most numerous were end scrapers ($n = 105$), side scrapers ($n = 34$), composite artifacts ($n = 34$), notches ($n = 18$), and knives ($n = 13$). In addition, 63 cores were recovered. The presence of natural filters with traces of use ($n = 69$) is also notorious. Bifaces ($n = 5$), bifacial blades with asymmetrical bevel ($n = 4$), and projectile points ($n = 10$) are under-represented; because they are the elements most commonly sought by collectors, it was possible to analyze more than 50 pieces from the TP with morphometric techniques (Heider & Rivero, 2018). The natural blades are chalcedony A or B ($n = 49$) and silica ($n = 15$), of very good quality for knapping. One further aspect to highlight is the presence of 12 artifacts made by pitting abrasion (e.g., reeds and grinding tools) in igneous and metamorphic rocks.

The cores were considered, following the proposal of Aschero (1975, 1983) and Franco (2004, 2014). A total of 63 cores were recovered in the TP. A general analysis verified the predominance of Chalcedony B ($n = 20$) and Chalcedony A ($n = 16$), and rhyolite ($n = 13$), quartz ($n = 11$), indeterminate rocks ($n = 2$), and silica ($n = 1$), were also recovered, although to a lesser extent. The variability is notorious in comparison with the neighboring sectors of the northern Pampa Seca (Heider, 2015), where the totality of the cores corresponds to Chalcedony A or quartz (Heider, 2016b). The shapes of the cores are varied. The highest variability of shapes corresponds to Chalcedony B (i.e., amorphous, polyhedral, and irregular discoidal) and rhyolite (i.e., amorphous, globular). Tested nodules were recorded in both Chalcedony A and B as well as rhyolite. Most of the quartz cores were depleted core bodies ($n = 10$). There are also cases of depleted cores in silica and Chalcedony A. Rhyolite is the only rock for which very large specimens were identified ($n = 8$). The remaining elements are distributed between large ($n = 5$), medium large ($n = 18$), medium small ($n = 29$), and small (3). In 75% ($n = 45$) of the samples, three or more extractions were identified, in some cases reaching eight or more ($n = 9$). It is noteworthy that only in Chalcedony B and rhyolite, three extractions or less were recorded.

As regards obsidian samples, two projectile points were recovered at the *La Maroma* site and two edge flakes at *Médano del Marcelo* and *Médano del Centenario* (Table 3). The results indicate TP connection vectors with southwestern Mendoza,

Table 3 Obsidian artifacts identified in the TP

Sitio	C.A.	T.A	E.	L	A	Esp	XRF	C.
La Maroma 1	Formatized tools	Point	En.	43	17	4	LM2	SLLM01
La Maroma 1		Point	Fr.	24	21	3	LM1	SLLM02
Médano Marcelo	Knapping products	Edge flake	Fr.	20	11	3	LM1	SLMM01
Centenario		Edge flake	Fr.	21	15	3	LM1	MZCE01

Glossary: *A.C.* Artifact class, *A.T.* Artifact type, *E* Condition, *En* whole, *Fr* Fractured, *L* Length in millimeters, *W* Width in millimeters, *Th* Thickness in millimeters, *XRF* Provenance in XRF, *C* Code

close to the Andes Mountains, where *Laguna del Maule* is located, with the subtypes *Laguna del Maule* 1 and *Laguna del Maule 2* (Barberena et al., 2019). These sources are located at a distance of approximately 450 km on/close to the border between Argentina and Chile (37°S). *Laguna del Maule 1* corresponds to a primary source located in the Andean Mountains (Barberena et al., 2019; Durán et al., 2004), while *Laguna del Maule 2* has been identified in the secondary deposits of the *Barrancas* (Barberena et al., 2019) and Colorado (Salgán et al., 2023) rivers (Fig. 1).

Discussion and Future Perspectives

In biogeographic terms, the TP is considered part of a marginal environment of central Argentina, with scarce water and vegetation availability (Heider et al., 2019). The archaeological record has low density and high spatial concentration. Given these conditions, a biogeographic perspective was combined with the Organization of Technology (sensu Nelson, 1991). Recently, the size or scale of the space that potentially functioned as an internodal space was evaluated. Internodes are understood as spaces between settlements or areas with comparatively low densities of stable population or lacking it (Nielsen, 2006) and occupied in a less intense and/or recurrent way (i.e., Barberena et al., 2017; Borrero & Barberena, 2006; Mizoguchi, 2009; Nielsen, 2006). Ethnohistoric and archaeological reports show that the TP was a space of low population density and difficult transit (Greslebin, 1961; Heider, 2015; Heider et al., 2019; among others). In this sense, Curtoni and collaborators (2022) showed that there were no stable populations in the TP during historical moments. Using historical cartography and geographic information systems, they proposed that *Rankülche* communities had an effectively occupied territory (in the north of La Pampa province) and a series of dominated territories. One of these, crossed by roads (“*rastrilladas*”), connected their enclaves with the spaces colonized by European and Creole communities on the edges of the Central Mountains of San Luis and Córdoba (Heider & Curtoni, 2019; Curtoni et al., 2022). These characteristics, typical of Internodal Space, persist even in the twenty-first century. In contrast, surrounding spaces had more stable populations, low-scale production of domestic resources, higher demographic load, and rocks suitable for knapping, among other characteristics (Berón, 2013; Borgo, 2020; Heider, 2016b; Heider et al., 2017; Sario, 2011; among many others).

In relation to the availability of lithic resources, the TP lacks rocks suitable for knapping. This must have constituted a risk factor, perhaps a limiting one, which hunter-gatherer groups had to face with strategies and technological decisions that left their mark on the archaeological record. On the other hand, there is an abundance of good-quality rocks in different nearby areas (nodal in our scale) (Fig. 3), where archaeological studies and petrographic analyses have been conducted on raw material sources (i.e., Borgo, 2020; Heider et al., 2017, 2020; Martínez et al., 2014; Sario, 2011). These studies provide a basis from which the geographic provenance of the artifacts identified at the TP sites can be addressed. We believe that they constitute a valid indicator of human circulation patterns and geographic vectors in the access to the internode. Moreover, we understand that the circulation of rocks implies a dynamics of circulation of information and hierarchization of space that can be observed in the TP. In view of the above, lithic technology is a central element of analysis and discussion, as it provides two lines of evidence: the vectors of entry of lithic resources and the technological trends.

The size of the tools recovered in the TP is a valuable indicator to address both lines of evidence. Very large or large pieces were generally made from metamorphic and igneous rocks, which were used to make active and passive grinding artifacts by the techniques of picking and abrasion. We believe that these were part of what Kuhn (1994) identified as site equipment where dunes were hierarchized in relation to the rest of the landscape (Heider et al., 2019). The cores are the pieces suitable for knapping of the largest size, generally large or medium large. In this set, rhyolite and Chalcedony B are the largest in size, which is related to the proximity of the sources (70 km linear distance between sites and source, Fig. 5) and to their presence in the landscape as very large nodules (Heider et al., 2017; Martínez et al., 2014). In both cases, their dimensions allowed obtaining good percussion platforms. The Chalcedony A cores have medium or smaller sizes and come from 60 km or more (Heider & Demichelis, 2015). In the rest of the rocks, the cores are small and depleted. The tools shaped on flakes vary notoriously in size, artifact types, and cortex percentage. In no case, do they have proportions equal to or greater than the cores. The overall trend is represented by small or very small medium-sized pieces, with low presence of cortex (with the exception of rhyolite and chalcedony) and clear signs of reactivation.

The results allow proposing that the lithic resources used are available in the areas or nodes surrounding the TP (Fig. 3). Rhyolites are available in the west (Martínez et al., 2014), Chalcedony B in the northwest (Heider et al., 2017), and Chalcedony A in the southwest (Heider & Demichelis, 2015), while quartz, microcrystalline silicas, igneous, and metamorphic rocks do not have specific petrographic studies. However, the most probable and closest sources are to the north, in Sierras Pampeanas Occidentales (Borgo, 2020; Borgo et al., 2019a, 2020). Finally, a set of obsidian pieces also of extra-regional origin (more than 450 km) come from the primary and secondary sources of Laguna del Maule (Heider et al., 2022a, b).

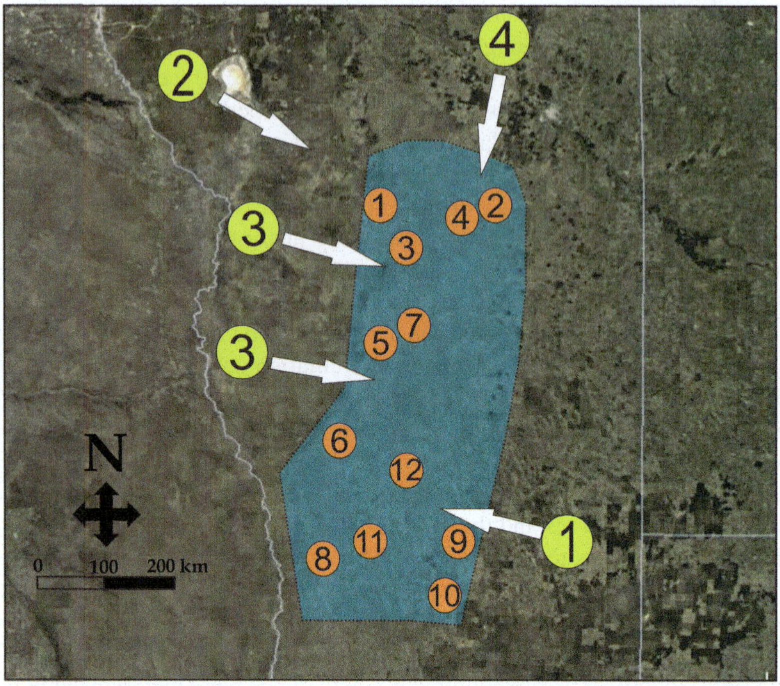

Fig. 5 Rock circulation in the TP. Numbers in yellow circles correspond to sources of lithic raw material near the TP. The white lines are the rock input vectors

Geochemical and technomorphological characterizations conducted in previous works (Heider, 2015; Heider et al., 2022a, b) allowed us to generate archaeological expectations regarding possible east–west connections at latitudes near parallels 34°S and 35°S, as well as to inquire about the strategies by which this raw material was incorporated into the area. In the geochemical results, three sources of obsidian provenance were recognized, all located in south-central western Argentina. A total of 12 artifacts were analyzed, including the four presented here, with 11 artifacts coming from the mountain range source Laguna del Maule 1 and Laguna del Maule 2, and one piece coming from the extra mountain range source El Peceño 1. The latter corresponds to the closest source, with a distance between 250 and 300 km from the sites, but with low representation in the record. In the TP, the artifacts correspond to finished instruments and two edge flakes, typical of the final manufacturing stages. This reinforces the idea that they were obtained as part of exchanges or infrequent visits with populations located in southern Mendoza, and not as a response to a technological need for rocks of excellent quality.

It is within this framework that we can propose that the lithic resources represented in the archaeological record of the TP come from distances between 100 and 150 km, corresponding to allochthonous rocks of very good quality, located near TP. The use of Chalcedony A and B, quartz, and, to a lesser extent, rhyolite, stands out. This indicates preponderant east–west and to a lesser extent north–south circulation vectors in addition to infrequent interactions with resources and populations located near the Andean Cordillera (source Laguna del Maule 1) and northern Patagonia (source *Laguna del Maule* 2), which have records in similar proportions in the Pampean and Central Mountain Regions (Berón et al., 2017; Giesso et al., 2008; Heider et al., 2022b; Stern & Aguerre, 2013).

The classic distinction between direct and indirect forms offers an additional point to understand the form of entry of the resources and their vectors. Direct access to Chalcedony A and B as well as to rhyolite can be assumed in view of our results. All three meet a series of archaeological expectations for direct access (Franco, 2004), such as high frequency of raw materials in the archaeological assemblages, representation of all stages of the artifact production sequence, high cortex index, use of raw materials not restricted to a particular type of artifact, and presence of cores, which is an excellent indicator of distance (spatial and/or social) to the source (Bayón & Flegenheimer, 2004; Franco, 2004, 2014; among others). In contrast, the remaining resources exhibit a smaller number of indicators that might support the direct access hypothesis.

The proximity to the sources of rhyolites and chalcedonies allowed us to recover a greater variability of artifacts in the archaeological record (from nodules to knapping debris). This includes the wide variability of forms in the cores and artifacts. However, regardless of the distance to the sources, a conserved technological strategy, in terms of Nelson, 1991, is observed. The absence of local rocks was a limiting factor that conditioned their use, maintenance, and discard. In this sense, we observed the presence of flakes from bipolar knapping, reactivation of artifacts, and natural filters with traces of use. The larger artifacts have longer and standardized blades. They are end scrapers, side scrapers, or projectile points. In general, they have longer technical series, with knapping, reworking, and retouching. Bifacial and thinning flakes are the most common in this group of pieces. Probably many of these elements entered the TP as finished artifacts or in advanced stages of manufacture. This is particularly noticeable in the obsidian projectile points of extra-regional origin.

The above analysis of lithic technology and how it entered the TP is expected to make an initial contribution to the studies of Ie in central Argentina. The TP dunes are eco-refuge, capable of sustaining human occupations in an environment lacking surface water (Heider et al., 2019). As mentioned by Veth (1989, 1995), the risks of deserts are real, and mitigation measures have been quite extreme on a global scale. Ultimately, it was also behavioral decisions that transformed the dunes into eco-shelters due to their capacity to retain water. This reoccupation shaped and transformed these spaces into places where human groups could safely obtain water, possibly as a result of an improvement in the conditions of the oasis, as a result of the positive feedback between the occupation and the increased availability of water in the dunes and lakes (Heider et al., 2019). As dunes were confined to a few specific points in the landscape, there probably was a redundancy in the occupation of these hierarchical spaces. It is proposed that these geographic and hydroecological characteristics formed the biogeographic basis for the generation of a space of "population convergence," which is supported by the recovery of rocks from different vectors. In the same line of argument, the TP can be included in a wide obsidian circulation circuit in southern Córdoba and San Luis (Heider & Salgán, 2021). In this sense, the presence of extra-regional rocks in the Central Mountain and Pampean Region nodes could have entered the record through the action of local people who obtained them directly in nearby nodes (Heider et al., 2022a). The convergence of people to the Ie undoubtedly created opportunities for the exchange of goods that, in turn, would be taken to other nodes where they would eventually be discarded (Nielsen, 2017).

Finally, regarding our initial question about the motives that led human groups to occupy this hostile place, the proposal of Nielsen (2017) makes sense, considering the fact that two kinds of occupation are expected in the internodes: transit and extractive. The above leads us to propose that the TP could have functioned as an extractive Ie of faunal resources. The open grassland could have offered hunter-gatherers game resources of great interest (guanacos and pampas deer) for the inhabitants of nearby nodes. In this sense, wildlife is a valuable resource, as more intense human activity in the nodes causes their retraction toward interstitial areas (Nielsen, 2017).

The results here reported constitute a starting point to find answers to such as the role of the TP in the dynamics of social interaction in central Argentina, what kind of control hunter-gatherers exercised over that territory, what technological innovations and ethnoecological knowledge were necessary to occupy the TP, how long the occupations extended in time, and what technological responses hunter-gatherers implemented to the thermal and nutritional stress generated by marginal environments. We believe that the study of sparsely populated or scarcely traveled spaces is strategic to understand human societies in terms of the circulation of information, objects, and people on large spatial scales. The PT is one of such spaces, with scarce archaeological records and population.

A field and cabinet agenda, already under development, raises questions and hypotheses complementary to the ideas presented here. Is it possible that the archaeological record is the result of a settlement/subsistence strategy specific to the TP and not the result of its use as an internodal space? Is it possible to differentiate in sites located in dunes the expectations of a strategy of high mobility with low population density and ephemeral sites versus internodal sites? Is the recurrence of grinding artifacts the result of the recurrence of occupations in the extractive internode? As Gerola (2022) mentions, the distinction and study of a space as nodal or internodal are useful for analytical purposes as long as it is linked to the specific dynamics and particularities of each of the groups under investigation. In short, this work is a concrete, initial attempt to explore the limits and possibilities of Internodal Archaeology for the case study. In this sense, a biogeographical view was added to the conceptual categories of node/internode to analyze the human occupations connected to their ecological landscape. The studies carried out raise new hypotheses: the TP as Ie was used jointly by human groups inhabiting surrounding nodes that, at the same time, created an exchange space that allowed dissipating social tensions. Further field and laboratory research will be aimed at obtaining relevant archaeological data to test this hypothesis and develop, in the long term, explanations based on a generalizable body of data at regional and macro-regional levels.

Acknowledgments We would like to thank the inhabitants of Southern Córdoba and San Luis for receiving us in their places and accompanying us in the field work, the members of the native communities for their interest in our work, and the governments of San Luis and Mendoza for the research permits. The field work was financed by CONICET. To the organizers of the IX Congreso de Arqueología de la Región Pampeana. We appreciate language revision by staff from the Instituto de Lenguas, Universidad Nacional de San Luis. Finally, we would like to thank the reviewers for their comments, which greatly enriched this work. We would also like to thank the editor, Dr. Mariano Bonomo. All written work is the responsibility of the authors.

References

Aldenderfer, M. (1998). *Montane Foragers, Asana and the South-Central Andean Archaic*. University of Iowa Press.

Álvarez, M., & Heider, G. (2019). Conocimiento tradicional y sus implicancias para la caza de ñandú y jabalí en comunidades campesinas del sur de la provincia de San Luis, Argentina. *Etnobiología, 17*(1), 5–19.

Andrefsky, W. (2008). The analysis of stone tool procurement, production, and maintenance. *Journal of Archaeological Research, 17*(1), 65–103.

Aschero, C. (1975). *Ensayo para una clasificación morfológica de artefactos líticos aplicada a estudios tipológicos interpretativos*. Unpublished report to CONICET. Manuscript.

Aschero, C. (1983). *Ensayo para una clasificación morfológica de artefactos líticos. Apéndices A y B*. Unpublished report to the Ergology and Technology course, Facultad de Filosofía y Letras. Universidad de Buenos Aires. Manuscript.

Aschero, C., & Hocsman, S. (2004). Revisando cuestiones tipológicas en torno a la clasificación de artefactos bifaciales. In A. Acosta, D. Loponte, & M. Ramos (Eds.), *Temas de Arqueología. Análisis Lítico* (pp. 7–25). Universidad Nacional de Luján.

Barberena, R., Romero Villanueva, G., Lucero, G., Fernández, M., Rughini, A., & Sosa, P. (2017). Espacios internodales en Patagonia septentrional: Biogeografía, información y mecanismos sociales de interacción. *Estudios Atacameños, 56*, 57–75.

Barberena, R., M. V. Fernández, A. Rughini, K. Borrazzo, R. Garvey, G. Lucero, C. Della Negra, G. Romero, V. Durán, V. Cortegoso, M. Giesso, C. Klesner, B. MacDonald y M. D. Glascock (2019). Deconstructing a complex obsidian 'source scape': a geoarchaeological and geochemical approach in northwestern Patagonia. Geoarchaeology. *An International Journal 34*, 30–41. http://dx.doi.org/10.31048/1852.4826.v12.n1.21865

Barberena, R., Menéndez, L., le Roux, P., et al. (2020). Multi-isotopic and morphometric evidence for the migration of farmers leading up to the Inka conquest of the southern Andes. *SciRep, 10*, 21171. https://doi.org/10.1038/s41598-020-78013-x

Bayón, C., & Flegenheimer, N. (2004). Cambio de planes a través del tiempo para el traslado de roca en la pampa bonaerense. *Estudios Atacameños, 28*, 59–70.

Berón, M. (2006). Relaciones interétnicas e identidad social en el registro arqueológico. In: B. Williams, & B. Alberti (Eds.), *Género y Etnicidad en la Arqueología Sudamericana, Serie Teórica* (Vol. 4(I), pp. 119–138). FACSO, UNICEN.

Berón, M. (2013). La arqueología del sector occidental de la región pampeana. Trayectoria y reposicionamiento respecto a la arqueología nacional. *Revista del Museo de La Plata, sección Antropología, 13*(87), 7–29.

Berón, M. (2016). Dunes, hills, waterholes, and saltpeter beds: Attractors for human populations in western Pampa, Argentina. *Quaternary International, 422*, 163–173.

Berón, M., Di Biase, A., Musaubach, M., & Páez, F. (2017). Enclaves y espacios internodales en la dinámica de poblaciones en el wall-mapu. aportes desde la arqueología pampeana. *Estudios Atacameños, 56*, 253–272.
Binford, L. (1979). Organization and formation processes: Looking at curated technologies. *Journal of Anthropological Research, 35*(3), 255–273.
Binford, L. (1980). Willow smoke and dogs' tails: Hunter-gatherer settlement systems and archaeological site formation. *American Antiquity, 45*(l), 4–20.
Borgo, M. (2020). *Estrategias de aprovisionamiento y explotación de recursos líticos en grupos cazadores-recolectores en la cuenca Alta y Media del río Quinto* [Unpublished doctoral thesis]. Facultad de Ciencias Sociales, UNICEN.
Borgo, M., Heider, G., Ramos, G., & Curtoni, R. (2019a). Primeros resultados de los estudios líticos en la localidad arqueológica Quebrada del Chiquero. *Revista Anales de Arqueología y Etnología, 74*(1), 39–56.
Borgo, M., Ramos, G., Heider, G., Chiesa, J., Ortíz Suarez, A., Gil, R., & Curtoni, R. (2019b). Análisis petrográficos de rocas silíceas en el centro-este de la provincia de San Luis. *Revista de Arqueología y Etnohistoria de Regiones Áridas y Semiáridas, XIII*, 121–135.
Borgo, M., Heider, G., Ramos, G., & Curtoni, R. (2020). Estudios preliminares en el área de canteras La Falla (San Luis). *Revista del Museo de Antropología, 13*(1), 191–196.
Borie, C., Power, X., Parra, S., Salinas, H., Rostan, P., Galarce, P., Peña, I., & Traverso, F. (2017). Tras la huella del sílice pampino. Nuevas metodologías para el rastreo de las áreas fuente de aprovisionamiento lítico en Taltal. *Estudios Atacameños, 56*, 103–131.
Borrero, L. (1994–1995). Arqueología de la Patagonia. *Palimpsesto Revista de Arqueología, 4*, 9–69.
Borrero, L., & Barberena, R. (2006). Hunter-gatherer home ranges and marine resources. An archaeological case from southern Patagonia. *Current Anthropology, 47*(5), 855–867.
Butzer, K. (1989). *Arqueología: una ecología del hombre*. Editorial Bellaterra.
Cabrera, A. (1976). Regiones fitogeográficas argentina. In: A. Cabrera, (Ed.), *Enciclopedia Argentina de Agricultura y Jardinería* (Vol. II(1), pp. 1–85). ACME SACI.
Cattáneo, R., Izeta, A., & Costa, T. (2013). *El patrimonio arqueológico de los espacios rurales de la provincia de Córdoba*. Universidad Nacional de Córdoba. Manuscript.
Chiavazza, H. (2013). No tan simples: pesca y horticultura entre grupos originarios del norte de Mendoza. *Comechingonia Virtual, 1*, 27–45.
Cortegoso, V., Neme, G., Giesso, M., Durán, V., & Gil, A. (2012). El uso de la obsidiana en el sur de Mendoza. In A. Gil & G. Neme (Eds.), *Paleoecología Humana en el Sur de Mendoza* (pp. 180–211). Sociedad Argentina de Antropología.
Curtoni, R., Heider, G., González Roglich, M., Houspanossian, J., & Mollo, N. (2022). Las rastrilladas del centro de Argentina como reflejo de la territorialidad indígena y de frontera. *Diálogo Andino, 68*, 35–45.
Dearing, J. (2006). Climate–human–environment interactions: Resolving our past: Climate of the Past. *European Geosciences Union, 2*, 187–203.
Dellafiore, C., Demaría, M., Maceira, M., & Bucher, E. (2001). Estudio de la distribución y abundancia del venado de las pampas en la provincia de San Luis mediante entrevistas. *Revista Argentina de Producción Animal, 21*, 137–144.
Demaría, M. (2008). *Cambios en la cobertura de pastizales pampeanos semiáridos en San Luis (Argentina), y su relación con variables climáticas y ambientales* [Unpublished doctoral thesis]. Universidad de Alcalá de Henares.
Dincauze, D. (2000). *Environmental archaeology. Principles and practice*. Cambridge University Press.
Durán, V., M. Giesso, M. Glascock, G. Neme, A. Gil y L. Sanhueza. (2004). Estudio de fuentes de aprovisionamiento y redes de distribución de obsidiana durante el Holoceno tardío en el sur de Mendoza (Argentina). *Estudios Atacameños 28*, 25–43.
Ebert, J. (1992). *Distributional archaeology*. University of New Mexico Press.
Ericson, J. (1984). Toward the analysis of lithic production systems. In J. Ericson & B. Purdy (Eds.), *Prehistoric quarries and lithic production* (pp. 1–9). Cambridge University Press.

Forman, S., Tripaldi, A., & Ciccioli, A. (2014). Eolian sand sheet deposition in the San Luis paleodune field, western Argentina as an indicator of a semi-arid environment through the Holocene. *Palaeogeography, Palaeoclimatology, Palaeoecology, 411*, 122–135.

Franco, N. (2004). La organización tecnológica y el uso de escalas espaciales amplias. El caso del sur y oeste del Lago Argentino. In A. Acosta, D. Loponte, & M. Ramos (Eds.), *Temas de Arqueología. Análisis Lítico* (pp. 101–144). Universidad Nacional de Luján.

Franco, N. (2014). Lithic artifacts and the information about human utilization of large areas. In P. Escola & S. Hocsman (Eds.), *Artefactos Líticos, Movilidad y Funcionalidad de Sitios: Problemas y Perspectivas* (pp. 116–127). British Archaeological Reports.

Gamble, C. (1993). People on the move: Interpretations of regional variation in Palaeolithic Europe. In J. Chapman & P. Dolukhanov (Eds.), *Cultural transformations and interactions in Eastern Europe* (Center for the Archaeology of Central and Eastern Europe. Monograph 1) (pp. 37–55). Ashgate Publishing Company.

Garreaud, R., Vuille, M., Compagnucci, R., & Marengo, J. (2009). Present-day South American climate. *Palaeogeography, Palaeoclimatology, Palaeoecology, 281*, 180–195.

Gerola, I. (2022). Paisajes dinámicos de la Cuenca Superior del Río Grande de San Juan (Jujuy, Argentina). Una aproximación internodal a la microrregión de Cusi Cusi y sus lógicas de habitar el paisaje. *Mundo de Antes, 16*(1), 269–230.

Giesso, M., Berón M. y Glascock M. (2008). Obsidian in Western Pampas, Argentina: Source Characterization and Provisioning Strategies. *International Association for Obsidian Studies Bulletin 38*,15–18.

Giesso, M., Durán, V., Neme, G., Glascock, M., Cortegoso, V., Gil, A., & Sanhueza, L. (2011). Study of obsidian source usage in the Central Andes of Argentina and Chile. *Archaeometry, 53*(1), 1–21.

Gil, A. (1997–1998). Cultígenos prehispánicos en el sur de Mendoza. Discusión en torno al límite meridional de la agricultura andina. *Relaciones, XXII-XXIII*, 295–318.

Gil, A., & Neme, G. (2010). Registro arqueológico en la cuenca media del Atuel: viejos y nuevos problemas; viejos y nuevos datos. In M. Zárate, A. Gil, & G. Neme (Eds.), *Condiciones Paleoambientales y ocupaciones humanas durante la transición Pleistoceno-Holoceno de Mendoza* (pp. 239–276). Sociedad Argentina de Antropología.

Gil, A., Giardina, M., Neme, G., & Ugan, A. (2014). Demografía humana e incorporación de cultígenos en el centro occidente argentino: explorando tendencias en las fechas radiocarbónicas. *Revista Española de Antropología Americana, 44*(2), 523–553.

Gil, A., Menéndez, L., Atencio, J., Peralta, E., Neme, G., & Ugan, A. (2018). Estrategias humanas, estabilidad y cambio en la frontera agrícola sur americana. *Latin American Antiquity, 29*(1), 6–26.

Gould, R., & Saggers, S. (1985). Lithic procurement in Central Australia: A closer look at Binford's idea of embeddedness in archaeology. *American Antiquity, 50*(1), 117–136.

Greslebin, H. (1924). *Fisiografía y noticia preliminar sobre arqueología de la región de Sayape: Provincia de San Luis*. Talleres Gráficos Ferrari.

Greslebin, H. (1928). La antigüedad del hombre en la región de Sayape, provincia de San Luis, República Argentina (nota preliminar). In *XXIII International Congress of Americanists* (pp. 305–312).

Greslebin, H. (1961). Interrogatorios Ranquelinos. *Cuadernos del Instituto Nacional de Investigaciones Folklóricas, 2*, 51–70.

Heider, G. (2015). *Los pueblos originarios en el norte de Pampa Seca. Una mirada arqueológica sobre los cazadores-recolectores del sur de Córdoba y San Luis, Argentina* [Unpublished doctoral thesis]. Facultad de Filosofía y Humanidades, Universidad Nacional de Córdoba.

Heider, G. (2016a). La gestión de recursos líticos en el Norte de Pampa Seca. *Relaciones de la Sociedad Argentina de Antropología, XLI*(2), 375–396.

Heider, G. (2016b). Un aporte a la dispersión de las ortocuarcitas pampeanas: el sur de Córdoba como una frontera. *Intersecciones en Antropología, 17*(3), 303–313.

Heider, G., & Curtoni, R. (2016). Investigaciones arqueológicas en la Provincia de San Luis: a 150 años de sus inicios, historia y perspectivas. *Revista del Museo de Antropología, 9*(1), 35–48.

Heider, G., & Curtoni, R. (2019). Chew Upültripalen Gualicho. Un análisis inicial de la cosmovisión Rankülche y su relación con el registro arqueológico de la Pampa Central. *TEFROS, 17*(2), 73–96.
Heider, G., & Demichelis, A. (2015). Loma de los Pedernales, a local raw material source in the North of Pampa Seca, Argentina. *Quaternary International, 375*, 3–12.
Heider, G., & López, L. (2016). El consumo de recursos vegetales silvestres en grupos cazadores-recolectores del Norte de Pampa Seca (San Luis y Córdoba, Argentina). *Mundo de Antes, 10*, 73–99.
Heider, G., & López, L. (2018). The South American agricultural frontier: The first direct evidence of maize consumption in San Luis, Argentina. *Antiquity, 92*(365), 1260–1273.
Heider, G., & Rivero, D. (2018). Estudios morfométricos aplicados a puntas de proyectil lanceoladas "Ayampitín" del Holoceno Temprano-medio en las Sierras y Llanuras Pampeanas Argentinas, Sudamérica. *Latin American Antiquity, 29*(3), 572–590.
Heider G. y L. Salgán. (2021). Nuevos estudios de procedencia d obsidianas en espacios internodales del Norte de Pampa Seca. IX In: Congreso de arqueología de la Región Pampeana Argentina. Universidad Nacional de Mar del Plata. Quintana, C., M. Álvarez; G. Bonnat, D. Mazzanti; M. Barros; V. Puente y M. Bonomo (eds.) . p. 78.
Heider, G., Rivero, D., & Baldo, E. (2015). Rocas de uso arqueológico en las Sierras Centrales. Fuentes de recursos líticos identificadas y potenciales en las provincias de Córdoba y San Luis, Argentina. *Revista de Antropología del Museo de Entre Ríos, 1*(2), 55–72.
Heider, G., Ortiz Suarez, A., Gil, R., Perino, E., Chiesa, J., Basaez, A., Bazán, C., & Díaz, I. (2017). Chert's source in Las Travesías from the center of Argentina. A case study in the arid zone. In *11th. International Symposium on knappable material, "From toolstone to stone tools"* (pp. 156–157). Buenos Aires and Necochea, (Argentina).
Heider, G., Jobbágy, E., & Tripaldi, A. (2019). Uso del espacio semiárido por poblaciones prehispánicas: El papel de los paisajes de dunas como eco-refugios en el Centro de Argentina. *Boletín de la Sociedad Geológica Mexicana, Volumen Especial Geoarqueología, 71*(2), 229–248.
Heider, G., Ortiz Suarez, A., Rivero, D., Baldo, E., Pastor, S., Ramos, G., Borgo, M., Gil, R., Chiesa, J., Costa, C., Recalde, A., Curtoni, R., Capriolo, J., & Muñoz, L. (2020). Estudios geoarqueológicos multiproxy de fuentes y canteras líticas de las Sierras Pampeanas y llanuras adyacentes. *Revista del Museo de Antropología, 13*(1), 31–36.
Heider, G., Curtoni, R., Borgo, M., Chiesa, J., Olivan, A., & Martínez Quirós, V. (2022a). New research on the Intihuasi Cave, San Luis, Argentina. Implications for the peopling of the southern cone of South America. *Latin American Antiquity, 33*, 857–862.
Heider, G., Salgán, L., Neme, G., & Gil, A. (2022b). Recursos líticos y vectores de interacción en las Pampas Centrales de Argentina: Un abordaje desde el uso de la obsidiana. *Revista del Museo de Antropología, 15*, 23–36.
Iniesta, L., Ots, M., & Manchado, M. (2020). Prácticas y tradiciones alimenticias prehispánicas y de la colonia temprana en Mendoza (centro oeste de Argentina). Un aporte desde la arqueología y la etnohistoria. *RIVAR, 7*(20), 46–66.
Iriondo, M. (1990). A late Holocene dry period in the Argentina plains. *Quaternary of South America and Antarctic Peninsula, 7*, 198–218.
Iriondo, M. (1999). Climatic changes in the South American plains: Records of a continents scale oscillation. *Quaternary International, 57–58*, 93–112.
Iriondo, M., & Kröhling, D. (1995). El Sistema Eólico Pampeano, Comunicaciones del Museo Provincial de Ciencias Naturales Florentino Ameghino. *Links, 5*, 1–68.
Jones, T., Brown, G., Raab, M., Mcvickar, J., Spaulding, W., Kennett, D., York, A., & Walker, P. (1999). Environmental imperatives reconsidered: Demographic crises in Western North America during the Medieval Climatic Anomaly. *Current Anthropology, 40*(2), 137–170.
Kelly, R. (2012). Obsidian in the Carson Desert: Mobility or trade? In R. Hughes (Ed.), *Perspectives on prehistoric trade and exchange in California and the Great Basin*. University of Utah Press.
Kuhn, S. (1994). A formal approach to the design and assembly of mobile toolkits. *American Antiquity, 59*, 426–442.

Lagiglia, H. (1982). El proceso de agriculturización en el sur de Cuyo: la cultura del Atuel II. In *Actas del V Congreso Nacional de Arqueología Argentina 2* (pp. 231–254). San Juan, (Argentina).

Lagiglia, H. (2002). Arqueología Prehistórica del sur Mendocino y sus Relaciones con el Centro Oeste Argentino. In A. Gil & G. Neme (Eds.), *Entre Montañas y Desiertos: Arqueología del sur de Mendoza* (pp. 43–64). Sociedad Argentina de Antropología.

Lalinde, V., Heider, G., Curtoni, R., Borgo, M., Chiesa, J., & Ramos, G. (2018). Late Holocene plant use in the Sierras Pampeanas of Argentina: Evidence from phytoliths and starch grains. *Journal of Archaeological Science: Reports, 21*, 413–421.

Leroy, S. (2006). From natural hazard to environmental catastrophe: Past and present. *Quaternary International, 158*, 4–12.

López, L., Berón, M., Prates, L., Medina, M., Heider, G., & Pastor, S. (2020). Alimentación vegetal entre pueblos originarios de la Diagonal Árida Argentina: Sierras Centrales, Pampa Seca y Norpatagonia. *Revista Iberoamericana de Viticultura, Agroindustria y Ruralidad, 21*(7), 81–102.

Martel, A., Zamora, D., & Lépori, M. (2017). Tráfico y movilidad caravanera en la Puna catamarqueña. Una mirada internodal. *Estudios Atacameños, 56*, 197–223.

Martínez, A., Heider, G., Baldo, E., & Merlo, M. (2014). Comparación petrográfica de las riolitas de la Fm. Cerro Varela con material arqueológico recolectado en el Sur de la provincia de Córdoba y el Centro-sur de la provincia de San Luis. In *XIX Congreso Geológico Argentino* (pp. T8–T32). San Miguel de Tucumán, (Argentina).

Medina, M., Pastor, S., & Recalde, A. (2016). The archaeological landscape of late prehispanic mixed foraging and cultivation economy (Sierras of Córdoba, Argentina). *Journal of Anthropological Archaeology, 42*, 88–104.

Medina, M., Balena, I., & Rivero, D. (2019). Proyectiles y procesos de intensificación: una aproximación desde Boyo Paso 2, 1500–750 AP (Sierras de Córdoba, Argentina). *Chungara, 51*(4), 517–529.

Mizoguchi, K. (2009). Nodes and edges: A network approach to hierarchisation and state formation in Japan. *Journal of Anthropological Archaeology, 28*(1), 14–26.

Morales, M., Barberena, R., Belardi, J., Borrero, L., Cortegoso, V., Durán, V., Guerci, A., Goñi, R., Gil, A., Neme, E., Yacobaccio, H., & Zárate, M. (2009). Reviewing human–environment interactions in arid regions of southern South America during the past 3000 years. *Palaeogeography, Palaeoclimatology, Palaeoecology, 28*, 283–295.

Nelson, M. (1991). The study of technological organization. In M. Schiffer (Ed.), *Archaeological method and theory* (pp. 57–100). University of Arizona Press.

Neme, G., & Gil, A. (2012). El Registro Arqueológico del sur de Mendoza en Perspectiva Biogeográfica. In G. Neme & A. Gil (Eds.), *Paleoecología Humana en el Sur de Mendoza: Perspectivas Arqueológicas* (pp. 254–279). Sociedad de Antropología Argentina.

Nielsen, A. (2006). Estudios internodales e interacción interregional en los Andes circumpuneños. In H. Lechtmann (Ed.), *Esferas de Interacción prehistóricas y fronteras nacionales modernas: los Andes sur centrales* (pp. 29–62). Instituto de Estudios Peruanos, Institute of Andean Research.

Nielsen, A. (2017). Actualidad y potencial de la arqueología internodal surandina. *Estudios Atacameños, 56*, 299–317.

Núñez, L., & Dillehay, T. (1995). *Movilidad giratoria, armonía social y desarrollo en los Andes Meridionales: Patrones de tráfico e interacción económica*. Universidad del Norte.

Oyarzábal, M., Clavijo, J., Oakley, L., Biganzoli, F., Tognetti, P., Barberis, I., Maturo, H., Aragón, R., Campanello, P., Prado, D., Oesterheld, M., & León, R. (2018). Vegetation units of Argentina. *Ecología Austral, 28*(01), 040–063.

Pastor, S., & Boixados, R. (2016). Arqueología y Etnohistoria: Diálogos renovados en torno a las relaciones entre las sociedades de los llanos riojanos y de las sierras noroccidentales de Córdoba (Períodos Prehispánico Tardío y Colonial Temprano). *Diálogo Andino, 49*, 311–328.

Pimentel, G., & Barros, A. (2020). La memoria de los senderos andinos. Entre huacas, diablos, ángeles y demonios. *Boletín Museo Chileno de Arte Precolombino, 25*(1), 201–225.
Renfrew, C., & Bahn, P. (1998). *Archaeology. Theories, methods and practice*. Thames and Hudson.
Rocchietti, A., & Ribero, F. (2015). Fechados radiocarbónicos y distribuciones arqueológicas en localidades del Sur de Sierra de Comechingones (Provincia de Córdoba). In A. Pifferetti & I. Dosztal (Eds.), *Metodologías científicas aplicadas al estudio de los bienes culturales. Datación, caracterización, prospección y conservación* (pp. 31–54). ASPHA.
Salgán, M.L., Pompei, M.P., Gil, A., Neme, G., Sruoga, P., Glascock, M.D. (2023). La conexión entre ambientes de tierras bajas y altas en el límite Cuyo Patagonia (Argentina): Un análisis sobre el transporte y uso de obsidiana Laguna del Maule. *Latin American Antiquity*. Published online 1–17. https://doi.org/10.1017/laq.2023.37
Sario, G. (2011). *Poblamiento humano en la provincia de San Luis: una perspectiva arqueológica a través del caso de la organización de la tecnología en Estancia La Suiza* [Unpublished doctoral thesis]. Universidad Nacional de Córdoba.
Servicio Meteorológico Nacional. (2021). *Estadísticas Climatológicas 1991–2021, Buenos Aires*. Disponible en http://www3.smn.gob.ar/
Soriano, A. (1992). Río de la Plata Grasslands. In R. Copeland (Ed.), *Ecosystems of the world, natural grasslands, introduction and western hemisphere* (pp. 367–407). Elsevier.
Stern, C. y Aguerre A. (2013). Obsidianas en sitios arqueológicos de la provincia de La Pampa, Argentina: su procedencia. *Intersecciones en Antropología 14*, 263–269.
Sturm, C., Clark, J., & Barton, L. (2016). The logic of ceramic technology in marginal environments: Implications for mobile life. *American Antiquity, 81*(4), 645–663.
Tapia, A. (2008). *Arqueología histórica de los cacicazgos ranqueles* [Unpublished doctoral thesis]. Facultad de Filosofía y Letras, Universidad de Buenos Aires.
Torrence, R. (1984). Monopoly or direct access? Industrial organization at the Melos obsidian quarries. In J. Ericson & B. Purdy (Eds.), *Prehistoric quarries and lithic production* (pp. 49–64). Cambridge University Press.
Tripaldi, A., & Forman, S. (2007). Geomorphology and chronology of Late Quaternary dune fields of western Argentina. *Palaeogeography, Palaeoclimatology, Palaeoecology, 251*, 300–320.
Tripaldi, A., Zárate, M., Forman, S., Badger, T., Doyle, M., & Ciccioli, P. (2013). Geological evidence for a drought episode in the western Pampas (Argentina, South America) during the early–mid 20th. century. *The Holocene, 23*(12), 1731–1746.
Veth, P. (1989). Island in the interior: A model for the colonization of Australia's arid zone. *Archaeology in Oceania, 24*(3), 81–91.
Veth, P. (1995). Aridity and settlement in northwest Australia. *Antiquity, 69*, 733–746.
Vilanova, I., Karsten, S., Geilenkirchen, M., Schäbitz, F., & Schulz, E. (2015). Last millennial environmental reconstruction based on a multi-proxy record from Laguna Nassau, Western Pampas, Argentina. *Neues Jahrbuch für Geologie und Paläontologie - Abhandlungen, Schweizerbart'sche Verlagsbuchhandlung, 277*(2), 209–224.
Vilanova, I., Tripaldi, A., Piovano, E., Forman, S., Chiesa, J., Jobbágy, E., Rojo, L., Heider, G., & Schittek, K. (2017). Vegetation and environmental changes related to hydroclimate regimes in Wester Pampas, Argentina, over the last 1.5 kyr. In *5th. Open science meeting. Global challenges for our common future: a paleoscience perspective* (p. 406). PAGES OSM.
Vilanova, I., Tripaldi, A., Karsten, S., Rojo, L., Piovano, E., Forman, S., Jobbagy, E., Heider, G., & Chiesa, J. (2022). Late Holocene environmental and hydro-climatic variability inferred from a shallow lake record, blowout dunes, Argentinian western Pampas, South America. *Journal of South American Earth Sciences, 116*, 103826.
Yacobaccio, H. (1994). Biomasa animal y consumo en el Pleistoceno-Holoceno Surandino. *Arqueología, 4*, 43–71.
Zárate, M., & Tripaldi, A. (2012). The aeolian system of central Argentina. *Aeolian Research, 3*, 401–417.

Evidence of Social Interaction in the Pampas: The Case Study of Hangar Site (Province of Buenos Aires, Argentina)

Cristian A. Kaufmann, María Clara Álvarez, Adriana M. Blasi, and María Paula Barros

Abstract The Hangar site is located in the Pampas region and has a high relative occurrence of silicified limestone, a raw material not available in the nearest outcrops usually exploited. The main objective of this chapter is to evaluate and discuss the frequency and use of this rock at Hangar. These data are compared and integrated with information from other sites where silicified limestone is represented to analyze its distribution in the region. We focus on the evidence of silicified limestone in the site by analyzing the *chaîne opératoire*, together with the petrographic study of five thin sections. We also conduct a bibliographic survey of the sites with silicified limestone in the Pampas. These data are used to construct fall-off curves by plotting the proportion of silicified limestone against the distance to the sources. Results indicate that the raw material found in Hangar comes from El Carancho, a source that is 340 km far from the site. These analyses allowed us to propose that during the final Late Holocene, there was an increased social interaction and a broadening of the territories of the hunter-gatherer bands that occupied the Pampas.

Supplementary Information The online version contains supplementary material available at https://doi.org/10.1007/978-3-031-55194-9_8.

C. A. Kaufmann (✉) · M. P. Barros
INCUAPA, UNCPBA-CONICET. Universidad Nacional del Centro de la Provincia de Buenos Aires (UNCPBA), Facultad de Ciencias Sociales, Olavarría, Buenos Aires, Argentina
e-mail: ckaufman@soc.unicen.edu.ar; pbarros@soc.unicen.edu.ar

M. C. Álvarez
Grupo de Estudios Ambientales, IMASL-CONICET, UNSL, San Luis, Argentina

Facultad de Ciencias Sociales, Universidad Nacional del Centro de la Provincia de Buenos Aires, Olavarría, Argentina
e-mail: malvarez@soc.unicen.edu.ar

A. M. Blasi
Comisión de Investigaciones Científicas de la Provincia de Buenos Aires (CIC) – División Mineralogía, Petrología y Sedimentología de la Facultad de Ciencias Naturales y Museo, Universidad Nacional de La Plata (UNLP), La Plata, Argentina
e-mail: ablasi@fcnym.unlp.edu.ar

G. F. Bonnat et al. (eds.), *Current Research in Archaeology of South American Pampas*, The Latin American Studies Book Series,
https://doi.org/10.1007/978-3-031-55194-9_8

Keywords Hunter-gatherers · Silicified limestone · Petrographic sections · Final late Holocene

Introduction

Movement for gathering information, goods, or social reasons is very important in hunter-gatherer societies. Besides utilitarian reasons, obtaining knowledge and social information is essential for small-scale bands (Whallon, 2011). Networks may link persons over wide geographic regions, and the movement of goods, persons, and information may exceed the traditionally considered kinship (Binford, 2006). This topic has been studied through different lines of evidence such as obsidian transfer distances (Pearce & Moutsiou, 2014), lithic point types (Buchanan et al., 2017), shell beads (Miller et al., 2018), ostrich eggs (Mitchell, 1996) and ceramic distributions (Quinn & Burton, 2016), among many others.

In the Pampas of Argentina, some studies were conducted to evaluate interaction through the analyses of the traveling of particular artifacts. Interaction networks have been explored through the analyses of marine shell distribution (Bonomo, 2007), ceramic motifs and manufacture (Catella, 2017; Di Prado, 2018), and lithic studies (Flegenheimer et al., 2003; Politis et al., 2003; Berón, 2006; Martínez, 2017; Martínez et al., 2017; Berón et al. 2015; Messineo et al., 2019; Messineo & Barros, 2021; among others). In particular, lithic artifacts have traditionally been used in interaction studies, as their source of origin can sometimes be easily traced (Belardi et al., 2006; Pallo & Borrero, 2015). Moreover, in addition to the excellent preservation of this type of materials, good quality raw materials are usually transported over long distances (Meltzer, 1989; Kuhn, 1995).

The recent excavation of the Hangar site in the Pampas region provides valuable information on social interactions. This site is dated about 1000 years BP and has a high relative frequency of silicified limestone, a raw material not available in the nearest outcrops usually exploited (e.g., Tandilia). In addition to the presence of this rock, other characteristics differentiate Hangar from other sites of earlier periods, such as the high number of projectile points, abundant ceramic sherds, and the presence of unusual tools in the regional context such as crutch-shaped drills. In this chapter, we evaluate all these aspects, focusing on the silicified limestone frequency, the *chaîne opératoire*, and the probable sources of the rocks. The main objective is to evaluate and discuss the frequency and use of this rock at Hangar. These data are compared and integrated with information from other sites where silicified limestone is represented to analyze its distribution in the region. We explore aspects related to territoriality and social interaction during the end of the late Holocene. For this period, it is proposed that, in a context of high human demography, the circulation of goods, people, and ideas among different sectors of the Pampas and neighboring regions would have increased (Berón, 2004, 2007; Curtoni, 2006; Mazzanti, 2006; Curtoni & Berón, 2011; Martínez et al., 2017; Messineo & Barros, 2021).

Regional Setting and Lithic Resources of the Pampas

The Pampas of Argentina is a large ecosystem of grasslands that have been inhabited by hunter-gatherers, at least since ca. 12,200 ^{14}C years BP (Politis, 2008). This extensive plain is located in the eastern part of Argentina between 30° and 39° south latitude and 57° and 65° west longitude. The Pampas are mainly humid temperate prairies covered by grass (Soriano et al., 1992). To the south and west of this region, there is also a large number of arboreal as *Prosopis caldenia* belonging to the southern sector of the Espinal phytogeographic province (Oyarzábal et al., 2018). The Pampas are characterized by an east–west moisture gradient and increasing continentality toward the northwest. Average temperatures are between 20 and 24 °C in summer and 7 and 10 °C in winter. The annual total rainfall increases toward the east; from 400 mm in the southwest to 1000 mm or more in the northeast (Burgos, 1968). Typical animals of the Pampas inhabit open grasslands (Ringuelet, 1961). Today, some mammals, such as guanaco (*Lama guanicoe*) and pampas deer (*Ozotoceros bezoarticus*), are almost absent from the region, although they were abundant in the past (Merino, 2003; Politis et al., 2011).

The Hangar site is located in the basin of the Salado Creek, in the humid Pampas subregion, province of Buenos Aires, in the depressed Pampa; from a phytogeographical point of view, it corresponds to the hygrophyte and halophyte grassland unit (Oyarzábal et al., 2018). Along this basin and the nearby shallow lakes, several archaeological sites have been recorded; most of them dated to the Middle and Late Holocene (Crivelli Montero et al. 1987, 1997; Eugenio, 1994; Kaufmann & González, 2013; Barros et al., 2018). These contexts are located in the plains between two hilly systems, named Ventania and Tandilia. This inter-hill grassland presents abundant shallow lakes and streams (Zárate & Rabassa, 2005) and has no rocks available for knapping.

In the Pampas region, there are several sources of good knapping quality rocks, which were used for making tools. In the northwestern sector of the Tandilia hilly system, there is a white orthoquartzite outcrop, recently discovered at the Don Guillermo locality (Fig. 1a) (Barros et al., 2021). Approximately 40 km east, different rocks have been found such as granite, orthoquartzite SBG (Sierras Bayas Group), silicified dolomite, chert, and quartz (Fig. 1b; Politis, 1984; Messineo et al., 2004; Messineo & Barros, 2015). In the central portion of the Tandilia hilly system, there are outcrops of orthoquartzite SBG, chert, silicified dolomite, diabase, and granitoid rocks (Fig. 1c; Flegenheimer, 1991; Flegenheimer et al., 1996; Flegenheimer & Bayón, 2002; Paulides, 2007; Colombo, 2011). In the southeastern portion, there are outcrops of quartzites and quartz of the Balcarce Formation (Fig. 1d; Mazzanti, 1997; Donadei Corada, 2019; Bonnat 2021). Outcrops of silicified tuff, quartzite sandstones, and coarse-grained quartzites have been detected in the plains located between the two hilly systems (Fig. 1e; Politis, 1984; Madrid & Salemme, 1991). A wide variety of siliceous rocks, basalt, and quartzite are available on the coast of Buenos Aires province in the form of coastal cobbles and pebbles (Fig. 1f; Bonomo, 2005).

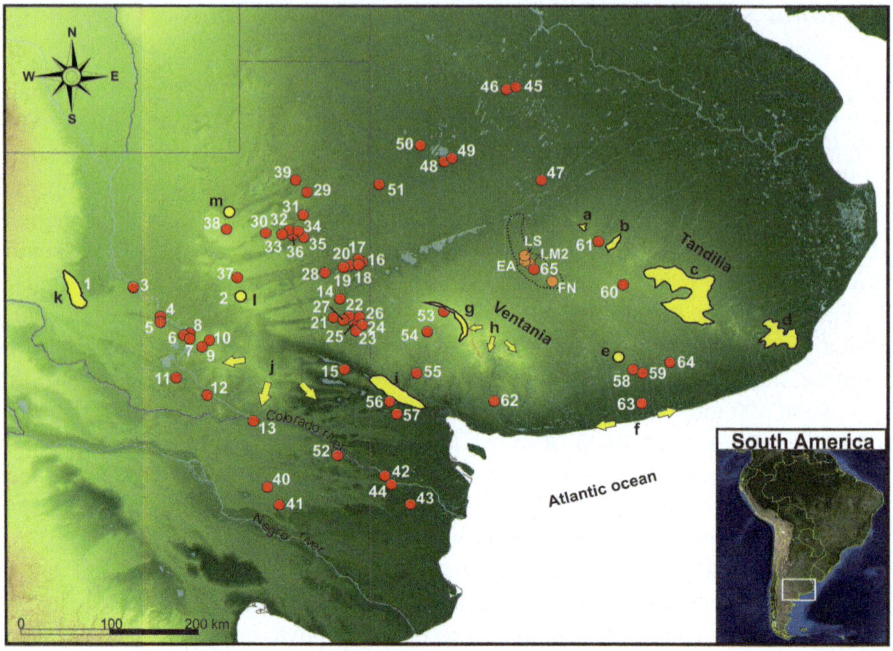

Fig. 1 Sites of the Pampas region with presence of silicified limestone and main lithic resources used in the regionReferences = (1) Meseta del fresco (source); (2) El Carancho 1 (source); (3) El Encuentro; (4) Médanos Lote 18; (5) Médanos Lote 24; (6) El Castillo; (7) Médanos de Villa; (8) Chenque 1; (9) La Casona; (10) El Molino; (11) Manantial Curacó; (12) Tapera Moreira 1; (13) Confluencia 2; (14) La Chola; (15) La Colorada Grande; (16) Médanos de Costilla; (17) Médanos de Phagouapé; (18) Médanos de Alduncín; (19) La Cascada; (20) Bajo de Atreucó; (21) Laguna La Tigra; (22) Laguna Guatraché; (23) Bajo de Marcelino; (24) Médano San Sixto; (25) Luis Otero; (26) Bajada la Minnesota; (27) El Carmel; (28) ChLillhué 1 y 3; (29) Médanos de Peñín; (30) Laguna Chapalcó; (31) Laguna de Montoya; (32) Manantial Naicó; (33) Laguna de Paisani; (34) Laguna de la Ruta; (35) Bajo del Medio; (36) Laguna del Fondo; (37) Laguna Traru Lauquen; (38) Don Isidoro 2; (39) Laguna de Rojo; (40) Laguna del Lote 4; (41) Laguna del Lote 15; (42) El Puma 3; (43) La Modesta; (44) Zoko Andi; (45) Laguna de los Pampas; (46) Laguna Giaccone; (47) Cabeza de Buey 2; (48) La Susana 1; (49) Laguna Chica; (50) Huencú Nazar; (51) Médano Santa Clara; (52) Médano La Enriqueta; (53) La Montaña sitio 1; (54) Los Chilenos sitio 2; (55) San Martín 1; (56) Laguna Chasicó; (57) Salinas Chicas; (58) Las Brusquillas 1; (59) Las Brusquillas 2; (60) La Barrancosa 2; (61) Arroyo Tapalqué 1; (62) Paso Mayor; (63) Las Toscas; (64) Laguna La Vizcacha; (65) Hangar; (a) orthoquartzite; (b) granite, orthoquartzite SBG, silicified dolomite, chert and quartz; (c) orthoquartzite SBG, chert, silicified dolomite, diabase and granitoid rocks; (d) quartzites and quartz; (e) silicified tuff, quartzite sandstones and coarse-grained quartzites; (f) siliceous rocks, basalt and quartzite; (g) rhyolite, metaquartzite and orthoquartzite; (h) metaquartzite, sandstone and quartz; (i) chalcedony, quartzite, xylopal, silicified limestones, schist and quartz; (j) basalt, chalcedony, siliceous rocks, andesite, quartzite, quartz, xylopal, rhyolite and obsidian; (k) silicified limestone (Meseta del Fresco); (l) silicified limestone (El Carancho); (m) silicified limestone (Valle Daza); EA) Escuela Agropecuaria; LS) Laguna Seca; LM2) Laguna Muscar 2; FN) Fortín Necochea. The dotted line indicates the micro-region of study

In the western sector of the Ventania hilly system, there are outcrops of rhyolite, metaquartzite, and orthoquartzite from La Mascota, La Lola, Trocadero, and Napostá Formations (Fig. 1g; Bayón & Zavala, 1997; Oliva & Moirano, 1997; Catella, 2014). Secondary pebble and cobble deposits of metaquartzite, sandstone, and quartz are also available in the fluvial courses that drain the southwestern flank of the Ventania system, like the Sauce Grande River that reaches the Atlantic Ocean coast (Fig. 1h; Bayón & Zavala, 1997; Bayón et al., 2006). In the Chasicó-Salinas Chicas basin, there are remnants of the Tehuelche gravel mantle, which include a great lithological variety with a predominance of volcanic and plutonic rocks and a smaller proportion of chalcedony, quartzite, xylopal, silicified limestones, schist, and quartz (Fig. 1i; Oliva & Moirano, 1997; Oliva et al., 2006; Martínez et al., 2009; Catella et al., 2010, 2013; Catella, 2014).

Secondary deposits of cobbles and pebbles of the Tehuelche mantle are concentrated in fluvial valleys, such as the Salado-Curacó, Colorado, and Negro rivers (Fig. 1j; Berón, 2006; Martínez et al., 2009; Santos Valero, 2015; Carrera Aizpitarte & Berón, 2021). These deposits are composed of rocks of different lithology, among which basalt, chalcedony, siliceous rocks, andesite, quartzite, quartz, xylopal, rhyolite, and obsidian are the most important (Berón et al. 1995; Berón, 2006; Carrera Aizpitarte, 2014).

Silicified Limestone Sources

The sources of silicified limestone are located in the Espinal and Monte phytogeographic provinces (Dry Pampas subregion) and are Meseta del Fresco, El Carancho, and Valle Daza localities (Fig. 1k, l, m; Charlin, 2002; Berón, 2006; Carrera Aizpitarte et al., 2015).

The limestones of the El Fresco Formation (facies V) were described by Melchor and Casadío (1999) as oolitic grainstones, bioclastic grainstones, calcareous arenite, and peloidal grainstone with dispersed gastropods with different degrees of silicification. According to the authors, the sediments of this Formation were deposited in a depressed area, episodically occupied by a freshwater or slightly saline lake. Rubin (1990) interprets outcrops close to those of the Meseta del Fresco as deposited in a tidal plain. Recently Carrera Aizpitarte et al. (2015) described the Meseta del Fresco peloidal and skeletal limestones as partially silicified, while in others, the silicification was total and prevented to recognize the compositional characteristics of the original limestone. At Meseta del Fresco (Fig. 1k), the silicified limestone is a high-visibility source of supply, with good quality for knapping (Berón & Curtoni, 2002; Curtoni et al., 2004; Berón, 2006). The rocks show high variability of colors and differences in the degree of silicification. The range of shades goes from white and beige to pink, red, and burgundy (Berón & Curtoni, 2002; Curtoni et al., 2004). Oolites and drusites are present in the rock sectors where the original limestone has not been completely silicified (Carrera Aizpitarte & Berón, 2021).

For the El Carancho locality, there are no geological antecedents; however, according to Berón and Carrera Aizpitarte (2019), the silicified limestone outcropping there would have been deposited in a continental shallow lake environment. The materials studied from El Carancho by one of the authors (AB) showed stromatolitic texture or biolamination, so they were associated with biogenic carbonates, as a result of biological processes in environments with the presence of microbial mats. The silicified limestone recognized in the southern margin of the El Carancho shallow lake (Fig. 1l) was identified as partially opalized clay (Berón, 1996–1998, 2004, 2006; Berón & Curtoni, 2002). This one has low visibility and good quality for knapping. The shades vary from light beige to dark gray and their macroscopic characteristics are similar to the rocks found in Meseta del Fresco. However, at the microscopic level, there are differences, since the rocks of El Carancho do not show skeletal carbonate grains (Carrera Aizpitarte & Berón, 2021).

Finally, a silicified limestone source was identified in Valle Daza (Fig. 1m). This is a low abundant and low visibility outcrop. The nodules are small and the quality of knapping is poor (Charlin, 2002). This source has not been characterized so far (Berón & Carrera Aizpitarte, 2019). Until now, there is evidence that the silicified limestone that outcrops in Meseta del Fresco and El Carancho were transported to various sites in the Pampas region (Berón, 2004, 2006; Messineo, 2008; Massigoge, 2011, 2012; Carrera Aizpitarte, 2014; Santos Valero, 2015, 2017; Santos Valero & Armentano, 2017; Barros et al., 2020; Berón & Carrera Aizpitarte, 2019). However, it is also important to note that there may be sources of this raw material that have not yet been identified.

The Hangar Site

The Hangar site is located at 37°14′37.40″S and 61°14′5.04″W, on a gentle slope in the Salado Creek. In 2017, we made a collection on a surface of 11,750 m^2 and we excavated 11 m^2. Two radiocarbon dates were obtained on guanaco bones at the *Laboratoire des sciences du climat et de l'environnement* (France): 835±30 years BP (GifA18041; median of 706 years cal BP) and 1335±30 years BP (GifA18042; median of 1226 years cal BP). These dates together with geomorphological characteristics, the faunal association, and the presence of abundant pottery inform that the occupations correspond to the end of the Late Holocene. The site has been impacted by the digging action of armadillos and agricultural livestock production, which have generated displacement and fragmentation of archaeological materials (Álvarez et al., 2020).

Lithic artifacts from the site include end scrapers, projectile points, side scrapers, drills, and knives. The presence of cores, flakes, and micro-flakes indicates the manufacturing and maintenance activities of the tools. The most used rocks were chert, orthoquartzite, and silicified limestone. Artifacts manufactured by abrasion and polishing were also recovered, such as bola stones and grinding tools mainly made on granitoid rocks. The detailed analysis of the projectile points (n = 46) shows that they were made, reactivated, recycled, and discarded at the site. These would have been part of weapon systems such as bows and arrows and darts (Barros et al., 2018).

Faunal remains recovered include guanaco, greater rhea (*Rhea americana*), vizcacha, pampas deer, Canidae, cow (*Bos taurus*), different species of armadillos, and micromammals. Some of these taxa such as guanaco, greater rhea, and vizcacha show evidence of human butchery (Barros et al., 2018).

The ceramic sherds found at the site correspond to vessels with simple contours and globular or sub-globular shapes. The firing of the pieces would have involved a long, stable process under non-oxidizing conditions (Borges Vaz et al., 2022). Smoothing, brushing, and painting were recognized as finishing techniques, while decorative techniques common to the Pampas region, such as incisions and impressions, are still absent (Politis et al., 2001; González & Frère, 2010; Di Prado, 2013). Some of the sherds showed traces of soot on the external surface, suggesting that some of the containers were used for food preparation and cooking (Borges Vaz et al., 2022).

The human remains found at the site were scattered on the surface and could have been exhumed by armadillo activity in conjunction with agricultural activities. These remains were not yet dated and belonged to infants and adults. The presence of red paint on the cortical surface of a few specimens indicates that at least some of the burials are secondary (Álvarez et al., 2020).

In sum, different analyses carried out so far indicate that the Hangar would have functioned as a base camp where various activities were carried out, such as the manufacture, replacement, and conditioning of lithic artifacts, along with the butchering and cooking of different prey (Barros et al., 2018).

Materials and methods

Part of the lithic materials analyzed are those collected in the surface context. The analyses followed a technological and techno-morphological perspective in order to assess the main characteristics of the assemblages as well as to build the *chaîne opératoire* (Aschero, 1975, 1983; Bellelli et al. 1985–1987; Pelegrin, 1991, 1995; Geneste, 1991; Perlès, 1991; Inizan et al., 1995). First, the artifacts were separated by raw material based on macroscopic criteria. Each group was classified by artifactual categories (cores, tools, knapping products). Quantitative (e.g., length, width, thickness, weight) and qualitative variables (e.g., blank, platform, typological group of the tools) were considered. In the case of modules, the criteria proposed by Bagolini (1968) and modified by Aschero (1975) were used. The *chaîne opératoire* was reconstructed for the most abundant rocks, although in this chapter we present the results only for the silicified limestone.

In the case of materials from excavation, the frequency of silicified limestone was calculated for excavation and screening materials.

Five silicified limestone artifacts were selected for thin sections. The pieces chosen were those that showed the greatest macroscopic differences among them. The petrographic study enabled a comparative and tentative analysis with previously studied silicified limestones from Meseta del Fresco and El Carancho. In the samples where the depositional texture of the limestone was recognized, before siliceous replacement, the classification of Dunham (1962) was used. It is important to note that there could be other limestone facies that have not been sampled. For this reason, it is not possible to have a correlation or similarity with the whole range of silicified limestone outcroppings in the two mentioned places.

We conducted a bibliographic survey of the sites with silicified limestone in the Pampas region. For this purpose, a database was built where the following variables were recorded: site, location, chronology, presence of silicified limestone, and percentage of this rock about the total raw materials. These data were used to construct fall-off curves by plotting the proportion of silicified limestone against the distance to the sources (Renfrew et al., 1968; Renfrew, 1975). We calculate the average of the values by grouping contexts into 50 km intervals from each source (Meseta del Fresco and El Carancho) to construct a summary distance-decay model. The shape of the curve allows us to examine the relative importance of the silicified limestone as a function of distance from sources and to infer possible foraging ranges. Moreover, we used the QGIS software version 3.22.2 to produce a map using the Hot Spot Analysis tool. This tool calculates the clustering of spatial phenomena. These maps allow easy visualization to display event density or occurrence.

Results

Lithic Assemblage and Silicified Limestone Artifacts in Hangar Site

In the surface lithic assemblage, the most represented raw material is chert ($n = 752$; 52.26%), followed by orthoquartzite SBG ($n = 399$; 27.73%), and silicified limestone ($n = 155$; 10.77%). Among the materials recovered in silicified limestone are tools ($n = 23$; 1.59%) and flakes ($n = 132$; 9.17%) (Table 1). In the excavation of the site, 3 artifacts and 166 micro-flakes on this raw material were also recovered.

The silicified limestone knapping objective was linked to the production of a wide variety of tools such as end scrapers ($n = 9$), projectile points ($n = 5$), crutch-shaped drills ($n = 2$), composite tools ($n = 2$), knives ($n = 2$), biface ($n = 1$), side scraper ($n = 1$) and retouched flake ($n = 1$) (Fig. 2; Table 1). For the manufacture of the end scrapers, interior flakes obtained in reduction stages after decortexing were selected. Except for one item, all end scrapers are complete and are small in size (in a range of 14.2 × 13.2 × 3.9 mm and 23.8 × 19.5 × 7.7 mm). Considering the moduli width/thickness, blanks are very thick and thick, and the length/width moduli indicate the presence of various categories such as very short, short, medium-normal, and medium-elongated. The delineation of the edges was carried out in a direct ($n = 7$) and inverse unifacial ($n = 2$) manner using marginal and summary

Table 1 Lithic artifacts recovered in the Hangar site

		Lithic raw material										
		Chert	Sil. dolomite	Quartz	Orthoquartzite SBG	Diabase	Rhyolite	Metaquartzite	Silica	Sil. limestone	Chalcedony	Undetermined
Tool	Scraper	45	–	1	12	–	–	–	–	9	–	1
	Projectil point	17	–	–	7	–	2	–	1	5	1	–
	Crutch-shaped drill	2	–	–	–	–	–	–	–	2	–	–
	Retouched flake	7	–	–	–	–	–	–	–	1	–	–
	Composite tool	1	–	–	7	–	–	–	–	2	–	1
	Utilized flake	6	–	–	2	–	–	–	–	–	–	–
	Und. fragment/ edge tool	6	–	–	1	–	–	–	–	–	–	–
	Side scraper	–	–	–	6	–	–	–	–	1	–	–
	Bola stone	–	–	–	–	2	–	1	–	–	–	–
	Piéce esquillé	2	–	–	–	–	–	–	–	–	–	–
	Biface	1	–	–	–	–	–	–	–	1	–	–
	Notch	2	–	–	–	–	–	–	–	–	–	–
	Knife	1	–	–	–	–	–	–	–	2	–	–
	Preform projectil point	–	–	–	1	–	–	–	–	–	–	–
	Burinated tools	–	–	–	1	–	–	–	–	–	–	–
	Undetermined	–	–	–	1	–	–	–	–	–	–	–
	Hammerstone	1	–	–	–	–	–	–	–	–	–	–
	Core	16	–	–	1	–	–	–	–	–	–	–

(continued)

Table 1 (continued)

		Lithic raw material										
		Chert	Sil. dolomite	Quartz	Orthoquartzite SBG	Diabase	Rhyolite	Metaquartzite	Silica	Sil. limestone	Chalcedony	Undetermined
K. product	Complete Flake	189	1	–	65	–	–	14	–	34	–	8
	Broken Flake	189	7	–	109	–	5	13	5	41	3	1
	Flake with no butt	267	2	–	179	–	–	23	2	46	1	13
	Ud. knapping product	–	–	1	7	–	–	11	1	11	8	4
	TOTAL (N)	752	10	2	399	2	7	62	9	155	13	28
	TOTAL (%)	52.26	0.69	0.14	27.73	0.14	0.49	4.31	0.63	10.77	0.9	1.95

References = *K* knapping, *Und* undifferentiated, *Ud* undetermined, *Sil* silicified

Fig. 2 Silicified limestone artifacts from the Hangar site: (**a**) crutch-shaped drill, (**b**) end scraper, (**c**) biface, (**d**) projectile point. Scale = 1 cm

retouching. About the projectile points, five fragments correspond to stemless artifacts. In three of them, only the base (straight and attenuated convex) is present, in one the mesial sector and the base (attenuated convex,) and in another the apex and the mesial sector. Due to their condition and degree of modification, it was not possible to characterize the selected blanks. The most common perimeter shape is the short triangular, followed by the elongated triangular, while the cross-section is asymmetrical biconvex and only in one case symmetrical biconvex. Bifacial work is important in manufacturing, which consists of extended knapping and marginal retouching (Barros et al., 2018). The analyses of the weapon system showed that two items correspond to arrows and one to a dart (Kaufmann et al., 2021).

Two crutch-shaped drills were also manufactured on silicified limestone. The larger item is a recycled projectile point in which the active part of the drill was shaped on the apical blade and the original attenuated convex base of the point was maintained. This is a medium-small size and medium-elongated module, which presents an oblique fracture in the active part (Fig. 2a). The second artifact has a longitudinal fracture on the active part and a longitudinal fracture on both sides of the body. Both drills were made by bifacial thinning. On the other hand, the composite tools have an end scraper edge and a notch, are complete, and were made on thick and very thick flakes, the length/width modules are represented by the category short-very wide and medium-elongated and are of small size. The delineation of the edges was made by micro-retouching and unifacial marginal retouching. Knives, with one complete artifact and one fractured artifact, were made on thin flakes and the module of length/width is medium-normal, while the size is medium-small. Edge delineation was performed by micro-retouching and bifacial retouching. The biface is complete, of medium-small size, and normal laminar module; it may correspond to the roughout of a tool. Lastly, the side scraper is fractured and made on a thick flake. Micro-retouching and unifacial retouching were used to delineate the edge. The retouched flake has a small size and short length/width module and presents summary retouching. The knapping products that are part of the assemblage ($n = 121$) are reduction flakes and micro-flakes, thinning flakes, formatization, and reactivation of the cutting edge, obtained by direct percussion and bipolar in a smaller proportion. These types of flakes are related to stages of finalization and maintenance of the tools.

Petrographic Thin Sections

The petrographic study made it possible to know the types of silicification present in the replaced limestone. It also allowed us to determine the presence of allochemical components and of siliciclastic clasts. The main results are described below.

Thin section # 1. Silicified wackestone. Scattered in the micrite are sparse carbonate grains (cg) and agglutinated grains (ag). Cryptocrystalline to microcrystalline quartz (mqz) and opal replace the micrite. There are also sectors of voids or fractures with bands of opal (op) and chalcedony (cl) to the center of the pore. There are also laminae of organic matter (Fig. 3).

Thin section #2. Silicified mudstone (fine limestone) brecciated with isolated siliciclastic grains, of medium sand-sized feldspars and quartz. Limestone has been replaced by cripto- to micro-crystalline (mqz) to crypto-crystalline quartz. Chalcedony (cl) or palisade quartz (cle), and feathery chalcedony (clp) have precipitated in the fractures and pores (Fig. 3).

Thin section #3. Silicified packstone. In the silicified micrite, some agglutinated carbonatic grains, non-skeletal carbonate grains with micritic coating, pellets and grains of quartz and feldspars are observed. The micrite was replaced by microcrystalline to crypto-crystalline quartz. Presents voids coated by bands of opal, chalcedony, or palisade quartz and sometimes mega-quartz have precipitated into the fractures (Fig. 3).

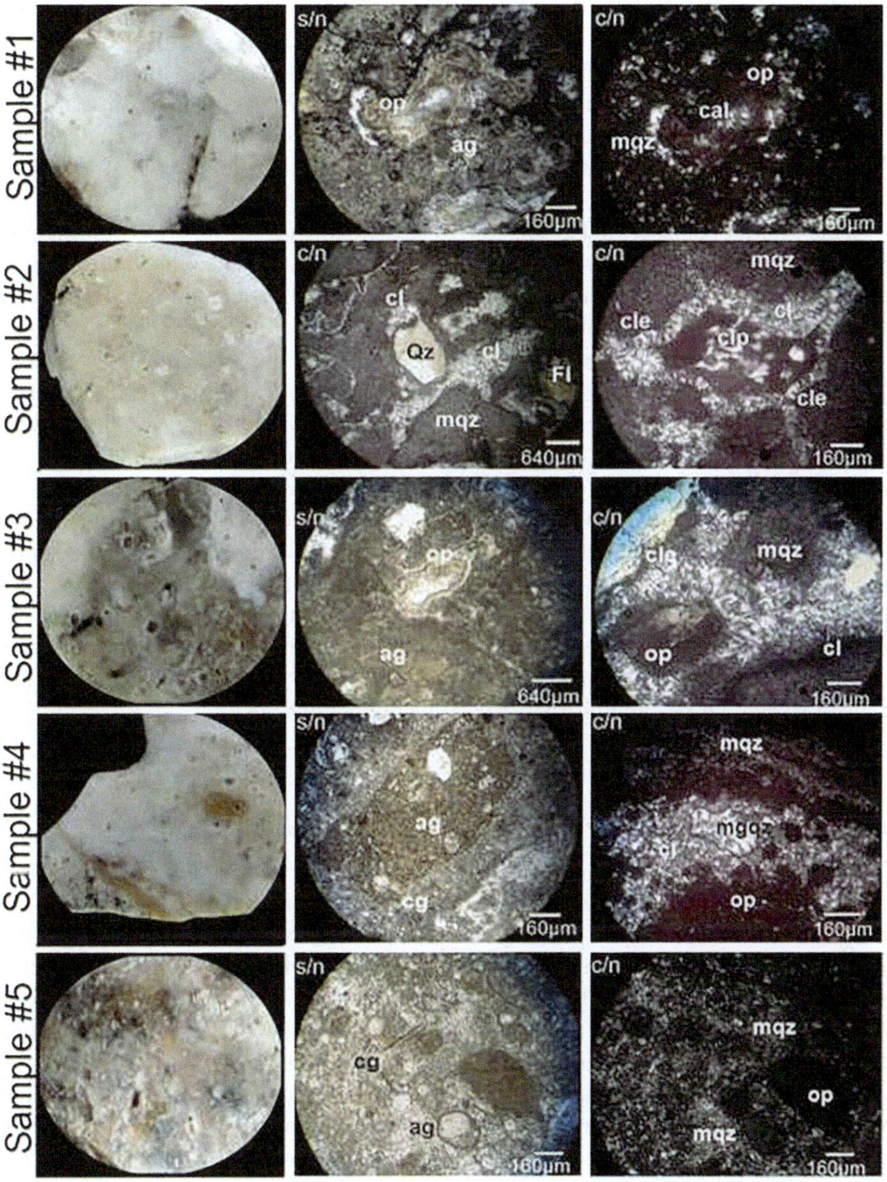

Fig. 3 Macroscopic view of the analyzed samples (left); microphotographs under petrographic microscope (center and right). References: carbonate grains = cg, agglutinated grains = ag, feldspars = F, quartz grains = Qz, microcrystalline quartz = mqz, opal = op, chalcedony = cl, palisade quartz = cle, plumose chalcedony = clp, megaquartz = mgqz, s/n = parallel nicol, c/n = crossed nicols

Thin section #4. Silicified packstone. Several sectors are observed, some with banding, others with some allochemical components such as subspherical non-skeletal carbonate grains with a micritic coating (possibly oolites with surface coating), agglutinated grains, and others with carbonate lithic clasts. The limestone is replaced by micro-quartz and the allochemical components by opal and chalcedony. In the banded sector, the replacement is by opal and micro-quartz. Opal, chalcedony, and meso-quartz have precipitated in the void or fractures (Fig. 3).

Thin section #5. Silicified grainstone to packstone with subspherical nonskeletal carbonate grains with a micritic coating (possibly surface oolites) and agglutinated grains. Limestones are replaced by micro-quartz and allochemical components by opal and chalcedony. No banded sectors were seen (Fig. 3).

The results of the thin-section analysis indicate that the silicified limestones of Hangar present biogenic carbonates, as a result of biological processes in environments with the presence of microbial mats. The observations made in the present work have not allowed us to find allochemical components such as ooids, skeletal fragments, peloids, etc., typical of the not totally silicified limestones of the Meseta del Fresco. On the other hand, simple dispersed carbonate clasts (with a superficial micritic envelope), carbonate aggregates gathered by biofilms, banding of organic matter, and carbonate layers (biolamination), which were later silicified, were recognized. These petrographic characteristics lead to postulate a biogenic origin of the materials associated with the development of microbial mattes. Based on the results obtained in this study, lithological and genetic similarity is found between the materials of El Hangar site and those of El Carancho locality.

Distribution and Fall-off Curves of Silicified Limestone

Sixty-five sites with 80 contexts with the presence of silicified limestone were recorded. The chronological range goes from 7400 years BP to the end of the nineteenth century. Most of the contexts ($n = 30$) could be assigned to the Late Holocene, while others were dated to the initial ($n = 10$) and final ($n = 7$) Late Holocene and historical period ($n = 3$). Six of the contexts correspond to the middle Holocene, one to the early Holocene, and in 23 cases, the chronology is unknown.

In the first place, two graphs were made with the percentages of silicified limestone in archaeological contexts in the Pampas region and their distance to the sources of Meseta del Fresco and El Carancho, respectively. When Meseta del Fresco is considered as the zero point (Fig. 4a), a low frequency of silicified limestone is observed in the contexts located less than 200 km from the source (Fig. 4a). Between 200 and 300 km, there are more sites and higher percentages of this rock. Then, the percentages and the number of sites with this raw material start to decrease. On the other hand, if El Carancho is considered as the zero point (Fig. 4b), the largest number of sites and highest percentages of silicified limestone are located in the first 250 km. Then, low frequencies are recorded, up to 500 km, which is the last context with this rock. The Hangar site, in both cases, is located in the

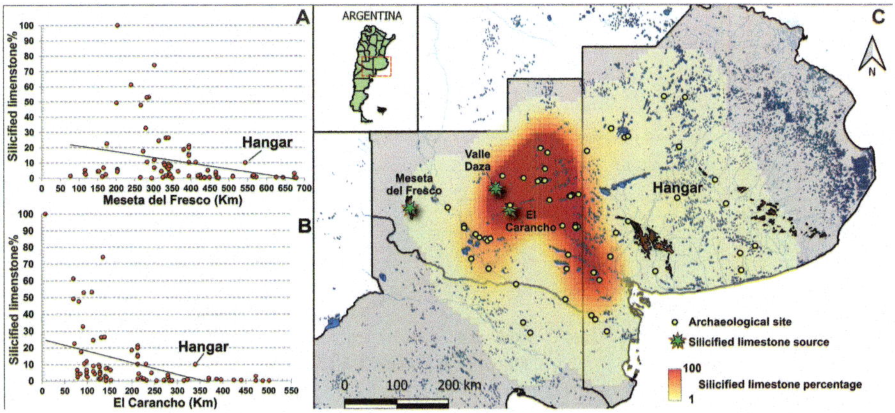

Fig. 4 Silicified limestone sites distribution and distance to the main sources. (**a**) Scatter graph and trend line showing the distance to Meseta del Fresco (x-axis) and the proportion of silicified limestone (y-axis); (**b**) Scatter graph and trend line showing the distance to El Carancho (x-axis) and the proportion of silicified limestone (y-axis), (**c**) Hot Spot Analysis Map

low-frequency zone of silicified limestone, but it stands out from the rest of the contexts due to its higher percentage of this rock.

Among the sites with high frequencies of silicified limestone (around 50% and higher) are Don Isidoro 2, Laguna Traru Lauquen, Laguna del Fondo, Laguna de Paisani, Laguna de Montoya, Médanos de Peñín, and Laguna Chapalcó (Charlin, 2002; Berón, 2004; Carrera Aizpitarte, 2007, 2014). Distances from these sites to El Carancho vary between 8 and 136 km (Supplementary Table 1). With percentages of around 20–30% are El Molino, Médanos de Phagouapé, Laguna La Tigra, El Carmel, Laguna de la Ruta, Médano Santa Clara, and the Laguna Chasicó sites (Berón, 2004; Carrera Aizpitarte, 2007; Catella, 2014; Berón et al., 2015; Páez, 2021; Santos Valero & Messineo, 2021). Distances from these sites to El Carancho vary between 72 and 213 km. The remaining sites, with lower percentages of silicified limestone, are found between 78 and 501 km from El Carancho (see Fig. 4c; Supplementary Table 1).

Second, the fall-off curves were constructed, again considering the distance to the two sources mentioned above. In the case of Meseta del Fresco (Fig. 5a), an atypical curve is observed, since in the first 200 km the percentages of silicified limestone are low and gradually rise to their maximum peak, in the range of 200–250 km. Then, the curve descends, although there is a slight increase that corresponds to the range of distances in which the Hangar contexts are found (500–550 km from Meseta del Fresco). Concerning El Carancho (Fig. 5b), a more expected curve is observed, given that the frequencies of silicified limestone are high in the first 100 km and begin to decrease until 250 km, where it stabilizes at low percentages, except for Hangar (300–350 km from El Carancho).

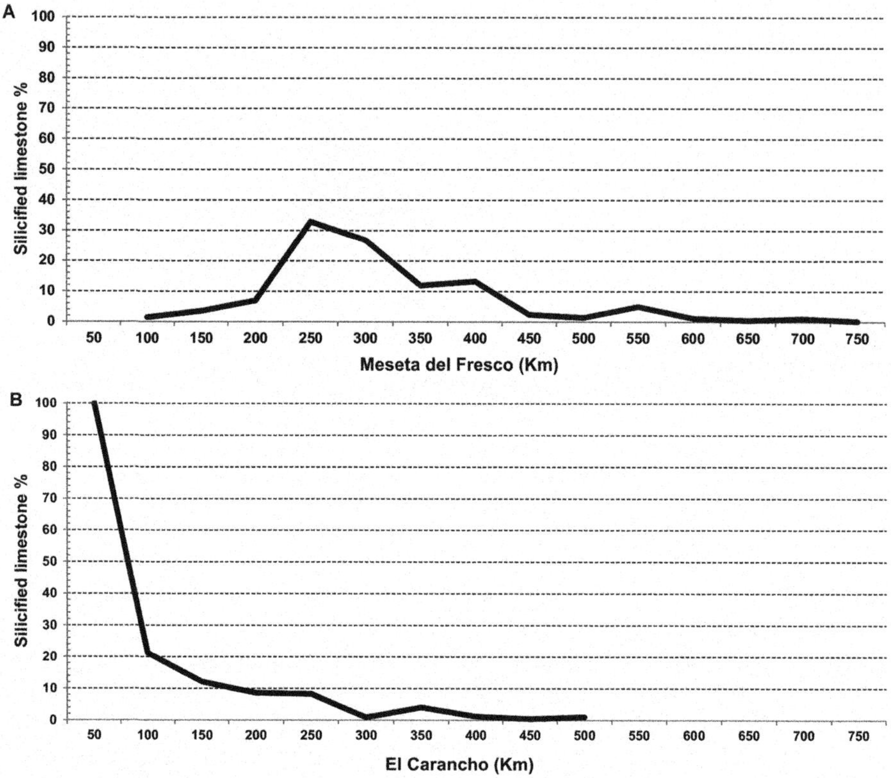

Fig. 5 Fall-off curves of silicified limestone in relation to the distance to the mains sources. (**a**) Distance to Meseta del Fresco, (**b**) Distance to El Carancho

Discussion

The Hangar is a site with low integrity as it has been affected by formation processes such as the action of tillage and animals of fossorial habits (Álvarez et al., 2020). However, it is characterized by a high temporal resolution (sensu Binford, 1981) with occupations concentrated at the end of the Late Holocene, which allows us to discuss some chronological trends. About the lithic assemblage, silicified limestone is the third most represented rock. The artifact assemblage made from this raw material includes tools, flakes, and micro-flakes. This rock could have entered the site in the form of prepared cores and bifaces (Kelly, 1988) from which flakes were extracted to manufacture tools such as end scrapers or were bifacially reduced to make projectile points. In some cases, these artifacts could also have entered the site already finished (e.g., projectile points) or in an advanced state of manufacture. Moreover, through bipolar reduction, it was sought the production of blanks on which several typological groups were made. However, evidence indicates that the *façonnage* was the main process for the manufacturing of projectile points, and end scrapers. Likewise, the presence of micro-flakes is mainly evidence of the maintenance of the edges of the tools. A case of recycling (sensu Schiffer, 1987) was also recorded on this rock; a drill was shaped on a projectile point that kept its original attenuated convex base.

Regarding the source of origin of the silicified limestone, petrographic analyses indicate that raw material found in Hangar can be linked with the studied sections of the outcrops in the El Carancho locality. The presence of agglutinated grains and non-skeletal carbonate grains, among other characteristics, indicates that these are biogenic carbonates, as a result of biological processes in environments with the presence of microbial mats. Still, it is not discarded that beyond these general trends, rocks from Meseta del Fresco and perhaps from sources such as Valle Daza or others that have not yet been identified may have been used. However, when considering the fall-off curves it is clear that Meseta del Fresco would not have been the main source of the rock for the studied assemblages. This is based on the low frequencies recorded at the sites closest to this source and the high increase near El Carancho, which would be the main place of supply. However, sampling biases cannot be discarded, given that survey in the Meseta del Fresco area is still under development.

The shape of the El Carancho fall-off curve shows a wide range of action, reaching up to 500 km from the source. The most intense use of the rock occurs in the first 100 km, where it is likely to be acquired as part of the groups' mobility circuits (Binford, 1979). At greater distances, where low frequencies are recorded, acquisition through interaction with other groups is more likely. These trends are observed diachronically, averaging occupations of different chronologies, most of which correspond to the Late Holocene. At this point, it is interesting to mention the case of Hangar, which stands out for its percentage of silicified limestone (10%) about the other sites in the Salado basin.

In the references on the study microregion, there is no data on the presence of silicified limestone in the sites, although two projectile points manufactured on this rock were found on the surface of the Escuela Agropecuaria site (see Supplementary Fig. 1). These points were found by local settlers and are not part of the analyzed collection from this site. It is different from what occurs in the Hangar site, where the frequency of silicified limestone is relatively high about the distance of its sources. This, together with the record of several crutch-shaped drills, is associated with a chronology of the final Late Holocene (from ca. 1000 years BP to the European conquest; Quintana & Mazzanti, 2001). As different researchers have stated, in the Pampas region and neighboring areas, during the last millennia, there was a sustained process of complex social interactions that triggered, among other aspects, the exchange of goods of various kinds. The dynamics of the complexification process generated mechanisms of social articulation such as exchange, parental or political alliances, and specific long-distance travel, which connected the different hunter-gatherer groups (e.g., Barrientos, 1997, 2001; Martínez, 1999; Madrid & Barrientos, 2000; Politis & Madrid, 2001; Pedrotta, 2005; Mazzanti, 2006; Oliva et al., 2006; Catella, 2014; Berón et al., 2015).

The use of lithic raw materials at the Hangar as well as other contextual evidence from the site allow us to propose some ideas about territories, social interactions, and access to outcrops in the Pampas region. It is important to emphasize that these are hypotheses that will have to be tested based on future analyses. The final Late Holocene occupations of Hangar, as seen in this chapter, present a frequency of silicified limestone that is above that observed for other sites in the area. On a micro-regional scale, this context differs from other sites in the Salado basin by registering very even frequencies of orthoquartzite and chert. In sites such as Escuela Agropecuaria (Crivelli Montero et al. 1987–88; 1997; Eugenio, 1994; Ríos Malan, 2022), Laguna Seca (Barros et al., 2018), Fortín Necochea (Silveira, 1992; Crivelli Montero et al., 1997), and Laguna Muscar 2 (Barros et al., 2018; Kaufmann et al., 2019) orthoquartzite is present with percentages of ca. 60%, followed by chert with ca. 30%. We propose that chronology would be a very important aspect to consider to understand these results. The Hangar occupations support the idea that in a context of high human demography, the circulation of goods, people, and ideas between different sectors of the Pampas and neighboring regions will have increased (Mazzanti, 2006; Berón, 2007; Curtoni, 2007a; Berón et al., 2015; Martínez et al., 2017; Messineo & Barros, 2021). Beyond the presence of western rocks in the east, it is important to note that the reverse trend has also been recorded. Several researchers have noted the presence of orthoquartzite tools, such as side scrapers, in the western sites (Carrera Aizpitarte, 2014; Berón et al., 2015; Heider, 2015, 2016).

In this context, it is proposed that during the final Late Holocene, there was a greater social interaction among the hunter-gatherer bands occupying the Pampas region; the west-central portion of the Pampas region may have been part of a shared territory. The material evidence and the typological groups represented in the sites are in accordance with this proposal. A characteristic item of all these is the projectile points. These are medium and small-sized, triangular, stemless, and with straight to convex bases. These points are predominantly manufactured with rocks from the Tandilia hilly system such as orthoquartzite and chert (Messineo & Barros, 2021). Some projectile points with similar characteristics are made with rocks from the Ventania hilly system and silicified limestone. These items are also found in Central Pampean Dune fields (Messineo et al., 2019) and the Dry Pampas (Berón, 2004; Heider, 2015). Crutch-shaped drills are another type of artifact that share designs and rocks in several sites in the Pampas region, generally associated with Late Holocene chronologies. These artifacts are present in different areas of the Pampas such as the inter-hill plains (Bórmida, n.d.; Eugenio, 1994; Pedrotta, 2005), southern Pampas (Torres, 1922; Bayón et al., 2006; Vecchi et al., 2013), western Pampas (Viani, 1930) and the Dry Pampas (Berón, 2004; Carrera Aizpitarte et al., 2013).

In general, it has been proposed that during the Holocene, in the inter-hill area, access to the most distant sources, such as silicified limestone, did not occur through direct access to the sources, but rather exchange strategies were developed between groups occupying different territories (Messineo, 2008). By the time of the final Late Holocene, evidence from the Hangar indicates a change in the management of lithic resources. The procurement of orthoquartzite, chert, and silicified limestone could have been carried out under the same mobility system. This is referred to by some authors as direct access, meaning the size of the territory or foraging range that a human group usually occupied, and which is associated with social interaction among hunter-gatherer societies (Bamforth, 1986, 1990; Kuhn, 1995; Mangado, 2006; Hughes, 2011; Pallo & Borrero, 2015; Seeman et al., 2020). This increase in the frequency of these rocks occurs even in sectors of the territory far from these sources. The presence of rocks from the Pampean Caldenal in the northern sector of the Inter-hill plains and the southwest of the dune field, as well as the presence of orthoquartzites from Tandilia in the Caldenal, show a strong social interaction between these human groups. The interaction between the xerophytic forest environments in the western Pampas and the northwest sector of the Inter-hill plains would have been generated through the natural corridors. One of the most important was formed by a series of interconnected shallow lakes called Las Encadenadas. This corridor functioned as a very important route in historical times. According to several historic sources, this was called *Camino a Salinas Grandes* or *Rastrillada de los Chilenos* (Road of the Chileans) (García [1810](1969)). This consisted of a group of routes used by indigenous people to head cattle from Argentina to Chile. This route was also used by Argentina to obtain salt from the shallow lake system known as Salinas Grandes. Some authors (Berón, 2006; Berón et al., 2017; Curtoni, 2007b) indicate that this route constituted one of the main nodes in the organization of indigenous movements connecting the Humid and occidental Pampas. We propose that this route would have functioned as a corridor for the circulation of information, people, and goods and that it would have had an important time depth, dating back at least a thousand years BP.

Final Remarks

In this chapter, we evaluated and discussed the existence of social interaction based on the distribution of silicified limestone in different sites of the Pampas region. The evidence studied allowed us to determine that this raw material entered the Hangar site in the form of prepared cores and bifaces as well as finished tools. Different activities were carried out with this rock and different types of artifacts were manufactured. Its acquisition could have been by interaction or direct access. Although the evidence does not allow us to differentiate between these two options, we can propose that there was access to the silicified limestone source. Moreover, the petrographic analysis showed this source would be linked to the outcrops in El Carancho locality, which are biogenic carbonates resulting from biological processes. The fall-off curve is in accordance with these data, showing an intense use of silicified limestone in the first 100 km from the source. At greater distances, lower frequencies are recorded, and the Hangar site stands out for its percentage of 10% at a distance of 340 km. According to this curve, the silicified limestone source of El Carancho would have been widely used by the groups that inhabited the Pampean Caldenal and its presence would have reached very distant areas of the Pampas, such as the plains located between the Tandilia and Ventania hilly systems.

Considering different lines of evidence such as lithic projectile point types and raw material percentages, it is proposed that during the final Late Holocene, the west-central portion of the pampas region could have been part of a shared territory. In these periods, there was a context of higher human demography, and consequently, the circulation of goods, people, and ideas increased. These data are in accordance with the results already discussed by other researchers, who proposed a scenario of social complexity for the Late Holocene periods. In this sense, some corridors that connected nodes would have functioned for the circulation of information much earlier than in already-known historical times.

Acknowledgments Financial support for this research was provided by the Agencia Nacional de Promoción Científica y Tecnológica, Argentina (PICT 2015-0235; PICT 2017-1969; PICT 2020-00789) and Mission Archéologique Française en Argentine. INCUAPA-CONICET and FACSO-UNCPBA provided institutional support. We would like to thank Emilio Eugenio for allowing us to use the material from the Escuela Agropecuaria site and Luciana Catella for allowing us to examine the collection of sillicified limenstone from her study area. To the Juarros family, who permitted us to conduct fieldwork on their land, and to Pedro Guillardoy for providing useful bibliography. To Pablo Messineo, for his valuable contributions and suggestions. Two anonymous reviewers improved this manuscript with their suggestions.

References

Álvarez, M. C., Rafuse, D. J., Bellinzoni, J. E., & Kaufmann, C. A. (2020). Armadillos as taphonomic agents of archaeological sites in the Pampas Region, Argentina. *Journal of Archaeological Science: Reports, 31*, 10102293. https://doi.org/10.1016/j.jasrep.2020.102293

Aschero, C. A. (1975). *Ensayo para una Clasificación Morfológica de Artefactos Líticos aplicada a Estudios Tipológicos Comparativos*. Report to the Consejo Nacional de Investigaciones Científicas y Técnicas (CONICET), Buenos Aires. Unpublished manuscript.

Aschero, C. A. (1983). *Ensayo para una Clasificación Morfológica de Artefactos Líticos aplicada a Estudios Tipológicos Comparativos. Apéndice A-C. Revisión 1983*. Cátedra de Ergología y Tecnología. Facultad de Filosofía y Letras. Universidad de Buenos Aires, Buenos Aires. Unpublished Manuscript.

Bagolini, B. (1968). *Ricerche sulle dimensioni dei manufatti litici non ritoccati*. Università di Ferrara.

Bamforth, D. B. (1986). Technological efficiency and tool curation. *American Antiquity, 51*, 38–50. https://doi.org/10.2307/280392

Bamforth, D. B. (1990). The settlement raw material and lithic procurement in the Central Mojave Desert. *Journal of Anthropological Archaeology, 9*, 70–104. https://doi.org/10.1016/0278-4165(90)90006-Y

Barrientos, G. (1997). *Nutrición y dieta de las poblaciones aborígenes prehispánicas del sudeste de la Región Pampeana* [Unpublished doctoral Thesis]. Facultad de Ciencias Naturales y Museo, Universidad de La Plata.

Barrientos, G. (2001). Una aproximación bioarqueológica al estudio del poblamiento prehistórico tardío del Sudeste de la Región Pampeana. *Intersecciones en Antropología, 2*, 3–18.

Barros, M. P., Heider, G., Álvarez, M. C., Kaufmann, C., & Bellinzoni, J. (2018). First results of the hunter-gatherer weapon system studies in the middle basin of the Salado creek (Pampas Region, Argentina). *Journal of Lithic Studies, 5*(2), 1–23. https://doi.org/10.2218/jls.2785

Barros, M. P., Bellinzoni, J., Colantonio, M. J., Álvarez, M. C., & Kaufmann, C. A. (2020). Aportes del sitio Hangar al estudio de la gestión tecnológica en la cuenca del Arroyo Salado (provincia de Buenos Aires) durante el Holoceno tardío. *Revista del Museo de Antropología, 13*(1), 237–242. https://doi.org/10.31048/1852.4826.v13.n1.24180

Barros, M. P., Álvarez, M. C., Kaufmann C. A., Gómez Peral, L., Rafuse, D., Politis, G. G. (2021). Una nueva cantera de cuarcita en el extremo occidental de Tandilia. Proceedings of the *IX Congreso de Arqueología de la Región Pampeana Argentina*: 139.

Bayón, C., & Zavala, C. (1997). Coastal site in south Buenos Aires: A review of Piedras Quebradas. In J. Rabassa & M. Salemme (Eds.), *Quaternary of South America and Antarctic Peninsula 10* (pp. 229–253). A. A. Balkema, Bookfield.

Bayón, C., Flegenheimer, N., Valente, M., & Pupio, A. (1999). Dime cómo eres y te diré de dónde vienes: La procedencia de rocas cuarcíticas en la región pampeana. *Relaciones de la Sociedad Argentina de Antropología, XXIV*, 187–235.

Bayón, C., Flegenheimer, N., & Pupio, A. (2006). Planes sociales en el abastecimiento y traslado de roca en la Pampa Bonaerense en el Holoceno Temprano y Tardío. *Relaciones de la Sociedad Argentina de Antropología, XXXI*, 19–45.

Belardi, J., Carballo Marina, F., & Espinosa, S. (2006). Cazadores recolectores en la cuenca media e inferior del río Coyle. In J. B. Belardi, F. Carballo Marina, & S. Espinosa (Eds.), *La cuenca del Río Coyle. Estado Actual de las Investigaciones* (pp. 97–128). Universidad Nacional de la Patagonia Austral.

Bellelli, C., Guráieb, A., & García, J. (1985–1987). Propuesta para el análisis y procesamiento por computadora de desechos de talla lítica (DELCO-Desechos líticos computarizados). *Arqueología Contemporánea, 2*(1), 36–53.

Berón, M. A. (1996–1998). Nuevos rumbos, viejos caminos. Cuenca Inferior del Chadileuvú (37° 33′ a 38° 11′ Lat. Sur). Primer trabajo de campo 1996. *Palimpsesto, 5*, 106–118.

Berón, M. A. (2004). *Dinámica poblacional y estrategias de subsistencia de poblaciones pre-hispánicas de la cuenca Atuel-Salado-Chadileuvú-Curacó* [Unpublished doctoral Thesis]. Facultad de Filosofía y Letras, Universidad de Buenos Aires.

Berón, M. A. (2006). Base Regional de recursos minerales en el occidente pampeano. Procedencia y estrategias de aprovisionamiento. *Relaciones de la Sociedad Argentina de Antropología, XXXI*, 47–88.

Berón, M. A. (2007). Circulación de bienes como indicador de interacción entre las poblaciones de la pampa occidental y sus vecinos. In C. Bayón, A. Pupio, M. I. González, N. Flegenheimer, & M. Frere (Eds.), *Arqueología en las Pampas* (pp. 345–364). Sociedad Argentina de Antropología.

Berón, M. A., & Carrera Aizpitarte, M. P. (2019). Materias primas y circuitos de movilidad en el Noroeste de Patagonia. Una aproximación al estudio de los conflictos sociales a partir de la evidencia arqueológica. *Revista del Museo de Antropología, 12*(1), 7–22. https://doi.org/10.31048/1852.4826.v12.n1.19812

Berón, M. A., & Curtoni, R. P. (2002). Propuestas metodológicas para la caracterización arqueológica de canteras y talleres de la Meseta del Fresco (La Pampa, Argentina). In D. L. Mazzanti, M. A. Berón, & F. Oliva (Eds.), *Del mar a los salitrales. Diez mil Años de Historia Pampeana en el Umbral del Tercer Milenio* (pp. 171–184). Sociedad Argentina de Antropología y Facultad de Humanidades, Laboratorio de Arqueología, Universidad Nacional de Mar del Plata.

Berón, M. A., Carrera Aizpitarte, M. P., & Páez, F. (2015). Arqueología en el área de Valles Transversales (provincia de La Pampa, Argentina). Caracterización y tendencias de los conjuntos arqueológicos. Implicancias sociales en la construcción del paisaje. *Relaciones de la Sociedad Argentina de Antropología, 40*(2), 549–587.

Berón, M. A., Di Biase, A. R. S., Musaubach, M. G., & Páez, F. N. (2017). Enclaves y espacios internodales en la dinámica de poblaciones en el Wall-Mapu: aportes desde la arqueología pampeana. *Estudios Atacameños, 56*, 253–272.

Berón, M., Migale, L., & Curtoni, R. (1995). Hacia la definición de una base regional de recursos líticos en el área de Curacó. Una cantera taller: Puesto Córdoba (La Pampa, Argentina). *Relaciones de la Sociedad Argentina de Antropología, 20*, 111–128.

Binford, L. R. (1979). Organization and formation processes: Looking at curated technologies. *Journal of Anthropological Research, 35*, 255–273. https://doi.org/10.1086/jar.35.3.3629902
Binford, L. R. (1981). *Bones: Ancient men and modern myths*. Academic.
Binford, L. R. (2006). Bands as characteristic of "mobile hunter-gatherers" may exist only in the history of anthropology. In F. Sellet, R. Greaves, & P. L. Yu (Eds.), *The archaeology and ethnoarchaeology of mobility* (pp. 3–22). University Press of Florida.
Bonnat, F. (2021). Análisis de la organización de la tecnológica lítica de los grupos cazadores-recolectores tempranos del área de Tandilia oriental. Sociedad Argentina de Antropología.
Bonomo, M. (2005). *Costeando las Llanuras. Arqueología del Litoral Marítimo Pampeano* [Colección de Tesis Doctorales]. Sociedad Argentina de Antropología.
Bonomo, M. (2007). El uso de los moluscos marinos por los cazadores-recolectores pampeanos. *Chungará, 39*(1), 87–102. https://doi.org/10.4067/S0717-73562007000100006
Borges Vaz, E., Álvarez, M. C., Barros, M. P., & Kaufmann, C. A. (2022). Tecnología cerámica de grupos cazadores-recolectores en el área Interserrana Bonaerense: el caso del sitio Hangar (Partido de General Lamadrid, Provincia de Buenos Aires). *Relaciones de la Sociedad Argentina de Antropología, 47*(1), 121–138. https://doi.org/10.24215/18521479e033
Bórmida, M. (n.d.). s/f. *Prolegómenos para una arqueología de la Pampa Bonaerense*. Edición oficial de la Provincia de Buenos Aires. Dirección de Bibliotecas, Museos y Archivos Históricos.
Buchanan, B., Hamilton, M., Hartley, J., & Kuhn, S. (2017). Investigating the scale of prehistoric social networks using culture, language, and point types in western North America. *Archaeological and Anthropological Sciences, 11*(1), 199–207. https://doi.org/10.1007/s12520-017-0537-y
Burgos, J. J. (1968). El clima de la provincia de Buenos Aires en relación con la vegetación natural y el suelo. In L. A. Cabrera (Ed.), *Flora de la Provincia de Buenos Aires Parte 1* (pp. 33–99). Colección Científica del INTA.
Carrera Aizpitarte, M. P. (2007). *Estrategias de aprovisionamiento lítico en el área centro-este de la provincia de La Pampa* [Unpublished Graduation Thesis]. Facultad de Ciencias Sociales, Universidad Nacional del Centro de la Provincia de Buenos Aires.
Carrera Aizpitarte, M. P. (2014). *Estudio de las estrategias de aprovisionamiento lítico en las áreas Curacó, Bajos sin Salida, Valles Transversales y Centro-este (provincia de La Pampa, Argentina)* [Unpublished doctoral Thesis]. Facultad de Ciencias Sociales, UniversidadNacional del Centro de la Provincia de Buenos Aires.
Carrera Aizpitarte, M. P., & Berón, M. A. (2021). Explotación de recursos líticos en dos canteras prehispánicas de la provincia de La Pampa (Argentina): Meseta del Fresco y Manto Tehuelche. *Estudios Atacameños, 66*, 7–34. https://doi.org/10.22199/issn.0718-1043-2020-0051
Carrera Aizpitarte, M. P., Luna, L., Aranda, C., & Berón, M. A. (2013). Médano La Enriqueta: un lugar de entierro de cazadores tardíos sobre el río Colorado (Departamento. Caleu Caleu, provincia de La Pampa). *Revista del Museo de La Plata, 13*(87), 167–184.
Carrera Aizpitarte, M. P., Berón, M. A., & Blasi, A. (2015). Study of siliceous outcrops of Meseta del Fresco, La Pampa, Argentina. *Quaternary International, 375*, 27–43. https://doi.org/10.1016/j.quaint.2014.11.061
Catella, L. (2014). *Movilidad y utilización del ambiente en poblaciones cazadoras recolectoras del sur de la Región Pampeana: la cuenca del Arroyo Chasicó como caso de estudio* [Unpublished doctoral Thesis]. Facultad de Ciencias Naturales y Museo, Universidad Nacional de La Plata.
Catella, L. (2017). La tecnología cerámica en la cuenca del arroyo Chasicó. Una aproximación a la movilidad de las sociedades del Holoceno tardío de Patagonia nororiental y de la región pampeana sudeste, Argentina. *Magallania, 45*(1), 137–163. https://doi.org/10.4067/S071822442017000100137
Catella, L., Moirano, J., & Oliva, F. (2010). Disponibilidad de materias primas líticas y su importancia para el análisis del uso del espacio y la organización de la tecnología en sociedades cazadoras recolectoras. In M. Berón, L. Luna, M. Bonomo, C. Montalvo, C. Aranda, & M. Carrera Aizpitarte (Eds.), *Mamül Mapu: pasado y presente desde la arqueología pampeana* (pp. 215–231). Libros del Espinillo.

Catella, L., Manassero, M., Moirano, J., & Oliva, F. (2013). Nuevos aportes al estudio del aprovisionamiento de cuarcita en la Región Pampeana, Argentina. *Cuadernos del Instituto Nacional de Antropología y Pensamiento Latinoamericano-Series Especiales, 1*(2), 200–215.

Charlin, J. (2002). Aprovisionamiento de materias primas líticas en el N.O. de la provincia de la Pampa a fines del Siglo XIX. In D. Mazzanti, M. A. Berón, & F. Oliva (Eds.), *Del Mar a los Salitrales. 10.000 años de Historia pampeana en el umbral del Tercer Milenio* (pp. 205–218). Laboratorio de Arqueología, Facultad de Humanidades, Universidad Nacional de Mar del Plata.

Colombo, M. (2011). El área de abastecimiento de las ortocuarcitas del grupo Sierras Bayas y las posibles técnicas para su obtención entre los cazadores y recolectores pampeanos. *Intersecciones en Antropología, 12*, 155–166.

Crivelli Montero, E., Eugenio E., Silveria, M. (1987–1988). El sitio Fortín Necochea (partido de General Lamadrid, provincia de Buenos Aires), estado actual de los Trabajos. *Paleoetnológica, 4*, 7–37. CAEA.

Crivelli Montero, E., Eugenio, E., Pardiñas, U., & Silveira, M. (1997). Archaeological investigation in the plains of the province of Buenos Aires, Llanura Interserrana Bonaerense. In J. Rabassa & M. Salemme (Eds.), *Quaternary of South American and Antarctic Peninsula* (Vol. 10 1994, pp. 167–209). A. A. Balkema, Bookfield.

Curtoni, R. P. (2006). Expresiones simbólicas, cosmovisión y territorialidad en los cazadores-recolectores pampeanos. *Relaciones de la Sociedad Argentina de Antropología, 31*, 133–160.

Curtoni, R. P. (2007a). *Arqueología y paisaje en el área centro-este de La Pampa* [Unpublished doctoral Thesis]. Facultad de Ciencias Naturales y Museo, Universidad Nacional de La Plata.

Curtoni, R. P. (2007b). Análisis e interpretación de las rastrilladas indígenas del sector centro-este de la provincia de La Pampa. *Revista de Arqueología Histórica Argentina y Latinoamérica, 1*, 65–92.

Curtoni, R. P., & Berón, M. A. (2011). Perception, identity and meaning in the social and ritual construction of the landscape. The Lihue Calel Hills, La Pampa, Argentina. *Revista Chilena de Antropología, 24*, 97–11. https://doi.org/10.5354/0719-1472.2012.18168

Curtoni, R. P., Barros, M. P., & Berón, M. A. (2004). Meseta del Fresco: análisis de canteras y talleres. In C. Gradín & F. Oliva (Eds.), *La Región Pampeana. Su pasado arqueológico* (pp. 287–296). Venado Tuerto.

Di Prado, V. (2013). Estudio de la alfarería del sitio Calera (partido de Olavarría, provincia de Buenos Aires, Argentina) desde la perspectiva del estilo tecnológico. *Revista del Museo de La Plata, 13*(87), 279–298.

Di Prado, V. (2018). Prácticas alfareras prehispánicas y procesos de interacción social en el centro-este de Argentina durante el Holoceno tardío. *Latin American Antiquity, 29*(3), 552–571. https://doi.org/10.1017/laq.2018.28

Donadei, J. P. (2019). Local and nonlocal rocks: Technological strategies and raw material management. Hunter-gatherer mobility for mid-Holocene groups of eastern Tandilia range (Argentina). *Journal of Archaeological Science: Reports, 24*, 264–275. https://doi.org/10.1016/j.jasrep.2019.01.011

Dunham, R. (1962). Classification of carbonate rocks according to depositional texture. In W. E. Ham (Ed.), (pp. 108–121). AAPG Memoir 1.

Eugenio, E. O. (1994). Recursos, Tecnología y Movilidad Territorial de los Cazadores de General La Madrid. In M. Silveira (Ed.), *10.000 AÑOS Prehistoria, Etnohistoria e Historia del Partido de General La Madrid* (pp. 47–66). Asociación Amigos del Complejo Cultural General Lamadrid.

Flegenheimer, N. (1991). La Liebre, un sitio cantera-taller. *Boletín del Centro, 2*, 58–64.

Flegenheimer, N., & Bayón, C. (2002). Cómo, Cuándo y Dónde? Estrategias de abastecimiento lítico en la Pampa Bonaerense. In D. Mazzanti, M. A. Berón, & F. Oliva (Eds.), *Del Mar a los Salitrales, Diez Mil Años de Historia Pampeana en el Umbral del Tercer Milenio* (pp. 231–241). Facultad de Humanidades. Universidad Nacional de Mar del Plata.

Flegenheimer, N., Kain, C., Zárate, M., & Barna, A. (1996). Aprovisionamiento de Cuarcitas en Tandilia, las Canteras de Arroyo Diamante. *Arqueología, 6*, 117–141.

Flegenheimer, N., Bayón, C., Valente, M., Baeza, J., & Femeninas, J. (2003). Long distance tool stone transport in the Argentine Pampas. *Quaternary International, 109–110*, 49–64. https://doi.org/10.1016/S1040-6182(02)00202-1

García, P. A. [1810](1969). Diario de un viaje a Salinas Grandes. In De Angelis P. *Colección de obras y documentos relativos a la historia antigua y moderna de las provincias del Río de la Plata,* T° IV (pp. 259–391). .

Geneste, J. M. (1991). L'approvisionnement en matières premières dans les systèmes de production lithique: la dimension spatiale de la technologie. In R. Mora, X. Terradas, A. Parpal, & C. Plana (Eds.), *Tecnología y cadenas operativas líticas, Treballs d'Arqueología 1* (pp. 1–36). Universidad Autónoma de Barcelona.

González, M. I., & Frère, M. M. (2010). *Diseños Prehispánicos de la Alfarería Pampeana.* Facultad de Filosofía y Letras, Universidad de Buenos Aires.

Heider, G. (2015). *Los Pueblos Originarios en el Norte de Pampa Seca. Una mirada arqueológica a los cazadores recolectores del Sur de las provincias de Córdoba y San Luis, Argentina* [Unpublished doctoral Thesis]. Facultad de Filosofía y Humanidades, Universidad Nacional de Córdoba.

Heider, G. (2016). Un aporte a la discusión sobre las ortocuarcitas del grupo Sierras Bayas: ¿el sur de Córdoba como una frontera? *Intersecciones en Antropología, 17*, 303–313.

Hughes, R. E. (2011). Sources of inspiration for studies of prehistoric resource acquisition and material conveyance in California and the Great Basin. In R. E. Hughes (Ed.), *Perspectives on Prehistoric Trade and Exchange in California and the Great Basin* (pp. 1–21). The University of Utah Press.

Inizan, M. L., Reduron-Ballinger, M., Roche, H., & Tixier, J. (1995). *Technologie de la pierre taillée* (p. 201). CREP, CNRS, Université Paris Nanterre X.

Kaufmann, C., & González, M. (2013). Rescate Arqueológico de Restos Óseos Humanos en el Sitio Laguna Seca 1 (Partido de General Lamadrid, Provincia de Buenos Aires). *Revista del Museo de La Plata, 13*(87), 125–136.

Kaufmann, C., Bellinzoni, J., Álvarez, M. C., González, M., Favier-Dubois, C., & Barros, M. P. (2019). Aportes del estudio de los procesos de formación de sitio para el análisis de los restos humanos del sitio Laguna Muscar 2 (Lamadrid, Buenos Aires). *Comechingonia, 23*(1), 325–348. https://doi.org/10.37603/2250.7728.v23.n1.25982

Kaufmann, C. A., Álvarez, M. C., Messineo, P. G., Barros, M. P., Bonomo, M., & Heider, G. (2021). Guanaco Hunting Strategies in the Southeastern Pampas during the Late Holocene. In J. B. Belardi, D. L. Bozzuto, P. M. Fernández, E. A. Moreno, & G. A. Neme (Eds.), *Ancient hunting strategies in Southern South America* (The Latin American studies book series) (pp. 83–112). Springer. https://doi.org/10.1007/978-3-030-61187-3_4

Kelly, R. (1988). The three sides of a biface. *American Antiquity, 53*(4), 717–734. https://doi.org/10.2307/281115

Kuhn, S. (1995). *Mousterian lithic technology. An ecological perspective.* Princeton University Press.

Madrid, P., & Barrientos, G. (2000). La estructura del registro arqueológico del sitio Laguna Tres Reyes 1 (Provincia de Buenos Aires): nuevos datos para la interpretación del poblamiento humano del Sudeste de la Región Pampeana a inicios del Holoceno tardío. *Relaciones de la Sociedad Argentina de Antropología, XXV*, 179–206.

Madrid, P., & Salemme, M. (1991). La ocupación tardía del sitio 1 de la Laguna Tres Reyes, Adolfo Gonzáles Chaves, Provincia de Buenos Aires. *Boletín del Centro de Registro del Patrimonio Arqueológico y Paleontológico, 3*, 165–179.

Mangado, J. (2006). El aprovisionamiento en materias primas líticas: hacia una caracterización paleocultural de los comportamientos paleoeconómicos. *Trabajos de Prehistoria, 63*(2), 79–91. https://doi.org/10.3989/tp.2006.v63.i2.18

Martínez, G. (1999). *Tecnología, subsistencia y asentamiento en el curso medio del Río Quequén Grande: Un enfoque arqueológico* [Unpublished docotral Thesis]. Facultad de Ciencias Naturales y Museo, Universidad de La Plata.

Martínez, G. (2017). *Arqueología de cazadores-recolectores del curso inferior del Río Colorado*. Serie Monográfica del INCUAPA Nro. 6, Facultad de Ciencias Sociales (UNCPBA).

Martínez, O., Rabassa, J., & Coronato, A. (2009). Charles Darwin and the first scientific observations on the Patagonian Shingle Formation ("Rodados Patagónicos"). *Revista de la Asociación Geológica Argentina, 64*(1), 90–100.

Martínez, G., Santos Valero, F., Flensborg, G., Carden, N., Stoessel, L., Alcaraz, A. P., Vaz, B., & Erika. (2017). Was there a process of regionalization in Northeastern Patagonia during the late Holocene? *The Journal of Island and Coastal Archaeology, 12*(1), 95–114. https://doi.org/10.1080/15564894.2016.1163756

Massigoge, A. (2011). Nuevas evidencias arqueológicas del Holoceno tardío en el área Interserrana: el sitio Las Brusquillas 2 (partido de San Cayetano, provincia de Buenos Aires, Argentina). *Cazadores Recolectores del Cono Sur, 5*, 179–195.

Massigoge, A. (2012). Las Brusquillas 1 (partido de San Cayetano, provincia de Buenos Aires): un nuevo sitio del Holoceno tardío del área Interserrana de la región pampeana. *Intersecciones en Antropología, 13*, 377–392.

Mazzanti, D. L. (1997). Excavaciones arqueológicas en el sitio Cueva Tixi, Buenos Aires, Argentina. *Latin American Antiquity, 8*, 55–62. https://doi.org/10.2307/971592

Mazzanti, D. L. (2006). La constitución de territorios sociales durante el Holoceno tardío. El caso de las sierras de Tandilia, Argentina. *Relaciones de la Sociedad Argentina de Antropología, XXXI*, 277–300.

Melchor, R., & Casadío, S. (1999). *Descripción geológica de la Hoja 3766-III "La Reforma" (1:250.000), Provincia de La Pampa*. Servicio Geológico Minero Argentino.

Meltzer, D. J. (1989). Was stone exchanged among Eastern North American Paleoindians? In C. Ellis & L. Lothrop (Eds.), *Eastern Paleoindian Lithic Resource Use* (pp. 11–39). Westview Press.

Merino, M. L. (2003). *Dieta y uso de hábitat del venado de las pampas, Ozotoceros bezoarticus celer, Cabrera 1943 (Mammalia-Cervidae) en la zona costera de Bahía Sanborombón, Buenos Aires, Argentina*. Implicancias para su conservación [Unpublished doctoral Thesis]. Facultad de Ciencias Naturales y Museo, Universidad Nacional de La Plata.

Messineo, P. G. (2008). *Investigaciones arqueológicas en la cuenca superior del arroyo Tapalqué (Partidos de Olavarría y Benito Juárez, Provincia de Buenos Aires)* [Unpublished doctoral Thesis]. Facultad de Ciencias Naturales y Museo, Universidad Nacional de La Plata.

Messineo, P. G., & Barros, M. P. (2015). Lithic raw materials and modes of exploitation in quarries and workshops from the center of the Pampean grassland of Argentina. *Lithic Technology, 40*(1), 3–20. https://doi.org/10.1179/2051618514Y.0000000007

Messineo, P. G., & Barros, M. P. (2021). Hunter-gatherer mobility and territories in the dunefields of center Argentina. *Journal of Lithic Studies, 8*(1), 1–19. https://doi.org/10.2218/jls.4313

Messineo, P. G., Barros, M. P., Poiré, D. G., & Gómez Peral, L. (2004). Características litológicas de los niveles de chert o ftanitas en las Sierras Bayas (Partido de Olavarría, Provincia de Buenos Aires). In G. A. Martínez, M. A. Gutiérrez, R. P. Curtoni, & M. A. Berón Madrid (Eds.), *Aproximaciones Contemporáneas a la Arqueología Pampeana. Perspectivas Teóricas, Metodológicas, Analíticas y Casos de Estudio* (pp. 307–319). FACSO-UNICEN.

Messineo, P. G., Barros, M. P., Pal, N., & Scheifler, N. A. (2019). Transporting rocks to an empty environment of lithic raw materials. The case of the Central Pampean Dunefield (Argentina). *Journal of Archaeological Science: Reports, 25*, 433–446. https://doi.org/10.1016/j.jasrep.2019.05.012

Miller, J. M., Sawchuk, E. A., Reedman, A. L. R., & Willoughby, P. (2018). Land Snail Shell Beads in the Sub-Saharan Archaeological Record: When, Where, and Why? *African Archaeological Review, 35*, 347–378. https://doi.org/10.1007/s10437-018-9305-3

Mitchell, P. J. (1996). Prehistoric exchange and interaction in southeastern southern Africa: marine shells and ostrich eggshell. *African Archaeological Review, 13*, 35–76.

Oliva, F., & Moirano, J. (1997). Primer Informe sobre Aprovisionamiento Primario de Riolita en Sierra de La Ventana. In M. A. Berón & G. Politis (Eds.), *Arqueología Pampeana en la Década de los 90* (pp. 137–146). Museo de Historia Natural de San Rafael.

Oliva, F., Ávila, J. D., Ayuso, M., Catella, L., & Gallego, M. (2006). Investigaciones arqueológicas en la cuenca del arroyo Chasicó y llanura adyacente, Sudoeste de la Provincia de Buenos Aires. *Comechingonia, 9*, 123–139. https://doi.org/10.37603/2250.7728.v9.n1.27619

Oyarzábal, M., Clavijo, J., Oakley, L., Biganzoli, F., Tognetti, P., Barberis, I., Maturo, H. M., Aragón, R., Campanello, P. I., Prado, D., Oesterheld, M., & León, R. J. C. (2018). Unidades de vegetación de la Argentina. *Ecología Austral, 28*, 40–63. https://doi.org/10.25260/EA.18.28.1.0.399

Páez, F. (2021). Laguna La Tigra. Un sitio de cazadores-recolectores del Holoceno tardío en el área de Valles Transversales (Provincia de la Pampa). *Comechingonia, 25*(2), 115–150. https://doi.org/10.37603/2250.7728.v25.n2.29011

Pallo, M. C., & Borrero, L. A. (2015). ¿Intercambio o movilidad?: una evaluación sobre el uso de escalas de análisis espaciales y curvas de declinación en Patagonia centro-meridional (Argentina). *Latin American Antiquity, 26*, 287–303. https://doi.org/10.7183/1045-6635.26.3.287

Paulides, L. (2007). Prepárate, vas a viajar. Sobre tendencias productivas en la formatización de núcleos en el complejo de canteras de arroyo Diamante, partido de Benito Juárez (provincia de Buenos Aires). In C. Bayón, A. Pupio, I. González de Bonaveri, N. Flegenheimer, & M. Frère (Eds.), *Arqueología de las Pampas* (pp. 233–249). Sociedad de Antropología Argentina.

Pearce, E., & Moutsiou, T. (2014). Using obsidian transfer distances to explore social network maintenance in late Pleistocene hunter-gatherers. *Journal of Anthropological Archaeology, 36*, 12–20. https://doi.org/10.1016/j.jaa.2014.07.002

Pedrotta, V. (2005). *Las sociedades indígenas del centro de la provincia de Buenos Aires entre los siglos XVI y XIX* [Unpublished doctoral Thesis]. Facultad de Ciencias Naturales y Museo, Universidad Nacional de La Plata.

Pelegrin, J. (1991). Les savoir-faire: une très longue histoire. *Terrain, 16*, 106–113.

Pelegrin, J. (1995). *Technologie lithique – Le Châtelperronien de Roc-de-Combe (Lot) et de La Côte (Dordogne)*. C.N.R.S Éditions (Cahiers du Quaternaire, 20).

Perlès, C. (1991). Économie des matières premières et économie du débitage: deux conceptions opposées? In *25 ans d'études technologiques en préhistoire: bilanet perspectives* (pp. 35–45). Juan -les-Pins A.P.D.C.A. Actes des XI e rencontres internationales d'Archéologie et d'Histoire d' Antibes.

Politis, G. G. (1984). *Arqueología del área Interserrana Bonaerense* [Unpublished doctoral Thesis]. Facultad de Ciencias Naturales y Museo, Universidad de La Plata.

Politis, G. G. (2008). The pampas and campos of South America. In H. Silverman & W. H. Isbell (Eds.), *The handbook of South American archaeology* (pp. 235–260). Springer.

Politis, G. G., & Madrid, P. (2001). Arqueología Pampeana: estado actual y perspectivas. In E. Berberian & A. Nielsen (Eds.), *Historia Argentina. Prehispánica* (pp. 737–814). Brujas.

Politis, G. G., Martínez, G., & Bonomo, M. (2001). Alfarería temprana en sitios de cazadores recolectores de la Región Pampeana (Argentina). *Latin American Antiquity, 12*(2), 167–181.

Politis, G. G., Bonomo, M., & Prates, L. (2003). Territorio y movilidad entre la costa Atlántica y el interior de la región pampeana (Argentina). *Estudos Ibero-Americanos, PUCRS, XXIX*, 11–35.

Politis, G. G., Prates, L., Merino, M., & Tognelli, M. F. (2011). Distribution parameters of guanaco (*Lama guanicoe*), Pampas deer (*Ozotocerosbezoarticus*) and marsh deer (*Blastocerus dichotomus*) in Central Argentina: Archaeological and paleoenvironmental implications. *Journal of Archaeological Science, 38*, 1405–1416. https://doi.org/10.1016/j.jas.2011.01.013

Quinn, P. S., & Burton, M. M. (2016). Ceramic distribution, migration and cultural interaction among late prehistoric (ca. 1300–200 B.P.) hunter-gatherers in the San Diego region, Southern California. *Journal of Archaeological Science: Reports, 5*, 285–295. https://doi.org/10.1016/j.jasrep.2015.11.022

Quintana, C. A., & Mazzanti, D. (2001). Selección y aprovechamiento de recursos faunísticos. In D. Mazzanti & C. A. Quintana (Eds.), *Cueva Tixi: Cazadores y Recolectores de las Sierras de Tandilia Oriental. Geología, Paleontología y Zooarqueología* (pp. 181–209). Laboratorio de Arqueología, Universidad Nacional de Mar del Plata, Publicación Especial 1.

Renfrew, C. (1975). Trade as action at a distance: questions of integration and communication. In J. A. Sabloff & C. C. Lamberg-Karlowsky (Eds.), *Ancient Civilisation and Trade* (pp. 3–59). University of New Mexico Press.

Renfrew, C., Dixon, E., & Cann, J. R. (1968). *Further analyses of Near Eastern Obsidians Proceedings of the Prehistoric Society* (Vol. 34, pp. 319–331).

Ringuelet, R. A. (1961). Rasgos fundamentales de la zoogeografía de la Argentina. *Phycis, 22*(63), 151–170.

Ríos Malan, M. (2022). Una primera aproximación al estudio de la manufactura de artefactos líticos de cuarcita en el sitio Escuela Agropecuaria (partido de General Lamadrid, provincia de Buenos Aires, Argentina). *Intersecciones en Antropología, 23*(1), 5–19. https://doi.org/10.37176/iea.23.1.2022.653

Rubin, O. E., 1990. *Estudio estratigráfico de las sedimentitas terciarias del centro y oeste del Departamento Puelén, región sudoeste de la provincia de La Pampa* [Unpublished Graduation Thesis]. Universidad de Buenos Aires.

Santos Valero, F. (2015). Contribuciones al conocimiento de la tecnología lítica en el valle inferior del Río Colorado: Sitio El Puma 3 (provincia de Buenos Aires). *Intersecciones en Antropología, 16*, 237–251.

Santos Valero, F. (2017). *Análisis tecnológico de sitios del Holoceno tardío en el valle inferior del río Colorado (Partidos de Villarino y Patagones; Prov. de Bs. As.)* [Unpublished doctoral Thesis]. Facultad de Ciencias Sociales, Universidad Nacional del Centro de la Provincia de Buenos Aires.

Santos Valero, F., & Armentano, G. M. (2017). La tecnología lítica. In G. Martínez (Ed.), Arqueología de Cazadores-Recolectores del Curso Inferior del río Colorado (Provincia de Buenos Aires, Argentina). Aportes al conocimiento de las ocupaciones humanas Pampeano-Patagónicas (pp. 121-145). Instituto de Investigaciones Arqueológicas y Paleontológicas del Cuaternario Pampeano-Consejo Nacional de Investigaciones Científicas y Técnicas-Universidad Nacional del Centro de la Provincia de Buenos Aires..

Santos Valero, F., & Messineo, P. G. (2021). Médano Santa Clara medio siglo después (partido de Salliqueló, Campo de Dunas del Centro Pampeano): nuevas perspectivas de análisis del conjunto lítico. *Revista del Museo de Antropología, 14*(3), 47–60. https://doi.org/10.31048/1852.4826.v14.n3.32273

Schiffer, M. B. (1987). *Formation Processes of the Archaeological Record*. University of New Mexico Press.

Seeman, M. F., Colucci, A. N., & Fulk, C. (2020). Hunter-gatherer mobility and versatility: A consideration of long-term lithic supply in the Midwest. *American Antiquity, 85*(1), 113–131. https://doi.org/10.1017/aaq.2019.72

Silveira, M. O. (1992). Etnohistoria y arqueología en Pampa Interserrana (provincia de Buenos Aires, República Argentina). *Palimpsesto, 2*, 29–50.

Soriano, A., León, R. J. C., Sala, O. E., Deregibus, V. A., Cauhépé, M. A., Scaglia, O. A., Velázquez, C. A., & Lemcoff, J. H. (1992). Río de La Plata grasslands. In R. Coupland (Ed.), *Natural grasslands. Introduction and western hemisphere* (pp. 367–407). Elsevier.

Torres, L. M. (1922). Arqueología de la Península San Blas (Provincia de Buenos Aires). *Revista del Museo de La Plata, 26*, 473–532.

Vecchi, R., Frontini, R., & Bayón, C. (2013). Paso Vanoli: una instalación del Holoceno tardío en valles fluviales del sudoeste bonaerense. *Revista del Museo de La Plata, 13*(87), 77–93.

Viani, J. L. (1930). Descripción de algunos ejemplares líticos de la antigua industria indígena Trenquelauquense (oeste de la provincia de Buenos Aires). In *La región pampeana: su pasado arqueológico*. Talleres Gráficos Ferrari Hnos.

Whallon, R. (2011). An introduction to information and its role in hunter-gatherer bands. In R. Whallon, W. A. Lovis, & R. K. Hitchcock (Eds.), *Information and its role in hunter-gatherer bands* (pp. 1–28). Cotsen Institute of Archaeology, University of California.

Zárate, M. A., & Rabassa, J. (2005). Geomorfología de la provincia de Buenos Aires. In R. E. de Barrio, R. O. Echeverri, M. F. Caballé, & E. Llambías (Eds.), *Geología y Recursos Minerales de la provincia de Buenos Aires* (pp. 119–138). Relatorio del 16° Congreso Geológico Argentino.

Technology and Use of Lithic Raw Materials from the Delta of the Paraná River and the Interior of the Province of Entre Ríos, Argentina. Preliminary Results of the Application of Micro-wear Analysis

Marcela Leipus

Abstract This paper presents the first results of the technological and functional analysis of some lithic assemblages from archaeological sites located in the Paraná Delta and interior of the province of Entre Ríos (Los Tres Cerros 1 and 2, Los Laureles, Tapera Vázquez, Cementerio de Los Indios, Laguna del Negro 1 and El Cerrito de Puerto Esquina). These assemblies are manufactured in various raw materials, widely prevailing different varieties of silicified sandstone and secondarily others such as silica, quartz, and silicified limestone. They come from different geological formations located within the Paraná River basin and along the Uruguay River basin. From a techno-morphological point of view, they are characterized by the wide variability of base forms, very little representation of retouched edges, a high frequency of flakes and fragments with natural edges, and potentially usable surfaces. The micro-wear analysis allowed us to characterize the uses of the artifacts and the presence of post-depositional alterations. The uses of the edges correspond to the work on several materials such as wood, non-woody vegetables, bones, and hide. Regarding the kinematics, traces developed by longitudinal and transverse actions with different working angles could be observed. These results provided relevant information about the selection of raw materials for the manufacture of artifacts with specific or versatile uses, as well as the relationship between morphology and function and the context of use, thus generating a deeper approach to the study of the dynamics of production and consumption of different resources.

Keywords Raw materials · Technology and use · lithic instruments · Micro-wear analysis · Traces of use · Paraná River Delta · Entre Ríos province · Northeastern Argentina

M. Leipus (✉)
División Arqueología del Museo de La Plata, Facultad de Ciencias Naturales y Museo
Universidad Nacional de la Plata, La Plata, Argentina
e-mail: mleipus@fcnym.unlp.edu.ar

G. F. Bonnat et al. (eds.), *Current Research in Archaeology of South American Pampas*, The Latin American Studies Book Series,
https://doi.org/10.1007/978-3-031-55194-9_9

Introduction

The processes of production and use of stone tools are conditioned by environmental and social factors, which involve choices and decisions that find their material correlation in specific aspects such as the differential use of raw materials, manufacturing techniques and use, their designs, the existence or not of reactivation and/or reuse, and the modes of disposal, among others (Álvarez, 2003; Castillo Bernal, 2004; Cueto, 2015; De Ángelis, 2016; Leipus, 2006, 2014a, b; Leipus & Mansur, 2007; Lynch et al., 2015; Mansur, 1999; Mansur et al., 2007; Marreiros et al., 2020; Pal, 2012, 2013–2015; Pawlik et al., 2020; Pedergnana & Ollé, 2020; Van Gijn, 2014).

The micro-wear analysis is a methodology that allows an approach to these aspects, through the characterization of the traces of use that develop on the edges in contact with different materials (e.g., hide, bone, and wood). This analysis provides direct information about the kinematics of the instrument and the material worked, but also about other aspects such as the selection of one or another raw material, the useful life of an artifact or edge, the intensity, and mode of use (reactivations, reuse, etc.). On the other hand, it allows characterizing microscopic technological and post-depositional alterations, which are extremely important when analyzing materials from archaeological contexts. In this way, the analysis allows inferring activities related to the context of use and deposition (Álvarez, 2003, 2004; De Angelis, 2016; Leipus, 2006, 2014a, b, 2021; Leipus & Mansur, 2007; Mansur, 1999; Mansur et al., 2007; Marreiros et al., 2020; Pal, 2012; Pal & Leipus, 2018, 2021; Pedergnana & Ollé, 2020; Van Gijn, 2014).

The aim of this paper is to present the results obtained from the application of the use-wear analysis methodology to a series of lithic assemblages from archaeological sites located in the Paraná River Delta and interior plains of the province of Entre Ríos, Argentina.

On the other hand, through the study of the practices of use, which encompass the ways of doing and using, it is proposed to contribute to the study of the lithic technology of the societies that occupied this territory. It is expected that the data and interpretations generated in this work will help to the knowledge of the strategies of use of lithic artifacts, strengthening a little explored line in the research area.

Archaeological Background of the Study Area

The first studies in the lower and middle Paraná River identified materials assigned to what are now known as archaeological entities in the region: Goya-Malabrigo and Guaraní. The first published references are from the first half of the nineteenth century, and it is from this moment that archaeological investigations increase Bonomo et al., 2019; Politis & Bonomo, 2018a, b). Since 2006, a multidisciplinary team has been carrying out systematic archaeological investigations in the upper

Paraná River Delta and the counties of Diamante, Victoria, and Gualeguay in the province of Entre Ríos, Argentina (Fig. 1). From these studies, more than 80 archaeological sites were detected, of which eight have been excavated (Bonomo et al., 2019). In the case of the occupations of the interior plains of Entre Ríos province, they would correspond to hunter-gatherer societies, that is, to others different from the Goya Malabrigo (Apolinaire, 2017).

The main foci of the research include: establishing the chronology of pre-Hispanic occupations, mortuary practices, the use of wild faunal and botanical

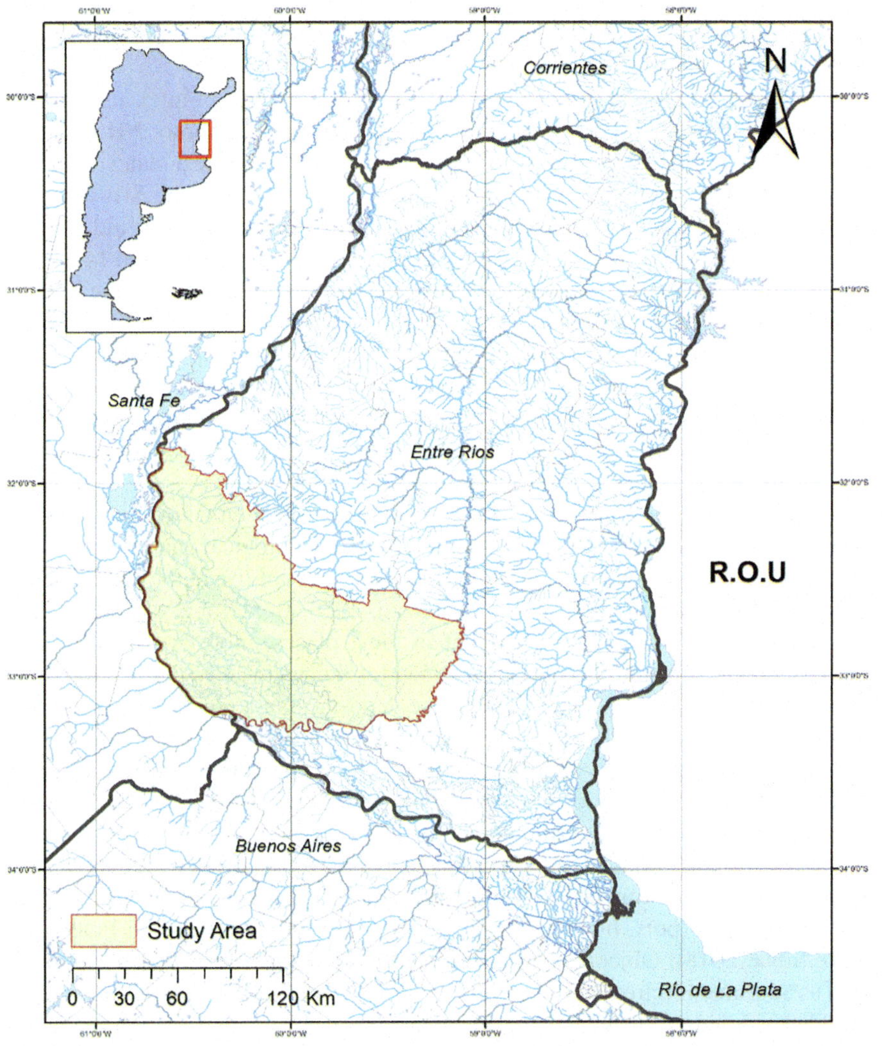

Fig. 1 Study area

resources, manufacturing practices and use of ceramic technology, formation processes—both anthropic and natural—of the mound structures and the horticulture of corn, beans, and squash (Bonomo et al., 2019; Politis & Bonomo, 2018a, b).

Likewise, the existence of a low level of social hierarchy was evaluated, already mentioned in the first chronicles of the sixteenth century (Bonomo et al., 2013, 2019) in the populations associated with the Goya-Malabrigo archaeological entity, which could be a derivation of the southern expansion of the Arawak people (Politis & Bonomo, 2012). The study of the dispersion of the Guaraní groups and the modes of interaction with local populations were also addressed (Bonomo et al., 2014). Briefly, the advances in research in the last decade show a more complex panorama that integrates local indigenous history with general adaptive processes that have occurred throughout the South American Lowlands (Bonomo, 2012; Bonomo et al., 2019; Politis & Bonomo, 2018a, b).

The landscape construction and the settlement pattern are complex and show different types of sites with different hierarchies (Gianotti & Bonomo, 2013; Bonomo et al., 2019; Politis & Bonomo, 2012, 2018a, b). The earliest dates for Goya-Malabrigo in the Paraná River are around 2050 years 14C BP at the Arroyo Aguilar II site, located in the middle course of this river. The Upper Delta would also have begun to be populated, according to the 43 radiocarbon ages obtained in 14 archaeological sites. The earliest evidence of the construction of mounds in the area was obtained in the Los Dos Cerros archaeological locality (LDC), which is characterized by the presence of two mound structures, one of which shows in its basal deposits the beginnings of the process constructive around 1950 years 14C BP. However, the genesis of elevated terrain surfaces may be a consequence of landscape modeling by natural agents (Bonomo et al., 2019; Politis & Bonomo, 2018a, b; Gianotti, 2021; Gianotti & Bonomo, 2013).

In recent years, the study of human bone remains has been taking center stage in the region and, based on the different lines of bioarchaeological analysis, it has been contributing to the reconstruction of pre-Hispanic ways of life in the area (Mazza & Loponte, 2012; Ramos van Raap & Scabuzzo, 2018, Bonomo et al., 2019). Thus, for example, it has been observed that in most places, mortuary spaces are not totally segregated from those places destined for domestic activities. An interesting and shared aspect among several sites is the deposition of some burials on layers of shells and ceramic sherds or covered by them. In addition to the delimitation of these burials by means of calcium carbonate concretions, the burials are characterized by a greater diversity of objects, which would indicate a differential status of the individuals buried there (Bonomo et al., 2019).

Pottery from the middle and lower reaches of the Paraná River is generally fragmented and in large quantities. They correspond to containers that have different sizes and shapes (pots, bowls, spoons, and kettles, among others; Di Prado et al., 2020; Silva, 2018). Objects were also made that were involved in other practices, such as adornment (circular and biconical earrings and beads), fishing (net weights) and fiber spinning ("*torteros*"). In this diverse repertoire, the flared pieces and the zoomorphic modeling that makeup vessels (although they generally appear isolated) stand out. Figurative references embodied in these appendices include mammals, mollusks, reptiles and birds; species that predominate both in the Middle and Lower Paraná Valley assemblages (Bonomo et al., 2019, 2021). To a lesser extent, there are anthropomorphic representations that show possible hairstyles, the use of headdresses, headbands and tattoos, or facial painting and piercings in the earlobes. The discovery of potsherds and clay masses with imprints of nets or baskets stands out (Gianotti & Bonomo, 2013) which, together with the spinners, show the weaving of fibers that have not been preserved in the record.

The systematic study of ceramic pieces attributed to Goya Malabrigo made it possible to define morphological categories that account for the variety of forms present in this entity (Bonomo et al., 2019; Di Prado et al., 2020; Politis et al., 2018a, b). Various aspects of the manufacturing process have been deepened based on the combination of experimental, techno-morphological and microscopic studies, and decoration has been analyzed from different perspectives (Bonomo et al., 2019 and bibliography cited there).

In this region, the diet was based on the exploitation of a wide range of resources by hunting mammals (coypu -*Myocastor coypus*-, capybara -*Hydrochoerus hydrochaeris*-, cuis -*Cavia* sp.- and cervids -*Blastocerus dichotomus*, *Ozotoceros bezoarticus* and *Mazama* sp.) and fishing mainly for *Characiformes* (*Hoplias* sp., *Prochilodus lineatus*, *Leporinus obtusidens*, *Salminus brasiliensis*, *Piaractus mesopotamicus*, among others) and *Siluriformes* (*Pimelodus maculatus*, *P. albicans*, *Rhamdia quelen*, *Pterodoras granulosus*). To a lesser extent, the use of freshwater birds, reptiles and molluscs is observed (Bastourre, 2014; Bastourre & Azpelicueta, 2020; Bonomo et al., 2019; Politis & Bonomo, 2018b). This diet was supplemented by harvesting "*algarrobo*", "*achira*", and wild rice, together with the cultivation of the widespread American triad of corn-beans-pumpkin. The results of investigations into biosiliceous micro-remains and starch grains have made it possible to verify the presence of wild and domesticated plant resources in the sediments of the "*cerritos*" *(small hills)* and also in ceramic artifacts that stored them (Bonomo et al., 2019; Sánchez et al., 2013).

Stable isotope studies (δ13C in collagen and apatite and δ15N in collagen) carried out on 22 samples of human bone remains from 15 archaeological sites in the Paraná River Delta and its adjacent plain, indicate that the majority of Goya-Malabrigo individuals have values isotopes consistent with a predominant consumption of C3 resources. This implies domesticated or wild plants and terrestrial herbivores that consumed this type of plant.

The presence of ribbed, semi-ribbed, and flat projectile points is observed, with and without peduncle and harpoon heads made with bones and antlers of hunted animals. In addition, for both cutting and grinding, there are few lithic artifacts made on various raw materials -siltstone, carbonate and silicified sandstone, xylopal, silicified limestone, vesicular basalt, quartzite, granite, and chalcedony- from the Paraná, Ituzaingó, Puerto Yeruá, Serra Geral and Botucatú, from the Martín García Complex and from the Uruguay River boulder deposits (Apolinaire, 2017; Apolinaire et al., 2019; Bonomo & Blasi, 2010; Bonomo et al., 2019).

The metal objects (e.g., copper), transferred from their sources and production workshops, were used to mark social and rank differences within the groups. Some people had unequal access to these imported goods, reflecting asymmetrical social relations as evidenced by grave goods, since they were buried with these objects. Through control of the exchange of products from distant geographical areas that transcended ethnic borders, these individuals had the ability to access these sophisticated goods which they could use to consolidate their social position, prestige, and leadership. This is supported by the first chronicles that identify with their own names at least three great indigenous chiefs of the Paraná region, using the adjectives "main chief", "main lord" or "*mayoral*", in reference to their government and command capacities (Bonomo et al., 2019: Politis & Bonomo, 2018a, b).

Finally, regarding the lithic material in the research area, the assemblages tend to be scarce compared to other materials such as pottery. That is why they have received little attention, however, systematic work on lithic technology in the southern region of northeastern Argentina has been increasing in recent years. The main problem of these works focused on topics such as: the production of artifacts and strategies implemented in manufacturing (Hocsman, 1999; Buc & Silvestre, 2006, 2011; Castro, 2013; Pérez et al., 2018; Silvestre & Capparelli, 2017), the analysis and delimitation of different sequences of manufacture (Apolinaire, 2017), or the study of primary and secondary raw materials sources (Apolinaire, 2018; Apolinaire et al., 2019; Bonomo & Blasi, 2010, Gascue et al., 2019a, b; Loponte et al., 2011; Sacur Silvestre et al., 2013; Silvestre & Capparelli, 2017). However, micro-wear approach has been little explored to date, despite the fact that it allows addressing such an important aspect of technology as the practices involved in the use of stone artifacts (Buc & Silvestre, 2006; Silvestre et al., 2020).

Techno-Morphological Analysis on the Lithic Assemblages of the Paraná River Delta and Interior Plains of Entre Ríos Province

The lithic assemblages analyzed in this paper come from surface and stratigraphic contexts (Table 1). The total of artifacts analyzed is 56 items, which add up to a total of 119 edges, points, bases, and surfaces, coming from the sites: Los Tres Cerros 1

Table 1 Totals of artifacts analyzed by site and context

Site	Number of artifacts/Context	Total artifacts recovered
Los Tres Cerros 1 (LTC1)	8 (stratigraphy)	25
Los Tres Cerros 2 (LTC2)	2 (stratigraphy)	Still under study
Laguna del negro 1 (LDN1)	27 (stratigraphy)	141
Cementerio de los Indios (CDLI)	4 (stratigraphy)	9
El Cerrito de Puerto Esquina (ECPE)	6 (surface)	17
Los Laureles (LL)	8 (surface)	24
Tapera Vázquez (TV)	1 4 (stratigraphy)	4
Total:	**56**	–

(LTC1 dates of the occupations between circa 1227 and 560 years BP, Scabuzzo et al., 2015), Los Tres Cerros 2 (LTC2, 920 ± 40 years BP), Laguna del Negro 1 (LDN 11990 + 90 years BP, Apolinaire, 2017), Cementerio de los Indios (CDLI, at the moment there are no dates), El Cerrito de Puerto Esquina (ECPE, at the moment there are no dates), Los Laureles (LL, at the moment there are no dates) and Tapera Vázquez (TP 650 ± 60 and 520 ± 60 years BP; Bonomo et al., 2014) (Fig. 2; Table 1).

For this analysis, the criteria proposed by Andrefsky (1998), Aschero (1983), and Mansur-Franchomme (1983) were taken into account. As for the selected rocks, the most represented is the sandstone of the Ituzaingó Formation ($n = 43$; 76.7%), followed by carbonate sandstones, ($n = 3$; 5.3%), quartz ($n = 3$; 5.3%), silicified sandstone of the Salto Chico Formation ($n = 2$; 3.5%), silicified limestone ($n = 2$; 3.5%) and in smaller proportions, chalcedony ($n = 1$; 1.8%), xylopal ($n = 1$; 1.8%) and gabbro ($n = 1$; 1.8%).

The sources of hard rocks suitable for manufacturing are concentrated along the two main river courses of the province of Entre Ríos (Apolinaire et al., 2019). However, there are also small discontinuous outcrops in specific sectors of various tributaries, such as in the Gualeguay River (Apolinaire, 2017, 2018; Apolinaire et al., 2019). In the first sector, the geological formations with potentially useful rocks are the Paraná Formation (sandstones and coquinas) and the Ituzaingó Formation (sandstones and xylopal). Along the course of the Uruguay River, the

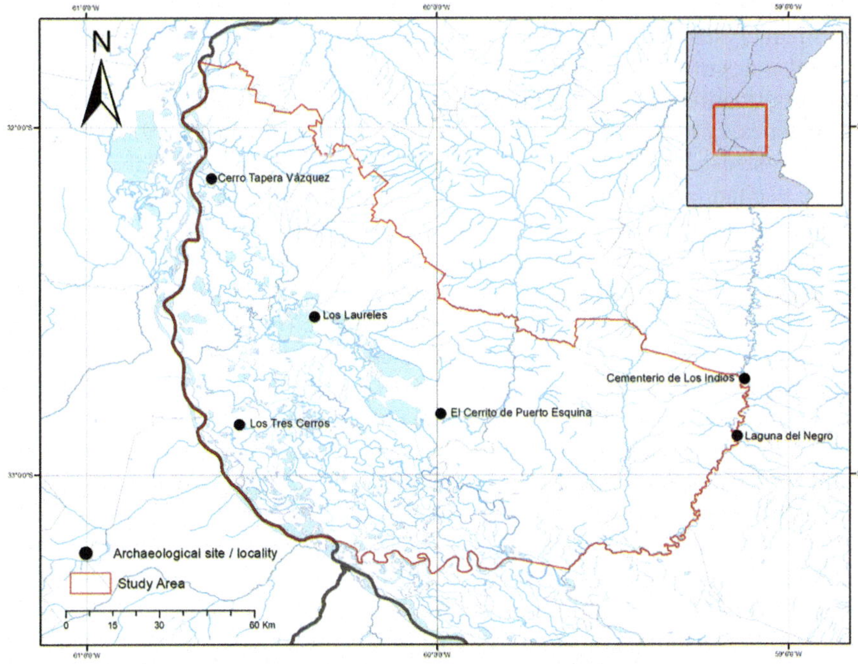

Fig. 2 Location of archaeological sites

Serra Geral Formation (massive and alveolar tholeiitic basalts), Puerto Yeruá (sandstones and silicified limestones), and El Palmar (silicified sandstones and xylopal) are located (Apolinaire, 2017, 2018; Apolinaire et al., 2019).

The sandstones of the Ituzaingó Formation generally have a yellowish-brown color and great diversity in terms of their knapping quality (Apolinaire, 2018; Apolinaire et al., 2019; Leipus, 2019; Leipus & Apolinaire, 2020). They comprise silicified sandstones, both with siliceous and carbonate cement, both being found recurrently in the lithic contexts of the area. In a hand sample, the former show greater toughness, a conchoidal fracture and a good degree of cohesion between the grains. In contrast, although they may exhibit good consolidation, the fracture planes are irregular, greater friability and rougher surfaces are observed (Leipus, 2019; Leipus & Apolinaire, 2020) (Fig. 3).

The sandstones of the Paraná Formation have high toughness and small clasts. Its size is difficult, but it can generate conchoidal fractures suitable for the manufacture of instruments. However, this raw material has been recovered very little in the lithic assemblages of the area and generally in the form of knapping debris. The siliciclastic rocks of the Salto Chico Formation consist mainly of silicified sandstones with very good quality in terms of presenting smaller clasts, tenacious siliceous cement, and crypto-crystalline characteristics (opal and chalcedony) (Apolinaire et al., 2019). Observed in an optical reflection microscope, the sandstones of the Ituzaingó and Salto Chico formations (both with siliceous and carbonate cement) have large quartz crystals and little foundation matrix (Leipus, 2019; Leipus & Apolinaire, 2020). On the other hand, the carbonate sandstones of the Paraná Formation have more small crystals and a greater amount of foundation matrix. In the reddish silicified sandstones of the Salto Chico Formation and in the reddish basalts of the Serra General Formation registered a different situation. Large crystals are also observed in sandstones, but siliceous cement is abundant (Leipus & Apolinaire, 2020).

The most used flake blanks as supports are undifferentiated flakes (n = 21; 37.5%). Second, angular flakes (n = 13; 23%) and pebbles (n = 6; 10.7%); and in smaller proportions: undifferentiated (n = 4; 7%), pebbles fragments (n = 3; 5.3%), edge flakes (n = 3; 5.3%), primary flakes (n = 1; 1.8%;), secondary flakes (n = 2; 3.5%), core fragment (n = 1; 1.8%), nodule (n = 1; 1.8%) and flake (n = 1; 1.8%). Regarding knapping, unretouched edges predominate (n = 39; 70%), followed by retouched edges (n = 10; 18%), polished and/or smoothed edges (n = 2; 3.5%), artifact formatted edge (n = 1; 1.8%), notch (n = 1; 1.8%), and a little axe (n = 1; 1.8%, formatted by chopping, abrasion, and polishing).

Regarding the presence of platforms, 38 artifacts preserve them (69%) and 17 do not (31%). The most represented types are flat (n = 22; 58%), followed by cortical (n = 11; 29%) and, in smaller proportions, dihedral (n = 3; 8%), a punctate (2.6%), and a filiform one (2.6%). Cortical platforms predominate in the various types of sandstones, more frequently in those of the Ituzaingó Formation, as well as the punctate, the filiform and the dihedral types. The flat ones, although they are mainly present in these sandstones, are also found in silicified limestones and in chalcedony. The presence of cortex is really very low and is shown in the set by few primary and secondary flakes, in addition to the cortical platforms.

Fig. 3 Sample of instruments manufactured on sandstone of the Fm. Ituzaingó from LTC1, LTC2 and LDN1 sites. (**a**) LTC2 Sondeo 1 N II Retouched edges. (**b**) LTC1 C5 NIII 132 Unretouched edge/polished surfaces. (**c**) LTC1 C4 NIII 73 Polished edges. (**d**) LDN1 C6 H226 Unretouched edges. (**e**) LDN1 C9 H138 Unretouched edges. (**f**) LDN1 Sondeo 1 N25-30 1 Unretouched edges

The most represented sizes in the pieces are medium (medium-small $n = 14$; 25% and medium-large $n = 14$; 25%), followed by very large ($n = 9$; 16%, including 8 unretouched edges and a gabbro little axe), large ($n = 5$; 9%), and small ($n = 8$; 14%). There is a high frequency of complete pieces ($n = 38$; 68%) compared to fractured ones ($n = 18$; 32%).

Evidence of heat treatment was identified on some LDN1 artifacts: LDN1 SONDEO 1 25-30 A: straight unretouched edges on a flake core fragment. LDN1 C4 H69: convex unretouched edge on undifferentiated flake. LDN1 C6 H162: straight retouched edge on primary flake. All three were manufactured on silicified sandstones of the Ituzaingó Formation (see section Micro-wear analysis).

In relation to the techno-morphological categories, the important scarcity of formatted instruments is observed, only some edges with marginal retouch or with two notches. Unretouched edges are the most frequent, followed by polished and/or smooth. It should be noted that flakes with unretouched edges are the most represented artifacts in other sites in the area (Castro, 2013; Hocsman, 1999). A particular case is a gabbro little axe from the LDN1 site manufactured from a pebble, with evidence of chipping and polishing at both ends. In summary, regarding the sharpening of the edges, the unretouched ones predominate widely and the presence of four polished ones was recorded (Tables 2 and 3). The sharpening of the edges by knapping presents marginal retouches and microretouches.

Artifacts with more than one edge prevail, either retouched or unretouched, with the most represented being those with between two (n = 20) and three edges (n = 17). In some cases, the presence of up to four edges in the same piece is seen, and in other cases, there are several probably active surfaces, as well as some ventral faces and a tip of an artifact with three edges. The pieces with a single edge add up to 18. The angles of the edges present a great variability: acute angles prevail between 25° and 55° (n = 70; 63%) and with less representation between 60° and 90° (n = 41; 37%), being the most frequent those of 55°, followed by those of 50°, 45°, and 40°.

In summary, from the technological and morphological point of view, the lithics assemblages are characterized by not presenting standardization and by the scarce variability of raw materials; by the wide diversity of base shapes, edges, and surfaces; a very scarce representation of retouched and/or polished edges and a high

Table 2 Classification by use. Presence/absence of wear on the analyzed edges, which are sorted into technological categories by number of edges (n = 119)

Technological categories	SU	PU	NDA	WU	Total
Unretouched edges	73	–	11	–	**84**
Retouched edges	18	–	–	2	**20**
Polished edges	3	–	–	–	**3**
Pebble	–	–	–	2	**2**
Notch	1	–	–	–	**1**
Axe	–	4	–	–	**4**
Surfaces	3	–	–	–	**3**
Point	1	–	–	–	**3**
Base	1	–	–	–	**3**
Total	**100**	**4**	**11**	**4**	**119**

References: *TC* technological category, *U E* unretouched edge, *P E* polished edge, *P S* polished surfaces, *R E* retouched edge, *P* pebble, *A* axe, *N* notch, *SU* sure use, *PU* probable use, *NDA* not determinable by alteration, *WU* without use

Table 3 General results of the use-wear analysis

Reference	RM	TC	N of edges	E M	Angle	E T	Determination	MW	K
LTC1 C12 N 21 GENERALES	CS	U E	2	S	75	LNE	SU	U	U
				S	80	LNE	SU	U	U
LTC1 C5 SE N4 N°182	CS	U E	3	S	75	LNE	SU	W	U
				S	85	LNE	SU	MV	T
				S	80	LNE	SU	HM	U
LTC1 C19 N9 N°75	Ch	U E	2	S	25	LNE	NDA	–	–
				CV	35	LNE	NDA	–	–
L3C1 C4 73	SS I	P E	2	CV	80	LPE	SU	SPM	L
				CV	45	LPE	SU	SPM	L
LTC1 C32 NIVELACIÓN	SS I	U E	1	CV	55	LNE	SU	U	U
LTC1 C32 3	SS I	U E	3	S	45	LNE	NDA	–	–
				S	45	LNE	NDA	–	–
				S	50	LNE	NDA	–	–
LTC1 C5 N3 N°132	SS I	U E + P S	1	S	55	SPE	SU	HM	U
			1 SU	SU	–	SU	SU	W	T
			1 SU	SU	–	SU	SU	HM	T
			1 SU	SU	–	SU	SU	W	L/T
LTC1 C17 N9 N°39	SS I	R E	2	S	50	RSE	SU	HM	U
				CV	55	LRE	SU	HM B?	T
LTC2 C2 N4 N°H33	Q	P	1	S	80	SNE	WS	–	–
LTC2 SONDEO 1 N2	SS I	R E	3	IR	45	LPE	SU	P	T/S
				S	55	LPE	SU	W	L
				IR	55	LPE	SU	W	T/A
LDN1 C2 N°190	SS I	R E + U E	2	S	65	LPE	SU	HM	U
				S	55	LNE	SU	HM	U
LDN1 C3 H411	SS I	U E	2	CV	65	LNE	SU	H/B	L
				CV	80	LNE	SU	H	L
LDN1 SONDEO 1 N15–30 N°1	SS I	R E + U E	2	CV	55	´LNE	SU	B	T
				CV	60	LRE	SU	B	L
LDN1 C2 H187	SS I	R E	2	CV	45	LRE	SU	H	T
				CV	40	LRE	SU	W	T
LDN1 C4 15–20 H35	SS I	R E + U E	3	IR	40	LNE	SU	W	U
				S	60	LNE	SU	U	L
				CV	75	LNE	SU	U	U
LDN1 C2 H148	SS I	U E	3	IR	40	LNE	SU	W	U
				CO	60	LNE	SU	B	U
				IR	85	LNE	SU	B	L

(continued)

Table 3 (continued)

Reference	RM	TC	N of edges	E M	Angle	E T	Determination	MW	K
LDN1 C4 H62	SS I	U E	3	IR	40	LNE	SU	HM	U
				IR	40	LNE	SU	H	T
				IR	25	LNE	WU	–	–
LDN1 C4 H143 125140	SS I	U E	2	CV	55	LNE	SU	HM	U
				S	40	LNE	SU	HM	U
LDN1 SONDEO 1 15–20	SS I	U E	2	S	45	LNE	SU	B	T
				CV	55	LNE	SU	W	U
LDN1 SONDEO 1 25–30 A	SS I	U E	3	S	45	LNE	SU	HM B?	U
				S	45	LNE	SU	HM	U
				S	55	SNE	SU	U	U
LDN1 C6 40–45 N°387	SS I	U E	2	CV	65	LNE	SU	PM	U
				S	65	LNE	SU	PM	T/S
LDN1 C4 H69	SS I	U E	1	CV	35	LNE	SU	HM	U
LDN1 C4 H365	SS I	U E	1	S	50	LNE	SU	W	T
LDN1 C6 H226	SS I	U E	3	S	45	LNE	SU	W	L
				S	50	LNE	SU	W	L
				S	75	LNE	SU	SPM	L
LDN1 C2 H205	SS I	U E	4	S	50	LNE	SU	U	U
				S	90	LNE	SU	U	U
				S	65	LNE	SU	HM	U
				S	55	LNE	SU	W	U
LDN1 C2 GRAL 3	SS I	U E	1	S	45	SNE	SU	B	T
LDN1 C6 H162	SS I	R E	1	S	70	LRE	SU	W	L
LDN1 C3 H177	SS I	R E	3	S	85	LRE	SU	W	T
				S	55	LRE	SU	W	L
				S	55	LNE	SU	W	L/T
LDN1 C3 H122	SS I	U E	2	CV	45	LNE	SU	HM	L
				S	35	LNE	SU	HM	U
LDN1 C4 H36	SS I	U E	1	CV	55	LNE	SU	HM	U
LDN1 C10 454	SS I	U E	2	S	50	LNE	SU	U	L
				CV	55	LNE	SU	SPM	L
LDN1 C1 100	SS I	U E	1	CV	75	LNE	SU	W	L
LDN1 C9 H138	SS I	U E	2	CV	70	LNE	SU	H	T
				CV	70	LNE	SU	W/B	T
			P	P	–	–	SU	W	U
			B	B	–	–	SU	H	T
LDN1 C9 NIVELACIÓN	SS I	U E	3	S	85	LNE	SU	HM	U
				S	75	LNE	SU	U	L
				S	70	LNE	SU	HM	U

(continued)

Table 3 (continued)

Reference	RM	TC	N of edges	E M	Angle	E T	Determination	MW	K
LDN1 C1 N°86	SL	U E	2	IR	25	LNE	NDA	–	–
				IR	40	LNE	NDA	–	–
LDN1 C3 H372	SS I	U E	3	IR	25	LNE	SU	U	U
				S	25	LNE	SU	U	U
				S	50	LNE	SU	U	U
LDN1 C3 H19	G	H	E	–	–	–	PU	–	–
			E	–	–	–	PU	–	–
			BO	–	–	–	PU	–	–
			BO	–	–	–	PU	–	–
CDLI 2 SONDEO 1 2530	SS I	U E	1	S	85	LNE	SU	U	T
CDLI 2 SONDEO 1 2025	SS I	U E	3	IR	50	LNE	NDA	–	–
				S	45	LNE	NDA	–	–
				S	50	LNE	NDA	–	–
CDLI2 SONDEO 1 N°1	SS I	U E	2	S	55	LNE	SU	H	T
				CV	55	LNE	SU	H	T
CDLI2 SONDEO 1 2025 N°2	SS I	U E	2	S	35	LNE	SU	HM	L
				CV	40	LNE	SU	U	L
EL CERRITO SUP N°14	SS I	U E	1	CV	80	SNE	SU	HM	T
EL CERRITO SUP N°13	CS	U E	3	S	45	LNE	US	U	T
				S	40	SNE	SU	H	T
				S	55	LNE	SU	U	U
EL CERRITO SUP N°19	SS I	P E	1	CV	80	LPE	SU	H	T
EL CERRITO SUP N°15	X	U E	1	S	55	LNE	NDA	–	–
EL CERRITO SUP N°16	SL	U E	1	CV	50	LNE	SU	B	L
EL CERRITO SUP 17	SS I	U E	3	S	40	LNE	SU	U	U
				IR	80	LNE	SU	HM	U
				IR	85	LNE	SU	PM	L
LOS LAURELES SECTOR 4 N°6	Q	U E	1	CV	80	SNE	WU	–	–
LOS LAURELES SECTOR 4 N°7	SS Ch	R E	1	S	50	LRE	SU	U	U
LOS LAURELES SECTOR 4 N°3	Q	P	1	S	85	SNE	WU	–	–
LOS LAURELES SECTOR 3 N°9	SS I	N	1	IR	40	LRE	SU	U	U

(continued)

Table 3 (continued)

Reference	RM	TC	N of edges	E M	Angle	E T	Determination	MW	K
LL SECTOR 4 N°2	SS I	R E + U E	3	S	50	RSE	SU	U	U
				IR	75	LNE	SU	W	T
				IR	75	LNE	SU	HM	U
LL SECTOR 4 N°5	SS I	U E	3	CV	55	LNE	SU	W	T
				S	55	LNE	SU	HM	T
				S	85	LNE	SU	W	U
LL SECTOR 3 N°7	SS Ch	U E	2	CV	60	LNE	SU	PM	L
				IR	50	LNE	SU	PM	T
LL SECTOR 3 5	SS I	U E	3	S	50	LNE	SU	W	U
				S	85	LNE	SU	B	L
				IR	65	LNE	SU	W	U
TAPERA VAZQUEZ C12 135–140 N°18	SS I	R E	1	CV	55	RSE	SU	U	U
Total: 56	**56**	**56**	**119**	**115***	**110****	**113*****	**119**	**100**	**100**

References: *RM* raw material, *SS I* silicified sandstone FM Ituzaingó, *CS* carbonate sandstone, *SS Ch* sandstone Salto Chico, *SL* silicified limestone, *Q* quartz, *X* xilopal, *Ch* chalcedony, *G* gabbro, *TC* technological category, *U E* unretouched edge, *P E* polished edge, *P S* polished surfaces, *R E* retouched edge, *P* pebble, *A* axe, *N* notch, *EM* edge morphology, *S* straight, *CV* convex, *SU* surface, *IR* irregular, *CO* concave, *P* point, *B* base, *E* extreme, *BO* borders, *ET* edge type, *LNE* large natural edge, *LPE* large polished edge, *SPE* short polished edge, *RSE* retouched short edges, *LRE* long retouched edges, *SNE* short natural edges, *SU* sure use, *PU* probable use, *NDA* not determinable by alteration, *WU* without use, *MW* material worked, *W* wood, *B* bone, *H* hide, *HM* hard material, *SPM* soft plant material, *PM* plant material, *U* undetermined, *K* kinematic, *L* longitudinal, *T* transverse, *L/T* longitudinal/transverse, *T/S* transverse/smoothing

*Do not count the extremes and the borders

**Do not count the point, the base, the extremes, the borders and surfaces

***Do not count the point, the base, the extremes and the borders

representation of unretouched edges and surfaces. There is also a high frequency of flakes with long unretouched edges and a variety of sizes, in which complete pieces predominate; as well as various edge morphologies, surfaces, and edge angles. Therefore, given the variability of the present flake blanks and the frequency of flat platforms, it is possible to propose a technological practice toward obtaining flakes without prior knapping of the cores and through the use of hard hammers (Fig. 3).

Micro-wear Analysis

Methodology

From the point of view of this analysis, the various varieties of sandstone have been studied very little, not only in the research area but also in other parts of the world, despite being one of the rocks, most used by societies in the past for the manufacture and use of artifacts by knapping. However, in recent years, there has been an increase in the micro-wear analysis of these raw materials, whether from methodological, experimental, and/or archaeological perspectives (Clemente, 2013; Laborda Martínez, 2018; Leipus, 2019, 2021; Leipus & Apolinaire, 2020; Masojć et al., 2021; Pedergnana & Ollé, 2020; Sade Martínez et al., 2021; Silvestre et al., 2020; Van Gijn & Houkes, 2006).

Within this framework, prior to the analysis of the archaeological material, an experimental collection was made with the same raw materials present in the archaeological artifacts (Leipus, 2019; Leipus & Apolinaire, 2020). The methodology used includes the characterization of rocks such as the various varieties of "heterogeneous coarse-grained" sandstones, made up of a matrix that can be microcrystalline, cryptocrystalline, or amorphous with variable composition and quartz crystals of different sizes cemented by this matrix (Mansur, 1999; Leipus, 2006, 2014b; Leipus & Mansur, 2007). This proposal consists of a mixed analysis in which the alterations that are produced by use in the fracture surface of the crystals and in the matrix are analyzed in a complementary way. It is analyzed taking the criteria developed for flint or chalcedony, while the fracture surface of the quartz crystals is observed taking into account the pattern of traces developed based on the analysis of hyaline quartz and obsidian (Mansur, 1999). When the fracture surfaces of the quartz crystals are observed fresh, they show the technological features very clearly: compression waves and striations that converge toward the point of percussion and/or pressure. From use, these technological traces begin to change, tending to disappear (Mansur, 1999). Although the analysis of the traces of use is carried out separately, the functional characterization is based on taking the traces formed on both surfaces (matrix and fracture surface of the crystals) as a whole (Álvarez, 2003; De Ángelis, 2016; Mansur, 1999; Leipus, 2006, 2014a, b, 2016, 2019; Leipus & Apolinaire, 2020; Leipus & Mansur, 2007; Flegenheimer & Leipus, 2007; Pal, 2012, 2013–2015).

The optical instruments used in the analysis belong to the Laboratorio ATMA "Análisis tecno-funcional de materiales arqueológicos" (CADIC-CONICET, Ushuaia, Argentina). A LEICA S6D binocular loupe with magnifications between 6X and 40X and two metallographic microscopes were used: Olympus BH2-UMA with magnifications between 50X and 500X and Leica DM 2700 MHRL with magnifications between 50X and 500X with a stand (column) 420 mm high and 24.4 by 37.4 mm mechanical stage. Observations at 200X were prioritized. A Leica Application System Version 4.9 image capture system was also used.

For the micro-wear analysis of archaeological artifacts, potentially usable edges, whether unretouched or retouched, whose length is equal to or greater than 1.5 cm, were analyzed. Due to the fact that some artifacts present more than one edge, all of them present were analyzed, as well as the points, bases, and surfaces.

For the identification of the kinematics, the activities were divided into longitudinal, transversal, and rotational; for functional inference, the methodological criteria of Alonso Lima and Mansur (Alonso Lima & Mansur, 1986/1990) were applied, with some modifications. The classification of edges was made from the identification of traces of use and taphonomic alterations, discriminated as follows:

- Edges with sure use: they present traces that allow to identify with certainty the type of material worked and/or the kinematics.
- Edges with probable use: they present traces of use accompanied by superficial alterations (taphonomic or technological) that modify them. Here are included those parts with little use time. Given these situations, it is not possible to clearly identify the material worked or the kinematics.
- Edges not determinable by alteration: they present taphonomic alterations that erase or obliterate the traces of use and that do not allow a functional inference to be made.
- Edges without use: they present clear signs of not having been used (e.g., fresh edges).

Post-depositional alterations (e.g., soil luster, patinas, sedimentary abrasion, striations) hinder the study of traces of use since they transform the surface of the rocks and highlight their representativeness. Therefore, their studies must be complementary to those of traces of use and technological traces (Mansur, 1999; Leipus, 2006, 2014a, b; Leipus & Mansur, 2007).

As mentioned above, a randomly selected sample of a total of 56 pieces, totaling 119 edges, surfaces, bases and point, was included in the analysis (Table 1; Fig. 3; Taking into account the high representativeness of artifacts with more than one edge or surface, the results of the micro-wear analysis are presented by total edges and surfaces. The base and the point are complementary to two edges of the same instrument (LDN1 C9 H138) and the surfaces, in turn, are complementary to an edge of the piece LTC1 C5 N3 N°132.

Micro-wear Results

In accordance with the stated objectives, the results obtained will be presented and discussed taking into account the entire sample, without discriminating by site, in order to record possible trends in general lines.

Of the sample of 56 artifacts, 50 showed traces of use (91%) and in 5 pieces (9.09%) intense post-depositional alterations were recorded (e.g., abrasion, striations with a random distribution pattern, etc.). In this case, it was not possible to observe traces of use, classifying as not determinable by alteration (2 pieces of sandstone, 1 of xilopal, 1 of chalcedony and 1 of silicified limestone). A particular case is that of the piece LDN1 C3 H 19 (a little axe manufactured in gabbro, of which the ends and edges were observed), in which micropolish was not observed but there were negative flakes and a certain degree of rounding, although it was not possible to determine with certainty if these traces were developed by the process of use (Table 3).

Of the total of 119 edges analyzed, 100 presented traces of use (84%) from which it was possible to identify the material worked with, the kinematics performed with the edge, or both, classifying without use. Traces of use were observed on the base and the point as well as on all three surfaces. In a total of four complementary edges, "fresh" technological traces (waves and percussion striations) were perceived and no traces of use were recorded, nor were there severe post-depositional alterations and, consequently, they were classified as unused. The total of edges with alterations is 11. (Table 2).

In relation to the materials worked, wood is the most represented ($n = 24$; 24%), and in second place, bone ($n = 10$; 10%) and hide ($n = 10$; 10%), followed by plant materials ($n = 5$; 5%) and soft plant material (e.g., grasses, rushes) ($n = 4$; 4%) (Figs. 4 and 5; Tables 3 and 4). In the particular case of piece LDN1.C3.H411 with two unretouched edges used, one was actually used on hide and the second probably on hide and bone and in the case of the piece LDN1 C9 H138 with one unretouched edge, one point and base used, the first presented traces of use that could be the result of wood or bone work. In some cases, it has only been possible to determine the origin or relative hardness of the material (e.g., hard material) ($n = 22$; 22%); and in others neither the origin nor the relative hardness could be determined, so they are classified as undetermined ($n = 22$; 22%) (Tables 3 and 4).

Regarding the kinematics performed with the edges, transverse ($n = 26$; 26%) and longitudinal ($n = 24$; 24%) actions were identified in similar proportions, with smoothing (transverse action) being the least represented ($n = 3$; 3%) and

Table 4 Results of microwear analysis by Material Worked and Kinematics

	Kinematics					
Material Worked	Transverse	Longitudinal	Smoothing	Indeterminate	L/T	Total
Wood	6	6	1	9	2	**24**
Bone	4	4		2	–	**10**
Hide	8	1	1	–	–	**10**
PM	2	2	1	1	–	**6**
SPM	–	4	–	–	–	**4**
HM	3	2	–	17	–	**22**
B and H	–	1	–		–	**1**
Indeterminate	2	4	–	16	–	**22**
W or B	1	–	–	–	–	**1**
Total	**26**	**24**	**3**	**45**	**2**	**100**

References: *PM* plant material, SPM soft plant material, *HM* hard material, *B* and *H* none and hide, *W* or *B* wood or bone, *L/T* longitudinal/transverse

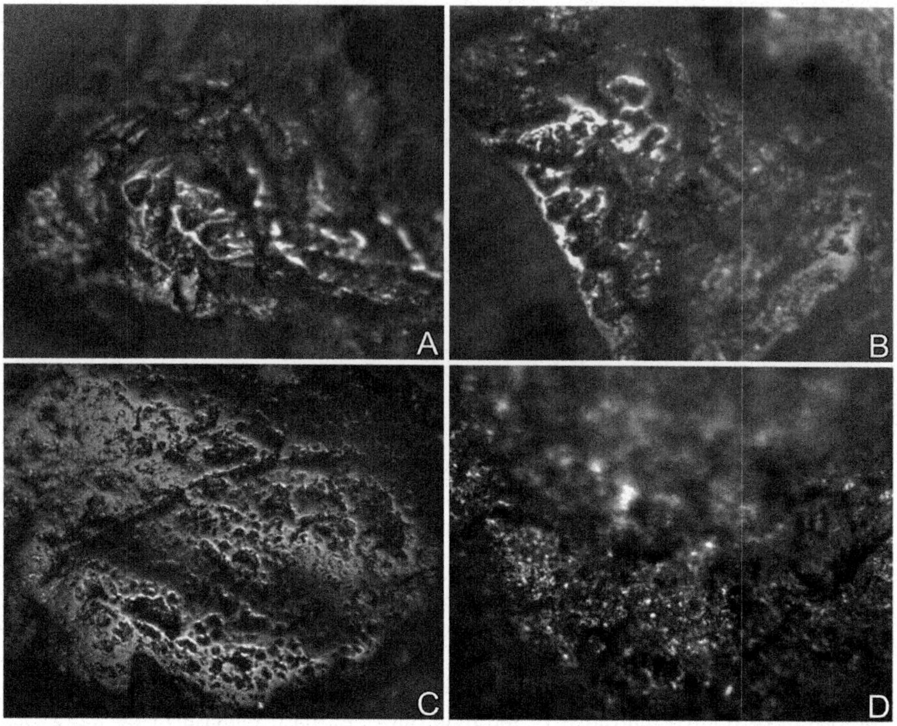

Fig. 4 Use Wear identified on artifact edges from sites LTC1 and LTC2 manufactured on Silicified sandstone Ituzaingó Formation. (**a**) LTC1.C4.73 Polished edges. Edge 1. Longitudinal work on soft plant material. 200X. (**b**) LTC1.C4.73 Polished edges. Edge 2. Longitudinal work on soft plant material. 200X. (**c**) LTC1.C5. NIII.132 Unretouched edge/polished surfaces. Surface 3. Longitudinal work on wood. 200X. (**d**) LTC2. Sondeo 1. NII Retouched edges. Edge 1. Transversal work (smoothing) on skin. 200X

bidirectional kinematics were recorded in two cases (2%). On the other hand, the representation of edges with indeterminate kinematics is high (*n* = 45; 45%) (Tables 3 and 4; Figs. 4 and 5).

Relating the kinematics and the materials worked (Figs. 4 and 5; Tables 3 and 4), it is observed that the general transversal kinematics is the most represented in hide processing (*n* = 8; 8%) and secondarily in wood (*n* = 6; 6%) and bone (*n* = 4; 4%). In the case of plant material, two edges were used transversally (2%); on hard materials, three edges (3%) and indeterminate materials two edges were used (2%). In relation to smoothing, this kinematics was used to process wood (*n* = 1; 1%), hide (*n* = 1; 1%) and plant material (*n* = 1; 1%). On the other hand, longitudinal kinematics was used in wood (*n* = 6; 6%) and bone (*n* = 4; 4%); also to process soft plant material (*n* = 4; 4%), plant material (*n* = 2; 2%), hard material (*n* = 2; 2%) and in an edge that worked both bone and hide (1%). A total of four edges were used with indeterminate kinematics. On wood, two edges (2%) were used with transversal/

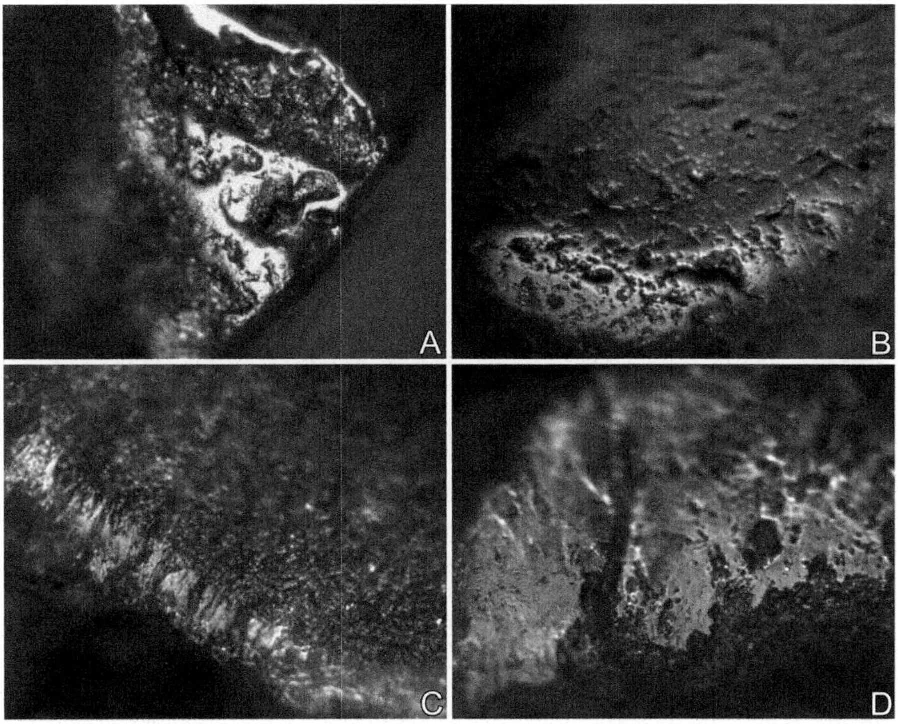

Fig. 5 Use Wear identified on artifact edges from sites LTC1 and LTC2 manufactured on Silicified sandstone. Ituzaingó Formation. (**a**) LDN1.C6.H226 Unretouched edges. Edge 3. 200X Longitudinal work on soft plant material. (**b**) LDN1.C3.H177 Unretouched edges. Edge 1. Transversal work on wood. 200X. (**c**) LDN1.Sondeo 1.N15–30.1 Unretouched edges. Edge 1. Transversal work on bone. 200X. (**d**) LDN1.C9.H138. Un retouched edges. Edge 1. Transversal work on skin. 200X

longitudinal kinematics (pieces LDN1.C3.H177, one point and LL Sector 4 Number 5, one surface).

In the cases in which the kinematics could not be determined but the worked material or its hardness/origin could, 9 edges (9%) were used on wood, two edges (2%) on bone, one on plant material (1%), and 17 edges (17%) on hard materials. On a total of 16 edges (16%) it has not been possible to determine either the kinematics or the substance worked.

The greatest variability of worked materials is observed in the sandstones of the Ituzaingó Formation, probably due to the fact that it is the most represented raw material. The edges of this rock were used to process: wood ($n = 23$), hard materials ($n = 21$), hide ($n = 9$), bone ($n = 9$), soft plant material ($n = 4$), plant material ($n = 3$), and hide and bone ($n = 1$) (Figs. 4 and 5). The indeterminate materials add up to a total of 17. In the case of the sandstone of the Salto Chico Formation, the edges were used on plant material ($n = 2$) and undetermined material ($n = 1$). Carbonatic sandstone ridges were used for wood ($n = 1$), hide ($n = 1$), plant material ($n = 1$), hard material ($n = 1$), and indeterminate ($n = 4$). A single edge of silicified limestone was used in bone processing (Table 3).

Regarding the relationship between the raw materials and the kinematics carried out, a trend similar to that observed in the worked materials is registered. The silicified sandstones of the Ituzaingó Formation present the greatest variability: longitudinal ($n = 22$), transversal ($n = 22$), smooth ($n = 3$), longitudinal/transverse ($n = 2$), transversal/longitudinal ($n = 1$), and indeterminate ($n = 39$). In the case of carbonate sandstones, their edges were used in transverse actions ($n = 3$) and indeterminate ($n = 5$); in the sandstones of the Salto Chico Formation: longitudinal ($n = 1$), transversal ($n = 1$) and indeterminate ($n = 1$); finally, in silicified limestone, only longitudinal kinematics were recorded (Table 3).

Arbitrarily, the edge angles were grouped into two: between 25° and 55°; between 60° and 90°. With the first group, the following were processed: wood ($n = 14$), hide ($n = 6$), bone ($n = 4$), hard materials ($n = 13$), soft plant materials ($n = 2$), material vegetable ($n = 1$) and indeterminate ($n = 15$). In the second group, the edges were used on: wood ($n = 7$), bone ($n = 4$), hide ($n = 3$), soft plant material ($n = 1$), hard materials ($n = 8$), plant material ($n = 5$), hide and bone ($n = 1$) wood or bone ($n = 1$) and indeterminate ($n = 7$). Taking into account the relationship between these two variables, there would seem to be a general trend toward a greater use of sharp edges between 25 ° and 55 °. However, we must bear in mind that this trend would not be reflected in the processing of any type of plant material. In order to relate these two variables, the angles of the edges were arbitrarily grouped into the same groups. In the first group, the following were used: transversal ($n = 13$), longitudinal ($n = 11$), smoothing ($n = 2$), longitudinal/transverse ($n = 1$), transversal/longitudinal ($n = 1$), and indeterminate ($n = 27$), and in the second group, for longitudinal ($n = 13$), transversal ($n = 8$), smoothing ($n = 1$) and indeterminate ($n = 16$) works (Table 3).

In the case of the types of edge and the materials worked, it is recorded that the most used are the long unretouched edges: to process wood ($n = 15$), bone ($n = 7$), hide ($n = 5$), vegetable materials ($n = 6$), soft plant materials ($n = 2$), wood and hide ($n = 1$), wood or bone ($n = 1$) hard materials ($n = 17$), and indeterminate ($n = 17$). Second, retouched long edges were used in the processing of wood ($n = 6$), bone ($n = 2$), hide ($n = 2$), hard material ($n = 1$), and indeterminate ($n = 2$). The natural short edges have processed: bone ($n = 1$), hide ($n = 1$), hard material ($n = 1$), and indeterminate materials ($n = 1$). The long polished edges were used to process: hide ($n = 1$) and soft plant materials ($n = 2$) and the single short polished edge was used on some hard material and the short tweaked edges were used on wood ($n = 1$) and undetermined materials ($n = 2$). Finally, the surfaces worked wood ($n = 2$) and hard material ($n = 1$), the base, hide ($n = 1$) and the point, wood ($n = 1$) (Table 3).

In the case of the relationship between the types of edge and the kinematics, a situation similar to that proposed for the type of edge and worked materials is observed: the most used are the long natural edges, which were used for longitudinal work, ($n = 18$), transverse ($n = 15$), smooth ($n = 1$), longitudinal/transverse ($n = 1$), and indeterminate ($n = 36$). Second, the long retouched edges were used in longitudinal ($n = 4$), transverse ($n = 4$), smoothing ($n = 2$), and indeterminate materials ($n = 3$) actions. In this case, it stands out that the same number of retouched long edges were used in transverse and longitudinal kinematics. The retouched short edges were used in transversal work (n = 3) and the natural short edges in transversal (n = 3) and indeterminate (n = 1). Long polished edges were used in longitudinal (n = 2) and transverse (n = 1) work and a single short polished edge with indeterminate kinematics. The surfaces were used in transverse (n = 2) and transverse/longitudinal (n = 1) work, the base with transverse kinematics and the point in some type of indeterminate action (Table 3).

Regarding the relationship between these two variables, it is observed that the straight edges are the ones that present the greatest variability of uses: woodworking (n = 12), hide (n = 2), bone (n = 4), material vegetal (n = 2), soft vegetal material (n = 1), hard materials (n = 13), and indeterminate (n = 16). The convex edges, although their quantity is less in relation to the straight ones, seem to present a considerable variability of uses: wood processing (n = 4), skin (n = 5), bone (n = 4), soft plant material (n = 3), plant material (n = 2), hide and bone (n = 1), wood or bone (n = 1), hard materials (n = 5), and indeterminate materials (n = 4). The irregular edges, little represented, also show a diversity of worked materials: wood (n = 5), hide (n = 2), bone (n = 1), plant material (n = 2), hard material (n = 3) and undetermined materials (n = 2). A single concave edge was used to process bone, the surfaces were used to process wood (n = 2) and hard material (n = 1), the point for wood and the base for hide.

Finally, regarding the relationship between these variables, the straight edges were used in longitudinal (n = 11), transversal (n = 10), smoothing (n = 1), longitudinal/transverse (n = 1) and indeterminate works (n = 27). The convex cutting edges were used in longitudinal (n = 11), transversal (n = 10), smooth (n = 1) and indeterminate (n = 8) work. Irregular edges, with longitudinal (n = 2), transverse (n = 3), smooth (n = 2), and indeterminate (n = 8) kinematics. The single concave edge and the point were used with indeterminate kinematics, the base in a transverse work and the surfaces in two transverse actions and one transverse/longitudinal (Table 3).

Discussion and Final Considerations

Although the results of the study presented here are preliminary made on a small sample of artifacts, we believe that the identification of the uses given to the lithic instruments of the study area, in particular the sandstone varieties, is vital for the interpretation of the archaeological assemblages, since the selection of these rocks is predominant in most of the sites in the region. Such identification can provide relevant information when done in conjunction with techno-morphological analysis. Thus, then, the discussion of the results obtained is presented below, relating technology, morphology and productive activities.

The predominant technological option was direct percussion by hand on various varieties of sandstone, mainly to produce flakes without preparation of the cores and, in a lesser way, the use of pebbles. Medium sizes were selected, obtaining long unretouched edges and, to a lesser extent, edges shaped by marginal retouching, or polished edges with straight and convex edges at acute angles. Although the presence of a great variability of edges was recorded (e.g., long unretouched, long retouched, short), the use of long edges in non-retouched flakes was prioritized, for various productive tasks and contexts of use.

With the sandstones of the Ituzaingó Formation processed the greatest variety of materials, with wood being the most represented, followed by bone, hide and, to a lesser extent, soft plant material, both in longitudinal and transverse actions. In the case of the Salto Chico sandstones, the carbonate sandstones and the silicified limestone, the cutting edges were used to work various materials with different kinematics (e.g., bone cutting in silicified limestone), but the frequency of use is relatively low.

A high percentage of processing of materials of plant origin is observed, mainly wood and, second, soft non-wood. The animal substances worked in smaller proportions, but similar to each other, are the bone and the hide. As already mentioned, direct and indirect evidence of the use of different resources, both plant and animal (e.g., LTR1, LDN1) has been recovered in the archaeological record (Bastourre, 2014; Bastourre & Azpelicueta, 2020; Bonomo et al., 2019; Brea et al., 2013; Colobig et al., 2015; Gianotti & Bonomo, 2013; Ramos et al., 2021). The characteristics present in the traces of use, mainly in the micropolish, correspond to the processing of these resources in a fresh condition and with a single type of kinematics. Only in one case (piece LDN1.C3.H411) was the use of the same edge identified in two work processes: to cut hide and bone. As for the kinematics, traces developed by longitudinal, transverse, and bidirectional actions with different working angles were recorded, which would also indicate the presence of various work processes: cutting or sawing wood, scraping hide, wood, and cutting soft plant materials.

In addition, the presence of traces of use should be highlighted, which only allowed us to characterize the hardness of the processed material. The representativeness of edges with indeterminate traces of use and with indeterminate kinematics is also high. This could be explained by several causes: the use of cutting edges on materials that develop undifferentiated traces of use regardless of their working time (e.g., meat, fish slaughter) or by the processing of substances that do develop diagnostic traces (e.g., wood, hide) but that have been used with little use time.

This situation has been observed in experiments carried out with various varieties of quartzite and sandstone (Leipus, 2006, 2014a, b, 2019; Leipus & Apolinaire, 2020). On the other hand, several experimental and archaeological analyses have shown that raw materials that we call "heterogeneous" coarse-grained, develop diagnostic traces slowly compared to flint, chalcedony, or silicified limestone (Alonso Lima & Mansur, 1986/1990; Álvarez, 2003; Clemente, 2013; De Ángelis, 2016; Mansur, 1999; Leipus & Mansur, 2007; Leipus, 2006, 2014a, b; Flegenheimer & Leipus, 2007; Pal, 2012, 2013–2015; Pedergnana & Ollé, 2020). The traces of use generated from experimentation on orthoquartzites are diagnostic after 30′ and 60′ of use. The experimental sandstones developed diagnostic traces at the same times of use as the orthoquartzites. These seem to reflect a similar use of the development model (Leipus, 2019; Leipus & Apolinaire, 2020). Thus, he slow rate of development of traces of use compared to finer-grained raw materials (e.g., homogeneous, such as flint, chalcedony, and silicified limestone) is related to the structure of heterogeneous raw materials (e.g., orthoquartzites, rhyolite, and sandstones) because they are made up of large crystals immersed in a foundation matrix (Alonso Lima & Mansur 1986/1990; Álvarez, 2003; Clemente, 2013; De Ángelis, 2016; Flegenheimer & Leipus, 2007; Leipus, 2006, 2014a, b, 2019; Leipus & Apolinaire, 2020; Mansur, 1999; Leipus & Mansur, 2007; Pal, 2012, 2013–2015; Pedergnana & Ollé, 2020). This situation makes the rate of development of traces of use slower than the models proposed for fine-grained raw materials.

Returning to the issue of undifferentiated traces of use, a particular case is the processing of fish, a highly exploited resource in the area of research. At the LTC1 site, two ichthyoarchaeological samples from the two excavated sectors were analyzed.

SC (grids 1, 10, and 15; n = 5767) and LE (grid 14, n = 26,261), finding 148 specimens with cut marks on bones (Bastourre & Azpelicueta, 2020). From micro-wear analysis, it is known that the processing of fish (e.g., scaling, cut of meat) develops very little traces of use even in fine-grained raw materials such as flint (Clemente Conte et al., 2020; Van Gijn, 1989; Hogberg et al., 2009; Rimkus, 2016), an exception would be the bone cut. For example, Van Gijn in his pioneering work on analysis of traces of use developed by fish processing in flint (1989), proposes that "these results lead to the conclusion that the processing of fish is largely invisible from the point of view of use-wear analysis. Certainly, fish-processing tools can be hidden among the tools interpreted as butchering implements, or among bone-working tools. Only the linear streaks of polish seem to be characteristic for fish-processing, but they are exclusively associated with scaling and are absent when scaling soft-scaled fishes, such as pike" (Van Gijn, 1989: 44). As mentioned above, in so-called "heterogeneous" coarse-grained raw materials such as orthoquartzites and sandstones, a longer use time is necessary for diagnostic traces to develop. Consequently, in the case of fish processing, the stage of development would not go beyond undifferentiated, since these raw materials develop traces of use more slowly. Therefore, it is proposed that some of the edges that show traces of indeterminate use could correspond to fish processing. However, this proposal must be contrasted through an experimental program aimed at processing this resource, which is already in process.

On the other hand, we propose as a hypothesis that the processing of fish may have been carried out with the edges of bone or shell instruments. It is proposed as a hypothesis, given that until now, there are few micro-wear analysis works on bone material in the area (e.g., Silvestre et al., 2020) and no evidence of the use of bone instruments has been recovered for scaling or processing fish.

There are ethnographic data that provide us with information about the processing of fish through the use of shell instruments by the Northwest Coast societies of North America (Van Gijn, 1989) or for the processing of other materials in the Amazon area (Prous & Pessoa Lima, 2019). In the research area, as well as in nearby areas, the record of mollusc shells is very common, either in discard areas interspersed with ceramic sherds, lithic remains and bone fragments (e.g., LDN1; Apolinaire, 2017; Bonomo et al., 2019), as food remains (Loponte et al., 2019) or in the form of personal adornments: necklace beads, "*tembetás*" in ritual and funerary contexts and as "*base*" concentrations for human burials (e.g., LTC1) (Acosta et al., 2015; Bonomo et al., 2019; Buc et al., 2019; Gascue et al., 2019a, b). From the archaeological point of view, Prous and Pessoa Lima (2019, Fig. 2) present evidence of instruments with active edges, retouched or not, manufactured on shells (e.g., knives and scrapers) which could be used to scale fish, scrape wood, etc. However, in the research area there is no archaeological data on the use of this raw material to manufacture instruments, nor on the application of micro-wear analysis; there is only antecedents from the experimental point of view to elaborate bone instruments (Buc et al., 2013). We consider a very good work proposal to consider the hypothesis formulated about the probable use of bone artifacts and shells.

Regarding the procurement strategies of the various sandstones and other raw materials used, both the surveys carried out and the macroscopic analyses of samples and thin cuts indicate that in the study area there are different geological formations and quarries that present the same type of rock. However, a formation can offer different kinds of raw materials (Apolinaire et al., 2019). In addition, the petrographic characteristics -which vary even within the same type of rock, particularly in the sandstones-, provide differential aptitudes for the manufacture and use of artifacts. The presence of lithics raw materials of various origins in the archaeological contexts of the research area indicates that interaction circuits existed in pre-Hispanic times (Bonomo & Blasi, 2010; Apolinaire et al., 2016). Raw materials entered whose sources of procurement are both local and medium and long distance, that is, between 10 and more than 100 km (Apolinaire, 2017).

Since silicified sandstones are the predominant rocks in most of the lithic assemblages, they can be considered a priority in procurement strategies. Although the thin section analyses did not allow to differentiate the silicified sandstones of the different formations present in the area, taking into account the distance to the sites (between 10 and 60 km), it is likely that they come from the local geological formations or immediately available from the Ituzaingó Formation, which emerges in the tributaries of the Gualeguay River (Apolinaire, 2017; Apolinaire et al., 2019).

Other widely used rocks are fluvial pebbles of diverse lithology, mainly chalcedony and silicified limestone and, to a lesser extent, quartzite, quartz and basalt that are available in the course of the Uruguay River. Certain artifacts on quartz and silicified limestone do not have a cortex and, consequently, their provenance is more difficult to determine; it could come from the deposits of pebbles or from long-distance primary sources, such as the Martín García Complex (quartz), the Botucatú Formation (quartzites), Puerto Yeruá Formation (silicified limestone) and/or Serra Geral Formation (quartz, quartzite, basalt). A similar situation was recorded in the artifacts made on xylopalo, whose origin may be local (Ituzaingó Formation), or at medium and long distance (Salto Chico and Concordia formations) (Apolinaire, 2017). Thus, it is possible to propose different forms of procurement: in the interior plains, medium- and long-distance non-locals raw materials were obtained mainly indirectly, involving intra-group exchange circuits, although the direct procurement is not ruled out in some cases. Most entered as nodules and were used to make artifacts at the sites, while a few entered as finished instruments, probably as part of the individuals transportable toolkit (Apolinaire et al., 2016). Indirect provisioning on the interior plains possibly involved intra-group exchange circuits. On the other hand, in the Upper Paraná Delta, although rocks from the middle distance (silicified sandstones) and some boulders were used to make artifacts at the sites, there was no evidence of manufacturing for the most majority of raw materials.

The pattern of use of lithic materials and their dispersion in the landscape differs significantly throughout the study region. Probably, the indirect procurement of these rocks involved the interaction with groups from the plains, who acted as intermediaries in the circulation routes of the rocks available in the course of the Uruguay River. Although the lowlands of the Upper Paraná River Delta and the high plains of the interior of Entre Ríos province functioned in an integrated manner in terms of procurement of lithics raw materials, the exploitation and circulation within both sectors respond to different categories, ways of doing things, and social representations (Apolinaire, 2017).

Regarding the use of the various varieties of sandstone for the manufacture of instruments, the results show that they are rocks that have edges that tend to be effective and durable for a variety of jobs in different contexts of use. This obtained use derives from the reactivation processes of the edges and the maximum degree of development of traces of use in many of them (which allowed the identification of the processed material, the kinematics or both). The various varieties of sandstones, particularly those of the Ituzaingó Formation, are the ones that showed the majority of well-developed and diagnostic traces of use.

Although it is well known that all materials from archaeological contexts present various types of alterations with different intensities, in this case, the incidence of post-depositional processes has not been a factor that strongly influenced the conservation of traces of use. The frequency of pieces in which diagnostic traces were observed is high, including those from superficial sites, which have presented traces with an intense degree of development in the same way as those from stratigraphy. The percentage of artifacts with intense post-depositional traces is similar to that recorded in other archaeological contexts analyzed by the author (Leipus, 2006, 2014a, b). Artifacts on fine-grained rocks, such as xylopal and chalcedony, are the ones that present the highest intensity of post-depositional traces, and no functional inference can be made. However, there are also pieces in the various varieties of sandstone that show traces of post-depositional processes. Although this situation could be due to their representation in the sample, it could also be explained by the structural and compositional characteristics of the sandstones: although they develop traces of use slowly, so is the incidence of taphonomic processes (Alvarez, 2003; Gibaja et al., 2002; Leipus, 2006 and bibliography cited there; Leipus, 2014a, b and bibliography cited there; Lynch et al., 2015; Mansur, 1999; Leipus & Mansur, 2007; Marreirros et al., 2020; Pal, 2012, 2013–2015; Pawlik et al., 2020; Pedergnana & Ollé, 2020).

The presence of heat treatment in some sandstones, regarding their knapping properties, allows us to observe a loss of homogeneity of the siliceous cement and the mineralogical transformation of iron minerals and clays. These changes weaken the cementum and lead to the fact that the fracture must surround the clasts of the skeleton instead of being produced by planes that pass through it. This causes less fracture control and makes rocks more brittle and crumbly. Likewise, the Ituzaingó and Salto Chico sandstones change their color quickly and markedly when subjected to temperatures lower than the thermal stress of quartz, so the thermal alteration recorded in the archaeological samples is possibly not related to the intention of improving the quality of the sandstone raw material, but it could be the result of an involuntary or accidental alteration or of the search to alter other types of rock properties (e.g., color) (Apolinaire et al., 2019). From the functional analysis, we can affirm that the thermal alteration was probably carried out accidentally prior to use, given the diagnostic degree of the traces of use.

Relating the results of the analysis of the aforesaid traces with the processed resources, it is highly probable that the lithic instruments have participated not only in the use of subsistence resources (e.g., butchering hunting preys, harvesting, horticulture, procurement of firewood), but also in the first stages of operational chains for the manufacture of other types of technologies. Such could be the case of wood, since it was processed in longitudinal works that include cutting and sawing, as well as transversal: scraping and smoothing the bark and the internal part, roughing, among others. With it, containers, weapons points, shafts, digging sticks, poles for rooms, canoes were made, but it was also used for the firewood, as in the case of cutting and sawing. For the work of non-woody vegetables, such as rushes and grasses, the tasks were aimed at cutting in the fresh state using only a multi-edged instrument, which could be related to its provisioning and the first stages of processing to obtain fibers, which would be used for the manufacture of basketry, cordage, fishing nets, among others. They could also have participated in collecting wild and/or cultivated plants in subsistence activities.

Bone work, both with longitudinal and transverse kinematics, could be associated with the processing of prey and/or its conditioning for raw material. This includes cutting, sawing and scraping, associated with the various stages of the operational chain of the abundant bone technology recovered in the research area, such as retouchers, points, weapons for fishing, needles for nets. On the other hand, the work on hide -since only transverse actions were identified- is most likely related to scraping and smoothing to then proceed to stretching and drying. Containers, clothing, bags, and *tientos* were manufactured with them.

The edges that presented traces of use developed by hard and indeterminate materials, in the first case could be the result of the processing of wood and/or bone and in the second, of materials that do not produce diagnostic traces, such as meat cutting. and fish processing, with little time use.

From the holistic approach addressed in this work and relating the raw material procurement strategies with the results obtained from our analysis, two options can be proposed. Regarding the relationship between the procurement strategies (indirect, through intergroup exchange networks), it is feasible to propose that the provision of the various raw materials, especially sandstone, involved an important effort that can be measured in terms of the travel planning and timing, selection, primary manufacturing, and transportation, etc. But it also includes the ways of doing, the social representations, and the different modes of control of the material circulation networks (Apolinaire, 2017). The various technological options for the provision of the raw material are normally explained in terms of their qualities for knapping activities. However, another important aspect that must be considered is that of the qualities of the edges when they are used in various use processes. The functional study of different varieties of sandstone allowed to characterize some qualities of the edges, particularly in terms of the criteria of effectiveness, stability, and durability (Álvarez, 2004; Flegenheimer & Leipus, 2007; Leipus, 2006 and bibliography cited there; Leipus, 2014a, b and bibliography cited there; Mansur, 1999; Leipus & Mansur, 2007; Pal, 2012; Pal & Leipus, 2018, 2021).

Indeed, our experimental work shows that the structural differences between the sandstones not only condition their qualities for knapping, but are also reflected in the functional aspect, both in terms of the speed of formation of traces of use and to the strength and durability of the edges during use (Leipus, 2019; Leipus & Apollinaire, 2020). The flakes have good unretouched edges, as well as the retouched instruments, both being effective and durable (no microscopic evidence of reactivation was found).

From this point of view, if the instruments made in the various varieties of sandstone were not intended for specific activities, one might wonder why there was such an emphasis on the selection of these rocks whose supply presents us with a complex panorama, which involves not only environmental factors but also social ones. The results obtained allow us to propose a hypothesis that derives from the combined analysis of morphometric and functional criteria. Indeed, in the analyzed assemblages, the selected flakes are of medium and large sizes, and in some of them long, straight, or medium convex edges were formed, with intermediate to high acute angles. This relationship is interesting because it suggests that the preference for sandstones, which may have an apparent technological cause, could actually respond to functional requirements. The technological cause could be that they are available in abundance and in good quality, but with a dispersed pattern. In addition, they allow medium and some large flakes to be obtained with long unretouched edges or for subsequent sharpening. The functional cause has been evidenced by experimentation; it implies the resistance and stability of the retouched and non-retouched edges in the various varieties of sandstone, but mainly in that of the Ituzaingó Formation. These are resistant edges in the various processes of use, without a tendency to splinter due to the toughness of the rock, at the same time that the cohesion of grains and foundation matrix means that too much shattering does not occur either. Consequently, they tend to remain stable, on a micro scale, throughout intense and/or prolonged use (Leipus & Apollinaire, 2020). In the archaeological materials, this observation is corroborated by the absence of evidence of reactivation and by the maximum degree of development of the traces of use.

The diversity of tasks to which the edges were assigned does not reflect functional specificity in terms of the relationship between raw material and use. Although transverse actions predominate, there are also longitudinal ones; and if woodworking prevails slightly, there is also use on hide, bones, and non-woody vegetables. On the contrary, such diversity would indicate that the sandstones were selected to make versatile artifacts with more than one usable edge (preferably long) destined to different tasks, with very effective acute angles over time. The search for supports (flake blanks) that would allow the production of medium-sized artifacts, edges with particular morphometric characteristics (edge shapes, edge lengths and angles) and relatively stable in functional terms was prioritized. This situation is the one registered in the sites analyzed, and for this reason, we propose it as a working hypothesis to be contrasted with information from other archaeological contexts.

Such a strategy would be well suited to the possibilities of a region where raw material of relatively good quality abounds, although it is spatially dispersed. Together, the upwelling and supply zones show significant internal variability, as well as evidence of significant regional and extra-regional social exchange networks. Thus, the maximization criteria would not only go through the modification and reuse of fractured instruments, but also through the making of versatile and durable artifacts as part of a conserved technological strategy, which includes a variability of contexts of use (Leipus, 2006, 2014a, b, 2016, 2021; Leipus & Mansur, 2007; Pal & Leipus, 2018, 2021). From this point of view, it is postulated that the various varieties of sandstone were the best possible material among those available in the region.

The results that were obtained allow us to affirm that the majority of artifacts with unretouched edges show diagnostic traces of use. These technological practices, which do not include the standardization of designs, were effective for the use and processing of various materials. In this way, artifacts with unretouched edges are no longer just knapping products and/or debris, but rather sought-after goods: artifacts with edges, points, bases, and usable surfaces. That is, true instruments.

Taking into account the technological and morphological variables related to the results of the microscopic analysis, we can conclude that, based on the evidence of a great variability of materials worked as well as kinematics, the edges have been part of a versatile strategy that includes various contexts of use. These results provide relevant information about the relationship between morphology and function, the selection of raw materials for the manufacture of artifacts with specific or versatile uses, and the context of use. In this way, it contributes to a deeper approach to the pre-Hispanic social dynamics of production and consumption of the different resources by the societies of the Paraná River Delta region and the interior of the province of Entre Ríos.

Acknowledgments This work is dedicated to the memory of Dr. María Estela Mansur, who was my Doctorate director, colleague and friend. Without his constant guidance I would have never able to do what I did. I also dedicate it to the memory of the archaeologist Patricia Madrid who was also my guide, colleague and friend. Both from their experiences encouraged my dedication to the study of technology and the use of stone tools. I only have words of thanks for them. They will always be in my memory. I wish to express my gratitude to Dr. Eduardo Apolinaire and Dr. María Teresa Iglesias for their critical review of the manuscript, which greatly helped to improve it. To the Archaeologist Diego Gobbo for his invaluable help with the images and to Professor Miriam Griszka for her final review of the English version. To the ATMA Laboratory "Análisis tecno-funcional de materiales arqueológicos" (CADIC-CONICET, Ushuaia, Argentina) who always makes their facilities, equipment and microscopes available to me. This paper is part of the project "Arqueología de ambientes acuáticos del Centro-este argentino" (Código 11/N885) de la Universidad Nacional de La Plata, directed by Dr. Mariano Bonomo and Dr. Luciano Prates.

Thanks to the anonymous reviewers whose suggestions helped to improve the manuscript. All opinions written here are the sole responsibility of the author.

References

Acosta, A., Buc, N., & Davrieux, N. (2015). Producción y uso de ornamentos en las tierras bajas de Sudamérica: el caso de las poblaciones humanas prehispánicas del extremo meridional de la cuenca del Plata (Argentina). *Munibe Antropologia-Arkeologia, 66*, 309–325.

Alonso Lima, M., &Mansur, M. E. (1986/1990). Estudo traceológico de instrumentos em quartzo e quartzito de Santana do Riacho (Minas Gerais). *Revista do Museu de Historia. Natural, 11*, 173–190, Belo Horizonte.

Álvarez, M. (2003). *Organización Tecnológica en el Canal Beagle. El caso de Túnel I (Tierra del Fuego, Argentina)*. Unpublished docotoral thesis Universidad de Buenos Aires, Argentina.

Álvarez, M. (2004). Estrategias tecnológicas en los grupos canoeros tempranos del área Fuego-Patagónica. *Magallania, 32*, 191–208.

Andrefsky, W. (1998). *Lithics: Macroscopic approaches to analysis*. Cambridge University Press.

Apolinaire, E. (2017). *Arqueología del suroeste de Entre Ríos: tecnología, subsistencia e interacción social en tiempos prehispánicos*. Unpublished doctoral thesis: Facultad de Ciencias Naturales y Museo, Universidad Nacional de La Plata.

Apolinaire, E. (2018). Caracterización de los depósitos secundarios de rodados fluviales del río Uruguay inferior: su aplicación en contextos arqueológicos de las llanuras interiores del nordeste argentino. *Boletim do Museu Paraense Emilio Goeldi, 13*(2), 433–457.

Apolinaire, E., Bastourre, L., & Costa Angrizani, R. (2016). Arqueología de las tierras altas de Entre Ríos: primeros resultados de las prospecciones en el interior del departamento Gualeguay. *Intersecciones en Antropología, 17*, 91–107.

Apolinaire, E., Castiñeira, C., & Bonomo, M. (2019). Nuevos aportes para la delimitación de la base regional de recursos líticos de la provincia de Entre Ríos: relevamiento de afloramientos primarios y caracterización microscópica de rocas siliciclásticas. *Revista Arqueología de la Facultad de Filosofía y Letras, 25*(2), 71–22.

Aschero, C. (1983). *Registro de códigos para atributos descriptivos aplicados a artefactos líticos*. Unpublished Report to CONICET. Manuscript.

Bastourre, L. (2014). Estudios arqueofaunísticos en el Delta Superior del Paraná: el Sitio Los Tres Cerros 1 (provincia de Entre Ríos, Argentina). *Revista Chilena de Antropología, 30*(2), 109–115.

Bastourre, L., & Azpelicueta, M. M. (2020). Del registro ictioarqueológico a las prácticas alimentarias: el caso de los Tres Cerros 1 (Delta Superior del Paraná, Entre Ríos). *Relaciones de la Sociedad Argentina de Antropología, XLV*(1), 13–57.

Bonomo, M., & Blasi, A. (2010). Base regional de recursos líticos del Delta del Paraná. Estudio petrográfico de artefactos y afloramientos en el sur de Entre Ríos. *Revista Cazadores Recolectores del Cono Sur, 4*, 17–41.

Bonomo, M., Politis, G., & Gianotti, C. (2013). Montículos, jerarquía social y horticultura en las sociedades indígenas del Delta del río Paraná (Argentina). *Latin American Antiquity, 22*(3), 297–333.

Bonomo, M., Castro, J., & Silva, C. (2014). Tecnología y subsistencia en el sitio arqueológico Cerro Tapera Vázquez (Parque Nacional Pre-delta, República Argentina). *Cadernos do LEPAARQ do Laboratório de Ensino e Pesquisa em Antropología da Univerdidade Federal de Pelotas, 11*(22), 53–81.

Bonomo, M., Di Prado, V., Silva, C., Scabuzzo, C., Ramos van Raap, M. A., Castiñeira, C., Colobig, M. M., & Politis, G. (2019). Las poblaciones indígenas prehispánicas del río Paraná Inferior y Medio. *Revista del Museo de La Plata, 4*(2), 585–620.

Bonomo, M., Politis, G., Bastourre, L., & Moreira, G. (2021). Humanized nature: Symbolic representation of Fauna in pottery from the Paraná River of South America. In M. Bonomo & S. Achila (Eds.), *South American contributions to world archaeology* (pp. 411–446). Springer.

Brea, M., Franco, M. J., Bonomo, M., & Politis, G. (2013). Análisis antracológico preliminar del sitio arqueológico Los Tres Cerros 1 (Delta Superior del rio Paraná), provincia de Entre Ríos. *Revista del Museo de La Plata, 13*(87), 345–360.

Buc, N., & Silvestre, R. (2006). Funcionalidad y complementariedad de los conjuntos líticos y óseos en el humedal del nordeste de la Provincia de Buenos Aires: Anahí, un caso de estudio. *Intersecciones en Antropología, 7*, 129–146.

Buc, N., & Silvestre, R. (2011). Distribución de artefactos líticos y óseos en el humedal del Paraná inferior. Relacionando conjuntos del norte y sur del Paraná Guazú. *Arqueología de cazadores recolectores en la Cuenca del Plata, 5*, 171–188.

Buc, N., Mucciolo, L., Pérez Jimeno, L., Coll, M., & Deluca, S. (2013). Series experimentales en obtención de formas base para la confección de instrumentos óseos en la cuenca del Paraná medio e inferior. *Cuadernos del Instituto Nacional de Antropología y Pensamiento Latinoamericano – Series Especiales, 1*(2), 185–199.

Buc, N., Acosta, A., & Loponte, D. (2019). Cuentas y tembetás malacológicos de los grupos cazadores-recolectores prehispánicos del humedal del Paraná inferior. *Comechingonia, 23*(1), 1–10. Versión On-line ISSN 1851-0027.

Castillo Bernal, S. (2004). Estudios microscópicos de huellas de uso en artefactos líticos: algunas observaciones teóricas y metodológicas. *Cuicuilco, 11*(32), 205–227.

Castro, J. (2013). Arqueología del centro de la provincia de Entre Ríos (Argentina). *Arqueología, 19*(2), 287–304.

Clemente, I. (2013). Análisis funcional de los restos líticos del abrigo de Benzú a lo largo de 180.000 años de ocupación: una aproximación a las actividades productivas a partir de un sondeo secuencial. In J. Ramos Muñoz, D. Bernal, E. Casasola, E. Vijan De Vila, & J. J. Cantillo Duarte (Eds.), *El abrigo y la cueva de Benzú. Memoria de los trabajos arqueológicos de una década en Ceuta (2002–2012)* (pp. 438–751). Universidad de Cádiz, Servicio de Publicaciones.

Clemente-Conte, I., Mazzucco, N., Soares, N., da Silva, C., Muñoz, J., Vila, E., & Montañés, M. (2020). Fish resource exploitation in the southern Atlantic coast of the Iberian Peninsula: A view from the traceological analysis of flaked stone tools (sixth-fourth mill. cal BCE). *Quaternary International, 541*, 141–151.

Colobig, M. M., Sánchez, J., & Zucol, A. (2015). Análisis de macrorrestos vegetales en el sitio arqueológico Los Tres Cerros 1 (isla Las Moras, Victoria, Entre Rios). *Revista del Museo de Antropología, 8*(1), 115–124.

Cueto, M. (2015). *Análisis de los procesos de uso de artefactos líticos en sociedades cazadoras recolectoras. Ocupaciones correspondientes a la transición Pleistoceno/Holoceno. Meseta Central de Santa Cruz* (International Series 2776). British Archaeological Reports.

De Angelis, H. (2016). *Arqueología de los cazadores-recolectores de la Faja Central de la Isla Grande de Tierra del Fuego*. Sociedad Argentina de Antropología.

Di Prado, V., Píccoli, C., Silva, C., Morosi, M., Politis, G., Bonomo, M., & Barboza, C. (2020). Communities of practice in pre-Hispanic middle and lower Paraná (Northeastern Argentina). Ceramic petrography contributions. *Journal of Archaeological Science Reports, 34*, 1–16.

Flegenheimer, N., & Leipus, M. (2007). Trabajar en un espacio reducido, Cerro El Sombrero Abrigo 1. In *Número especial de la Revista Pacarina* (pp. 441–444). Facultad de Humanidades y Ciencias Sociales, Universidad Nacional de Jujuy.

Gascue, A., Bortolotto, N., Loponte, D., Acosta, A., Borges, C., Fleitas, M., & Fodrini, A. (2019a). Contextos geomorfológicos y tecno-económicos del registro arqueológico del bajo río Uruguay (margen izquierda). Resultados preliminares de nuevas prospecciones. *Arqueología, 25*(3), 87–117.

Gascue, A., Scarabino, F., Bortolotto, N., Clavijo, C., & Capdepont, I. (2019b). El rol de los moluscos en las poblaciones prehispánicas de Uruguay. *Comechingonia, 23*(1), 55. *versión On-line* ISSN 1851-0027.

Gianotti, C. (2021). Environment transformation and landscape domestication in the lowlands of northeast of Uruguay. Earthworks as technology for the management of flood ecosystems. In M. Bonomo & S. Achila (Eds.), *South American contributions to world archaeology* (pp. 283–318). Springer.

Gianotti, C., & Bonomo, M. (2013). De montículos a paisajes: procesos de transformación y construcción de paisajes en el sur de la Cuenca del Plata. *Comechingonia, 17*, 129–163.

Gibaja Bao, J. J., Clemente, I., & Mir, A. (2002). Análisis funcional en instrumentos de cuarcita: el yacimiento del paleolítico superior de la Cueva de la Fuente del Trucho (Colungo, Huesca). In I. Clemente, R. Risch, & J. Gibaja (Eds.), *Análisis Funcional: su aplicación al estudio de sociedades prehistóricas* (BAR international series) (Vol. 1073, pp. 79–86). Archaeopress.

Hocsman, S. (1999). *Tecnología lítica prehispánica en la cuenca inferior del arroyo Las Conchas (Depto. de Paraná, Pcia. de Entre Ríos): El sitio VU 4 como caso de estudio*. Unpublished doctoral thesis: Facultad de Ciencias Naturales e Instituto Miguel Lillo, Universidad Nacional de Tucumán, Tucumán.

Högberg, A., Puseman, K., & Yost, C. (2009). Integration of use-wear with protein residue analysis—A study of tool use and function in the south Scandinavian early Neolithic. *Journal of Archaeological Science, 36*, 1725–1737.

Laborda Martínez, M. A. (2018). Análisis traceológico de una muestra de hematites y ocres con estigmas de utilización recuperados en la cueva de Zatoya (Abaurrea Alta, Navarra). Aproximación a los modos de uso y a su posible utilidad. *Cuadernos de Arqueología de la Universidad de Navarra, 26*, 129–182.

Leipus, M. (2006). *Análisis de los modos de uso prehispánicos de las materias primas líticas en el Sudeste de la Región Pampeana: Una aproximación funcional*. Unpublished doctoral thesis: Facultad de Ciencias Naturales y Museo, UNLP.

Leipus, M. (2014a). El análisis funcional de materias primas heterogéneas y su aplicación a diferentes variedades de cuarcitas de la región Pampeana (Argentina): resultados experimentales y arqueológicos. In M. E. Mansur, M. Alonso Lima, & Y. Maigrot (Eds.), *Traceology today: Methodological issues in the old world and the Americas* (British archaeological reports 2643) (pp. 43–54).

Leipus, M. (2014b). Análisis funcional de base microscópica de los instrumentos líticos manufacturados por talla de las unidades estratigráficas Y, S y Z. In G. Politis, M. A. Gutierrez, & C. Scabuzzo (Eds.), *Estado actual de las investigaciones en el sitio Arroyo Seco 2* (Serie Monográfica 4) (pp. 229–274). INCUAPA-UNICEN. Olavarría.

Leipus, M. (2016). Variabilidad tecnomorfológica y funcional de las raederas en la región Pampeana (Argentina). *Revista de Antropología del Museo de Entre Ríos, 2*(2), 47–67.

Leipus, M. (2019). Identificación y caracterización de rastros de uso experimentales en materias primas líticas de la provincia de Entre Ríos. In *VIII Encuentro de Discusión Arqueológica del Nordeste*. Libros Pórtico.

Leipus, M. (2021). Relación entre materias primas, tecnología y contexto de uso en conjuntos líticos del Delta del Paraná e interior de la provincia de Entre Ríos. In *IX Encuentro de Discusión Arqueológica del Nordeste*. Libros Pórtico.

Leipus, M., & Apolinaire, E. (2020). Análisis funcional de base microscópica aplicado a materias primas de la provincia de Entre Ríos: identificación y caracterización de rastros de uso experimentales. *Revista del Museo de Antropología, 12*(3), 161–170.

Leipus, M., & Mansur, M. E. (2007). El análisis funcional de base microscópica aplicado a materiales heterogéneos. Perspectivas metodológicas para el estudio de las cuarcitas de la región Pampeana. In C. Bayón, A. Pupio, M. I. González, N. Flegenheimer, & M. Frere (Eds.), *Arqueología de las pampas* (pp. 179–200). Sociedad Argentina de Antropología.

Loponte, D., Tchilinguirian, P., & Sacur Silvestre, R. (2011). Caracterización de afloramientos de calizas silicificadas de la provincia de Entre Ríos (Argentina) y su vinculación con los circuitos de abastecimiento prehispánico. In M. R. Feuillet Terzaghi, M. B. Colasurdo, J. I. Sartori, & S. Escudero (Eds.), *Avances y perspectivas en la Arqueología del Nordeste* (pp. 125–139). Santa Fé.

Loponte, D., Parisi, F., Liotta, J., Wagner, M., & Acosta, A. (2019). Los moluscos fluviales en la subsistencia de los grupos cazadores-recolectores del humedal del Paraná inferior. *Comechingonia, 23*(1), 12–56. versión On-line ISSN 1851-0027.

Lynch, V., Fernández, M., & Miotti, L. (2015). Estudio experimental sobre la dinámica post-depositacional en rastros de uso: integridad artefactual del Componente I de Cueva Maripe (Santa Cruz, Argentina). *Magallania, 43*(1), 211–229.

Mansur, M. E. (1999). Análisis funcional de material lítico: Problemas de formación y deformación de rastros de uso. *Actas del XII Congreso Nacional de Arqueología Argentina, I*, 355–366.

Mansur-Franchomme, M. E. (1983). *Traces d'utilisation et technologie lithique: exemples de la Patagonie.* Unpublished doctoral dissertation, Universidad de Bordeaux I.

Mansur, M. E., Leipus, M., & De Angelis, H. (2007). Perspectivas sobre la conservación del material lítico a partir del análisis microscópico. In Número especial de la Revista Pacarina (pp. 263–269). Facultad de Humanidades y Ciencias Sociales/Universidad Nacional de Jujuy.

Marreiros, J., Calandra, I., Gneisinger, W., Paixao, E., Pedergnana, A., & Schunk, L. (2020). Rethinking use-wear analysis and experimentation as applied to the study of past hominin tool use. *Journal of Paleolithic Archaeology, 3*, 475–502.

Masojć, M., Kim, J. Y., Krupa-Kurzynowska, J., Sohn, Y. K., Ehlert, M., & Michalec, G. (2021). The oldest *Homo erectus* buried lithic horizon from the eastern Saharan Africa. EDAR 7 – An Acheulean assemblage with Kombewa method from the Eastern Desert, Sudan. *PLoS One, 16*(3), 1–53.

Mazza, B., & Loponte, D. (2012). Las prácticas mortuorias en el humedal del Paraná inferior. *Arqueología Iberoamericana, 13*, 3–21.

Pal, N. (2012). *Tendencias temporales en las estrategias de explotación y uso de los materiales líticos recuperados en la cuenca superior del arroyo Tapalqué (partidos de Benito Juárez y Olavarría): una perspectiva desde el análisis funcional.* Unpublished doctoral thesis: Facultad de Ciencias Sociales, Universidad del Centro.

Pal, N. (2013–2015). Identificación y caracterización de rastros de uso: programa experimental sobre rocas cuarcíticas y ftanita (provincia de Buenos Aires, Argentina). Boletín de Arqueología Experimental, 10, 161–173.

Pal, N., & Leipus, M. (2018). *Modos de producción y uso en la Subregión Pampa Húmeda durante el Holoceno: un enfoque integral de la tecnología lítica de los grupos cazadores-recolectores.* I Congreso Argentino de Estudios Líticos en Arqueología.

Pal, N., & Leipus, M. (2021). *Análisis del diseño de los instrumentos líticos utilizados en el procesamiento de vegetales: subregión Pampa Húmeda.* X Congreso de Arqueología de la Región Pampeana Argentina.

Pawlik, A., Olle, A., Márquez, B., Longo, L., & Skakum, N. (2020). Functional studies of prehistoric artefacts and their socio-economic meaning. *Journal of Archaeological Science: Reports, 33*, 102467.

Pedergnana, A., & Ollé, A. (2020). Use-wear analysis of the late middle Pleistocene quartzite assemblage from the Gran Dolina site, TD10.1 subunit (Sierra de Atapuerca, Spain). *Quaternary International, 569–570*(20), 181–211.

Pérez, M., Sacur Silvestre, R., & Buc, N. (2018). Tecnología de grupos guaraníes en las cuencas alta y baja de los ríos Paraná y Uruguay. *Revista de Antropología del Museo de Entre Ríos, 4*(2), 41–65.

Politis, G., & Bonomo, M. (2012). La entidad arqueológica Goya-Malabrigo (Ríos Paraná y Uruguay) y su filiación Arawak. *Boletín de la Sociedade de Arqueología Brasileira, 25*(1), 10–46.

Politis, G., & Bonomo, M. (Eds.). (2018a). *Goya-Malabrigo: arqueología de una sociedad indígena del noreste argentino.* UNICEN.

Politis, G., & Bonomo, M. (2018b). Estado actual y perspectivas de Goya-Malabrigo, una sociedad indígena del Noreste argentino. In G. Politis & M. Bonomo (Eds.), *Goya-Malabrigo: arqueología de una sociedad indígena del noreste argentino* (pp. 9–44).

Prous, A., & Pessoa Lima, A. (2019). O registro de moluscos nos vestigios arqueológicos pré-históricos do Brasil. *Comechingonia, 23*(1), 21. Versión On-line ISSN 1851-0027.

Ramos van Raap, M. A., & Scabuzzo, C. (2018). Estudios bioarqueológicos de la colección del Museo Arqueológico Manuel Almeida (provincia de Entre Ríos). *Relaciones de la Sociedad Argentina de Antropología, XLIII*(2), 185–205.

Ramos, S., Franco, M. J., Brea, M., Bonomo, M., & Politis, G. (2021). The use of wood during prehispanic times in the upper Paraná Delta revealed through analysis of ancient charcoal. *Vegetation History and Archaeobotany, 30*, 193–212.

Rimkus, T. (2016). Microliths in fisheries? Use-wear and experimental study of composite tools of the mesolithic South Lithuania. *Arheologija un Etnografija, 29*, 31–45.

Sacur Silvestre, R., Buc, N., Acosta, A., & Loponte, D. (2013). Estrategias de captura de presas y sistemas de armas de los cazadores-recolectores que habitaron el humedal del Paraná inferior: una aproximación experimental y arqueológica. *Comechingonia, 17*, 27–57.

Sade Martínez, K., De Angelis, H., & Castañeda Carrasco, F. (2021). Configuración de micro-rastros de uso en artefactos experimentales de lutita, basalto, toba y arenisca del yacimiento de Punta del Monte 1 (Aysén, Chile). *Boletín De Arqueología Experimental, 15*, 5–20.

Sánchez, J., Colobig, M. M., Zucol, A., Politis, G., Bonomo, M., & Castiñeira, C. (2013). Primeros resultados sobre el uso prehispánico de los vegetales en el sitio arqueológico Los Tres Cerros 1 (Victoria, Entre Ríos, Argentina): Análisis del registro biosilíceo. *Darwiniana, 1*(2), 201–219.

Scabuzzo, C., Ramos Van Raap, A., Bonomo, M., & Politis, G. (2015). Estudios bioarqueológicos en el sitio Los Tres Cerros 1 (Delta Superior del río Paraná, Entre Ríos, Argentina). *Boletim do Museu Paraense Emílio Goeldi, 10*(2), 509–535. https://doi.org/10.1590/1981-81222015000200015

Silva, C. B. (2018). Formas cerámicas Goya-Malabrigo: variabilidad y tipos morfológicos. In G. Politis & M. Bonomo (Eds.), *Goya-Malabrigo: arqueología de una sociedad indígena del noreste argentino* (pp. 73–87). UNICEN.

Silvestre, R., & Capparelli, A. (2017). La tecnología lítica de grupos guaraníes prehispánicos en la cuenca inferior del río Paraná. *Pesquisas, 70*, 53–77.

Silvestre, R., Buc, N., & Montero, R. (2020). Use-wear analysis of lithic and bone artifacts coming from Guaraní archaeological assemblages from the Paraná River basin, Argentina. *Journal of Archaeological Science: Reports, 31*, 102292.

Van Gijn, A. L. (1989). The wear and tear of flint: Principles of functional analysis applied to Dutch neolithic assemblages. *Analecta Praehistorica Leidensia, 22*, 23–50.

Van Gijn, A. L. (2014). Science and interpretation in microwear studies. *Journal of Archaeological Science, 48*, 166–169.

Van Gijn, A. L., & Houkes, R. (2006). Stone, procurement and use. In K. L. P. Louwe & P. F. B. Jongste (Eds.), *Schipluiden: A neolithic settlement on the Dutch north sea coast c. 3500 cal BC* (pp. 167–193). Faculty of Archaeology/Leiden University. Retrieved from https://hdl.handle.net/1887/32627

Between Shell Layers. Processes Involved in the Formation of Don Enrique Archaeological Site (Punta Indio, Buenos Aires Province, Argentina)

Naiquen Ghiani Echenique, Mariel S. Luengo, Enrique E. Fucks, María C. Paleo, and Emiliana Bernasconi

Abstract Don Enrique archaeological site is located in the northern coastal plain of Samborombón Bay (Punta Indio, Buenos Aires province, Argentina). The aim of this investigation is to interpret the stratigraphic context of the site and the geomorphological processes involved in its origin and formation, through the integration of geomorphological, micropaleontological, and archeological analyses. Two exposed sedimentary profiles from the quarries 'Cantera 1' and 'Cantera 2' were analyzed. Numerous ceramic fragments, along with other archaeological materials were collected from the surface and within the sedimentary profiles. The characteristics of these remains revealed that the site is associated with hunter-gatherer groups. We interpret the unit with archaeological materials in 'Cantera 1' as a paleosoil, constituted over a beach ridge and buried by bioclastic deposits of a high-energy beach environment. In 'Cantera 2', the soil continued being formed by current climatic conditions, while in 'Cantera 1' it started a new pedogenesis process over the shell deposit. We propose that the observed differences between both sectors are related to the paleoshoreline position and its topographic configuration. Also, micropaleontological analysis suggests changing environmental conditions from an estuarine environment with higher marine influence to the establishment of more restricted estuarine conditions as a result of coastal progradation processes. Don Enrique represents the first surveyed archaeological site in the northern sector of Samborombón Bay, where archaeological materials were found in situ, specifically in a buried paleosoil. Two radiocarbon dates obtained from *Mactra isabelleana* indicate that the site was formed during the Late Holocene, ~1300–1200 cal years BP.

N. Ghiani Echenique (✉)
Laboratorio de Análisis Cerámico, Facultad de Ciencias Naturales y Museo, Universidad Nacional de La Plata – Consejo Nacional de Investigaciones Científicas y Técnicas (CONICET), La Plata, Buenos Aires, Argentina
e-mail: naiquen@fcnym.unlp.edu.ar

G. F. Bonnat et al. (eds.), *Current Research in Archaeology of South American Pampas*, The Latin American Studies Book Series,
https://doi.org/10.1007/978-3-031-55194-9_10

Keywords Hunter-gatherers · Late Holocene · Archaeology · Geomorphology · Micropaleontology · Palynological analysis · Foraminiferal analysis · Samborombón Bay

Introduction

Don Enrique archaeological site is located at the northern coastal plain of Samborombón Bay (Punta Indio, Buenos Aires province, Argentina). It covers a series of quarries generated by shell exploitation over a beach ridge at 4 m a.s.l., about 350 m SE of the intersection of the Provincial Routes N°11 and N°36 (35°35′44.74″S and 57°15′53.91″W) and 1 km from the coast of the Río de la Plata estuary, separated from it by a marsh environment (Fig. 1).

During a prospecting pollen field trip, ceramic sherds were recorded in the area. The site was identified and registered through an archaeological prospection framed within the regional archaeological approach of the southern sector of Punta Indio, which contributed to forming a deeper knowledge of the hunter-gatherer groups that inhabited it during the Late Holocene (Ghiani Echenique, 2021).

The site is exposed in four extraction points and its survey shows an evident distribution of archaeological materials in an area of approximately 9000 m^2. According to the observations in the satellite images available on *Google Earth*, which were taken since the year 2003, the site exposure is recent. The extractions

M. S. Luengo
Centro de Estudios Integrales de la Dinámica Exógena (CEIDE), Facultad de Ciencias Agrarias y Forestales/Facultad de Ciencias Naturales y Museo, Universidad Nacional de La Plata – Consejo Nacional de Investigaciones Científicas y Técnicas (CONICET), La Plata, Buenos Aires, Argentina
e-mail: msluengo@fcnym.unlp.edu.ar

E. E. Fucks
Centro de Estudios Integrales de la Dinámica Exógena (CEIDE), Facultad de Ciencias Agrarias y Forestales/Facultad de Ciencias Naturales y Museo, Universidad Nacional de La Plata, La Plata, Buenos Aires, Argentina
e-mail: efucks@fcnym.unlp.edu.ar

M. C. Paleo
Laboratorio de Análisis Cerámico, Facultad de Ciencias Naturales y Museo, Universidad Nacional de La Plata, La Plata, Buenos Aires, Argentina
e-mail: mcpaleo@fcnym.unlp.edu.ar

E. Bernasconi
Instituto de Investigaciones en Biodiversidad y Medioambiente (INIBIOMA), Consejo Nacional de Investigaciones Científicas y Tecnológicas-Universidad Nacional del Comahue, San Carlos de Bariloche, Río Negro, Argentina
e-mail: bernasconi@comahue-conicet.gob.ar

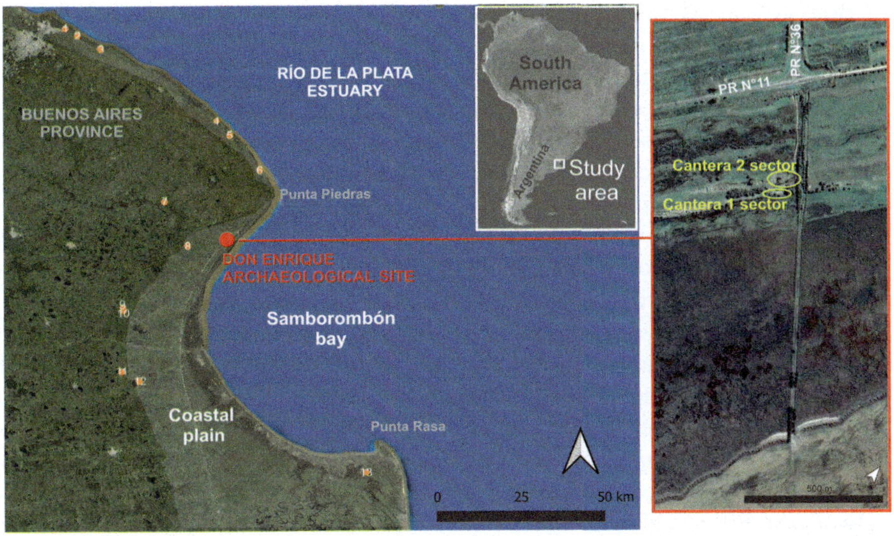

Fig. 1 Don Enrique archaeological site, in the coastal plain of the Samborombón Bay (Punta Indio, Buenos Aires province, Argentina). The coastal plain is represented (shaded surface), as well as other archaeological sites and localities mentioned in the text: (1) Palo Blanco, (2) La Maza I, (3) Arroyo Zapata, (4) Barrio San Clemente, (5) Las Marías, (6) Los Tres Ombúes, (7) Corral del Indio, (8) El Puesto, (9) La Guillerma, (10) San Ramón, (11) La Salada, (12) Los Molles, (13) El Divisadero Monte 6. (Modified from *Google Earth* 2022)

were observed during the first archaeological prospection made at the end of 2017, and they are also visible in the available images corresponding to 2019 and 2021.

Particularly, two extraction fronts, which are adjacent to the littoral ridge and separated by 18 m, were analyzed during the prospection. These were called 'Cantera 1' and 'Cantera 2'. A disposition of archaeological remains that differed from those observed in other coastal archaeological sites in the region was found in 'Cantera 1'. The stratigraphy showed that archaeological remains were under a shell layer, while in the other sites, they were in the soil formed over the shell deposits.

Don Enrique constitutes the first surveyed archaeological site in the northern sector of the Samborombón Bay, while south of the Salado river numerous sites were registered, such as La Salada, Los Molles and El Divisadero Monte 6. These sites correspond to hunter-gatherers that inhabited the coastal wetland from approximately 2000 years BP to post-Hispanic-indigenous contact, after 500 years BP (Aldazábal, 2004; Aldazábal et al., 2004; Aldazábal & Eugenio, 2008; among others). To the north, investigations centered in Barrio San Clemente archaeological locality and Las Marías archaeological site indicated occupations of hunter-gatherers and hunter-gatherer-fishers since nearly 1800 years BP to post-contact times (~300 years BP) (Paleo & Pérez Meroni, 1999, 2004, 2005/2006, 2007; Balesta et al., 1997; Pérez Meroni & Paleo, 1999; Paleo et al., 2018, 2002; Pérez Meroni et al., 2010; García et al., 2011; Ghiani Echenique et al., 2013; Day Pilaría, 2018;).

Both coastal sectors have similar characteristics in the archaeological record, also shared with the adjacent continental plain that belongs to the Salado river lower basin, like La Guillerma and San Ramón archaeological localities (González, 2005; González de Bonaveri et al., 2006; González & Frère, 2009; among others). These contexts mainly represent prolonged and/or recurrent occupations, with abundant material evidence of ceramics and, in less proportion, lithic artifacts, numerous archaeofaunistic remains and presence of allochthonous resources. The sites are located in high places where the *talar*, a plant community which might have been important to human groups because of protection and resources, developed (Balesta et al., 1997; Pérez Meroni & Paleo, 1999; Paleo et al., 2002; Aldazábal et al., 2004; González, 2005; González & Frère, 2009). It is worth mentioning that, from an archaeological landscape approach, the surveyed sites in the southern sector of Punta Indio, both in the coastal plain (Los Tres Ombúes and Don Enrique) and in the Samborombón river lower basin (Corral del Indio and El Puesto), were interpreted as part of a *talares* and wetlands landscape with a sense of identity for their inhabitants, who carried out their daily practices and constituted social interaction networks at different scales (Ghiani Echenique, 2021).

The referred archaeological contexts can be classified as shallow sites, according to the denomination established by Zárate et al. (2000/2002) for interfluvial Pampean environments. In these, the remains are buried near the surface in the horizon A of the current soil, and the action of pedogenesis predominates, with less incidence of sedimentation or erosion processes. However, northern to these sites, in the same coastal plain, archaeological remains included in the shell deposits were also registered, particularly in Palo Blanco, La Maza I and Arroyo Zapata archaeological sites (Cigliano, 1963, 1966; Ceruti & Crowder, 1973; Salemme et al., 1985; Salemme et al., 1989). They were interpreted considering its formation process by Salemme et al. (1989), as redeposits of sediment and archaeological materials in a high-energy environment. These observations clearly show that archaeological remains and shell deposits should be analyzed together in an interdisciplinary context to allow a proper interpretation.

Unlike the aforementioned archaeological contexts, archaeological remains in Don Enrique were found in a level situated between shell layers. This peculiarity implies the need to recognize the processes involved in the site formation, and, based on this, to interpret the nature of the record. That is why the aim of this investigation is to interpret the stratigraphic context of Don Enrique archaeological site and the geomorphological processes involved in its origin and formation, through the integration of the archaeological study and the information provided by geomorphological and micropaleontological analyses.

From the paleoenvironmental point of view, considering that the sediments represent a historical archive of climatic variations, then macroremains, microfaunistic communities (foraminifera and ostracods) and the vegetal composition will reflect changes in the ecological conditions through variations in their distribution, abundance, and diversity. Therefore, in this case, through the analysis of the distribution of foraminifera and plant composition, it will be possible to approach the paleoenvironmental and paleoclimatic fluctuations that occurred in the study area during the involved period.

Holocene Sea-Level Fluctuations and Regional Geomorphological Configuration

Erosion and accumulation processes that accompanied the sea-level changes that occurred during the Holocene (MIS 1) generated a multiplicity of deposits and geomorphological features which indicate the paleoenvironmental conditions of the wide Río de la Plata estuary.

The increase of the sea level associated with the Holocene warmer event is widely represented in the coastal plain located east of Buenos Aires province. In this geomorphological unit, parallel to the coastline, littoral transgressive-regressive deposits predominate, with altitude of less than 5 m a.s.l (Fidalgo et al., 1973; Codignotto & Aguirre, 1993; Cavallotto, 1995, 2002; Fucks et al., 2010, among others). The limit with the continental plain is represented by a sharp paleo cliff or a poorly marked surface, dominated by erosion. These littoral processes reworked loessic and littoral Pleistocene deposits, which incorporated silt sandy sediments and carbonatic concretions (Colado et al., 1995), as well as other concretions transported by the tributaries of the estuary and those generated in situ, mainly associated with mollusks remains.

Based on radiocarbon dates obtained for different depositional environments, numerous sea-level curves were proposed for the Río de la Plata estuary and adjacent coastal plains for the Holocene (e.g. Isla, 1989; Aguirre & Whatley, 1995; Cavallotto, 1995; Cavallotto et al., 2004). Recently, Prieto et al. (2017) formulated a sea-level curve for this region, from a wide database. The study revealed that during the transgressive phase, the sea level was similar to the current level near ~7000 cal years BP, and the transgressive maximum was reached at ~6000–5500 cal years BP, with an approximate altitude of +4 m. After these ages, it started a gradual decline to the current position.

The mentioned sea-level fluctuations conformed to the geomorphological unit called coastal plain, which includes environments with littoral bars, shell ridges, tidal flats, coastal lagoons, dunes and beaches, generated by the accumulation of marine sediments next to the continental plain (e.g. Fidalgo et al., 1973; Tricart, 1973; Violante & Parker, 2000; Violante et al., 2001; Cavallotto, 2002; Cavallotto & Violante, 2005; Schnack et al., 2005; Fucks & Deschamps, 2008; Fucks et al., 2010). In the Río de la Plata estuary, the formation of the coastal plain of the Buenos Aires province was conditioned by the presence of Punta Piedras, a coastal ledge which constitutes the northern end of the Samborombón Bay. It worked as a divergence point for the littoral streams, generating the Rio de la Plata coastal plain to the northwest, and the Ajó coastal plain to the southwest (Violante et al., 2001), of which the Samborombón Bay is a part.

These littoral streams were dominated by winds from the east and east-southeast, which still have a predominant effect on the coastal processes (Cavallotto, 2002). It is worth mentioning that strong winds with such orientation are associated with climatic anomaly events known as *sudestadas* (i.e. winds from the South-East) which imply maximum levels in the estuary water and are frequently accompanied by precipitations of varying intensity (Fucks et al., 2017). These events could be linked to anomalies in the low levels of atmospheric circulation that occur during the whole year (Escobar et al., 2004).

Samborombón Bay Coastal Plain

Geology and Geomorphological Characterization

The Samborombón Bay coastal plain is located between Punta Piedras and Punta Rasa, as seen in Fig. 1. In the proximity of Don Enrique archaeological site, the coastal plain presents a poor development because of the proximity to Punta Piedras, and the most noticeable feature is a succession of shell ridges that covers 2.5 km width. Its slightly different lineups indicate little changes in the littoral drift characteristic of functional coastlines along the time (Cavallotto, 2009). West of littoral ridges, there is a plain that represents an ancient tidal flat (Tricart, 1973), a wide sector which contacts the higher continental plain inland that widens in a southerly direction. It constitutes a muddy environment with alternation of plain sectors and meandering tidal paleochannels that are clogged in different grades (Vargas Gil et al., 1972; Sánchez et al., 1976) and are distinguishable by hydrophilic vegetation (Vervoorst, 1967). According to Fucks et al. (2010), this littoral environment from the Holocene marine ingression was grouped as Canal 18 Member from Canal de Las Escobas Formation. East of littoral ridges, there is an ancient intertidal plain that gradually turned into a marsh characterized by daily and extraordinary tides. Because the low energy of the waves conditions a predominantly fine texture, there are low and swampy grounds, called *cangrejales* (i.e. a sedimentary muddy plain) Vervoorst, 1967; Volpedo et al., 2005). The Samborombón Bay coastal plain presents a variable width of 0.5 km in the north up to 6 km approximately in the south, and it has abundant tidal channels, paleochannels and *cheniers*, little ridges of a shelly sandy composition. According to Fucks et al. (2010), the intertidal plain development implied the incorporation of intertidal muddy sediments to the tidal plain, as evidenced by the current dynamics (Carol et al., 2013; Volpedo et al., 2005), in alternation with *cheniers* formation periods, which indicate high energy episodes occurred during the last 3000 years (Luengo et al., 2018).

Sedimentary deposits related to the Holocene marine ingression in the center zone of Samborombón Bay were grouped as the Canal de Las Escobas Formation (Fidalgo et al., 1973, reformulated by Fucks et al., 2010) (Fig. 2). The Destacamento

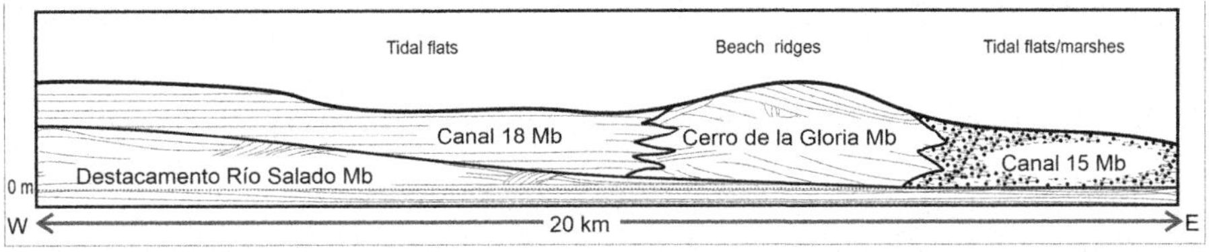

Fig. 2 Stratigraphic units related to the Holocene marine ingression in the center of Samborombón Bay, that correspond to the Members (Mb) which constitute the Canal de Las Escobas Formation

Río Salado Member (Mb) corresponds to sandy clay dark sediments situated at the base of the unit, and it can only be observed in ravines and stratigraphic cuts. The overlying sediments were classified as Canal 18, Cerro de la Gloria, and Canal 15 Members. The first is associated with tidal flats of an open environment that were deactivated since the formation of the shell ridges. The Canal 15 Member implies tidal flats that graduated to marshes, with the incorporation of current intertidal sediments and *cheniers* formation near the Salado river mouth. The Cerro de la Gloria Member includes beach ridges conformed by marine shells and sand sediments, which constitute the most conspicuous geoforms of the coastal plain. They are parallel to subparallel lineups, easily identifiable because of the higher position in relation to its surroundings, due to being separated by flooded depressions and the plant community called *talar* that is developed on them. These geoforms are composed of skeletal remains of mollusks, subordinated pebbles, and sands, constituting layers with different thicknesses and variable lengths, with a cross-stratification of low angle. The malacological fossiliferous content corresponds predominantly to bivalves. Close to 90% of the remains are assignable to *Mactra isabelleana* (d'Orbigny, 1846) and the other 10% to *Noetia bisculata* (Lamarck, 1819), *Brachidontes rodriguezi* (d'Orbigny, 1842), *Ostrea* sp., *Pitar rostratus* (Philippi, 1844), *Anomalocardia brasiliana* (Gmelin, 1791) and *Corbula patagonica* (d'Orbigny, 1846). Another taxon represented are marine gastropods typical of a shallow littoral zone, as *Zidona drufesnei* (Donovan, 1823), *Pachycymbiola brasiliana* (Lightfoot, 1786), *Olivella tehuelcha* (Duclos, 1835), *Crepidula* sp., *Urosalpinx* sp., *Olivancillaria* sp. and *Buccinanops* sp. (Fucks et al., 2010).

Various radiocarbon dates performed on materials from the Canal de Las Escobas Formation resulted in absolute ages of ~7000–6000 14C years BP for the Destacamento Río Salado Member and of ~6700–5500 14C years BP for the Canal 18 Member (see a synthesis in Fucks et al., 2010:33). In addition, ages of ~4800–4200 14C years BP were obtained in the main ridge of the Samborombón Bay central area, which corresponds to the Cerro de la Gloria Member by Gómez et al. (1988), while Codignotto and Aguirre (1993) dated in ~3000–2500 14C years BP the littoral ridges located northerly, near the Samborombón and Salado river mouths, and in ~3300–1600 14C years BP the most barrier spits identified in Punta Rasa. It is also worth mentioning the dates of ~3600–2600 14C years BP made by Luengo (2019) for the *cheniers* formed in the marsh environment near the Salado river mouth.

Soils and Vegetation

Imbellone et al. (2014) characterized the calcareous soils of the coastal plain, scarcely developed and with A-C and A-AC-C sequences, whose content of lithogenic calcium carbonate was contributed mainly by biogenic materials. In the Samborombón Bay, Haprendoles (Molisols Order) were identified on beach ridges, with a greater representation of Hapludolls (Molisols Order) in dunes and Udipsamments (Entisols Order) in the abundant sand mantles in the area (Sánchez et al., 1976). Also, an inland-coastal (W-E) vegetation gradient can be verified in the bay which encompasses the Pampas grassland, the *talar*, the *pajonal* and the intertidal and subtidal environments up to the estuary water (Volpedo et al., 2005).

The *talar* is a plant community which presents an important biodiversity and a variable composition. It is located in eastern Buenos Aires province, and it is dominated by *Celtis tala* Gillies ex Planch (Parodi, 1940; Vervoorst, 1967; Arturi & Goya, 2004; Torres Robles, 2009). As the cited authors refer, the *talar* integrate mosaics of microenvironments in which they combine with grasslands, *pajonales* and littoral communities. In the coastal plain North of Punta Piedras, the *talares* are distributed in variable width strips parallel to sub-parallel to the coast (50–100 m) and with surfaces between 1 and 30 ha (Goya et al., 1992). This distribution corresponds to the shell ridges separated by inter-ridges, low areas where the trees do not develop due to drainage limitations associated with the plane or concave-plane relief (Arturi & Goya, 2004). Toward the continent, the forest is also observed in lunettes of shallow lakes and fluvial levees (Parodi, 1940; Vervoorst, 1967).

Regarding the *talar* plant composition, according to Parodi (1940) throughout the distribution area, there is *C. tala* and to a lesser extent species, such as *Sambucus australis*, *Phytolacca dioica* L., *Jodina rhombifolia* Hook et Arn., *Acacia caven* (Molina) Molina, *Schinus fasciculatus* (Griseb.) IM Johnst., and *Schinus longifolius* (Lindl.) Speg. var. *longifolius*, *Scutia buxifolia*. This plant formation includes shrubby species such as *Baccharis trimera* (Less.) DC., *Baccharis articulata* (Lam.) Pers. and *Pavonia sepium* A. St.-Hil., climbers such as *Cayaponia bonariensis* (Mill.) Mart.Crov., *Passiflora caerulea* L. and *Araujia sericifera* Brot., epiphytes of the *Tillandsia* genus and many other native and exotic herbaceous plants from the undergrowth and associated grasslands. Different naturalized exotic species can also be observed. Parodi (1940) and Vervoorst (1967) mention the decrease in composition richness of the south of Punta Piedras *talares*, where they also present less extension. For the coastal plain of Samborombón Bay, León et al. (1979) indicated the presence of forests near the coast with a large predominance of *C. tala*.

The first pollen analysis of archaeological sequences from the littoral zone of the Río de la Plata corroborated the association of the *talar* forest with human occupation of the Barrio San Clemente archaeological locality (Páez et al., 1999). Paleo et al. (2002) delve into the formation and dynamics of the coastal environment. Considering the location of the archaeological sites in the beach ridges adjacent to the coast and the previous conformation of the innermost zones, the authors suggest that the formation of the soil on these landforms could have begun at ~3000 14C years BP. They interpret the development of the predominant gramineous-herbaceous steppe and also of the "*talar*" for the beginnings of human occupation ~1500 14C years BP, indicated by pollen from *Celtis* sp. and *Jodina* sp. Between ~1500 and ~200–300 14C years BP, the development of grassland and *talar* forest is simultaneous to the first hydrophilic elements (Cyperaceae and *Typha*). Therefore, during human occupation, this forest was a predictable and available resource, although minor variations such as lower *talar* diversity were recorded at ~900 14C years BP. From ~340 14C years BP to the present, there would have been a recomposition of the structure of the plant communities, becoming a more diverse and tri-stratified forest.

Additionally, pollen studies carried out by Fernández (1995, in Aldazábal, 2004) at the central sector of Samborombón Bay, corroborated the development of *talar* forests ~2000 14C years BP. Recent studies (Luengo et al., 2021) showed an earlier development, since ~6500 cal years BP for the central zone of the bay.

Materials and Methods

Field Methodology

In the Don Enrique archaeological site, 'Cantera 1' sector is located 1 km away from the coast and has an approximate area of 200 m^2. The 'Cantera 2' sector, of an innermost location, presents a deeper and wider excavation, of approximately 3800 m^2. In both quarries, archaeological materials were recovered from exposed profiles resulting from shell extraction and on the surface. Also, sedimentary and stratigraphic units from the exposed profiles were identified. It should be noted that in the geological and geomorphological studies carried out in the area, it is common to resort to quarries, as well as channel cuts, for the observation of outcrops that allow detailing the lithological and fossiliferous attributes of the different sedimentary units (Fucks et al., 2010). Therefore, the quarries present excellent conditions for observation. The recognition of attributes such as color (using Munsell's Sediment Color Charts), texture, sedimentary (Carver, 1971) and biogenic structures, contacts, presence of fossils, position of the fauna, presence of concretions and archaeological materials, among others, allow us to distinguish different units and to interpret the paleoenvironmental conditions represented by the record.

For the micropaleontological analysis, representative samples were taken from each of the stratigraphic units identified in the 'Cantera 1' sector. In particular, a total of four samples (every 5 cm intervals) were obtained from the unit containing the archaeological materials.

Archaeological Analysis

The recovered remains were counted and registered. Ceramic and archaeofaunistical materials were analyzed with special consideration of traces which could be related to post-depositional processes.

Ceramic Technology

The study was framed in the technological style perspective, interested in practices and knowledge linked to pottery (Stark, 1999; Gosselain, 2000; among others). In that sense, the technological choices carried out in pottery production and use were of interest. Different analysis scales were considered, understanding sherds as constitutive parts of vessels (Paleo & Pérez Meroni, 2005/2006; Cremonte & Bugliani, 2006/2009). The methodological guidelines established by Shepard (1956), Rice (1989), Balfet et al. (1992), and Convención Nacional de Antropología (1966) were followed. For a proper observation of macroscopic features, the analysis was conducted on ceramic fragments that exceed 2 cm in length. Particularly a subsample of 40 sherds recovered in 'Cantera 1' and 'Cantera 2' sectors were analyzed. Macroscopic and sub-macroscopic observations (10X and 40X magnifications) of the sherds surfaces were carried out to inquire about different aspects and fresh cuts were analyzed to characterize the ceramic paste (Orton et al., 1997; Cremonte & Bugliani, 2006/2009). In a complementary way, a petrographic study of two selected sherds allowed a deeper paste characterization.

With the purpose of linking the pottery assemblage characteristics to the place where they were recovered and inquiring about the processes linked to the site formation, the analysis was focused on the ceramic alteration, which refers to different processes that generate addition, loss or deformation of the ceramic material (Skibo, 1992). Some of these modifications are mechanical, while others are chemical reactions between ceramic and matrix (Schiffer & Skibo, 1989; Skibo, 1992; Sanhueza Riquelme, 1998). Because the traces registered in fragments could be related to the vessel used as same as to different agents acting in the archaeological record formation (Skibo, 1992; Sanhueza Riquelme, 1998), it is necessary to consider the context of the sherds to link them to post-depositional processes. In that sense, different traces were inspected on surfaces and edges of sherds: mineral deposits, surface jumped, surface attrition, cracked, delamination, root marks, edge rounding, fresh fractures and rodent marks (Schiffer & Skibo, 1989; Sanhueza Riquelme, 1998; Silva, 2017; Galligani et al., 2018).

Zooarchaeological Record

The archaeofaunistical analysis of the recovered assemblage consider the different stages of transformation of animals into social products, including prey acquisition, processing, preparation of foods, discard and taphonomic processes that could affect the remains (Mengoni Goñalons, 1999; Day Pilaría, 2018; among others).

The sample was classified into specimens, which refers to each bone, teeth, shell or fragment of them that constitutes the minimum unit of analysis (Grayson, 1984; Mengoni Goñalons, 1999, 2006/2010). Systematic determination of specimens was based on the biological systematic criteria (Salemme et al., 1991), while aspects like bone morphology, thickness and curvature, location and shape of articular surfaces, foramina and fossae, among others, were considered for the anatomic assignation (Berwick, 1975). Osteological atlas contributed to the identification, as well as reference collections that belong to the Facultad de Ciencias Naturales y Museo (Universidad Nacional de La Plata). The remains that could not be assigned to a taxonomic category were classified as indeterminate. For the Mammalia class, categories based on size and weight ranges were used: small (<15 kg), medium (15–50 kg), and large (>50 kg), in addition to indeterminate Mammalia. Some assigned remains were potentially linked with the regional archaeofaunistical record.

To determine and quantify assemblage composition, the following measures of taxonomic abundance estimations were used: NSP (total number of specimens), NISP (number of identified specimens per taxon) and MNI (minimum number of individuals) (Payne, 1975; Mengoni Goñalons, 1988; Grayson 1991). Moreover, skeletal parts abundance was estimated with the MNE index (minimum number of elements) and the general fragmentation degree with the NSP/NISP index and also could be evaluated the fragmentation of some taxa with the NISP/MNE index (Lyman, 1994; Mengoni Goñalons, 1999).

To evaluate the action of natural and cultural agents that could influence the records formation, the modification of the bone surfaces was analyzed both macroscopically and with a binocular microscope (10X and 40X magnifications). The remains were classified according to the weathering stages proposed by Behrensmeyer (1978), taking into account cortical alterations such as cracks and exfoliations. Mineral precipitations and carnivores, rodents and roots marks (e.g. Binford, 1981; Lyman, 1994; Mengoni Goñalons, 1999, 2006/2010; Gutiérrez et al., 2015) were observed, registered, and analyzed, as well as different cultural modifications, like thermo-alterations, fractures and cut marks (Binford, 1981; Medina & Teta, 2010; Mengoni Goñalons, 1999, 2006/2010; Day Pilaría, 2018).

Micropaleontological Analysis

The micropaleontological assemblage of the unit that contained the archaeological material in the 'Cantera 1' sector was analyzed. It includes palynological material (pollen and non-pollen palynomorphs (NPPs), such as moss and ferns spores, cysts of dinoflagellate, algae, among others) and calcareous foraminifera of four samples from Unit 1.3 (from top to bottom: 1.3.4, 1.3.3, 1.3.2 and 1.3.1).

Palynological Analysis

For the recovery of the palynological material, a volume of 1–2 cm^3 of sediment per sample was processed. To remove coarse particles, the samples were washed using sieves of 150 and 250 μm mesh. Standard physicochemical palynological techniques proposed by Faegri and Iversen (1989) and Mudie and Harland (1996) were used. The processing included: potassium hydroxide (KOH) 10% for deflocculation and neutralization of humic acids; removal of carbonates with hydrochloric acid (HCL) 10%, removal of silicates with hydrofluoric acid (HF) 48%, hydrochloric acid 25% hot to remove the silica gel product of the previous step and acetolysis to remove organic remains. The material was preserved in pure carbolic glycerin and stored in Kahn tubes. Then, it was observed under an optical microscope at 400X and 1000X magnification by placing an aliquot of the residue on a slide and sealing it with paraffin. The identification of the palynomorphs was made according to different atlases and published illustrations such as Heusser (1971) and Markgraf and D'Antoni (1978), among others.

Microfaunistic Analysis

Four samples were washed through a mesh screen with size sieve 63 μm (Tyler System n° 230) in order to eliminate the mud fraction, followed by air-drying. All foraminifera and ostracods were picked from 10 g of dried sediment. Systematic analysis of the fauna was based on Loeblich and Tappan (1992) and Sen Gupta (1999) and for ostracods was based on Moore (1961). The abundance and species richness of foraminifera were determined. Also, the proportions of hyaline, porcelaneous, and agglutinated individuals were recorded following Murray (2006). The Shannon-Wiener (H) and Fisher alpha (α) indexes were calculated according to the Buzas and Gibson (1969) method, in order to recognize environmental variations in the time involved.

Chronology

To establish a precise time frame for the archaeological site, two radiocarbon dates (C14) were obtained from biogenic material (shells) that corresponded to two samples recovered from different stratigraphic units. The bivalve *M. isabelleana* was selected because of its abundance in the parautochthonous shell concentrations (from the beach ridge unit) that are interpreted as high energy deposits (storm deposits). Also, most of the mentioned dates on the regional scale were made from biogenic material of this species. The better-preserved shells were chosen, considering they indicated a low degree of rework and retransport according to the evaluation of taphonomic attributes (fragmentation, abrasion, corrosion, and bioerosion).

The samples were sieved in the laboratory with 2.80 mm, 1.40 mm, and 0.354 mm meshes, to separate the different fractions and to wash the sandy matrix. The dating was made in the Laboratorio de Tritio y Radiocarbono, Centro de Investigaciones Geológicas, Consejo Nacional de Investigaciones Científicas y Técnicas (LATYR-CIG-CONICET) in La Plata, Argentina, and they were calibrated with CALIB.8.2 (Stuiver et al., 2020), using the *Marine20* calibration curve (Heaton et al., 2020). This curve represents the global marine reservoir effect, standardized in 400 years C14, according to the global ocean average (Stuiver & Brazuinas, 1993), and in consequence without considering local variations. In the Buenos Aires province littoral only one study was made in that sense, with no conclusive results (Gómez et al., 2008), because of that we prefer to calibrate the obtained ages using the global standard.

Results

Stratigraphy of Don Enrique Archaeological Site

The 'Cantera 1' sector of Don Enrique archaeological site has a 30 m long extraction front (Fig. 3a), in which a 1.75 m sedimentary profile (Profile A, 35°35′45.20″S, 57°15′53.50″W) was analyzed (Fig. 3b). Four stratigraphic units were determined and are described from the base to the top:

Unit 1.4: 0.55 m thick unit characterized by the alternation of fossiliferous layers from entire bivalve shells with a chaotic orientation to layers with unidentifiable fragmented shells, with sand lenses.

Unit 1.3: 0.35 m thick unit with fine sandy texture and dark brown color linked to a high organic matter content (10 YR 5/3 Brown at the base of the unit grading to 10 YR 5/2 grayish brown, to 10 YR 3/2 very dark grayish brown at the top of the unit). This unit presents a few fragmented shells distributed along the unit, biological structures, bones, small charcoal fragments and ceramic potsherds.

Unit 1.2: This unit overlies unit 1.3 in net contact. It is a 0.45 m thick compact fossiliferous layer conformed by mollusk shells, bioclast-supported with a sandy texture matrix. It presents a variation between entire to slightly fragmented valve layers (with mainly chaotic and nesting orientation), and fragmented to highly fragmented valve layers, and a slight inclination in a southeast direction, toward the interior of the profile.

Unit 1.1: The sequence culminates with the Horizon A of the current soil, 0.40 m of thickness. It is characterized by a very dark homogeneous structure and silty sandy texture with shell remains distributed along the unit. At the top, it presents a few shells and scarce vegetation.

The described units were observed along the entire outcropping extraction front. Notably, the unit 1.3 presented a variable thickness, between 0.30 and 0.35 m, with the top at a depth of 0.85 m and the base at 1.15–1.20 m. Several ceramic fragments were recovered at various georeferenced points (Fig. 3c).

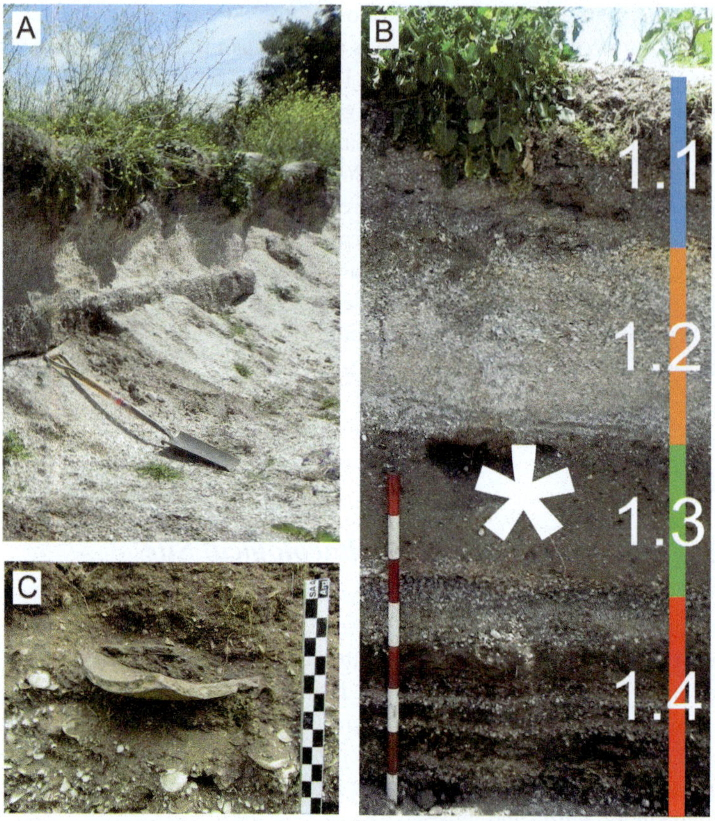

Fig. 3 Don Enrique archaeological site, 'Cantera 1' sector. (**a**) View of the shell extraction front; (**b**) Profile A, of 1.75 m potency. The asterisk indicates the location of the archaeological material, and the bar represents the stratigraphic units 1.1, 1.2, 1.3 and 1.4. (**c**) Potsherd in situ to a depth of 0.85 m, in correspondence with unit 1.3

In the 'Cantera 2' sector, the explored profiles showed archaeological material buried at 0.20–0.60 m. Remains buried at a depth of 0.35–0.60 m were recorded on the extraction front of about 50 m long parallel to 'Cantera 1' (Fig. 4a). For instance, ceramic fragments were recovered at 0.20–0.40 m depth, in the soil that corresponds to unit 2.1 in profile G (35°35′44.58″S, 57°15′53.71″W), which is 1 m long. Below this, a fossiliferous shell layer is developed (Unit 2.2) (Fig. 4b, c). However, the sequence has different characteristics from those described above, in particular, the absence of the upper shell layer observed in the 'Cantera 1′ profiles (Unit 1.3).

Fig. 4 Don Enrique archaeological site, 'Cantera 2' sector. (**a**) View of the shell extraction front; (**b**) Profile G, of 1 m potency. The asterisk indicates the location of the archaeological material, and the bar represents the stratigraphic units 2.1 and 2.2. (**c**) Associated potsherd in situ at 0.20–0.25 m deep, in correspondence with unit 2.1

Archaeological Materials Analysis

The recovered materials, including those found on the surface as well as in the exposed profiles of both site sectors, are 176 potsherds and 149 faunistic remains that corresponded to fishes, reptiles and mammals, some of which have some traces related to anthropic action. Also, eight little charcoal fragments that disintegrated when recovered and three lithic elements were registered. Two of these were elaborated in *tosca*, a local calcareous conglomerate, and the other in diorite, an allochthonous rock probably from the Tandilia mountain system, in the center-south of the Buenos Aires province. Table 1 details the ceramics and archaeofaunistical remains recovered in the exposed profiles and on the surface in the 'Cantera 1' and 'Cantera 2' sectors, which were exhaustively analyzed to characterize the post depositional processes that were involved in the archaeological record of the site.

Table 1 Don Enrique archaeological site. Ceramic and archaeofaunistical materials recovered in exposed profiles and on the surface of 'Cantera 1' and 'Cantera 2' sectors

		Cantera 1		Cantera 2	
		Profiles	Surface	Profiles	Surface
Materials	Ceramic fragments	26	22	58	32
	Archaeofaunistical remains	111	21	23	1

The analyzed ceramic subsample included 25 body sherds and 15 rim sherds, which were between 26 and 118 mm long, with a mean of 50 mm. A minor variation is registered in the width (18–64 mm), with a mean of 35 mm. The thickness was between 3 and 19 mm and the mean is 5.5 mm. Fragments that predominate in the assemblage had dark brown and gray surfaces, with smoothed surfaces, even though some of them had polished surfaces. Both surface treatments left linear marks, mainly on the external side. Other traces were color stains on both surfaces, which could be related to the manufacture as well as the use of the vessel over the fire, and soot deposits, which indicate the last mentioned practice, mainly situated in the external surface (n = 16).

The ceramic pastes of this assemblage presented a great homogeneity. Fresh fractures (n = 17) evidenced a sandy compact paste with homogeneous size inclusions and a low representation of large-sized individuals as grog temper and ferromagnesian concentrations that are frequent in the regional ceramic record (e.g. Paleo & Pérez Meroni, 2004, 2005/2006; González & Frère, 2009; Ghiani Echenique, 2021). Most sherds present a gray/black homogeneous fracture edge (n = 11), while in a minor proportion, differences were identified between core and margins, which refer to a variability related to cooking characteristics of vessels and/or parts of them. The observation of two thin sections allowed us to confirm the paste homogeneity. In these, moderated and low cavities were registered, as well as abundant inclusions of fine sand texture, angular to sub-rounded, well sorted. Quartz and feldspar are very abundant, while a low quantity of volcanic glass and shreds were identified in one of the thin sections, with a very low representation of other components as opaque minerals and pumice fragments.

Decorative techniques were registered in 15 sherds of the subsample. Red paint was registered on internal and/or external surfaces (n = 8), as well as incise geometric designs (n = 4), mainly situated on the external side and exceptionally on the internal or on the lip. These designs are simple, mainly made with full lines, while some fragments show a combination of incisions and paint (n = 3). The identified rims are straight (n = 8), inverted (n = 5) and everted (n = 1), with convex, beveled, bevel-rounded, rounded and flat lips. In two cases the represented profile allowed us to estimate the vessel morphology. In one case, a big inverted rim with incisions on the lip, corresponds to a restricted piece of 140 mm height. The other case, a rim that indicates a 150 mm mouth diameter, corresponds to a non-restricted small piece with a simple outline and 120 mm of estimated height. According to the categories formulated by Balfet et al. (1992) they correspond to a pot and a glass respectively.

Abundant ceramic alteration traces were observed on the external surface of the 40 analyzed potsherds, as well as on the external surface and on the fracture edges of most of them. There is a high frequency of calcium carbonate deposits, white

Table 2 Archaeofaunistical assemblage of Don Enrique archaeological site. NISP (number of identified specimens per taxon), MNE (minimum number of elements) and MNI (minimum number of individuals) per taxon

Taxon	NISP	MNE	MNI
Siluriformes	32	11	2
Scienidae	13	9	–
Pogonias cromis	2	1	1
Salvator merianae	1	1	1
Mammalia indeterminate	1	1	–
Mammalia medium-large	5	3	–
Mammalia medium	2	2	–
Ozotoceros bezoarticus	1	1	1
Ctenomys talarum	1	1	1
Cavia aperea	2	2	1
Myocastor coypus	2	1	1
Total NISP	*62*	*33*	*8*
Indeterminate	24	–	–
Total NSP	*86*	–	–

adherences that react with hydrochloric acid, present in 87.5% of the analyzed subsample, mainly on both surfaces (75%). Other surface alterations are less represented with percentages between 22.5% and 2.5%, such as surface jumping, surface attrition, slight cracked, delamination, and roots marks, which are linked to the referred carbonated adherences. On the fracture edges, rounding (22.5%), fresh fractures (17.5%) and rodent marks (10%) were identified. Particularly, edges with fresh fractures correspond to sherds recovered on the soil surface.

The archaeofaunistical assemblage comprises 86 bone specimens of which 73.9% were assigned to some taxonomic category. Fishes of Siluriformes order, Scienidae family, and *Pogonias cromis* (Linnaeus, 1766), as well as the reptile *Salvator merianae* (Duméril and Bibron, 1839) were identified. The most represented taxa of the sample are mammals of species *Ozotoceros bezoarticus* (Linnaeus, 1758), *Ctenomys talarum* (Thomas, 1898), *Cavia aperea* (Erxleben, 1777) and *Myocastor coypus* (Molina, 1782), in addition to Mammal indeterminate, medium and medium-large categories (Table 2).

Generally, the assemblage presents a good preservation, with a 2–3 weathering stage according to Behrensmeyer (1978), even though 14.1% of specimens have rough, cracked and peeled surfaces. Abundant root marks (28.2%) and rodent marks (6.5%) were registered, while carnivore marks were not observed. An interesting characteristic is the presence of surface carbonated adherences on some specimens (17.4%), which partially cover them, similar to those registered in the ceramics. Also it is worth mentioning that most of the specimens assigned to *Pogonias cromis* were articulated vertebrae.

Thermal alteration was observed on remains assigned to Siluriformes and *Ozotoceros bezoarticus* (11.9%) and intentional fractures in specimens of *Cavia aperea* and Mammal medium size category, which are considered anthropic modifications that indicate the faunal exploitation. Most of the thermalalteration elements have brown and black color and are indeterminate specimens with less than 3 cm length. Generally, the archaeofaunistical materials that were recovered in profiles and on the surface present similar percentages of natural and anthropic traces.

Palynological Analysis

The palynological results are presented in a qualitative manner, and discussed according to the paleoenvironmental and paleobiological implications of the record, as well as the foraminiferological results. Two samples of Unit 1.3 resulted in low pollen concentration, while the others resulted in barren. The following palynological types were identified: Poaceae, Chenopodiaceae, Asteraceae; *Baccharis* type, *Ambrosia* type, *Celtis*, Apiaceae; *Eryngium* type, Ranunculaceae, *Ephedra tweediana*, Bryophytes: *Phaeoceros* sp. and scarce dinoflagellate cyst of *Operculodinium* sp. The level 1.3.4, at the top of the unit, was characterized by the dominance of Asteraceae, Poaceae, Celtis and *Eryngium* type (Apiaceae) and were accompanied by Chenopodiaceae, *Ambrosia* type, *Baccharis* type, Ranunculaceae, *Ephedra tweediana* and *Phaeoceros* sp. Also, scarce fragments of *Operculodinium* were present. The level 1.3.3 resulted in a low diversity and was characterized by the dominance of Chenopodiaceae accompanied by Poaceae, *Baccharis* type and *Celtis*. The lower levels of the unit (1.3.2 and 1.3.1) resulted in barren.

Foraminiferological Analysis

We recognized seven species corresponding to five genera of foraminifera (one with open nomenclature) (Table 3), while ostracods were absent. All foraminifera individuals were calcareous hyaline specimens that belong to the Rotaliida order. The assemblages were mostly represented by *Criboelphidium gunteri* (Cole, 1931), *Criboelphidium poeyanum* (d'Orbigny, 1839) and *Elphidium galvestonense* (Kornfeld, 1931). Other species such as *Ammonia parkinsoniana* (d'Orbigny, 1839), *Buccella peruviana* (d'Orbigny 1839) and *Cibicidoides dispars* (d'Orbigny, 1839) were present in minor proportions. The abundance registered at each sample varied between 3 and 612 individuals, whereas the richness (S) ranged between 4 and 6 species. The Shannon-Wiener (H) diversity index varied between 0.4 and 1.4 and the Fisher index (α) between 0.7 and 0.9.

Table 3 Foraminifera fauna distribution, Abundance (A), Species richness (S), Shannon-Wiener (H) and Fisher index (α)

Sample	*Criboelphidium gunteri*	*Criboelphidium poeyanum*	*Elphidium galvestonense*	*Ammonia parkinsoniana*	*Buccella peruviana*	*Cibicidoides dispars*	*Elphidium* sp.	*A*	*S*	*H*	*α*
1	209	81	61	41	14	–	–	405	5	1.3	0.9
2	196	74	20	34	–	–	–	324	4	1.1	0.7
3	294	162	29	–	29	29	–	544	5	1.2	0.8
4	290	10	–	–	10	–	10	320	4	0.4	0.7

Table 4 Radiocarbon dates obtained for the stratigraphic Units 1.2 and 1.4 of the 'Cantera 1' sector of Don Enrique archaeological site

Sample	Material	Radiocarbon age	Calibration (1σ) (*Marine20* curve)
DEN U4 (LP- 3727)	Shells (*Mactra isabelleana*)	1880 ± 70 years 14C BP	1178–1359 BP
DEN U2 (LP- 3729)	Shells (*Mactra isabelleana*)	1890 ± 60 years 14C BP	1199–1365 BP

Chronology

Radiocarbon dates of 1880 ± 70 years 14C BP were obtained for unit 1.4 (LP-3727) of 'Cantera 1' sector, and of 1890 ± 60 years 14C BP for unit 1.2 (LP-3729). These ages, considering the global marine reservoir effect, correspond to the age ranges of 1178–1359 cal years BP for the former and 1199–1365 cal years BP for the latter (Table 4). The statistical comparison made with the program indicates that the radiocarbon ages obtained are the same, with a confidence interval of 95%.

Archaeological Characterization of Don Enrique Archaeological Site and Observations Related to Post-depositional Processes

The characteristics of the analyzed archaeological materials allowed us to consider their correspondence to Late Holocene hunter-gatherer groups. The abundance of potsherds and archaeofaunistical remains is coincident with the identified sites of the period in a regional scale. The zooarchaeological assemblage represents a variety of taxa including fishes, reptiles and mammals, which were recognized at other archaeological context in the region, as Barrio San Clemente archaeological locality and Las Marías site (Day Pilaría, 2018). Even so, taxonomic diversity is relatively important considering the NSP. The pottery also shows similarities with the regional record, with representation of the same types of surface treatment (smoothed and polished) and decorative modalities that include red paint and incisions forming geometric designs, even though these are simple and with little technical variation. Also, the analyzed pastes show an evident homogeneity, with low representation of ferromagnesian concentrations and grog temper, which are abundant in nearby sites, such as Los Tres Ombúes and Corral del Indio (Ghiani Echenique, 2021). It is possible to consider the local origin of pottery, because of the abundance of clay in the zone, even if the record does not show evidence of the manufacturing process, such as clay masses. In the future, compositional analysis will be carried out to inquire about this possibility.

The Don Enrique record is also integrated with a few small charcoals which evidence the fire used by the inhabitants of the place. They are similar to those registered in Los Tres Ombúes archaeological site, which shows the use of firewood obtained in the *talar* and the coast (Ghiani Echenique et al., 2020). According to the regional archaeological knowledge, the fire was linked to the vessel manufacture and use, as color stains and soot deposits evidence. Also the thermal alteration of zooarchaeological remains could indicate the cooking of different dams as fishes and cervids. Other elements that signal faunal resources processing are the fractures identified in remains of medium mammal and the rodent *Cavia aperea*. The mentioned taxa also present processing evidence in other archaeological sites of the area (Day Pilaría, 2018; Ghiani Echenique, 2021). Regarding the thermo-altered specimens of less than 3 cm long, the discard in stoves practice can be considered. In addition, the site record has representation of *tosca* artifacts, which indicate local resource exploitation for tool manufacture, as well as lithic artifacts from an allochthonous resource (diorita), which come from at least 300 km away the site.

Post depositional alterations present in the ceramic assemblage are mainly linked to the high incidence of calcium carbonate over the materials, which coincides with the shell abundance in the stratigraphic profile. Considering the average maximum length of sherds as indicator of the ceramic assemblage fragmentation, Don Enrique (50 mm) has a minor grade than the identified in nearby sites as Los Tres Ombúes (38.5 mm) and El Puesto (26 mm). The edge rounding is not broadly represented, which could indicate the lack of direct incidence of coastal processes in most of these materials, according to the representation of polished surfaces and soot deposits, likewise a slight degree of surface attrition, jumping and delamination in the assemblage. In coincidence with it, is the 2–3 weathering stage of the archaeofaunistic specimens. Most ceramic and faunistic remains present a high incidence of calcium carbonate depositions, which probably indicate pedogenic carbonation, and they have no indicators linked to the hydric transport. Whereas in sites that have archaeological remains mixed with shells, these materials, for example, rounded sherds, show an intense abrasion, which suggests transport by a hydric high-energy agent (Cigliano, 1963, 1966; Salemme et al., 1989). Moreover, fresh fractures registered in Don Enrique ceramic assemblage mainly affected surface potsherds, which evidence an in situ fragmentation under depositional conditions, as those observed during the prospection. In that sense, we could affirm the recent character of the fragmentation, which would have happened once the material was exposed because of the soil alteration caused by the mineral extraction, after the year 2016.

Geomorphological and Paleoenvironmental Evolution of the Site

At ~1300–1200 cal years BP, according to the stratigraphical and geomorphological evidence, the alternation of high energy (probably storm events) with low energy processes occurred in a beach environment in the northern sector of the Samborombón bay, as represented in the 'Cantera 1' sector of Don Enrique archaeological site.

Such change is suggested by the alternation of fossiliferous bivalve shells (mainly *Mactra isabelleana*), complete or fragmented, with sand lenses in unit 1.4. This environment corresponded with the paleoshoreline position, 1 km inland from its current position, which constituted a beach ridge geomorphological unit.

In accordance with that, the micropaleontological evidence from unit 1.3 allowed us to infer an estuarine environment with high marine influence. The lower part of the study unit contains a marine restricted benthic foraminifera assemblage such as *Buccella peruviana,* a typical species of the inner shelf, found with very low relative frequencies. While in the superficial levels of the unit 1.3 *Ammonia parkinsoniana,* which lives in muddy-sandy sediments, typical of brackish marshes and lagoons, and marine environments from inner shelf to upper bathyal waters (Murray, 2006), was found. This genus is able to tolerate high salinity, even though it was thus registered in sites with large salinity variations, including sites with very low salinity (Márquez, 2019; Ballesteros-Prada et al., 2021). According to the palynological results, the lower levels of the unit were barren. It might be due to the sandy texture (being fine-textured sediments more favorable to the preservation of palynomorphs) which evidence a high energy environment, as revealed by previous palynological results from the Samborombón Bay coastal plain (Luengo, 2019). The distance between the depositional environment and the vegetation parent source could be another factor. These results are in agreement with an environment with high marine influence, evidenced by the foraminiferal record.

In this level, it is also evident a gradual change in the composition of microfossil assemblages that shows an increasing relative abundance of widely known opportunistic and marginal marine benthic foraminifera species such as *Buccella* sp., *Elphidium* spp. and *Ammonia* sp. It suggests changes in the environmental conditions from an estuarine environment with higher marine influence to the establishment of more restricted estuarine conditions as a result of coastal progradation processes.

These environmental changes are also reflected in the palynological record. The presence of Chenopodiaceae and Poaceae (most likely representing *S. perennis* and *Sporobolus* sp.) in the top levels of Unit 1.3, which correspond to the soil organic horizon, suggest a salt marsh vegetation environment. This halophytic vegetation is currently developed in the coastal area of Samborombón Bay and extends from the coastline up to at ca. 100 m eastward from the study site. As a recent multi-proxy study for the Samborombón Bay central area (Luengo et al., 2018) has shown, the displacement of this tidal environment responded to the coastal progradation during the Late Holocene until reaching its present position. These environmental changes reflected by the micropaleontological evidence could respond to this process.

Based on the stratigraphic evidence, we consider that the unit 1.3, which contains archaeological remains, corresponds to a paleosoil developed in a supratidal environment (beach ridge unit). The position of the archaeological materials and the presence of articulated bones and fragile elements as small charcoals, indicate the correspondence with a stabilized deposit favorable to the pedogenetic processes action.

Also, the record of the trees and shrubs *Celtis* sp. and *E. tweediana* accompanied by Asteraceae at the top of this unit, indicates the change from a tidal influence environment (saltmarsh vegetation) to the establishment of the *talar* forest near the site, which evidence more stable environmental conditions and pedogenic processes. These processes are also evidenced by abundant content in organic matter in the top of the unit.

Overlying the paleosoil, in 'Cantera 1' profiles, a shell deposit with a slight stratigraphic inclination in a south-eastern direction (Unit 1.2), could indicate a high energy event in a beach environment and generated by the estuarine dynamics. This bioclastic deposit would have occurred during one or more storm events, possibly *sudestada* extraordinary events. Local geomorphological changes in the environment propitiated the change of this specific site from supratidal to intertidal conditions, burying the soil where the record of anthropic activity remained.

Although the ages of both shell deposits have statistically equal chronologies, the presence of the soil developed at the expense of the lower deposit and buried by the upper one, exposes a period of time in which pedogenetic processes were active generating this soil.

Regarding the 'Cantera 2' profile, according to the stratigraphical evidence, unit 2.1 could be correlated with the 'Cantera 1' stratigraphic sequence unit 1.3. Particularly, in this sector, the absence of the shell deposit overlying the paleosoil that contains archaeological materials in 'Cantera 1' is noted. We consider that differences between both sectors could be explained in terms of the paleoshoreline position and topographical configuration.

In summary, we propose that Don Enrique archaeological site presents buried soil developed on a beach which was progressively deactivated by the coastal progradation, where the hunter-gatherers inhabited and left their remains (Fig. 5a). It was developed after 1300–1200 cal year BP, during a time interval of stable environmental conditions. After that, the deposition of bioclastic material occurred during one or more extraordinary storm events, forming a dense stratified shell concentration with evident chaotic and nesting orientation, which were generated by wave and current action (Martínez et al., 2006). It covered the mentioned soil in the 'Cantera 1' sector, which corresponds to the coastal front (Fig. 5b). Regarding the northern sector of the Samborombón Bay, extraordinary events like this can be considered as *sudestadas*, given the incidence on the coast in SE-NW direction. This climatic phenomena of importance in the dynamics of the current Río de la Plata estuary, would also have occurred in the recent past. In the Don Enrique archaeological site, the *sudestada* events would not have had an impact on the innermost sectors of the beach ridge such as 'Cantera 2', where the soil continued being formed by current climatic conditions, while in 'Cantera 1' started a new pedogenic process over the shell deposit (Fig. 5c).

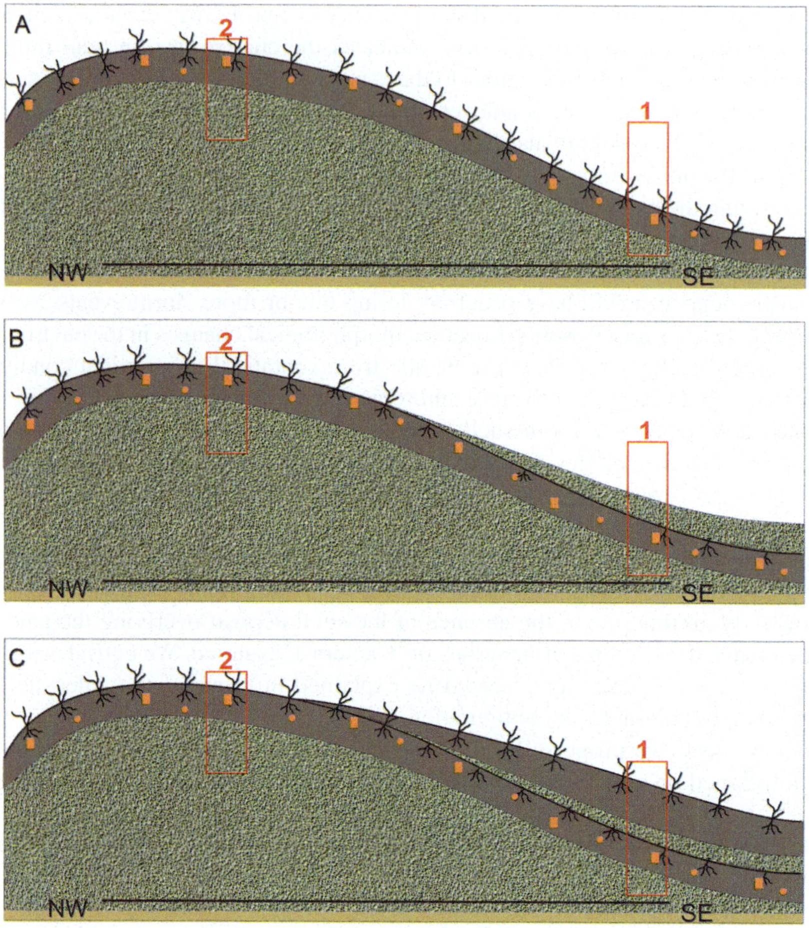

Fig. 5 Don Enrique archaeological site. Schematic NW-SE profile which indicates 'Cantera 1' (1) and 'Cantera 2' (2) sectors. The altitude difference is exaggerated. (**a**) Incorporation of the archaeological remains into the soil in formation; (**b**) Deposition of bioclastic material during a storm event; (**c**) In 'Cantera 2' sector the soil continued being formed, while in 'Cantera 1' it started a new pedogenic process

Final Considerations

We consider that both the shell ridge formation where the Don Enrique archaeological site is located and the events of shell deposition that buried the paleosoil that contains archaeological remains, correspond to the coastal progradation event linked to the Holocene regressive phase. The obtained radiocarbon dates show that these processes, and in consequence the site formation, occurred after ~1300–1200 cal years BP. It should be noted that the shell deposits correspond to one of the more recent moments in relation to the previous dates obtained for the

Canal de las Escobas Fm at the northern and central zone of Samborombón Bay, which allows us to consider a greater span of availability of biogenic material in the Río de la Plata estuary and hence, of the constitution on this Formation.

Don Enrique constitutes a hunter-gatherer archaeological site whose record shows some of the daily practices that the human groups carried forward in that place, as well as similarities with nearby sites that could reflect narrow social relations, and some particularities too. In this site the material remains were found in situ, specifically in a buried soil that indicates the incipient pedogenesis in a shell deposit of a deactivated beach in which the "*talar*" was starting to develop. Then, in the coastal front, this soil was covered by a new shell deposit linked to new extraordinary storm events, similar to the *sudestadas* that characterize the Río de la Plata estuary current dynamics, in response to local geomorphological changes. That is because the site went from being a stabilized area outside the reach of the waves and tides, to being a site directly affected by the waves, favoring accumulation caused by storms.

It is of great interest to continue studying the site, including an archaeological excavation, to deepen these considerations and establish the relations between this site and other archaeological places of the Río de la Plata estuary and the nearby areas during the Late Holocene. New dates obtained in archaeological materials could bring a more precise temporal frame to the human occupation of the site.

References

Aguirre, M. L., & Whatley, R. C. (1995). Late quaternary marginal marine deposits from North-Eastern Buenos Aires Province, Argentina: A review. *Quaternary Science Reviews, 14*, 223 251.

Aldazábal, V. (2004). *La ocupación humana en el sector centro oriental de la Pampa Deprimida*. (Unpublished doctoral thesis). Universidad de Buenos Aires.

Aldazábal, V., & Eugenio, E. (2008). La colonización humana en el Holoceno tardío del Humedal de la Bahía Samborombón. In L. Borrero & N. Franco (Eds.), *Arqueología del Extremo Sur del Continente Americano, resultados de nuevos proyectos* (pp. 55–81). Instituto Multidisciplinario de Historia y Ciencias Humanas-Consejo Nacional de Investigaciones Científicas y Técnicas.

Aldazábal, V., Weiler, N. E., & Eugenio, E. (2004). Una perspectiva geoarqueológica para comprender la ocupación humana en la costa central de la provincia de Buenos Aires. *Intersecciones en Antropología, 5*, 29–38.

Arturi, M. F., & Goya, J. F. (2004). Estructura, dinámica y manejo de los talares del NE de Buenos Aires. In M. F. Arturi, J. L. Frangi, & J. F. Goya (Eds.), *Ecología y manejo de los bosques de Argentina* (pp. 1–24). Editorial de la Universidad Nacional de La Plata.

Balesta, B., Paleo, M. C., Pérez Meroni, M., & Zagorodny, N. (1997). Revisión y Estado actual de las investigaciones arqueológicas en el Parque Costero Sur (Partido de Magdalena; Province of Buenos Aires). In M. A. Berón & G. G. Politis (Eds.), *Arqueología pampeana en la década de los '90* (pp. 147–160). Museo de Historia Natural de San Rafael -Facultad de Ciencias Sociales, Universidad Nacional del Centro de la Provincia de Buenos Aires.

Balfet, H., Fauvet-Berthelot, M. F., & Monzón, S. (1992). *Normas para la descripción de vasijas cerámicas*. Centre D'Études Mexicaines et Centroaméricaines.

Ballesteros-Prada, A., Luengo, M., Vilanova, I., Fucks, E., & Bernasconi, E. (2021). Foraminiferal paleodiversity and paleoenvironments at the NE coastal plain of Buenos Aires province

(Argentina) during the Mid-Holocene sea level highstand. *The Holocene, 31*(1), 108–120. https://doi.org/10.1177/0959683620961511
Behrensmeyer, A. (1978). Taphonomic and ecologic information from bone weathering. *Palaeobiology, 4*, 150–162.
Berwick, D. (1975). Valoración del análisis sistemático de los restos de fauna en sitios arqueológicos. *Chungara, Revista de Antropología Chilena, 5*, 125–140.
Binford, L. R. (1981). *Bones: Ancient men and modern myths*. Academic Press.
Buzas, M., & Gibson, T. (1969). Species diversity: Benthonic foraminifera in western North Atlantic. *Science, 163*, 72–75.
Carol, E. S., Kruse, E. E., Braga, F., & Tosi, L. (2013). El rol de la evapotranspiración en la geohidrología del humedal de la Planicie Costera de Ajó (provincia de Buenos Aires). In N. González, E. E. Krause, M. M. Trovatto, & P. Laurencena (Eds.), *Agua Subterránea, Recurso Estratégico* (Vol. I, pp. 72–77). Editorial de la Universidad Nacional de La Plata.
Carver, R. (1971). *Procedures in sedimentology*. Wiley-Interscience.
Cavallotto, J. L. (1995). *Evolución geomorfológica de la llanura costera ubicada en el margen sur del Río de la Plata*. (Unpublished doctoral thesis). Universidad Nacional de la Plata.
Cavallotto, J. L. (2002). Evolución holocena de la llanura costera del margen sur del Río de la Plata. *Revista de la Asociación Geológica Argentina, 57*(4), 376–388. http://www.scielo.org.ar/scielo.php?script=sci_arttext&pid=S0004-48222002000400003
Cavallotto, J. L. (2009). Caracterización e historia evolutiva del sustrato sobre el que se desarrollan los talares del noreste bonaerense. In J. Athor (Ed.), *Parque Costero del Sur. Naturaleza, conservación y patrimonio cultural* (pp. 66–81). Fundación de Historia Natural Félix de Azara.
Cavallotto, J. L., & Violante, R. A. (2005). Geología y geomorfología del Río de la Plata. In E. De Barrio, R. O. Etcheverry, M. F. Caballé, & E. Llambias (Eds.), *Geología y recursos minerales de la provincia de Buenos Aires. Relatorio XVI Congreso Geológico Argentino* (pp. 237–254). Asociación Geológica Argentina.
Cavallotto, J. L., Violante, R. A., & Parker, G. (2004). Sea-level fluctuations during the last 8,600 years in the de la Plata river (Argentina). *Quaternary International, 114*, 155–165.
Ceruti, C., & Crowder, R. (1973). La presencia de cerámica en los cordones conchiles litorales de la provincia de Buenos Aires (Argentina). Un sitio nuevo. In *Anales I Congreso Nacional de Arqueología Uruguaya and II Encuentro de Arqueología del Litoral* (pp. 1–48). Fray Bentos.
Cigliano, M. E. (1963). Arqueología del NE de la provincia de Buenos Aires. *Anales de la Comisión de Investigación Científica de la provincia de Buenos Aires, 4*, 473–511.
Cigliano, M. E. (1966). La cerámica temprana en América del Sur. El yacimiento de Palo Blanco (Partido de Berisso, provincia de Buenos Aires, Argentina). *Ampurias, 28*, 163–170.
Codignotto, J. O., & Aguirre, M. L. (1993). Coastal evolution in sea level and molluscan fauna in northeastern Argentina during the Late Quaternary. *Marine Geology, 110*, 163–175.
Colado, U., Figini, A., Fidalgo, F., & Fucks, E. (1995). Los depósitos marinos del Cenozoico superior aflorantes en la zona comprendida entre Punta Indio y el Río Samborombón, Provincia de Buenos Aires. In *Actas de las 4° Jornadas Geológicas y Geofísicas Bonaerenses* (Vol. 1, pp. 151–158). Junín.
Convención Nacional de Antropología. (1966). *Primera Convención Nacional de Antropología*. Facultad de Filosofía y Humanidades, Universidad Nacional de Córdoba.
Cremonte, M. B., & Bugliani, M. F. (2006/2009). Pasta, forma e Iconografía. Estrategias para el estudio de la cerámica arqueológica. Xama, 19–23, 239–262. https://www.researchgate.net/publication/306353180_Pasta_forma_e_iconografia_Estrtegias_para_el_estudio_de_la_cerámica_arqueológica
Day Pilaría, F. A. (2018). Gestión de los recursos faunísticos en sociedades cazadoras, recolectoras y pescadoras. Análisis arqueozoológico en sitios del litoral del Río de la Plata (partidos de Magdalena y Punta Indio, provincia de Buenos Aires). (Unpublished doctoral thesis). Universidad Nacional de la Plata. http://sedici.unlp.edu.ar/handle/10915/73571
Escobar, G., Vargas, W., & Bischoff, S. (2004). Wind tides in the Rio de la Plata estuary: Meteorological conditions. *International Journal of Climatology, 24*(9), 1159–1169. https://doi.org/10.1002/joc.1026

Faegri, K., & Iversen, J. (1989). Finding the grain. Laboratory technique. In K. Faegri, P. E. Kalana, & K. Krzywinski (Eds.), *Textbook of pollen analysis* (pp. 69–90). Wiley.

Fidalgo, F., Colado, U. R., & De Francesco, F. O. (1973). Sobre ingresiones marinas cuaternarias en los partidos de Castelli, Chascomús y Magdalena (provincia de Buenos Aires). In *Actas del 5° Congreso Geológico Argentino* (Vol. 4, pp. 225–240). Asociación Geológica Argentina.

Fucks, E. E., & Deschamps, C. (2008). Depósitos continentales cuaternarios en el noreste de la Región Pampeana. *Revista de la Asociación Geológica Argentina, 63*(3), 326–343.

Fucks, E. E., Schnack, E. J., & Aguirre, M. L. (2010). Nuevo ordenamiento estratigráfico de las secuencias marinas del sector continental de la Bahía Samborombón, provincia de Buenos Aires. *Revista de la Asociación Geológica Argentina, 67*(1), 27–39. http://www.scielo.org.ar/scielo.php?pid=S0004-48222010000400004&script=sci_arttext&tlng=es

Fucks, E. E., D'amico, G. M., Pisano, M. F., & Nuccetelli, G. (2017). Evolución geomorfológica de la región del Gran La Plata y su relación con eventos catastróficos. *Revista de la Asociación Geológica Argentina, 74*(2), 141–154.

Galligani, P., Balducci, F., & Sartori, J. (2018). Análisis del registro cerámico desde una perspectiva tafonómica: el caso de los sitios arqueológicos Familia Primón y Los Bañados (Centro-Este de Santa Fe, Argentina). *Mundo de Antes, 12*(1), 107–133. http://mundodeantes.org.ar/pdf/revista12-1/03-Galligani.pdf

García, M. S., Paleo, M. C., & López, L. (2011). Análisis del material lítico del sitio Las Marías. *Arqueología Rosarina Hoy, 3*, 115–126.

Ghiani Echenique, N. (2021). *Paisaje y dinámica social en el sector meridional del partido de Punta Indio (provincia de Buenos Aires) durante el Holoceno tardío* (Un abordaje arqueológico (Unpublished doctoral thesis)). Universidad Nacional de la Plata. http://sedici.unlp.edu.ar/handle/10915/130296

Ghiani Echenique, N., Uvietta, A. R., & Gambaro, R. (2013). Alfarerías tubulares en el noreste de la provincia de Buenos Aires: Caracterización y distribución. *Revista del Museo de La Plata, Sección Antropología, 13*(87), 299–314.

Ghiani Echenique, N., Valencia, C., & Paleo, M. C. (2020). De buena madera. Análisis antracológico de restos leñosos carbonizados procedentes del sitio Los Tres Ombúes (partido de Punta Indio, provincia de Buenos Aires). *Intersecciones en Antropología, 21*(1), 72–84. https://doi.org/10.37176/iea.21.1.2020.531

Gómez, G., Figini, A., & Fidalgo, F. (1988). Secuencia vertical de edades 14C en la las Escobas, en la localidad de Cerro de la Gloria, Bahía de Samborombón, provincia de Buenos Aires, Argentina. In *Actas 2° Jornadas Geológicas Bonaerenses* (pp. 277–281). Bahía Blanca.

Gómez, E., Borel, M., Aguirre, M., & Martínez, D. (2008). Radiocarbon reservoir age and hard-water effect for the Northeastern coastal waters of Argentina. *Radiocarbon, 50*(1), 119–129.

González, M. I. (2005). *Arqueología de alfareros-cazadores y pescadores pampeanos*. Sociedad Argentina de Antropología.

González de Bonaveri, M. I., Frère, M. M., & Escosteguy, P. (2006). El Sitio San Ramón 7, curso inferior del río Salado, provincia de Buenos Aires. *Relaciones de la Sociedad Argentina de Antropología, 31*, 187–199. http://www.saantropologia.com.ar/wp-content/uploads/2015/01/Relaciones%2031/08-%20Gonzalez%20et%20al.pdf

González, M. I., & Frère, M. M. (2009). Talares y paisaje fluvial bonaerense: arqueología del río Salado. *Intersecciones en Antropología, 10*, 249–265. http://www.scielo.org.ar/scielo.php?pid=S1850-373X2009000200006&script=sci_arttext&tlng=en

Gosselain, O. P. (2000). Materializing identities: An African perspective. *Journal of Archaeological Method and Theory, 7*(3), 187–217. https://doi.org/10.1023/A:1026558503986

Goya, J. F., Placci, L. G., Arturi, M. F., & Brown, A. (1992). Estructura y distribución de los talares de la reserva de biosfera Parque Costero del Sur. *Revista de la Facultad de Agronomía de La Plata, 68*, 53–64.

Grayson, D. K. (1984). *Quantitative zooarchaeology*. Academic Press.

Gutiérrez, M. A., Kaufmann, C. A., González, M. E., Scheifler, N. A., Rafuse, D. J., Massigoge, A., & Álvarez, M. C. (2015). The role of small carnivores in the movement of bones: Implications for the pampas archaeofaunal record, Argentina. *Archaeological and Anthropological Sciences, 8*(2), 257–276. https://doi.org/10.1007/s12520-015-0272-1

Heaton, T. J., Köhler, P., Butzin, M., Bard, E., Reimer, R. W., Austin, W. E., et al. (2020). Marine20—The marine radiocarbon age calibration curve (0–55,000 cal BP). *Radiocarbon, 62*(4), 779–820. https://doi.org/10.1017/RDC.2020.68

Heusser, C. J. (1971). *Pollen and spores of Chile: Modern types of the Pteridophyta, Gymnospermae, Angiospermae*. University of Arizona Press.

Imbellone, P. A., Giménez, J. E., & Mormeneo, M. L. (2014). Suelos calcáreos del litoral noreste de la provincia de Buenos Aires, Argentina. In P. A. Imbellone (Ed.), *Suelos con acumulaciones calcáreas y yesíferas de Argentina* (pp. 57–93). Instituto Nacional de Tecnología Agropecuaria.

Isla, F. I. (1989). Holocene coastal evolution in Buenos Aires Province, Argentina. *Quaternary of South America and Antarctic Peninsula, 11*, 297–321.

León, R. J. C., Burkart, S. E., & Movia, C. P. (1979). La vegetación de la República Argentina. Relevamiento fitosociológico del pastizal del norte de la Depresión del Salado (partido de Magdalena y Brandsen, provincia de Buenos Aires). *Serie Fitogeográfica, 17*, 11–93.

Loeblich, A. R., & Tappan, H. (1992). Present status of foraminiferal classification, in studies in benthic foraminifera. In Y. Takayanagi & T. Saito (Eds.), *Proceedings of the fourth international symposium on benthic foraminifera* (pp. 93–102). Tokai University Press.

Luengo, M. S. (2019). *Vegetación y ambientes del Holoceno en el sector continental de la Bahía Samborombón, provincia de Buenos Aires: cambios del nivel del mar* (Unpublished doctoral thesis). Universidad Nacional de la Plata. http://sedici.unlp.edu.ar/handle/10915/74246

Luengo, M., Bernasconi, E., Vilanova, I., & Fucks, E. (2018). Past environmental changes during the late Holocene sea-level fall (last 2.7 ka) at Bahía Samborombón, NE coastal plain (Argentina). *Journal South American Earth Science, 85*, 362–373. https://doi.org/10.1016/j.jsames.2018.05.012

Luengo, M. S., Ballesteros-Prada, A., Vilanova, I., Bernasconi, E., & Fucks, E. (2021). Environmental changes related to the Holocene Sea-level transgression and successive high-stand (7–5.5 kyr) at the NE coastal plain of Buenos Aires (Argentina). *Quaternary International, 602*, 4–14. https://doi.org/10.1016/j.quaint.2021.03.006

Lyman, R. L. (1994). *Vertebrate taphonomy*. Cambridge University Press.

Markgraf, V., & D'Antoni, H. L. (1978). *Pollen Flora of Argentina. Modern spore and pollen types of Pteridophyta, Gymnospermae, and Angiospermae*. University of Arizona Press.

Márquez, M. (2019). Holocene foraminifera of the southeast of the Pampa coastal plain, Argentina. *Journal of South American Earth Sciences, 94*. https://doi.org/10.1016/j.jsames.2019.102215

Martínez, S., Rojas, A., Ubilla, M., Verde, M., Perea, D., & Piñeiro, G. (2006). Asociaciones de moluscos del Holoceno marino de Uruguay: composición, geocronología e indicadores paleoambientales. *Ameghiniana, 43*(2), 385–397. http://www.scielo.org.ar/scielo.php?script=sci_art text&pid=S0002-70142006000200009&lng=es&tlng=es

Medina, M. E., & Teta, P. (2010). Alteración térmica y consumo de roedores caviomorfos en Quebrada del Real 1 (Pampa de Achala, Córdoba): Una aproximación experimental. Presented in the XVII Congreso Nacional de Arqueología, Mendoza.

Mengoni Goñalons, G. (1988). Análisis de materiales faunísticos de sitios arqueológicos. *Xama, 1*, 71–120.

Mengoni Goñalons, G. (1999). *Cazadores de guanacos de la estepa patagónica*. Sociedad Argentina de Antropología.

Mengoni Goñalons, G. (2006/2010). Zooarqueología en la práctica: algunos temas metodológicos. Xama, 19/23, 83–113. http://repositorio.filo.uba.ar/handle/filodigital/12683

Moore, R. C. (1961). *Treatise on invertebrate paleontology part Q. Arthropoda 3. Crustacea, Ostracoda*. Geological Society of America and University of Kansas Press.

Mudie, P. J., & Harland, R. (1996). Aquatic quaternary. In J. Jansonius & D. C. McGregor (Eds.), *Palynology, principles and applications* (pp. 843–877). American Association of Stratigraphic Palynologists Foundation.

Murray, J. (2006). *Ecology and applications of benthic foraminifera*. Cambridge University Press.

Orton, C., Tyers, P., & Vince, A. (1997). *La Cerámica en Arqueología*. Editorial Crítica.

Páez, M. M., Paleo, M. C., Pérez Meroni, M., & Pastorino, S. (1999). El bosque de tala como recurso potencial: interpretación arqueológica y palinológica. In *Libro de resúmenes del XIII Congreso Nacional de Arqueología Argentina* (pp. 369–370).

Paleo, M. C., & Pérez Meroni, M. (1999). Nuevos aportes a la arqueología de Punta Indio. In C. Diez Marin (Ed.), *Actas del XII Congreso Nacional de Arqueología Argentina* (Vol. III, pp. 165–169). Universidad Nacional de La Plata.

Paleo, M. C., & Pérez Meroni, M. (2004). Problemáticas vinculadas a las estrategias de subsistencia de la localidad arqueológica Barrio San Clemente. In C. J. Gradin & F. Oliva (Eds.), *El Área Pampeana. Su pasado arqueológico* (pp. 311–319). Rosario.

Paleo, M. C. and Pérez Meroni, M. (2005/2006). Dimensión social de la tecnología cerámica en sociedades cazadoras-recolectoras. Revista do Museu de Arqueología e Etnología, 15/16, 73–85.

Paleo, M. C., & Pérez Meroni, M. (2007). Primeros resultados del sitio "Las Marías", Partido de Magdalena. Provincia de Buenos Aires. In F. Oliva, N. de Grandis, & J. Rodríguez (Eds.), *Arqueología Argentina en los inicios de un nuevo siglo* (Vol. I, pp. 275–286). Laborde Editor.

Paleo, M. C., Páez, M. M., & Pérez Meroni, M. (2002). Condiciones ambientales y ocupación humana durante el Holoceno tardío en el litoral fluvial bonaerense. In D. Mazzanti, M. Berón, & F. Oliva (Eds.), *Del Mar a los Salitrales* (pp. 365–376). Universidad Nacional de Mar del Plata.

Paleo, M. C., Pérez Meroni, M., Menestrina, F., & Romero, L. (2018). Indagando sobre prácticas alimentarias en la costa del Río de la Plata a través del estudio de residuos orgánicos en tiestos cerámicos. *Comechingonia, 22*(2), 135–165.

Parodi, L. R. (1940). Distribución geográfica de los talares de la Provincia de Buenos Aires. *Darwin, 4*, 33–56.

Payne, S. (1975). Partial recovery and sample bias. In A. T. Clason (Ed.), *Archaeozoological studies* (pp. 7–17). North Holland Publishing Company.

Pérez Meroni, M., & Paleo, M. C. (1999). La utilización del espacio por grupos del litoral bonaerense. In J. López Mazz & M. Sans (Comps.), *Arqueología y bioantropología de Tierras Bajas* (pp. 165–172). Facultad de Humanidades y Ciencias de la Educación, Universidad de la República.

Pérez Meroni, M., Paleo, M. C., Pochettino, M. L., & Lema, V. S. (2010). Procesamiento y consumo de vegetales por grupos cazadores-recolectores del Holoceno tardío, en los partidos de Magdalena y Punta Indio, provincia de Buenos Aires. In M. Berón, L. Luna, M. Bonomo, C. Montalvo, C. Aranda, & M. Carrera Aizpitarte (Eds.), *Mamül Mapu: Pasado y presente desde la arqueología pampeana* (pp. 215–226). Editorial Libros del Espinillo.

Prieto, A. R., Mourelle, D., Peltier, W. R., Drummond, R., Vilanova, I., & Ricci, L. (2017). Relative Sea-level changes during the Holocene in the Río de la Plata, Argentina and Uruguay: A review. *Quaternary International, 442*, 35–49. https://doi.org/10.1016/j.quaint.2016.02.044

Rice, P. (1989). *Pottery analysis. A sourcebook*. University of Chicago Press.

Salemme, M., Tonni, E. P., & Ceruti, C. N. (1985). Los materiales faunísticos del sitio arqueológico "La Maza I" (partido de Berisso, provincia de Buenos Aires, Argentina): revisión crítica. *Revista de la Asociación de Ciencias Naturales del Litoral, 16*(2), 169–178.

Salemme, M., Miotti, L., & Aguirre, M. (1989). Holocene settlement in the Río de La Plata Littoral (Argentina): A methodological approach. *Geoarchaeology, 4*(1), 69–80.

Salemme, M., Miotti, L., & Tonni, E. (1991). La determinación sistemática de los mamíferos en el análisis arqueofaunístico. In N. Ratto & A. Haber (Eds.), *De Procesos, Contextos y Otros Huesos* (pp. 65–73). Facultad de Filosofía y Letras, Universidad de Buenos Aires.

Sánchez, R. O., Ferrer, J. A., Duymovich, O. A., & Hurtado, M. A. (1976). Estudio pedológico integral de los partidos de Magdalena y Brandsen. *Anales del LEMIT, serie, II*, 1–127.

Sanhueza Riquelme, L. (1998). Antecedentes y proposición metodológica para el estudio de huellas de alteración en cerámica. *Conserva, 2*, 69–79.

Schiffer, M. B., & Skibo, J. M. (1989). A provisional theory of ceramic abrasion. *American Anthropologist, 91*, 102–116.

Schnack, E. J., Isla, F. I., De Francesco, F. O., & Fucks, E. E. (2005). Estratigrafía del Cuaternario marino tardío en la provincia de Buenos Aires. In E. De Barrio, R. O. Etcheverry, M. F. Caballé, & E. Llambias (Eds.), *Geología y Recursos Minerales de la Provincia de Buenos Aires. Relatorio del XVI Congreso Geológico Argentino* (pp. 159–181). Asociación Geológica Argentina.

Sen Gupta, B. K. (1999). Systematics of modern foraminifera. In B. K. Sen Gupta (Ed.), *Modern Foraminifera* (pp. 1–371). Kluwer Academic Publishers.

Shepard, A. O. (1956). *Ceramics for the archaeologist*. Carnegie Institution of Washington.

Silva, C. (2017). Evaluación de los procesos de formación de sitio desde la alfarería: el caso de Laguna de los Gansos 1 (Departamento Diamante, Entre Ríos). *Comechingonia Revista de Arqueología, 21*(2), 145–174. http://www.scielo.org.ar/scielo.php?script=sci_arttext&pid=S1851-00272017000200007&lng=es&nrm=iso&tlng=es

Skibo, J. M. (1992). *Pottery function: A use-alteration perspective*. Plenum Press.

Stark, M. (1999). Social dimensions of technical choice in Kalinga ceramic traditions. In E. S. Chilton (Ed.), *Material meanings* (pp. 24–43). University of Utah Press.

Stuiver, M., & Brazuinas, T. F. (1993). Modeling atmospheric 14C influences and 14C ages of marine samples to 10.000 BC. *Radiocarbon, 35*(1), 137–189.

Stuiver M., Reimer P. J., & Reimer R.W. (2020). *CALIB 8.2*. http://calib.org. Accessed 15 December 2021.

Torres Robles, S. S. (2009). *Variación geográfica de la composición y riqueza de plantas vasculares en los talares bonaerenses y su relación con el clima, sustrato, estructura del paisaje y uso*. (Unpublished doctoral thesis). Universidad Nacional de la Plata. http://sedici.unlp.edu.ar/handle/10915/55171

Tricart, J. L. F. (1973). *Geomorfología de la Pampa Deprimida*. Instituto Nacional de Tecnología Agropecuaria.

Vargas Gil, J. R., Scoppa, C. O., & Iñíguez, A. M. (1972). Génesis de los suelos de la región norte de la Bahía de Samborombón. *Revista de Investigaciones Agropecuarias, 3*(9), 29–52.

Vervoorst, F. (1967). Las comunidades de la Depresión del Salado, provincia de Buenos Aires. In *Serie Fitogeográfica 7. La vegetación de la República Argentina*. Instituto Nacional de Tecnología Agropecuaria.

Violante, R., & Parker, G. (2000). El Holoceno en las regiones marinas y costeras del nordeste de la provincia de Buenos Aires. *Revista de la Asociación Geológica Argentina, 55*(4), 337–351.

Violante, R., Parker, G., & Cavallotto, J. L. (2001). Evolución de las llanuras costeras del este bonaerense entre bahía Samborombón y la laguna Mar Chiquita durante el Holoceno. *Revista de la Asociación Geológica Argentina, 56*(1), 51–66.

Volpedo, A. V., Yunes Nuñez, T., & Fernández Cirelli, A. (2005). El humedal mixohalino de Bahía Samborombón: Conservación y perspectivas. In J. Peteán & J. Cappato (Comps.), *Humedales Fluviales de América del Sur. Hacia un manejo sustentable* (pp. 89–110). Ediciones Proteger-International Union for Conservation of Nature.

Zárate, M., González de Bonaveri, M. I., Flegenheimer, N., & Bayón, C. (2000/2002). Sitios arqueológicos someros: El concepto de sitio en estratigrafía y sitio de superficie. Cuadernos del Instituto Nacional de Antropología y Pensamiento Latinoamericano, 19, 635–653. https://revistas.inapl.gob.ar/index.php/cuadernos/article/viewFile/566/335

The Peri-urban Concept in Historical Archaeology. An Approach from the Nineteenth-Century Southern Frontier of Argentina

Hernán Tomassini and Adolfo Eliges

Abstract Peri-urban spaces are usually understood as transition areas, with both urban and rural characteristics, where different economic and social practices are developed, mainly linked to agricultural production. The notion of peri-urban spaces has been widely used in different sciences, such as geography and economics, referring to heterogeneous cultural landscapes in which activities typically assignable to both rural and urban environments are carried out. This concept was also applied in studies related to the rural history of the Río de la Plata area (Buenos Aires Province, Argentina) between the eighteenth and nineteenth century; however, it has not yet been used in studies on historical archaeology in Argentina, a research line that continues to classify archaeological sites in a dichotomous way by their association with urban or rural contexts. In this work, we propose an archaeological approach to the concept of peri-urban applied to the study sites related to the southern frontier of the Pampean and Northern Patagonia regions from the last 25 years of the nineteenth century. In this frame, the following is analyzed: (1) the process of construction of the landscape and the material record present in the site "Boliche Lucanera," a commercial establishment located at Bahía Blanca city (Buenos Aires Province), and (2) the settlement called "Pueblo Avellaneda" (Río Negro Province), founded in the framework of the military campaigns of territorial conquest carried out by the state of Argentine against the original populations. Finally, both cases are compared and the relevance of the peri-urban concept for archaeological studies is discussed.

Keywords Peri-urban · Borders · Nineteenth century · Argentina · Pampa · Patagonia

H. Tomassini (✉)
Departamento de Humanidades, Universidad Nacional del Sur-CONICET, Bahía Blanca, Argentina

A. Eliges
Universidad Nacional de La Plata-CONICET, La Plata, Argentina
e-mail: adolfo.eliges@fcnym.unlp.edu.ar

G. F. Bonnat et al. (eds.), *Current Research in Archaeology of South American Pampas*, The Latin American Studies Book Series,
https://doi.org/10.1007/978-3-031-55194-9_11

Introduction

The interactions between the countryside and the city, with their consequent territorial manifestations, have been analyzed and studied from different perspectives within the Social Sciences. The idea of an intermediate peri-urban space arose in the first decades of the twentieth century to refer to transformations in the urban environment by the advance of industrial capitalism during the nineteenth century, industrialization, and migration from rural areas, a process developed mainly in the so-called central countries (Western Europe and United States) (see Avila Sanchez, 2005; Puebla, 2009/2010; Feito & Barsky, 2020). These reflections allowed thinking about the social, economic, and productive processes that take place there.

In Argentina, the peri-urban concept began to be used in the 1980s from Rural Geography as a tool to study the transformations that occurred in areas close to large urban centers such as Buenos Aires, Rosario, and Córdoba. In these cities, from the end of the nineteenth century, and with higher intensity in the first decades of the twentieth century, important territorial changes occurred. With the consolidation of the agro-export model[1] overseas migrations, population growth, and large-scale urbanization, the industrial and agricultural belts were consolidated on the outskirts of the main urban centers that responded to the needs (food supply and labor) of this model (Barsky, 2005; Feito, 2018). This process continued with nuances throughout the twentieth and twenty-first centuries, generating the interest of researchers and the public sector through the territorial transformations and the management of territorial planning by the State, expanding the variety of topics addressed (De Marco, 2020).

In the case of History, the researches involving peri-urban spaces are previous although, in most cases, without making use of that concept. In the rural history of the Pampean region, since the second half of the twentieth century, studies refer to the economic practices developed in the "ejido" lands[2] and the mechanisms of access to their property. During the 1980s, new approaches allowed to understand the role of the farms that grew in the surroundings of Buenos Aires city and its port in the colonial and post-independence period in supplying agricultural and livestock products, and studies on the different forms of alienation of land resources in the countryside of Buenos Aires (see Miguez, 1986; Bandieri & Blanco, 1996; Barcos, 2007; Gelman, 2007; Lanteri, 2018; among others). In these investigations (e.g., Banzato, 2011; Barcos, 2019), the peri-urban term was used as synonymous of "ejido" land, referring to the area of parceled land that surrounded the towns

[1] This is the name given to the economic model developed since, approximately, the middle of the nineteenth century and accentuated in the following decades, characterized by a productive and exporting profile consisting of the sale, mainly, of meat and cereals destined for European markets (See Barsky & Gelman, 2009).

[2] According to Melisa Pesoa (2014), the ejido is the urban-productive unit of the city and is made up of the urban nucleus and the concentric crop areas that surround it (chacras and quintas) destined for agriculture. For a historical development of the concept in Ibero-America and the current perspectives of analysis, see Lanteri (2020).

destined to establish a population, cereal cultivation, and fruit-horticultural, generally intended for family self-support and local supply (Barcos & Martirén, 2019).

Some recent works applied the concept to refer to the agricultural space used to produce primary consumer goods and supply nearby cities in towns of the Buenos Aires Province (Ciliberto, 2007; Principi & Ciliberto, 2009). In these studies, the peri-urban is considered as the "*área rural localizada en las inmediaciones al núcleo urbano, un universo productivo y mercantilmente definido en y por sus múltiples relaciones con la ciudad*[3]" (Ciliberto & Principi, 2014: 265).

In this work, we analyzed the peri-urban concept from an approach oriented to Historical Archaeology, considering two study areas. Based on our objectives, we synthesize the most relevant information of the Pampean and Northern Patagonian regions of Argentina. Initially, the studies focused on some of the oldest colonial sites in the country, such as the cities of Buenos Aires, Santa Fe, and Mendoza. This Urban Archaeology specialized in the study of different sites in the historic center and in different aspects of the daily life of the social groups that inhabited the city since the seventeenth century (Schávelzon, 1991, 1999). Following the definition given by one of its pioneers.

> *la arqueología urbana es la arqueología de la ciudad moderna [...] el objeto de estudio somos nosotros mismos, no son otros que nos precedieron, es el proceso de transformación del hábitat que vemos y hacemos a diario, haya comenzado la ciudad cuando sea que lo haya hecho, importa que llegue a hoy [...] es la arqueología de la ciudad y no dentro de la ciudad' [...] Es el cruce entre varias especialidades y por lo tanto es multidisciplinar: confluyen la historia urbana, la de la arquitectura, la conservación patrimonial, los estudios del hábitat, la antropología y obviamente la arqueología histórica y la de la arquitectura*[4] (Schávelzon, 2020: 13).

In the beginning of 1990s, research began to encompass the study of interethnic contact spaces in border and rural areas. Initially, those carried out on military structures belonging to the borderline from the late-colonial period until the culmination of the process called "Conquest of the Desert" stand out (Gómez Romero & Spota, 2006; Bagaloni & Pedrotta, 2018). Gradually, work on productive establishments, commercial facilities, and other elements that make up the rural landscape[5] were incorporated (Landa et al., 2018). In this way, a new area of research was established, that of Rural Historical Archaeology, delimitated:

> *por los temas y objetivos de estudio específicos, y no sólo en relación al ámbito o tipo de sitio en donde dichos estudios se desarrollan (estancias, chacras, pulperías,* etc.*) (Brittez,* 2004*). Las diversas temáticas abordadas dentro de este campo de conocimiento pueden*

---

[3] "Rural area located in the vicinity of the urban nucleus, a productive and mercantile universe defined in and by its multiple relationships with the city"

[4] "Urban archaeology is the archaeology of the modern city [...] the object of study is ourselves, it is not others who preceded us, it is the process of transformation of the habitat that we see and do on a daily basis, has started the city whenever it has, it matters that it reaches today [...] it is the archaeology of the city and not within the city' [...] It is the intersection between several specialties and therefore it is multidisciplinary: history converges architecture, heritage conservation, habitat studies, anthropology and obviously historical archaeology and architecture."

[5] Such as stalls, roads, and towns.

> *caracterizarse por estar incluidas en dos grandes núcleos de investigación: a- el paisaje (generalmente desde perspectivas constructivistas) y b- los procesos de cambio social (específicamente el que se fue dando desde un mundo de frontera o espacio fronterizo a uno rural moderno, incorporado al sistema Mundial entre mediados del siglo XIX e inicios del XX*[6] (Landa et al., 2018: 198).

According to the abovementioned, the investigations in Historical Archaeology were defined in two conceptualized areas (both physically and theoretically): the urban and the rural. However, the analysis of the material record and spatial transformation was not addressed, considering that these concepts do not cover the entire territorial organization, specifically those transitional zones called peri-urban areas. Those areas, with their own and distinct characteristics, began to consolidate towards the end of the nineteenth century and the beginning of the twentieth in the Pampas-Patagonian space. In the cases in which the peri-urban concept was applied, it was without a specific theoretical reflection that discusses its implications, or it was grouped with what is considered urban, but without considering its specificities.

Finally, we understand Historical Archaeology as "aquella disciplina que trata del estudio global y comparativo de procesos sociales pasados que involucran a una gran variedad de sectores sociales y grupos étnicos acerca de los cuales existe algún tipo de registro escrito, sea producido por ellos mismos o por otros[7]" (...that discipline that treats the global and comparative study of the past social processes that involve a great variety of social sectors and ethnic groups about which there exists some kind of written records, being it produced by themselves or by other people…; free translation by the authors) (Pedrotta & Bagaloni, 2021: 369). Following this position, far from proposing "archaeology of the peri-urban"or "peri-urban archaeology" different from those already mentioned, our objective is to discuss the present record and the spatial transformations in sites located in peri-urban contexts/areas and the occupation of that space and compare it with other similar sites. Thus, returning to questions raised since the beginning of the discipline, we ask ourselves, how is the process of change experienced by rural and urban areas in the decades between centuries expressed archaeologically? And in the case of peri-urban spaces, does registration change in those places? Is it the same? Are they more like one than the other? Based on this perspective, we seek to contribute to the understanding of the processes that arose after the Argentine state advance on the Pampean and Northern Patagonian regions.

---

[6] "because of the specific themes and objectives of the study, and not only in relation to the scope or type of site where said studies are carried out ("estancias" -extended ranches-, farms, "pulperías" -rural bars and stores-, etc.) (Brittez, 2004). The various topics addressed within this field of knowledge can be characterized by being included in two large research nuclei: a- the landscape (generally from constructivist perspectives) and b- the processes of social change (specifically the one that was taking place from a world from a border or border area to a modern rural one, incorporated into the World System between the mid-nineteenth and early twentieth centuries).

[7] "that discipline that deals with the global and comparative study of past social processes that involve a great variety of social sectors and ethnic groups about which there is some type of written record, whether produced by themselves or by others"

We present and discuss here some results of the work carried out in two sites located in this area (Fig. 1): (a) the construction process of the landscape and the material record present in the "Boliche Lucanera," a rural commercial establishment that operated since the end of the nineteenth century in the town of General Daniel Cerri, a border area within the city of Bahía Blanca (Buenos Aires Province, Argentina), and (b) "Pueblo Avellaneda" (Río Negro Province, Argentina), the first settlement established within the framework of the military campaign of territorial conquest carried out by the Argentine state between 1879 and 1885 against the indigenous populations. The town was built in the middle valley of the Río Negro, in a territory known as Choele Choel, whose domain was considered strategic by both indigenous groups and the European society. Based on the use of the peri-urban category and the idea of border, these processes are compared with the development of two neighborhoods in the city of Buenos Aires, inserted within their territory since the second half of the nineteenth century: the towns known today as San José de Flores and Belgrano.

Fig. 1 Study area with details of the cases analyzed and their comparison with the sites of the city of Buenos Aires

Theoretical Contributions to the Notion of Peri-urban

Peri-urban areas have been defined and interpreted, mainly from Geography, in different ways. Although urban and rural spaces were traditionally thought of in terms of opposition and dichotomy, the renewed social studies allow broader approaches from various disciplines, in which the links between these two concepts are considered in terms of interdependence and complementarity. However, the peri-urban concept is expressed in its specificity and maintains its identity, different from both the urban and the rural (Attademo & Ringuelet, 2008). De Marco (2020) defined them as productive territories (mainly oriented to primary rural exploitation), residential and services located on the outskirts of cities that, as they grow, consolidate the demand for certain products. This definition is close to that proposed by Barsky (2010), who considers them as territorial complexes in an interface situation between two types of apparently well-differentiated land uses, the countryside, and the city, and in whose space activities of both rural and urban territories are developed. In this frame, their characteristics are difficult to determine and study as it is a transitional area in constant transformation.

Respect to the notion of border, there are not a precise limit between two different spaces in their composition, but it can be used as a conceptual tool that allows us to understand the distinction of the uses of space, whether residential and/or productive, between urban, rural and peri-urban areas (Matossian & Vejsbjerg, 2020). Historically, the idea of border was associated with a territory of low demographic density that surrounds urban centers (hinterland) and that produced orchards and farms to supply the requirements of the city. This territory, influenced by classical sociology perspective, has been described as a kind of demographic gradient in concentric circles (Sorokin and Zimmerman, 1929, in Cardoso, 2020), which ranges from the urban to the rural, passing through the sub-urban, peri-urban and rur-urban (Fig. 2a). This perspective has typically been applied to explain the expansive phenomenon of large cities since the beginning of the twentieth century using different variables, such as population size, social diversification and stratification, proportion of farmers and agricultural land use, etc. Subsequently, Cardoso (2020) conceptualized this process as an interface between the city and the countryside, which would not result in a mixture but in a third territory, built in a process of reterritorialization, where the actors materially and symbolically appropriate the territory through their practices. The morphology of this *third territory* would then be made up of "perímetros superpuestos de tamaños y duraciones variables dependientes de las prácticas socio-espaciales: [resultando] múltiple, plural, de coexistencia de diferentes trayectorias, heterogéneo [y] ensamblado[8]" and ensambled; (...superposed perimeters of size and variable duration depending multiple, plural, coexisting with different trajectories, heterogeneous… and assembled; Cardoso, 2020:

[8] "overlapping perimeters of variable sizes and durations depending on socio-spatial practices. [... Thus, this third territory is] multiple, plural, of coexistence of different trajectories, heterogeneous [and] assembled"

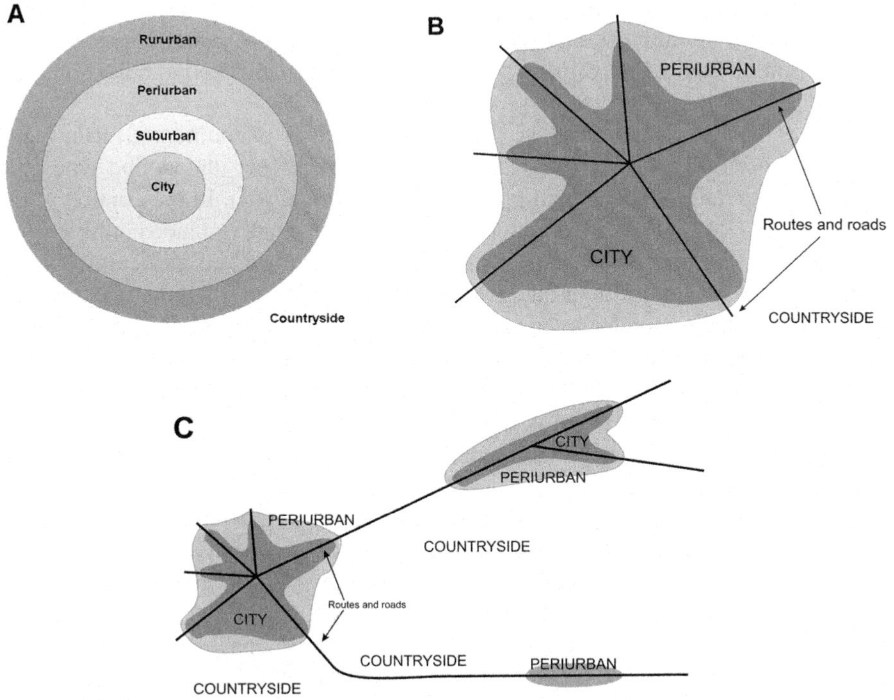

Fig. 2 Descriptive models of peri-urban development (modified from Cardoso, 2020): (**a**) A model based on precepts of traditional sociology, with a descending gradation centered on demographics and social complexity. (**b**) Territorialization processes in the socio environmental practices of the agents become the third territory with its characteristics. (**c**) Development of model B, applied to the agricultural colonies of the second half of the century century

405). From an historical perspective, situated in the process of reterritorialization experienced in the territory that surrounded the city of Buenos Aires towards the middle of the nineteenth century, "la primera etapa del mecanismo de transformación que se operó en la superficie agraria fue el proceso de periurbanización, entendido como la división de explotaciones ganaderas en parcelas pequeñas y medianas, para su venta posterior [...][9]" ("the first phase of the transformation mechanism that operated in the agrarian surface was the peri-urbanization, understood as the subdivision of cattle exploitations in small and middle sized parcels so as to be sold later on" (Svetlitza de Nemirovsky, 2010: 7; free translation by the authors). We propose that this urban expansion occurred radially, from the development of communication ways (routes, trains, etc.), and the use and reinforcement of structures linked to the logistics of the circulation of goods located around and

[9] "the first stage of the transformation mechanism that took place in the agrarian area was the process of peri-urbanization, understood as the division of livestock farms into small and medium plots, for later sale [...]"

perpendicular to these roads, which began to outline the orthogonal morphology of the towns. Concomitant with this process, there were substantial changes in the distribution of land ownership, with more limited and delimited subdivisions and the appearance of spaces reserved for public use (Fig. 2b).

This extension of the cities to the countryside is a phenomenon intrinsically linked to the model of capitalist expansion and, therefore, understood as a geographical correlate of this global phenomenon (Pesoa, 2014). In the context of the expansive process of the territorial model of the Nation-State produced in Argentina, fundamentally in the second half of the nineteenth century and particularly in the case of Pampean and northern Patagonian regions, we believe that the phenomenon should be associated with the concept of colonization (sensu Lanteri & Martirén, 2020) and the process of massive European immigration that occurred synchronously. In this sense, the notion of peri-urban should not only be understood as a dynamic edge in expansion, or as peninsulas that are inserted in the rural space but also as urban enclaves consolidating and extending over a territory imagined in the past as a desert (Fig. 2c).

In this frame, the archaeological correlation of peri-urban sites since the second half of the nineteenth century, should reflect this temporal-spatial dynamic of consolidation of a colonizing urban nucleus that, eventually, configures an expanding edge, transforming a rural or unexploited territory in a hybrid and, in turn, a specific expression characterized by transformation processes in spaces with low occupational density and oriented to the satisfaction of urban needs. Thus, the appearance of peri-urban sites would represent a chronological indicator, both in stratigraphy and in their dispersion concerning the cities involved.

The notion of landscape is a very useful category to enrich the approach and treatment of various issues related to spatiality. Here, the landscape is understood as a social construction that allows identifying the dynamic and interdependent relationships that are established between the agents participating in this process with the physical, social, and cultural dimensions of their environment through time and space, and how their actions and decisions continually modify it (Criado Boado, 1993; Ingold, 1993, among others). This concept contains material and ideological elements through which human groups assume it as their own (Curtoni, 2000). Taking into account these aspects, landscape archaeology allows the study of human spatiality from materiality, by examining how it has been shaped and organized by various economic, social, religious, symbolic, and cultural processes. Thus, the landscape becomes the object and aim of the research, being the result of the interaction of elements of a sociocultural and natural order, material and conceptual, past, and present. This interaction has led to the formation of a landscape that is the result of historical processes and must be historically understood (Gianotti Garcia, 2005).

The Perri-urban of the City of Buenos Aires IRES Around the Nineteenth· Century

As it has been already said, we understand the phenomenon of peri-urbanization as a process deeply anchored in the historical specificity of the expansion of national States. This process happened over a spatiality in dispute and closely linked to a set of territorial practices developed by both, individual and collective actors, eminently committed with its belonging to a state agency over territories imagined and assumed as part of its sovereign right. Thus, before testing the concept of peri-urban in our case studies, as it has been described, we will refer to the process of expansion and territorial incorporation that Buenos Aires city experienced towards its hinterland, which will provide a well-documented and almost synchronous reference regarding the forms that this process has adopted regionally and in which it is possible to identify the articulation between the concepts of peri-urban and border in the sense that we have described. The selection of these cases for comparison is due not only to the similarities in terms of their historical development but also to the varied archaeological representation that manifests their transformations in time and space as they were territorially and legally integrated into the capital city.

Towards the middle of the nineteenth· century, the present neighborhoods of San José de Flores and Belgrano, both integrated into the city of Buenos Aires in 1888, formed part of its hinterland, presenting a very low demographic density and an occupation linked to the use of natural resources, such as small farms and orchards or lime extractive enterprises, such as the case of Belgrano (Camino, 2014). Both territories were crossed by routes that formed part of the networks for the circulation of goods and people between the big city and distinct locations in the interior of the country and were located at a distance that was then perceived as that between the city and the countryside (Piñeiro, 1983).

San José De Flores

The current terrain of the San José de Flores neighborhood was acquired by Juan Diego Flores in 1776 to dedicate them to the lease and exploitation of wheat and firewood for the recently declared capital of the "Virreinato del Río de la Plata" ("Viceroyalty of the Rio de la Plata" (Ciliberto, 2007; Camino, 2011). This terrain was located at a strategic point for the communications and logistics of the city of Buenos Aires since the "Camino Real" (the current Rivadavia Avenue down-town Buenos Aires) ran through it to the west (Camino et al., 2012). In 1806, the San José parish was established, and as the place was known as "the lands of Flores," it began to be called San José de Flores. The owner donated land for the church, a square, and other public facilities and, 2 years later, the rest of the property was divided into blocks of 16 lots and put up for sale. The new subdivisions, perpendicular to the "Camino Real," were shaping the morphology of what would be the layout of a town and an obligatory stopover for wagons heading west. In 1811, the Cabildo of Buenos Aires granted it the institutional status of a district, which contributed to increasing its importance as a logistics point (Melero García et al., 1980). The town was configured in conjunction with the development of a set of farms and villas that supplied the big city. With the arrival of the railway, in 1857, the area began to be chosen for the establishment of recreational houses for the Buenos Aires bourgeoisie, a process that was strengthened as the new means of transport that linked it with the city increased and European immigration grew. Finally, in 1888 Flores was formally integrated into the "ejido" of the city of Buenos Aires (Camino, 2011).

Camino et al. (2012) present seven sites that testify the development of San José de Flores, first as a hinterland provider of food for the city, then as a satellite town, and, finally, as a neighborhood integrated into the big city. The authors linked the archaeological record to three historical periods: (a) beginning and growth of San José de Flores (1806–1857), characterized by the development of farms and country houses to supply the city; (b) consolidation and demographic explosion (1857–1888), in which there was an important boom in new means of transport that stimulated the transfer of goods and people between Flores and Buenos Aires, at a time when an explosive growth of the town center of the town; and (c) the period characterized by the incorporation of Flores to the city of Buenos Aires, between 1888 and 1914, is manifested in the almost total urbanization of the old fields and by the enormous migratory flood that the City of Buenos Aires received.

The archaeological record before 1857 corresponds to the Rodríguez-Visillac site (Camino, 2014), and it is represented by a construction (floor and a brick wall) with a very low density of findings, most of which are simple construction materials of local origin and some few industrialized materials of European manufacture (some fragments of English earthenware, one of Spanish majolica and a coin from the Buenos Aires Province, minted in England in 1822). Concerning the second period, a much higher density of findings was reported in a particular sector (La Moyosa residence), with construction materials mostly of European origin (ceramic tiles, roof tiles, and wall tiles from France, coatings of British origin, and German drainage pipes), and only locally made bricks. In the ceramic registry, British stoneware, earthenware, and bottles predominate, as well as some others imported from France, Germany, and Belgium. The third period includes materials dated between 1890 and 1911 and is mainly represented by the Corralón de Floresta and Plaza Pueyrredón sites, both characterized by the presence of fill materials from the incineration of waste from the city of Buenos Aires (Camino, 2011). Camino characterizes this moment as an "almost total" integration of Flores into the City. Slabs of British origin continue to predominate, followed by German and French. The vitreous set presents pieces of national, British, and American origin. As for construction materials, all the slabs and tiles are of French origin.

Regarding the archaeofaunistic remains found at the Rodríguez-Visillac and La Moyosa sites, it is argued that the diet "during the last 150 years" (Camino, 2014: 63) consisted mainly of *Bos taurus*, followed by *Ovis aries* and *Gallus*, in the first case, and *Bos taurus*, *Ovis aries*, and *Sus scrofa*, in the second case. *Gallus pavus*,

Ansus domesticus, *Lagostomus maximus*, and a group of unidentified fishes and snails were also found to a lesser extent.

Belgrano

The lands on which the current neighborhood of Belgrano is located, towards the end of the eighteenth century were used for raising and grazing cattle, a function for which they were considered very propitious. Those lands were crossed by the Camino Real to the north (current Cabildo Avenue), which linked Buenos Aires with Lima (Peru), so it was a busy and important route (Casella de Calderón, 1984). There was a chapel and a lime factory operated by a group of Franciscans religious (Padula & Silveira, 2015), so very few structures could be found; in addition, there was a small cemetery, a few scattered houses of fishermen, and small-scale lime exploitation (Schávelzon, 2020). In 1855, by request of the San José de Flores residents, the governor of the Buenos Aires Province founded the town of Belgrano on the land of Calera (Melero García et al., 1980), which, by 1864 was already a place of rest for the wealthy classes (Giunta, 2004), especially after the inauguration of the railway, which occurred a couple of years before. Towards the end of 1887, the Province of Buenos Aires ceded the districts of San José de Flores and Belgrano to the Nation, with which this district became a new neighborhood of the city.

A series of works for the enhancement of the public space known as Barrancas de Belgrano, between the years 2017 and 2018, allowed a group of archaeologists belonging to the Centro de Interpretación de Arqueología y Paleontología del Gobierno de la Ciudad de Buenos Aires undertook an intervention in a sector corresponding to the Chapel of the Franciscans. Near the chapel, an oval structure corresponding to a bucket well or cistern was found, made up of fractured bricks joined with mud mortar, a little over 3 m deep. On the northern limit, another structure of solid bricks with a different construction method was superimposed, interpreted as the remains of a possible wall arranged in a southwest-northeast direction over the oval structure and partially collapsed. Inside, two well-defined primary discard levels were identified, one with cultural materials corresponding to the last decades of the eighteenth century and the first decades of the nineteenth, and the other with discards located between the last quarter of the nineteenth century and the beginning of the twentieth (Silveira et al., 2019). The lower subset is distinguished by the presence of cultural materials with great chronological homogeneity and a high rate of remounting. The upper subset, on the other hand, is characterized by the presence of very varied cultural materials, with high fragmentation, a low rate of reassembly, and low chronological coherence between them. Considering the existence of a structure built above the well, which is currently collapsed, and the tendency to fill negative reliefs with materials removed from municipal waste treatment plants (burns), from other parts of the city (Schávelzon, 2020), is very likely that these latest residues recovered from the upper part of the well come from the aforementioned practice.

Bahía Blanca and Choele-Choel Region in the Framework of the Expansion of the Frontiers (Nineteenth-Century)

The City of Bahía Blanca

Bahía Blanca city was developed after the installation of a fort (Fortaleza Protectora Argentina) in 1828 to expand the border and secure rural farms. During most of the nineteenth century, Bahía Blanca had poor communication with other parts of the Buenos Aires Province. This state of isolation resulted in a very slow rate of growth. Both most of the population and the agricultural or livestock production extended a few leagues from the fortress and the main activity was trade with the indigenous people settled in the town or nearby and with those groups that periodically came to trade. For this reason, some authors called it, like the town of Carmen de Patagones, a factory (Gelman & Santilli, 2002; Ratto & Santilli, 2004; Ratto, 2008). Since 1870, when the process of expansion of the borderline in the Pampean region was accentuated, some of its characteristics were modified. Although agriculture maintains its marginal character, destined above all for local markets and practiced in "ejido" farms, livestock gains importance. Being first a wool area, refined beef gradually began to proliferate and its possibilities of export are taking shape with the sale of live cattle to England in the 1890s (Costantini, 2019). After the conquest of the Pampean-Patagonian territories, Bahía Blanca underwent profound changes with the arrival of the railway (1884), the agricultural colonization in its surroundings, the opening of commercial houses and export companies, the installation of cold-storage industries, and the inauguration of the port (1885) among other aspects. The acceleration of the development of the productive forces allowed direct communication of the local economy with the international economic system and the consideration by the ruling elite. In this frame, between the last years of the nineteenth century and the first decade of the twentieth century, it was consolidated as an important urban center, formally declared a city in 1895, from which the rest of the population centers that surrounded it were configured.

One of these spaces is the town of General Daniel Cerri[10] belonging to the Bahía Blanca district and located in the closeness of the local estuary, near the Sauce Chico river (Fig. 3). Since the establishment of the Fortaleza Protectora Argentina in 1828, this space was included within its limits, considered part of the "ejido", although state control did not materialize until the last decades of the century. For this reason, the occupation of the Creole society was limited and only a few occupations by the settlers and "indios amigos"[11] (friendly aboriginals) groups were

[10] The area where the town is located was traditionally called Cuatreros (which means "rustlers"). After 1943, it was changed to its current name (Pupio & Perriere, 2013).

[11] During the period of conquest, the success and maintenance of European colonial settlements depended, at times, on the military assistance provided by native groups against other groups of native populations. This phenomenon is known as "ethnic soldiers" and is widely represented in the American continent (González, 1990; Whitehead, 1990). In the case of the province of Buenos

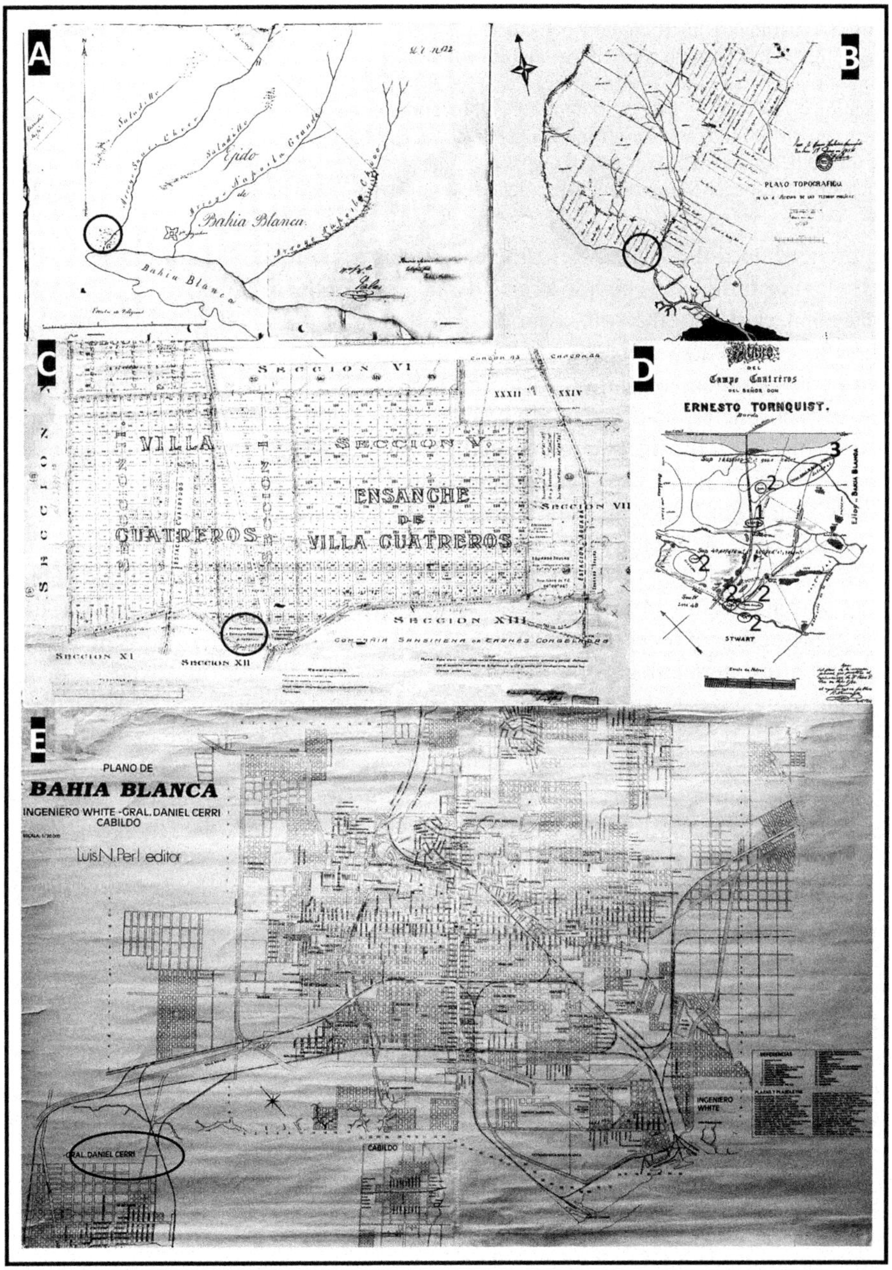

Fig. 3 Evolution of the Bahía Blanca and surrounding area based on the cartographic record: (**a**) Map of Bahía Blanca in 1856. The circle represents the current location of General Daniel Cerri on the Sauce Chico river (Museo y Archivo Histórico de Bahía Blanca). (**b**) Topographic map with the distribution of properties since 1870. The circle indicates the highest occupation of the sector (Plano Topográfico de la cuarta sección de las tierras públicas, Dirección de Goedesia, MOSP).

Fig. 3 (continued) (**c**) Boundary plan of the land currently occupied by General Daniel Cerri, made in 1913. A total division of the land is observed. The outlined area indicates the former location of the fort. Modified from Pupio & Perriere, 2013. (**d**) Map of the year 1884 with details of the location of the fort (1), posts and huts (2), and location of the old road to Patagones (3) that crossed the town. (Modified from Pupio and Perriere (2013). E) Map of Bahía Blanca from the first decades of the 20th. century. The location of the study area is indicated with its current name and the access routes to this space (Map library, Biblioteca Rivadavia))

referenced until the 1850s. Nevertheless, it was a frequent place of transit used by Creoles and Indigenous people as a crossing point for the so-called road to Patagones that connected with the southernmost population of the province of Buenos Aires and the rest of Patagonia. The Sauce Chico river was also the place of entry for the numerous indigenous parties that recurrently approached the incipient town to trade. From the second half of the century, the first official attempts at occupation are verified with the installation of an agricultural-military colony in a place called Nueva Roma, located near Bahía Blanca, and the sanction of different laws from 1860, which intensified the sale of properties to individuals and orders about the area. In 1870, the sector between Bahía Blanca and Nueva Roma was populated with ranches and houses (Pupio & Perriere, 2013). Since this decade, the state presence was manifested in a more forceful way with the installation of several forts and the construction of a defense ditch in 1876 to prevent the entry of raids in a period in which the national government began to develop the recent advances on the indigenous border (Bayón & Pupio, 2003). By 1877, the military commander of Bahía Blanca reported that three fortlets had already been built: one in the "Paso de los Chilenos" called Fortín Palao, another in "Los Saladillos" called Fort Borges, and another in the "*camino a Patagones*" called Fort Cuatreros (letter from military commander of Bahía Blanca, Daniel Cerri, to the justice of peace Pedro Hugony, 1877, emphasis added).

After the territorial conquest by the Argentine national state, some of the transformations that took place in Bahía Blanca were reproduced and have an impact on General Daniel Cerri. In the last decades of the nineteenth century, the practice of family-type horticultural activity aimed at local supply began to develop, in which the Italian immigrant contribution played a dominant role (Lorda, 2006). In the first years of the last century, urban growth intensified, facilitated by the installation of different railway stations in the sector (Villa Olga in 1891, Cerri in 1897, and Aguará in 1911). Additionally, the first local industries were developed with the installation of the Compañía Sansinena de Carnes Congeladas together with an annexed port (1903) and the Soulas et Fils skin washer and peeler (1905), intended for the export of primary goods, in line with the dominant economic model at the national level

Aires, during the government of Juan Manuel de Rosas and afterward, it refers to the name given to those indigenous groups that, negotiating with the state in exchange for rations, settled near the recently founded towns and provided an auxiliary militia service (Ratto, 1994, 2003). However, its meaning may vary regionally and temporally. For example, for the Chilean case during Colonial and Republican period see Ruiz-Esquide Figueroa (1993), Casanova Guarda (1996), Zavala Cepeda (2008, 2012) and Obregón Iturra (2008), among others.

(Kraser, 2017). All these transformations, in approximately 30 years, produced an enormous impact on the territorial configuration of the locality. Indeed, a territorial system was consolidated in the town of General Daniel Cerri that grew and developed in close ties with the city of Bahía Blanca, but with a significant degree of autonomy and its own dynamic (Lorda, 2006).

Boliche Lucanera: Archaeological Record

The Boliche Lucanera site (38°42′44.6″ S -62°24′38.7″ W) is a residential garbage pit, of both a house and a business, and is in the courtyard of a building erected in the first decade of the twentieth century. According to documents obtained from municipal offices, it was determined that the construction ended in 1908 and it was used from the beginning for the sale of alcoholic beverages. The commercial guides of the first decades of the twentieth century indicate that the Italia trade was established in the place, functioning as a warehouse, café, bar, billiards, and, around 1922, as a wine warehouse. This establishment continued as a warehouse until the 1970s. At present, the construction is abandoned.

Its discovery took place after the extraction of a tree by an excavating machine that uncovered industrial (i.e., glass, stoneware, earthenware, metal) and archaeofaunistic remains. This generated a well with irregular limits with an approximate area of 29 m^2 with the presence of remains in stratigraphy (Fig. 5a). A sequence of 23 stratigraphic units located from the surface to 120 cm deep was determined (Table 1). The partial classification indicates that most of the material corresponds to bones and vitreous remains (Tables 2 and 3). To a lesser extent, earthenware, stoneware, ceramics, and metals were found. Other scarce remains include fragments of kaolin pipes and mother-of-pearl, and plastic buttons. The studies carried out on the vitreous materials and remains of stoneware bottles (Vecchi & Tomassini, 2015), as well as the earthenware fragments (Tomassini & Weimann, 2020), allowed establishing the chronology of the site. On the one hand, alcoholic beverage containers were identified as square gin bottles of different brands: Schiedam and v. Hoytema and Co. from the Netherlands and Llave from the Jurgen Peters factory-made of Argentina. In addition, bottles of Hesperidina, "Cutler, Palmer and Co." brand whiskey, wine, and beer were recognized. The latter is presented in cylindrical bottles, mostly green and brown, used to protect the content from sun exposure. Three fragments of a light green cylindrical bottle bear inscriptions that allowed it to be identified as absinthe liqueur, an alcoholic beverage imported on a massive scale since approximately 1850 and present in most of the rural businesses in the region (Schávelzon, 1991; Carrera, 2014). Among the non-alcoholic beverages, an entire bottle of a "Ross's Belfast" brand carbonated Ginger Ale was recovered. They were used to drink alone or to dilute the alcoholic beverage. The rounded base prevented it from being placed in a vertical position, so the cork was permanently in contact with the liquid, preventing it from drying out and leaking. It is a torpedo or ballast-type bottle manufactured by the firm of W. A. Ross of Ireland used a hinged

Table 1 Boliche Lucanera site. Material distribution by stratigraphic unit. The grids and units with the most number of findings are highlighted

Garbage pit related grids	1.5 W-5S	1.5 W-6S	2.5 W-6S	2.5 W-5S	3.5 W-6S	3.5 W-5S	0.5E-8S	0.5E-7S	0.5E-6S	0.5E-5S	0.5 W-7S	0.5 W-6S	0.5 W-5S	Total
Base	–	–	–	–	–	–	3	–	2	–	1	1	–	7
0–60 cm	–	–	–	–	–	–	–	–	–	–	–	–	–	0
U 1–9	–	–	–	–	–	–	1	5	–	–	–	–	–	6
U 10	2	2	–	–	1	1	–	10	–	5	25	14	–	60
U 11–17	–	–	–	–	–	3	–	–	–	–	–	–	–	3
U 18	–	–	–	–	–	–	–	–	–	–	–	–	–	0
U 19	–	1	–	–	–	–	–	–	–	–	–	–	1	2
U 20	–	–	–	–	–	–	–	–	–	–	–	–	–	0
U 21	23	472	97	12	33	45	77	324	231	22	874	992	323	3513
U 22	3	–	–	7	–	–	–	15	11	9	3	22	–	52
U 23	–	–	–	–	–	–	–	–	2	1	–	–	–	3
Total	*28*	*475*	*97*	*0*	*34*	*49*	*81*	*354*	*235*	*37*	*903*	*1029*	*324*	*3646*

Table 2 Boliche Lucanera site. Number of fragments by type of material

Material	*N*
Ceramic	6424
Fauna	2292
Glass	1269
Metal	150
Others	10
Total	*4363*

Table 3 Boliche Lucanera site. Representation by type of container (vitreous elements)

Glass	*N* (fragments)
Cylindrical bottle of beer	272
Bottle of wormwood	3
Cylindrical soft drink/tonic bottle	1
Cylindrical bottle of sauce (mustard)	12
Cylindrical bottle of Vermouth (Bitter)	2
Cylindrical bottle of Hesperidina	107
Cylindrical wine bottle	400
Gin case bottle	140
Cosmetic/medicine/perfume bottle	22
Glass or cup	27
Tulips	63
Flat glass	199
Other	21
Total	*1269*

mold and manual termination type "Oil Finish'' and was manufactured between 1865 and the first decades of the twentieth century. According to the advertisements in the newspapers, it is known that by 1889 it was already being distributed in Bahía Blanca. Another product identified was a "Secrestat" brand bitter, manufactured by Maison J. Hré Secrestat Aîné in Bordeaux in 1851 and referenced in different frontier sites in the last quarter of the nineteenth century in the Pampean region (Tapia & Pineau, 2001; Pedrotta, 2002; Dosztal, 2014).

In the case of the earthenware remains, crockery elements such as plates of different sizes, cups, bowls, and dishes, corresponding to the whiteware and pearlware types, were mostly identified. Among the plates and bowls, the presence of four brands was determined, being the most representative Petrus Regout and Co. of Maastricht, produced from the last quarter of the nineteenth century in Holland. Several plates and saucers bear the "Ironstone China" seal of the J. and G. Meakin Hanley factory of England and include the phrase "Dieu et Mon Droit": translated as "God and my right." This stamp indicates a commitment to the British imperialist project on the part of the manufacturing company, as well as a claim to prestige. On the other hand, the word "England" indicates a temporary ascription to the last quarter of the nineteenth century (Majewsky & O'Brien, 1987). In addition, a bowl with blue floral decoration from the French brand "Opaque de Sarreguemines" was recovered. This company manufactured continuously between 1790 and 2007 and by the stamp, it can be ascribed to the last quarter of the nineteenth century (Puebla, 2017). Also, eight fragments of a yellow plate with the stamp "Rosario Boulogne" in the closest levels to the current ground were found. The stoneware elements belong to gin and beer containers. Among these there are monochromatic and bichromatic containers, corresponding to cylindrical and sinusoidal bottles (Schávelzon, 1991). One of the identified brands is from the H. Kennedy factory, located in Barrowfield, Glasgow, which operated from 1866 to 1929 and was used in Argentina as a beer container. Finally, 52 red-paste ceramic fragments with glazing on their inner face were recovered. These elements appeared both in the stratigraphic units near the discard area and in the pit made by the machine. Due to its color and type of pasta, it was identified as a cooking element (pot). These accessories were very common in the border areas due to their artisanal production and were manufactured by both Creole and indigenous populations. Despite this, no elements that could be attributed to indigenous materiality were recovered.

Choele Choel

The Río Negro valley constituted a scenario of multiple interactions, characterized by high mobility of people and the circulation and exchange of goods and ideas (Palermo, 1991; Nacuzzi, 2005). The intensity of these interactions increased with the presence of Europeans (Prates et al., 2016). Starting in the sixteenth century, the establishment of urban centers, particularly in the center of the Republic of Chile and in the north of Buenos Aires province, and the introduction of technologies, goods, and livestock, hitherto unknown to indigenous groups, prompted important changes in their mobility patterns and their material culture (Pinto Rodríguez, 1996; Martínez Sarasola, 2013). At the end of the eighteenth century, with the foundation of the "Fuerte de Nuestra Señora del Carmen," currently the Carmen de Patagones District (Buenos Aires Province), the intensification of these exchanges impacted the transformation of the North Patagonian and Pampean landscapes, which, at the beginning of the nineteenth century, was already crossed by an extensive network of roads that connected the Chilean Araucanía region with the Salinas Grandes, Buenos Aires, and Patagones (Palermo, 1991; Bechis, 2010). The main axis in the relations between indigenous and Hispano-Creole settlers was trade, in whose exchange, the bulk of the income of the aboriginal groups was constituted by cattle, initially wild, captured in the Pampean region (De Jong & Cordero, 2017). The intensification of this trade and the growth of livestock enterprises, because of the expansion of the Creole occupation of the pampas, led to the disappearance of wild cattle, and the development of a complex system of acquisition, transportation, and marketing of livestock. This situation involves numerous social groups and a violent history of conflict and the development of an intricate web of political relationships, intra- and interethnic.

In this context, the island of Choele Choel[12] located in the middle valley of the Río Negro, constituted a strategic center of the network of indigenous roads in historical moments, since the routes that led to Buenos Aires, Patagones, Valcheta, Las Manzanas, and Araucanía converged there. It provided important facilities for the transit and stay of the numerous contingents of travelers and the merchandise that circulated there, since it had drinking water and pastures for animals, constituting a fattening center for cattle before their commercialization in Chile; on the other hand, it was one of the few places where it was feasible to cross the Negro river to the southern bank, which also gave it a high defensive value against possible attacks by opposing indigenous or state groups (Bechis, 2010). It was also a space valued by the white, European society, due to the recognition of its strategic importance for cattle trafficking by indigenous groups, its central position for the military domination of Patagonia and the expectation it generated in various European powers the possibility of finding a possible navigable route that would join the two oceans, avoiding the dangers involved in crossing the Strait of Magellan or surrounding Cape Horn (Eliges, 2017a).

Despite the multiple interests that the different groups had over this territory, it was not until the irruption of the state troops at the end of May 1879 that their domain was guaranteed to the white, European society. Until then, different groups had disputed control of Choele Choel, although none had managed to monopolize it. On the contrary, the centrality of Choele Choel as a space for exchange allows us to understand the intricate cosmopolitan character of the type of relationships that were established there and the polysemy with which the different groups signified the territory (Palermo, 1991). In this context, the district of Patagones, located in the extreme south of the province of Buenos Aires, based its economy on the trade of salt, wheat, hides, salted meats and products from its exchanges with the indigenous economy, which had its epicenter in Choele Choel. That situation required the maintenance of a complex diplomatic system that would allow its population access to these resources which, at the beginning of the nineteenth century, only were obtained in a territory under the indigenous rule. During the term of the "Negocio Pacífico de Indios[13]"(i.e., the "Pacific Trade with the Indians), between 1829 and 1852, Patagones experienced a great development of commerce, which resulted in exponential growth of its population, which went from 505 inhabitants in 1823 to about 2000 in 1832 (Ruffini, 2000: 72).

[12] Strictly speaking, it is an archipelago, but this denomination is chosen because it is the way in which traditional historical literature has referred to this territory.

[13] The "Negocio Pacífico de Indios" (Pacific Business of Indians) was a border and internal security policy developed between 1829–1852, during Juan Manuel de Rosas' administration as governor of the province of Buenos Aires, which elaborated and complexly articulated interethnic relations between the different indigenous factions that disputed the territory of Pampa and Patagonia. This policy guaranteed a long period of harmony, which enabled the incorporation of Buenos Aires to the main cattle trade networks, by significantly reducing the number of indigenous irruptions against the productive undertakings of white society (Ratto, 1994, 2003).

After the fall of the Rosas government in 1852, the treaties with the indigenous groups lost validity, and violence once again predominated in their relations with white society. Towards the second half of the nineteenth century, the province of Buenos Aires was in a situation of permanent conflict with the indigenous factions that wiped out the countryside located on its southern border. The need to establish lasting peaceful relations that would allow productive agricultural development, at a time of strong integration of Argentina with the global economy, led to the establishment of peace treaties with the main indigenous leaders. In the case of the negotiations carried out with Cacique (i.e., the tribal chief) Yanquetruz, the army's main negotiator, the commander of the Patagones fort was aware of the need for Patagones to achieve "peace at all costs" (Levaggi, 2000: 285; Lanteri & Pedrotta, 2018). The agreement signed in 1857, in addition to including a series of mutual acknowledgments and economic compensation to the indigenous bias, guaranteed free trade to all tribes and Indians dependent on the cacique Yanquetruz in the town of Patagones or any other in the province of Buenos Aires, and assistance to government troops, should the governorate decide to fortify Choele Choel Island (Levaggi, 2000: 288–289; Lanteri & Pedrotta, 2018).

The signing of this treaty transformed Patagones relations with indigenous groups, which translated into a significant expansion of agricultural exploitation in the Negro River valley and a notable increase in trade with indigenous groups, which, just as it happened in Bahía Blanca, placed both cities as important "factories" linked to interethnic trade, turning out to be a source of significant prosperity for the city (Ruffini, 2000). In compliance with its clauses, the towns of Guardia Mitre and Fortín Conesa were founded in 1862 and in 1869, respectively, to the west of the city. There was also the settlement of "friendly tribes" on both banks of the Negro river, consolidating the establishment of peaceful relations between the white, European society and the indigenous groups that signed the agreement (Municipalidad de Carmen de Patagones, 2022). All this resulted in a transformation of the landscape that radiated from the city-port, upriver, towards the mountain range. After the arrival of the Argentine Army troops in Choele Choel, in May 1879, and the violent eviction of all the aboriginal groups, this process was dramatically accelerated, with the founding of numerous towns in the old indigenous whereabouts and the implementation of agricultural and livestock colonies, promoting the settlement of numerous contingents of European immigrants. By the end of the nineteenth century and the beginning of the twentieth, all the land captured in the Río Negro valley had already been distributed among the new private owners, with new zoning for public spaces in towns and neighborhoods and orthogonal territorial schemes (Fig. 4e).

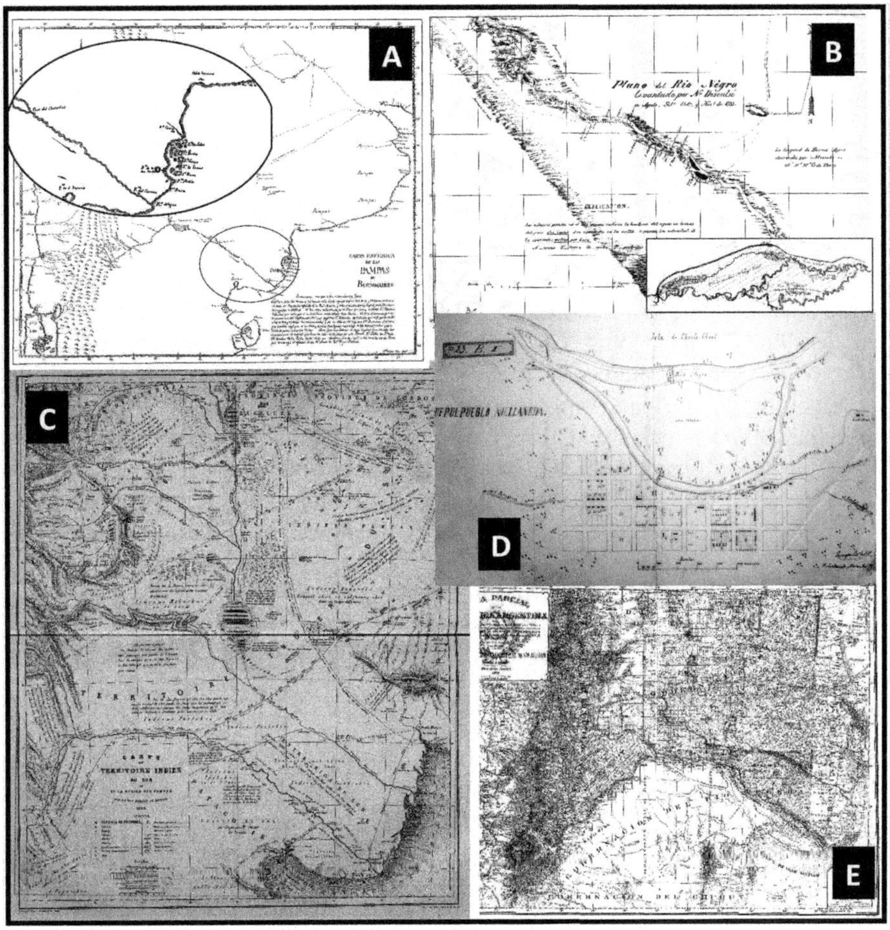

Fig. 4 Evolution of the Carmen de Patagones and Choele Choel area, between 1805 and 1889, based on the cartographic record. (**a**) Modified from the "Spherical Chart of the Pampas of Buenos Aires," made by Sebastián Undiano in 1805 (Archivo General de la Nación: Sala IX, Colonia, Exp. 1, legajo 39–5-5), the only reference is made to the " Fuerte del Carmen" [of Patagones] and the "Paso del Cholechel"; (**b**) Modified from sheets 1 and 5 of the "Plano del Río Negro" made by Nicolás Descalzi in 1833 (Archivo General de la Nación: Sala X, legajo 27–5-2). Sheet 1 shows the "Population of Carmen de Patagones" and a set of references to "houses" and "taperas" about 40 km to the west, with the names of their respective owners. Sheet 5 presents the "Isla de Chuleechel," with several references to military camps and the port "Encarnación". (**c**) 1865: "Carte du Territoire Indien du Sud et de la région des Pampas,'' by Martín De Moussy (Dehais, 2006: 178). This work collected all the cartographic knowledge gathered up to the time, in the case of Patagones, it only repeats the information provided by Descalzi three decades earlier. (**d**) 1879: "Croquis del Pueblo Avellaneda,'' by Jorge Rohde (Eliges, 2017b). The only map of the town, made a few months after the military conquest of the territory by the Argentine State. The orthogonal layout of the town is distinguished around the indigenous rake that connected Patagones with the mountain range and other towns to the north, and information on the adjudication of public and private land, the zoning of the town and a set of topographical references. (**e**) 1889: "Mapa Parcial de la República Argentina'', by Jorge Rohde (Dehais, 2006: 218). Detail of the towns and cities founded from the conquest of the indigenous territory and its division into cadastral and administrative jurisdictions. It also accounts for the existence of an "Old Camp of Choele Choel "(the Avellaneda Town), and a new one (the current city of Choele Choel)

Pueblo Avellaneda: Archaeological Record

Pueblo Avellaneda was the first urban settlement founded in the context of the "Desert Campaign," a name used to refer to the process of the Argentine state borders advancing over indigenous ancestral lands between 1879 and 1885. After a failed attempt, the town was raised by the expeditionary troops on a property now known as "Pampa de los Molinos" or "Zanjón de los Loros" (39°15′57 S–65°40′25 W), where it continued to be occupied for a little over two years before being moved to the location of the current city of Choele Choel, 2.5 km to the south. It had a network of streets and blocks structured around the main road (former indigenous road or "rastrillada"), residential, commercial, and administrative districts, and spaces reserved for different public services.

According to original cartography from the time of its foundation, the town had more than 30 hectares, of which approximately a third has been and continues to be occupied by different establishments. In the rest of the property, which we have defined as the "Pampa de Los Molinos" archaeological locality, a few structures have been built after the abandonment of the town; the remains generally are disposed of on the surface in their primary deposit position. The terrain has been affected by intense erosive processes that have marked it with numerous gullies that flow into large canyons up to 10 m deep; also, to a lesser extent, by low-intensity livestock farming. It is noteworthy that a large part of the archaeological record nowadays forms part of numerous private collections and local museums, as result of more than a century of action by private collectors.

Although most of the systematically collected material has been found on the surface, the testimony of some old residents, who have participated in spontaneous collections, refers to the existence of numerous accumulations of waste with little development with materials from the old town, which would be compatible with the practice of home burial of garbage. There are no written or cartographic testimonies that account for the existence of sites dedicated to the treatment or communal burial of waste or its massive abandonment in negative reliefs. The passage of the French expeditionary Henry de la Vaux, during the years 1896 and 1897, left the testimony of a desert town, just temporarily occupied by some indigenous families who would soon be evicted again, and where it was possible to find all kinds of objects on the surface: "Everywhere you can see abandoned ranches. You see pots and pans thrown everywhere" (De la Vaux, 1901: 39).

The study of the material culture of Pueblo Avellaneda was carried out on both the materials that have been extracted in our interventions and the main local collections. We have not found considerable evidence of indigenous materiality in any of these records, neither in the archaeological site nor in private collections. This aspect can be explained from the micro-local specificity, since at the regional level there is numerous evidence of indigenous presence (Nacuzzi, 2005; Prates, 2008; Prates et al., 2016; among others).

The archaeological record includes a large amount of glass, especially containers for alcoholic beverages, tonics, and cosmetics, ceramic materials (porcelain, stoneware, earthenware, kaolin, and bricks), metal (particularly a large number of nails and tin cans), textiles, and faunistic remains (Fig. 5b). As we can see in Table 4, the number of metallic parts is much greater than that of other materials represented; however, this arises from a notable bias in the collection criteria present in the collections, in which there are many metal nails but very few fragments of glass containers, among which only whole pieces or larges fragments have been selected. The proportions that emerge from this bias are contradictory to those found on the field, where the presence of vitreous fragments is significantly higher.

All kinds of industrialized objects are covered, such as food containers, especially tin cans. Construction materials also abound, predominantly nails and local construction bricks, which are scattered throughout the property. Ceramic materials (porcelain insulators) have also been found linked to the laying of telegraph lines and metal pipes, possibly belonging, in some cases, to wind musical instruments. Other highly represented items include military objects, accounting for an important variety of ammunition, parts of weapons, and fragments of uniforms; in particular, many buttons and hardware and even fragments of fabric from jackets and badges, and leather trimmings from the footwear of the uniformed men (Table 5). There are also parts of iron tools, pots, kettles, bulbs format, and cutlery of varied design and origin. Human bones were found, some thermo-altered, kept in bags cataloged as "hearth remains."

Although the materials collected in the field have not yet been analyzed, glass pieces are the most abundant. To date, 906 fragments have been analyzed, from which a minimum number of bottles (NMB) of 111 emerges; 52 of these fragments correspond to pyramidal bottles, among which the brands of gin Jürgen Peters and v. Hoytema and Co are of Dutch origin, and the local brands Llave, owned by Peters Hermanos, and Real Hollands, owned by William Paats and Co, both bottled gin imported in bulk from Holland; In general, all these bottles became popular in the country from the second half of the nineteenth century to the end of that century, although some may be somewhat older. Among the cylindrical bottles ($n = 31$), have been found Bordeaux and Champagne wine bottles, English beer, and wine bottles, all of which were in mass circulation in the second half of the nineteenth century. Bottles from Spanish Jerez were also found, manufactured since the beginning of the nineteenth century, and of local drinks, also mass-produced from the 1860s. Among the specimens preserved in local museums, some specimens of bottles of Turin Vermouth of the "Francesco Maletti Comp" and "Soprafino Martini Sola e Cia," the latter produced between 1863 and 1879. Other identified bottles correspond to the type of beer "India Pale Ale" from the Scottish firm "William Younger and Co's." Sinusoidal and cylindrical stoneware containers of local, British, and Dutch origin were also found. The typical beer porrons used to be matte white, although they could also have a glossy coating or have a decoration with colored baths on the upper half, typically called "chocolate bath." One of these specimens, a sinusoidal white bottle with a glossy finish, features an embossed stamp from Henry Kennedy's Barrowfield Pottery factory, which operated in Glasgow, Scotland, between 1866 and 1929. A large part of the bottles they exported contained beer of ginger ("Ginger") elaborated by the firm also Scottish, Tennent Limited of Glasgow.

Fig. 5 Representation of the elements found in the analyzed case studies. (**a**) Boliche Lucanera site. (**b**) Pueblo Avellaneda site

Table 4 Choele Choel: Collections and museums (totals by material)

Choele Choel: Collections and Museums					
Totals by material					
Material	CSM	CCM	CMG	CDL	Total
Ceramic	62	89	132	8	*291*
Fauna	0	70	0	1	*71*
Glass	46	45	0	16	*107*
Human skeletal remain	0	41	0	104	*145*
Metal	455	580	184	274	*1493*
Textile	2	25	0	0	*27*
Other	2	2	4	4	*12*
Total	*613*	*897*	*320*	*423*	*2253*

References	
CSM	Salatino Mazzulli Collection
CCM	Casa Maldonado Collection
CMG	Mony Gundín Collection
CDL	Collection donated to the laboratory by Mr. Salatino Mazzulli

Note: The Textile category includes leather articles that are part of clothing

Table 5 Choele Choel: Collections and museums (totals by function)

Collections and Museums: Totals by function					
Function or use	CSM	CCM	CMG	CDL	Total
Armament/ammunition	34	213	2	27	*276*
Beverage container	325	113	0	133	*571*
Construction material	22	0	0	0	*22*
Dress	0	14	4	67	*85*
Food container	82	265	194	17	*558*
Furniture	28	41	4	1	*74*
Hardware/ironwork	0	0	0	0	*0*
Indeterminate industrialized object	12	4	0	0	*16*
Inkwell	0	6	1	0	*7*
Rural/urban infrastructure	4	3	3	0	*10*
Saddle part	10	0	0	0	*10*
Smoking pipe	0	48	12	8	*68*
Tableware	13	18	38	12	*81*
Toner/perfume container	3	7	1	13	*24*
Tool	1	2	8	6	*17*
Trousseau	17	72	47	15	*151*
Utensil	1	1	5	0	*7*
Other	15	45	1	108	*169*
Total	*567*	*852*	*320*	*407*	*2146*

Empty bottles were also sold, as many regions brewed their beer; this is the case of the "Santa Rosa" brewery, owned by the German Juan Buhler. One of the stoneware fragments recovered shows the legend "J. Buhler and Co.," along with the phrase "Bolívar 480." The brand's historical references indicate that the factory was founded in 1852, but it operated at 480 Bolívar Street, city of Buenos Aires, between 1865 and 1890.

Though the "official" chronology of Pueblo Avellaneda runs between the years 1879, when it was founded, and 1882, when it was transferred, the relative dates obtained from some diagnostic materials would indicate the presence of materials dated between the second half of the nineteenth century and the first decades of the twentieth century. For example, numerous specimens have been recovered, some intact, of bottles of the Hesperidina brand, manufactured by the firm Bagley; it is an alcoholic beverage of local origin that has been changing its container over time, with a good part of the findings corresponding to the model marketed only in 1866. There are also copies corresponding to the first bottle models of the Argentine beer brands Quilmes and Palermo, founded respectively in 1888 and 1897, and a siphon to make carbonated water of the "Sparklets" type, patented in London in 1896. The Salatino Mazzulli collection houses a cardboard box of ten Remington model Argentine cartridges, caliber 0.43 Spanish, corresponding to the war arsenal of the Argentine Army and dated with a seal in October 1892. In addition to the numerous fragments of this ammunition, bullets of other calibers have been found, including some with a lateral spike system (Faucheux), considered ancient for the time in which the town worked.

Other very abundant elements are the tin food containers, popularized in Argentina from the 1880s. The semi-desert conditions present in the site allowed the conservation of materials that normally suffer great deterioration; however, sardine cans appear very frequently in almost all the archaeological localities. In private collections, a copy was identified with the inscription in relief "71 – Importe do Portugal – Packed in Portugal" on one of its covers, together with a bas-relief in the shape of an opening key. Cylindrical tin cans with welded lids and square and cylindrical friction lid cans were also found, one of these with the inscription "Manufactura de tobacco Hebra" on its lid. The fact that the discovery of cans of canned sardines is so frequent suggests that the acquisition of this type of food should not have been onerous at this time, so it could be expected that it would have happened later in the 1880s, considering that it was at that time that its importation increased substantially (Tapia et al., 2010). The finding of numerous riveted barrel bands, abundantly distributed in certain places, seems to account for the widespread practice of trading alcoholic beverages in bulk.

Discussion

The formation of the frontier spaces between the State and the indigenous societies in the Pampean and northern Patagonian regions of Argentina was a dynamic process that spanned almost the entire nineteenth century, from the post-independence period to the military campaigns carried out towards the end of the 1870s. During this time, different national and provincial governments developed strategies to occupy territories that were in control of the indigenous populations. One of the most common strategies was the foundation of forts and associated towns to constitute population centers as the borderline advanced, added to the spontaneous advance of the population.

In most cases, around these forts, a specific area called "ejido," or "panllevar" land, was designated for the cultivation of cereals, fruits, and vegetables for the livelihood of the population (Barcos & Lanteri, 2018). In the case of the Fortaleza Protectora Argentina (Argentine Protective Fortress), this sector extended a few leagues from the fort for fear of indigenous incursions, so that more remote areas, such as the current location of General Daniel Cerri, were scarcely occupied during the first half of the nineteenth century, except for some rural establishments and isolated posts. However, the place was frequently visited by travelers, merchants, authorities, and indigenous groups from colonial times since it was the obligatory step to continue the journey to the town of Carmen de Patagones in southernmost Buenos Aires Province. For this reason, it was common to install posts and places located along the paths, which formed part of a large system of communication between the different spaces. Carmen de Patagones, initially built by the Spanish Crown as a port enclave near the mouth of the Negro river in the Atlantic Ocean, after the independence process had to entrust its survival to the peace and trade agreements with the indigenous groups, from which could gradually develop its hinterland and project its urban territory towards Choele Choel around the axis configured by the old indigenous roads.

The expansion of the controlled territory in each of these nuclei was not always a linear process or developed unilaterally, because of government decisions. In most cases, it was the product of negotiations with the other predominant actors, the indigenous groups. In the Bahía Blanca area, the sector located near the Sauce Chico river was part of the department's official territory since 1865, and the river itself served as the boundary. However, until the moments before the military conquest, the authorities recommended it to circulate with special permission even though the territory was occupied and populated by members of the Hispanic/Euro Creole society. The famous Argentinian naturalist Francisco Moreno must have received a "passport" drawn up by the military commander of Bahía Blanca asking Chief Namuncurá, one of the main indigenous leaders, not to impede his march or harm him in any way (Moreno, [1879] 1997). For Bahía Blanca, the end of the military campaigns with the inauguration of the railway, and the port meant a boost in terms of the development of rail-port infrastructure based on what was required by the national economic model based on the export of primary products to the large industrial centers, mainly Great Britain. These factors, together with the arrival of European immigrants, generated demographic growth and a change in the urban and rural fabric. These spaces were modernized based on the aforementioned requirements, so the development of roads and railway lines was vital to accompany growth.

The territorial development of Carmen de Patagones towards its hinterland had three peaks during the nineteenth century. The first coincided with the economic boost generated by the implementation of the Pacific Business of Indians, applied by the then governor of Buenos Aires, Brigadier Juan Manuel de Rosas between 1829 and 1852; with his fall, dialogues policies towards indigenous groups were interrupted and, therefore, a return to times of inter-ethnic violence and economic recession. The second occurred once the Buenos Aires Province was able to make peace with Cacique Yanquetruz, in 1857, and Carmen de Patagones returned to economic growth, which translated into a new impulse in the development of its territorial influence towards the west. The peace treaty included the recognition of the desire of Buenos Aires to advance on Choele Choel, an issue that greatly interested the inhabitants of Carmen de Patagones, both for their intentions to develop their agricultural businesses and for the need to continue with the fruitful trade with indigenous groups, whose goods were exported by ship to the city of Buenos Aires. The third peak coincided with the military advance on the territory, which occurred in 1879. Despite the occurrence of the "Desert Campaign"'drastically changed the relationship of Carmen de Patagones with the territory, the subsequent urban and rural development towards the mountain range took place on the axis determined by the old indigenous roads and whereabouts. This process was strengthened by the ephemeral navigation of steamships on the Negro river and, later, by the arrival of the railway.

In this context, the sites studied herein and in comparison with other sites near the city of Buenos Aires (i.e., San José de Flores and Belgrano), allow interpreting a process of development of the urban edges produced in favor of their hinterlands, with special intensity from the second half of the nineteenth century. Despite the specifics, this process occurred in all cases in a radial way, from paths that linked these cities with other centers of social relevance and because of the socio-territorial practices of the subjects that inhabited them and not only because of the implementation of state policies. It has correlation with the trend that occurred, especially in the first half of the nineteenth century, by the fragmentation of the enormous estates into increasingly limited lots, especially in the areas that bordered the big cities, and their orthogonal reorganization around those routes. The daily life of these rural regions, increasingly committed to urban life, was gradually oriented towards the roads, which resulted in the demographic concentration, first, around the cart posts located next to the roads and, later, to railway stations as the means of communication developed. These areas of higher demographic density were transformed into towns, then into districts, and, in some cases, into neighborhoods that were territorially united to the cities, to the extent that these overflowed and the hinterlands were eliminated, a process that had much to do with European immigration and with the creation of recreational spaces for the local bourgeoisie (Camino, 2014; Schávelzon, 2017).

The peri-urbanization processes analyzed present many similarities. However, it is necessary to highlight that each one of them was affected by the conditions of its historicity, its actors, and its contingencies. The comparison to Buenos Aires city aims to distinguish a set of coincidences regarding how certain socio-spatial practices associated with those contexts had an impact on its hinterland. Considering this, we seek to rehearse an interpretation of the transformations experienced around the growing cities that transfigured the Pampas and Northern Patagonian border space in the second half of the nineteenth century. The radical change experienced in state policies towards indigenous territory from the last quarter of a century represented a break in its spatial evolution. The annexation of large extensions of land and its subsequent private sale generated the formation of large estates.

From this moment, the characteristics of these spaces (Bahía Blanca/General Daniel Cerri and Choele-Choel/Patagones) were transformed due to the consolidation and intensification of economic logic linked to the development of the agro-export model and the insertion of Argentina into the world market as a supplier of raw materials. In the case of the peri-urban territories of the city of Buenos Aires, we observe that their development was more linear due to their proximity to said city and their early incorporation into commercial circuits.

We consider that the archaeological correlation of this process is manifested through modes of accumulation of low artifactual density and high coherence and reassembly in the early moments of occupation of the site, turning towards less differentiated forms and typical rubbish of the twentieth century. In this sense, in the Boliche Lucanera garbage pit two aspects should be pointed out. On one hand, the discovery in the upper levels of eight fragments of a cream-colored earthenware plate-like creamware, of the Boulogne brand, manufactured in the city of Rosario (Santa Fe Province, Argentina) in the 1950s indicates the use of housing-commerce until well into the twentieth century. This also indicates the maintenance, for more than 70 years, of "traditional" disposal practices within spaces already considered "modern." These practices had a long tradition and had been established in everyday life since colonial times and are still used in rural areas (Schávelzon, 1999; Guillermo, 2004). In Bahía Blanca, by the first half of the century, it was still common to find landfills located in front of houses or backyards, as well as domiciliary wells, reservoirs, and cisterns, in neighborhoods, completely inserted into the urban space (Becher, 2012). The growth of the city, which went from being a small village to a large city, generated consequences in terms of garbage management and public hygiene. Policies related to the treatment and management of solid waste began to be implemented in the second half of the nineteenth century, although it was not until 1907 that the establishment of a communal service for the incineration or cremation of garbage and the sanitization of downtown streets with specific machinery was recorded (Becher, 2012). In Carmen de Patagones, the process of expansion towards the west, including the Choele Choel area, that had been developed in line with the businesses held by the indigenous groups was interrupted in 1879 by the military campaign that the Argentine State directed against them and was replaced by a new state-led colonizing model.

One last aspect to consider is the type of archaeological record present in other sites in the region, both rural and urban, for the period analyzed. In general, there are no significant differences in the type of materials founded, their brands and origin. This situation had already been highlighted by other researchers when comparing archaeofaunal and ceramic materials. Regarding the fauna, there are no great differences in the type of fauna of rural and urban contexts of the city of Buenos Aires, nor between what was recovered in productive and military settlements of the campaign (Brittez, 2009; Lanza, 2015). It is worth noting the clarification that diversity is greater in urban sites and that an important consumption of river fish was verified. Analyzing a typically rural site from the late nineteenth and early twentieth centuries in the province of La Pampa, Pineau and Andrade recorded that this space was fully stocked with a wide variety of products thanks to the presence of the railroad (Pineau & Andrade, 2018). This is verified from the study of stoneware and earthenware fragments that present a wide variability in decoration and types of containers and correspond to the same trademarks and stamps that have been recovered in other cities and towns for the same period.

In the sites studied herein (and the other sites used for comparison; San José de Flores and Belgrano), we also found high similarities in the type of material. For example, in the case of alcoholic beverages, the presence of gin case bottles of different brands (i.e. "v. Hoytema and Co." and "Jürgen Peters"), mainly from the Netherlands, is very common. The Hesperidina bottle is another highly represented element in all the cases described in a period that covers at least 50 years. In the case of stoneware, the beer bottles from the H. Kennedy Barrowfield factory in Glasgow (Scotland) is another commonly represented element. These aspects show how, despite the different stages and forms of development that each space was going through, certain consumption practices and habits in the ways of eating and drinking were installed both in cities and in border towns, and deserve to be pointed out.

Conclusions

This work is presented as a novel contribution to analyze and discuss the transformation of space in frontier areas of the Pampean and northern Patagonian regions during the nineteenth century. Our main goal was to propose a comparison from casuistry between rural and urban sites to identify correspondences and breaks in the archaeological record; however, the lack of specific interest in the peri-urban phenomenon in previous works made it impossible to establish reference points to generate an analytical matrix applicable to these cases. For this reason, we focused our comparisons with sites already studied from a diachronic perspective.

We consider it important to understand the historical evolution of the territories by analyzing all the stages of its transformation. In this sense, the notion of peri-urban and borders allows making the analysis more complex by incorporating forms of the spatial organization usually not considered. On the other hand, the interdisciplinary perspective of Historical Archeology is accompanied by the contributions of Geography, making it possible to understand how the population centers of the border areas developed. On this point, the morphology of the built space, whether rural, urban or peri-urban, reflects, in addition to the economic organization, the social organization, the political structures, and the objectives of the social groups, the daily transformations carried out by the different actors in their environment. That may be the result of their intentional praxis, the impact of their socio-spatial practices, disputes over the structuring of social space, or the mere iteration of the ways of exercising their trades and customs in the construction and recreation of the landscape. All of the above are practices that configure the different local spaces.

Acknowledgments We acknowledge the institutional support of CONICET, Universidad Nacional de Buenos Aires, Universidad Nacional del Sur and Universidad Nacional de la Plata. We would also like to thank Sol Lanteri for the suggestions and the critical reading of the manuscript. Thanks to Rodrigo Tomassini and Romina Frontini for the formal revision of the English texts. Thanks are also extended to the two anonymous reviewers and editors, whose comments and suggestions greatly improved this paper. Finally, we thank Juan Francisco Jimenez for the bibliographic suggestions related to "Indios Amigos" and Victoria Pedrotta for her constant support.

References

Attademo, S., & Ringuelet, R. (2008). Presentación. Estudios del mundo rural periurbano. *Mundo Agrario, 9*(17).

Avila Sanchez, H. (2005). Introducción. Líneas de investigación y el debate en los estudios urbano-rurales. In H. Avila Sánchez (coord.), *Lo urbano-rural, ¿nuevas expresiones territoriales?* (pp. 19–58). UNAM.

Bagaloni, V., & Pedrotta, V. (2018). Frontiers and Fortlets at the Pampa region, Argentina. *The Society for Historical Archaeology, 52*, 348–371.

Bandieri, S., & Blanco, G. (1996). La historia agraria Argentina en los siglos XIX y XX: Una síntesis pendiente. *Historia Agraria, 11*, 133–150.

Banzato, G. (2011). La periferia de "la gran transformación": el mercado de tierras en la provincia de Buenos Aires. Los casos de Chascomús y Junín en perspectiva comparada, 1780-1912. *Mundo Agrario, 12*(23), 00.

Barcos, M. F. (2007). Los ejidos de los pueblos de campaña: ocupación y acceso a la propiedad legal en Monte, 1829–1865. *Mundo Agrario, 7*(14).

Barcos, M. F. (2019). Pueblos y agricultura a fines del siglo XIX: los ensanches de ejidos en la provincia de Buenos Aires. *Estudios Rurales, 9*(17), 1–39.

Barcos, M. F., & Lanteri, S. (2018). Entre la antigua ocupación y la modernización liberal. Una primera mirada de conjunto sobre pueblos, ejidos y campos criollos e indígenas en la frontera sur de Buenos Aires (siglo XIX). *Estudios del hábitat, 16*(1), e033.

Barcos, M. F., & Martirén, J. (2019). La metamorfosis de una economía agraria en la pampa Argentina. Buenos Aires y Santa Fe entre las décadas de 1850 y 1890. *Anuario de Estudios Americanos, 76*(2), 585–614. https://doi.org/10.3989/aeamer.2019.2.07

Barsky, A. (2005). El periurbano productivo, en espacio en constante transformación. Introducción al estado del debate con referencias al caso de Buenos Aires. *Scripta Nova, 94*(36). http://www.ub.edu/geocrit/sn/sn-194-36

Barsky, A. (2010) La agricultura de "cercanías" a la ciudad y los ciclos del territorio periurbano. Reflexiones sobre el caso de la Región Metropolitana de Buenos Aires. In A. Svetlitza de Nemirovsky (coord.), *Globalización y agricultura periurbana en Argentina. Escenarios, recorridos y problemas* (pp. 15–29). Flacso Argentina.

Barsky, O., & Gelman, J. (2009). *Historia del agro argentino. Desde la Conquista hasta comienzos del siglo XXI*. Sudamericana.

Bayón, C., & Pupio, A. (2003). La construcción del paisaje en el sudoeste bonaerense (1865–1879): una perspectiva arqueológica. In R. Mandrini & C. Paz (compilers), *Las fronteras hispanocriollas del mundo indígena latinoamericano en los siglos XVIII- XIX. Un estudio comparativo* (pp. 345–374). Universidad Nacional del Centro de la Provincia de Buenos Aires-Universidad Nacional del Sur.

Becher, P. (2012). *Patrimonio urbano y arqueología histórica en Argentina. Reflexiones y perspectivas en torno a la vida cotidiana de la ciudad de Bahía Blanca (fines XIX-principios del XX)*. Editorial Académica Española.

Bechis, M. A. (2010). *Piezas de etnohistoria y de antropología histórica*. Sociedad Argentina de Antropología.

Brittez, F. R. (2004). Arqueología Rural en el partido de Coronel Brandsen, provincia de Buenos Aires. In C. Gradín & F. Oliva (Eds.), *La Región Pampeana –su pasado arqueológico* (Actas del I Congreso de Arqueología de la Región Pampeana Argentina) (pp. 211–222). Laborde Editor.

Brittez, F. R. (2009). Zooarqueología, tafonomía y procesos de formación de sitios rurales pampeanos: estado de la cuestión y expectativas para momentos tardíos. *Revista de Arqueología Histórica Argentina y Latinoamericana, 3*, 47–68.

Camino, U. (2011). San José de Flores Un lugar en el mundo. *Comechingonia, 14*, 173–189.

Camino, U. (2014). San José de Flores, de pueblo rural a barrio porteño. *Urbania. Revista latinoamericana de arqueología e historia de las ciudades, 3*, 59–73.

Camino, U., Traba, A., & Coloca, F. (2012). San José de Flores entre el pacto y el bondi. Una arqueología de pueblo y ciudad. *Revista del Centro de Estudios de Arqueología Histórica, 1*(1), 79–87.

Cardoso, M. M. (2020). Interfase. In A. Benedetti (Dir.), *Palabras clave para el estudio de las fronteras* (págs. 399–406). Teseo Press Design.

Carrera, J. (2014). Vinos y aguardientes en las pampas rioplatenses 1770–1850. *RIVAR. Revista Iberoamericana de Viticultura, Agroindustria y Ruralidad, 1*(2), 21–37.

Casanova Guarda, H. (1996). La Alianza hispano-pehuenche y sus repercusiones en el macro espacio fronterizo sur andino (1750–1800). In J. P. Rodríguez (Ed.), *Araucanía y Pampas. Un mundo fronterizo en América del Sur* (pp. 72–92). Ediciones de la Universidad de la Frontera.

Casella de Calderón, E. (1984). Belgrano: La ciudad yuxtapuesta. *Buenos Aires nos cuenta, 6*, 2–79.

Ciliberto, M. V. (2007). Patrimonio y producción en los entornos rurales de Buenos Aires. San José de Flores, 1800–1875. *Mundo Agrario, 8*(15).

Ciliberto, M. V., & Principi, A. (2014). Lo urbano-rural en la historiografía agraria rioplatense. Del final de la Colonia al inicio del siglo XIX. *Anuario Colombiano de Historia Social y de la Cultura, 41*(2), 261–290. https://doi.org/10.15446/achsc.v41n2.48788

Costantini, F. (2019). Capitales británicos y producción agraria en espacios de frontera. Bahía Blanca, Argentina (1860-1900). *Estudios fronterizos, 20*, e033.20. https://doi.org/10.21670/ref.1912033

Criado Boado, F. (1993). Límites y posibilidades de la Arqueología del Paisaje. *SPAL, 2*, 9–55.

Curtoni, F. (2000). La percepción del paisaje y la reproducción de la identidad social en la región pampeana occidental, Argentina. Paisajes Culturales Sudamericanos. In C. Gianotti García (coord.), *De las Prácticas Sociales a las Representaciones* (pp. 115–125). Santiago de Compostela.

De Jong, I., & Cordero, G. (2017). El malón en contrapunto: dinámicas de la diplomacia, el comercio y la guerra en la Frontera Sur (s. XVIII y XIX). In S. Serulnikov & G. Di Meglio (comps.),

Los saqueos en la historia Argentina. Variaciones de una acción colectiva (pp. 63–90). Siglo XXI.

De La Vaux, H. (1901). *Viaje a la Patagonia, 1896*. Librería Hachette.

De Marco, C. (2020). Los extramuros productivos. Un balance en el estudio de espacios periurbanos en Argentina. In J. M. Cerdá & G. Mateo (coord.), *La ruralidad en tensión* (pp. 147–182). Teseo Press.

Dehais, F. J. (2006). *Contribución a la cartografía de Patagonia o Chica entre 1519 y 1900*. Edición del autor.

Dosztal, I. (2014). El estudio de los bienes de consumo de origen vítreo del sitio arqueológico Casa de la Administración Alexandra Colony, siglos XIX y XX. *Santa Fe, Argentina. Cuba Arqueológica, 8*(1), 65–75.

Eliges, A. C. (2017a). La monumentalización del "pueblo Avellaneda" (Choele Choel, Provincia de Río Negro, 1879 a 1882). In J. G. (Comp.), *X Jornadas de Arqueología de la Patagonia* (p. 65). IDEAUS-CONICET.

Eliges, A. C. (2017b). La representación cartográfica del "segundo pueblo Avellaneda"- Choele Choel (Río Negro, 1879–1882). In J. G. (Comp.), *X Jornadas de Arqueología de la Patagonia* (p. 89). IDEAUS-CONICET.

Feito, M. (2018). *Problemas y desafíos del periurbano de Buenos Aires* (pp. 1–19). Estudios socioterritoriales.

Feito, M., & Barsky, A. (2020). Periurbano (Argentina, 1985–2020). In A. Salomón & J. Muzlera (Comp.), *Diccionario del agro iberoamericano* (pp. 897–808). Teseo Press Design.

Gelman, J. (2007). La historia económica Argentina, 1984–2004. Un balance con luces y sombras. *América Latina en la Historia Económica, 28*, 7–24.

Gelman, J., & Santilli, D. (2002). Una medición de la economía rural de Buenos Aires en la época de Rosas. Expansión ganadera y diferencias regionales. *Revista de Historia Económica, 1*, 81–107.

Gianotti Garcia, C. (2005). Arqueología del Paisaje en Uruguay. Origen y desarrollo de la arquitectura en tierra y su relación con la construcción del espacio doméstico en la prehistoria de las tierras bajas. In L. Mameli & E. Muntañola (Eds.), *América Latina: realidades diversas* (pp. 104–123). Instituto Catalán de Cooperación Iberoamericana-Casa de América, Colección AmerandCat.

Giunta, R. (2004). Revolución industrial y barrio. In L. Barela & M. Sabugo (Dir.), *Buenos Aires. El libro del barrio. Teorías y definiciones* (pp. 107–124). Instituto Histórico de la Ciudad de Buenos Aires.

Gómez Romero, F., & Spota, J. C. (2006). Algunos comentarios críticos acerca de 15 años de arqueología en los fortines pampeanos. *Relaciones de la Sociedad Argentina de Antropología, 23*, 161–185.

González, N. L. (1990). From Cannibals to Mercenaries: Carib Militarism, 1600–1840. *Journal of Anthropological Research, 46*(1), 25–39.

Guillermo, S. (2004). El proceso de descarte de basura y los contextos de depositación presentes en la ciudad de Buenos Aires. *Intersecciones en Antropología, 5*, 19–28.

Ingold, T. (1993). The temporality of the landscape. *World Archaeology, 25*(2), 152–174.

Kraser, M. B. (2017). *Paisaje cotidiano de General Cerri: Relatos, nostalgias y deseos sobre el patrimonio local*. EdiUNS.

Landa, C., Pineau, V., Montanari, M., & Doval, J. (2018). A través de una década de arqueología rural en el norte pampeano: pulperías, caminos, puestos y poblados (fines del siglo XIX -principios del siglo XX). In C. Landa, V. Pineau, M. Montanari, & J. Doval (Eds.), *Paisajes de la campaña pampeana (siglos XIX y XX). Abordajes desde la Arqueología rural en Argentina* (pp. 197–237). Archaeopress Publishing LTD.

Lanteri, S. (2018). La RER en el marco de la Historia Rural: un balance de más de veinte años. In *Nuevas miradas sobre el siglo XIX rural. Jornada de homenaje a Jorge Gelman*. http://ravignani.institutos.filo.uba.ar/rer

Lanteri, S. (2020). Ejido (Río de la Plata, Iberoamérica, siglo XIX). In A. Salomón & J. Muzera (Eds.), *Diccionario del agro iberoamericano* (pp. 479–490). Teseo Press.
Lanteri, S., & Martirén, J. (2020). Colonización. In A. Benedetti (dir.), *Palabras clave para el estudio de las fronteras* (pp. 127–137). Teseo Press Design.
Lanteri, S., & Pedrotta, V. (2018). Tierras, armas y política en la frontera sur bonaerense durante la década de 1850. Los "indios amigos", Maicá y Villa Fidelidad. *Anuario del Instituto de Historia Argentina, 18*(1), e066. https://doi.org/10.24215/2314-257Xe066
Lanza, M. (2015). Fauna menor en sitios históricos. *Archaeofauna, 24*, 342–362.
Levaggi, A. (2000). *Paz en la frontera. Historia de las relaciones diplomáticas con las comunidades indígenas en la Argentina (Siglos XVI–XIX).* Universidad del Museo Social Argentino.
Lorda, M. A. (2006). El desarrollo local, estrategia de gestión ambiental de la actividad agrícola en espacios próximos a la ciudad de Bahía Blanca. *Revista Universitaria de Geografía, 15*, 201–2017.
Majewsky, T., & O'Brien, M. (1987). The use and misuse of nineteenth-century English and American ceramics in archaeological analysis. *Advances in Archaeological Method and Theory, 11*, 97–209.
Martínez Sarasola, C. (2013). *Nuestros paisanos, los indios*. Editorial del nuevo extremo.
Matossian, B., & Vejsbjerg, L. (2020). Borde. In A. Benedetti (dir.), *Palabras clave para el estudio de las fronteras* (pp. 65–75). Teseo Press Design.
Melero García, R., Barela, L., González, L., de Molina, G. M., Pagani, E., Pérez, V., & Sprovieri, S. (1980). *Manual informativo de la ciudad de Buenos Aires*. Instituto Histórico de la Ciudad de Buenos Aires.
Miguez, E. J. (1986). La expansión agraria de la pampa húmeda (1850–1914). Tendencias recientes de su análisis histórico. *Anuario IEHS*, 89–119.
Moreno, F. (1997 [1879]). Reminiscencias del Perito Moreno. : El Elefante Blanco.
Municipalidad de Carmen de Patagones. (accessed: March 20, 2022). *Patagones Municipio. Historia: Períodos históricos.* Obtained from: https://patagones.gob.ar/es/periodos-historicos
Nacuzzi, L. R. (2005). *Identidades impuestas. Tehuelches, aucas y pampas en el norte de la Patagonia.* Sociedad Argentina de Antropología.
Obregón Iturra, J. (2008). Aproximación crítica al pensamiento dicotómico "indios amigos" *versus* "indios enemigos" bajo el gobierno del Marqués de Baides, Chile, 1639–1646. *Revista CUHSO, 15*(2), 25–30.
Padula, H., & Silveira, M. (2015). Restos malacológicos hallados en sitios históricos de la Ciudad de Buenos Aires. In H. Hammond & A. Zubismendi (Eds.), *Arqueomalacología. Abordajes metodológicos y casos de estudio en el Cono Sur* (pp. 145–174). Fundación de Historia Natural Félix de Azara.
Palermo, M. A. (1991). La compleja interacción hispano-indígena en el sur argentino y chileno durante el Período Colonial. *América indígena*, 153–192.
Pedrotta, V. (2002). Arqueología histórica en el Arroyo Nieves (Partido de Olavarría). Resultados preliminares de los primeros trabajos de campo. *Intersecciones en Antropología, 3*, 125–129.
Pedrotta, V., & Bagaloni, V. (2021). Un cuarto de siglo de arqueología histórica en las fronteras y espacios rurales de Argentina. *Latin American Antiquity, 32*(2), 368–384.
Pesoa, M. (2014). Sembrando ciudades: la fundación de ciudades en la provincia de Buenos Aires (1810–1916). *Quaderns de Recerca en Urbanisme, [en línia], 4*, https://raco.cat/index.php/QRU/article/view/311300 [Consulta: 25-03-2022]
Pineau, V., & Andrade, A. (2018). Los fragmentos de gres del sitio Mariano Miró. Análisis de su contenido y procedencia (La Pampa, siglo XX). *Revista de Antropología del Museo de Entre Ríos, 4*(2), 111–121.
Piñeiro, A. G. (1983). *Barrios, calles y plazas de la Ciudad de Buenos Aires: origen y razón de sus nombres*. Instituto Histórico de la Ciudad de Buenos Aires.
Pinto Rodríguez, J. (1996). Integración y desintegración de un espacio fronterizo. La Araucanía y las Pampas, 1550–1900. In J. Pinto Rodriguez (Ed.), *Araucanía y pampas: un mundo fronterizo en América del Sur* (pp. 11–46). Ediciones Universidad de la Frontera.

Prates, L. (2008). *Los indígenas del río Negro: un enfoque arqueológico*. Sociedad Argentina de Antropología.

Prates, L., Serna, A., Mange, E., & De Jong, I. (2016). Expresión material de la interacción entre indígenas y criollos en un sitio funerario del siglo XIX de Norpatagonia (Sitio Chimpay, Río Negro). *Intersecciones en Antropología, 17*, 35–48.

Principi, A., Ciliberto, M. V. (2009). Negocios e inversiones entre la campaña y la ciudad: los pulperos en las cercanías de Buenos Aires (primera mitad del siglo XIX). *Segundas Jornadas Nacionales de Historia Social*. Memoria Académica. Disponible en: http://www.memoria.fahce.unlp.edu.ar/trab_eventos/ev.9739/ev.9739.pdf

Puebla, G. (2009/2010). Caracterización del periurbano en países centrales y periféricos a través de cuatro autores. Breve recopilación y análisis bibliográfico. Breves Contribuciones del. *I.E.G*, 21, 135–155.

Puebla, L. (2017). *Análisis de sellos de lozas arqueológicas del área fundacional como contribución al estudio de la economía y sociedad de Mendoza (1800–1960)*. (Unpublished Graduation Thesis), Universidad Nacional de Cuyo. http://bdigital.uncu.edu.ar/9959. Access: 22 de octubre, 2019.

Pupio, A., & Perriere, H. (2013). *Malones, fortines y estancias en la identidad de General Daniel Cerri*. EdiUNS.

Ratto, S. (1994). Indios amigos e indios aliados. Orígenes del negocio pacífco en la provincia de Buenos Aires (1829–1832). *Cuadernos del Instituto Ravignani, 5*, 5–34.

Ratto, S. (2003). Una experiencia fronteriza exitosa: el Negocio Pacífico de Indios en la provincia de Buenos Aires (1829–1852). *Revista de Indias, LXIII*(227), 191–222.

Ratto, S. (2008). Allá lejos y hace tiempo. El fuerte de Carmen de Patagones en la primera mitad del Siglo XIX. *Quinto Sol, 12*, 45–72.

Ratto, S., & Santilli, D. (2004). De factoría a poblado agropecuario. La evolución del partido de Bahía Blanca hacia 1869. *Cuadernos del Sur-Historia, 33*, 47–78.

Ruffini, M. E. (2000). La ocupación de la tierra en la margen Sur del Río Negro (1820–1878). In M. E. Infesta (Coord), *Estudios de Historia Rural*, V (pp. 65–106). FaHCE.

Ruiz-Esquide Figueroa, A. (1993). *Los indios amigos en la frontera araucana*. Santiago de Chile.

Schávelzon, D. (1991). *Arqueología Histórica de Buenos Aires. La cultura material porteña de los siglos XVIII y XIX*. Corregidor.

Schávelzon, D. (1999). *Arqueología de Buenos Aires*. Emecé Editores.

Schávelzon, D. (2017) El Bajo Belgrano como borde urbano. una historia de rellenos y basurales. *Anales del IAA, 47*(1), 82–98.

Schávelzon, D. (2020). *Manual de Arqueología Urbana: técnicas para excavar Buenos Aires*. Centro de Arqueología Urbana (CAU).

Silveira, M., Padula, H., Orsini, R., & Bernat, E. (2019). Análisis zooarqueológico de los restos hallados en una estructura sanitaria vinculada a momentos tempranos del actual barrio de Belgrano. *Teoría y Práctica de la Arqueología Histórica Latinoamericana., 8*(9), 71–87.

Sorokin, P., & Zimmerman, C. (1929). Principles of rural-urban Sociology. New York: H. Holt.

Svetlitza de Nemirovsky, A. (2010). Introducción. In: A. Svetlitza de Nemirovsky (coord.), *Globalización y agricultura periurbana en Argentina. Escenarios, recorridos y problemas* (pp. 7–12). Flacso Argentina.

Tapia, A.H. and Pineau, V. (2001). Precisando la cronología de un sitio ranquel a través de los fragmentos vítreos. In F. Oliva, N. de Grandis & J. Rodríguez (comps.), *Arqueología Argentina en los inicios de un nuevo siglo* (pp. 473–483). Publicación del XIV Congreso Nacional de Arqueología Argentina.

Tapia, A. H.; De Rosa, H.; Landa, C.,Montanari, E. (2010). Identificación de jerarquización y prácticas sociales a través de artefactos metálicos. Fortín La Perra (1883–1885), La Pampa, Argentina. In *Actas del 53° Congreso Internacional de Americanistas, México (edición en CD-Rom)*.

Tomassini, H., & Weimann, J. (2020). Análisis de los materiales de loza del sitio Boliche Lucanera (General Daniel Cerri, partido de Bahía Blanca). *La Zaranda de Ideas. Revista de Jóvenes Investigadores, 18*(2), 40–58.

Vecchi, R., & Tomassini, H. (2015). Arqueología de rescate en un basural del siglo XIX. Sitio "Boliche Lucanera", General Daniel Cerri (partido de Bahía Blanca). Paper presented in the *VI Congreso Nacional de Arqueología Histórica Argentina*. Facultad de Filosofía y Letras, Universidad Nacional de Cuyo, Argentina.

Whitehead, N. L. (1990). Carib ethnic soldiering in Venezuela, the Guianas, and the Antilles, 1492–1820. *Ethnohistory, 37*(4), 357–385.

Zavala Cepeda, J. M. (2008). *Los Mapuches del Siglo XVIII: Dinámica Interétnica y Estrategias de resistencia*. Editorial Universidad Bolivariana.

Zavala Cepeda, J. M. (2012). La presencia mapuche en los Andes según las fuentes coloniales chilenas. *Estudios Transandinos, 17*(1), 119–134.

Challenges in the Management and Research of Cultural Heritage. The Patrimonialization of the EB 1789 Canyon from the Eighteenth Century in San Antonio De Areco, Province of Buenos Aires, Argentina

Verónica J. Acevedo and Laura Staropoli

Abstract In this chapter, several aspects of cultural heritage management are presented based on the experience of the enhancement and value revitalization of the EB 1789 cannon from the eighteenth century, in the town of San Antonio de Areco, Province of Buenos Aires (Argentina). These aspects raise the issue of negotiations and challenges in decision-making regarding cultural heritage. The complexity of patrimonialization also relates to the necessity and difficulty of the articulation process between the interests of the different social actors involved. Our objective during the management was to promote good practices for the protection of cultural heritage through research, conservation, and dissemination of strategies, concrete and experiential, that would help to promote heritage awareness in the community. We investigate the local social and cultural context, develop a good dialogue with the different social actors, and offer viable alternatives when issues arise while respecting the needs of the local community and political decisions. As we progressed, we deeply considered each decision and reflected upon each new situation, allowing this project to become implemented.

Keywords Argentina · Cultural heritage · Management · Revalorization process · Good practices · Research · Conservation · Dissemination

---

V. J. Acevedo (✉) · L. Staropoli
ArqueoLab-UBA, Instituto de Arqueología, Facultad de Filosofía y Letras, Universidad de Buenos Aires, Buenos Aires, Argentina
e-mail: vacevedo@filo.uba.ar; lstaropoli@filo.uba.ar

G. F. Bonnat et al. (eds.), *Current Research in Archaeology of South American Pampas*, The Latin American Studies Book Series,
https://doi.org/10.1007/978-3-031-55194-9_12

Introduction

This chapter outlines diverse aspects of cultural heritage management based upon due consideration of the enhancement and value revitalization of the EB 1789 cannon from the eighteenth century in the town of San Antonio de Areco, Province of Buenos Aires (Argentina). The town is located 110 km from the "Ciudad Autónoma de Buenos Aires," the capital of Argentina (Fig. 1). The patrimonialization objective was to give visibility to EB 1789 cannon and San Antonio de Areco history and to promote good practices for the protection of cultural heritage through specific research, conservation, and dissemination strategies within the framework of Participative Science and Visible Thinking (Acevedo & Staropoli, 2021; Acevedo et al., 2020).

This work was developed by Project ArqueoLab-UBA, a laboratory specialized in archaeometric, technological, and virtual analysis and studies for the research, conservation, dissemination, enhancement, and value revitalization of cultural heritage. It was created by a group of social scientists from the University of Buenos Aires. Their aim is to contribute to the knowledge of the human past through the application of procedures and techniques from other disciplines, not only to resolve archaeological and historical problems but also conservation ones related to

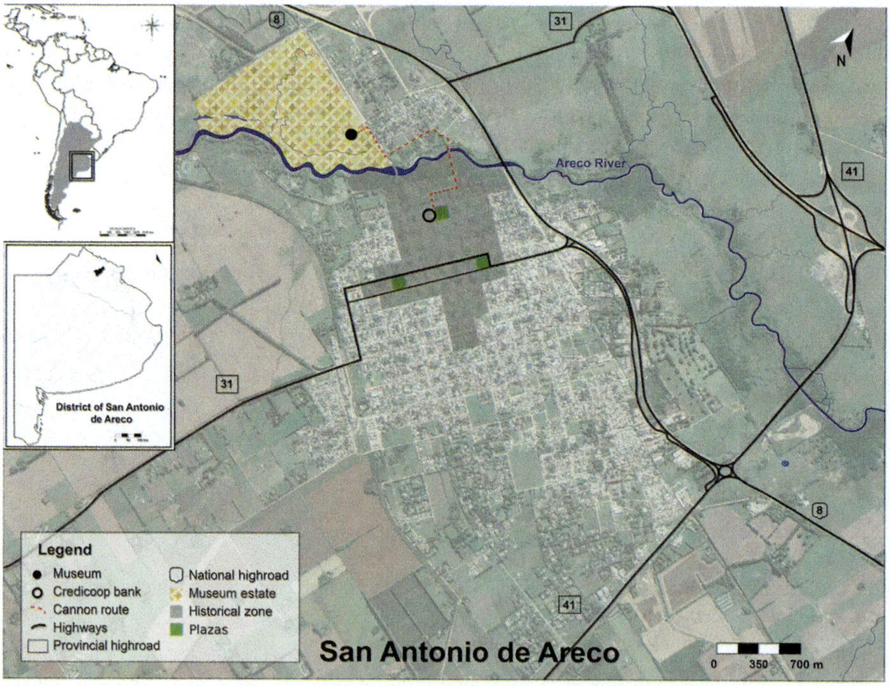

Fig. 1 Map of San Antonio de Areco, Province of Buenos Aires, Argentina

different types of cultural materials (Acevedo, 2019; Acevedo & Staropoli, 2021; Acevedo et al., 2020).

Our research group was specially invited to do this work due to the collaboration and cooperation with the institutional policy and the Urban and Rural Heritage Protection and Rehabilitation Program (*Programa de Protección y Rehabilitación del Patrimonio Urbano y Rural*) from the Secretary of Planification (*Secretaría de Planificación*), and Creole Park and Gauchesco Museum Ricardo Güiraldes (*Parque Criollo y Museo Gauchesco Ricardo Güiraldes*) from the Secretary of Social Development, Production, Tourism and Culture (*Secretaría de Desarrollo, Producción, Turismo y Cultura*), both of them under the dependency of the town of San Antonio de Areco authorities (Acevedo et al., 2020). This program aims to improve the social, cultural, and economic use, and enjoyment of cultural heritage by recovering the value of use and exchange. Furthermore, it gives orientation on how to proceed with rehabilitating cultural heritage by means of adequate behavior in open and built social spaces. In line with these objectives, the program is part of the provincial Territorial Ordering Code (*Código de Ordenamiento Territorial*) existing since 2014 and takes into account the Sustainable Development Goals (*Objetivos de Desarrollo Sostenible, ODS*), to which the district of San Antonio de Areco subscribed to (Arellano, 2017).

The project of enhancement and value revitalization was born when a historical document about the cannon was found at the Gauchesco Museum Ricardo Güiraldes (*Museo Gauchesco Ricardo Güirlades*), immediately after, the program authorities were informed. From there forward, a bibliographic and document investigation started to inquire and ascertain facts regarding the cannon and its history. Also, historians and researchers were consulted. The new information recovered increased the interest in this artillery piece which had been forgotten at the museum property for decades. As a result, the district authorities developed the enhancement and value revitalization project of the cannon with the goal to highlight a national historic event in which the local community played an important role. The descendants of the residents who took part in those historical facts still live in San Antonio de Areco, so this project also allows them to recover their identities, symbolized and materialized by the EB 1789 cannon (Acevedo et al., 2020).

As the cannon was lying in the museum park, it was necessary to establish a dialogue and negotiate with the director and personnel for the practical implementation of the project (Fig. 2). Firstly, the enhancement arose from the dialogue between the parties, whose intention was to move the cannon from the museum park to the original place where it was found buried, where the Credicoop Bank now is with the objective to give visibility to the object (Fig. 1).

The museum was founded in 1938, and it was declared a national historical monument in 1999, which made it a tourist and cultural attraction for the locality. The institution was named *Don Ricardo Güiraldes*, a renowned writer, and it focuses on featuring everything related to his personal life and literature pieces, particularly the book *Don Segundo Sombra*. As such, the museum also exhibits historical past expressions related to the Argentinian *gaucho* figure, his customs, traditions, belongings, and way of life. Moreover, the museum constitutes itself as an open

Fig. 2 Cannon in the *Parque Criollo and Museo Gauchesco Ricardo Güiraldes* (June 2019)

space where diverse cultural activities are carried out including workshops organized by our project to recover past technology. Instead, this is a space for the graduation and capacity of the students of cultural heritage management careers at the National University of San Antonio de Areco (*Universidad Nacional de San Antonio de Areco, UNSAdA*) (Acevedo et al., 2020).

Brief Review of the EB 1789 Cannon

The cannon is an artillery piece made of melted iron engraved with the denomination EB 1789 in the gun trunnions. The initial EB refers to the Ehrendals Bruk brand manufactured by the Ehrendals factory, Gnesca district in Sweden, the number points likely to the year of fabrication (Fig. 2). The object originally was found lying under the present property of the Credicoop Bank in the historical center of the town, in front of the Ruiz de Arellano plaza and was recovered during an excavation at the beginning of the twentieth century (Acevedo et al., 2020; Arellano, 2017).

The reason why it was buried and concealed is still unknown. Nevertheless, information available on the territorial register was made during the eighteenth and early nineteenth centuries suggests that nailing cannons down and using them as milestones was a frequent social practice in Spain as well as in its colonies, as in the province of Buenos Aires at that time. They were also used as boundary stones for urban zones or remelted to fabricate ammunition. Another line of reasoning mentions that funding of cannons completely sedimented can be related to their own collapse after functional obsolescence aggravated by environmental factors (Archivo Histórico de la Dirección de Geodesia, Catastro y Tierras de la Provincia de Buenos Aires, 2022).

It is known that in 1927 the cannon was donated by the local inhabitants to the Provincial Museum Complex Enrique Udaondo (*Complejo Museográfico Provincial Enrique Udaondo*) at Luján, a district in the province of Buenos Aires, for its exhibition. Once the Creole Park and Museum Ricardo Güiraldes (*Parque Criollo y Museo Gauchesco Ricardo Güiraldes*) was inaugurated in 1938, it requested a return to the new museum where it remained for 80 years. Additionally, letters exchanged by the authorities of both districts stored at Creole Park and Museum Ricardo Güiraldes (*Parque Criollo y Museo Gauchesco Ricardo Güiraldes*), detail all activities of extraction, cleaning, and donation of the cannon as well as its return (Arellano, 2017).

An inquiry at the Belgrano Popular Library indicates that the EB 1789 cannon was originally posted at Areco Fort in 1727. Also, an article published in the local journal La Gaceta N°40 on November 7 of 1937, titled "A gaucho type museum will be created at the Park Güiraldes." It mentions that two small cannons will be placed on the embankment together with the old cannon from the Areco Fort (Arellano, 2017).

On the other hand, various archaeologists visited the historical area and other historical points of interest in San Antonio de Areco, the Creole Park and Museum Ricardo Güiraldes (*Parque Criollo y Museo Gauchesco Ricardo Güiraldes*), as part of their research on historical artillery pieces. The technical data collected indicates that the EB 1789 cannon has similar manufacturing and material characteristics such as a hundred iron and bronze pieces found in the abovementioned museums and which correspond to a collection of objects from the seventeenth to nineteenth century. This set stands out for including muzzle-loading cannons of different calibers, designs, and manufacturing marks engraved on the body and trunnions from Spain, England, and Sweden. In addition, they represent several fire mouths of possible naval use that would have been used in various war contexts in the province of Buenos Aires (Ciarlo, 2017).

Regarding the artillery piece, the cannon was manufactured in Europe through the implementation of a technology that is characteristic of a historical moment in which the French Revolution took place. In fact, all the artillery that was used in the context of the Río de La Plata was brought from Europe to be used in the colonies (Ciarlo, 2017).

Thanks to the research carried out, it is important to highlight that the cannon as a material object represents a part of the history of San Antonio de Areco, linked to the defense of the Areco Fort and is part of a set of artillery pieces used by the neighbors and local militia who participated in war actions in the province of Buenos Aires in historical times.

It is necessary to mention that the town was constituted as an obligatory post on the Camino Real to Alto Perú, which linked the port of Buenos Aires with the port of Lima, Peru. The Areco post was on the direct road to the city of Córdoba (Central Argentina) which was central on the route linking the Viceroyalty of Río de La Plata with the Viceroyalty of Alto Perú. The post was established as an important commercial and civil route and for the movement of militias and military troops (Garavaglia, 2009). San Antonio de Areco is thus of special interest to the national history.

In this regard, there is a series of historical documents archived at the Hipólito Vieytes Municipal Historical Archive (*Archivo Histórico Municipal Hipólito Vieytes*) that testify to the local historical events that occurred, as well as the names of the neighbors who were part of the militias and who served as the basis for the investigation to the enhancement of the object (Arellano, 2017). In this sense, the cannon has a symbolic meaning with respect to the specific context of use.

In addition, the cannon is a symbolic referent, with a material and immaterial heritage value for the local inhabitants, which in turn becomes a tangible relic that unites the present generation with past residents of San Antonio de Areco, protagonists of historical events and laid the town foundations for the formation.

The Local Context

The Project ArqueoLab-UBA has compiled a large amount of information from all the work carried out over the last few years, which includes reports presented in the county, documentary research, contribution of historians, tasks carried out by professionals archaeologists, as well as fieldwork, diffusion activities, and meetings with authorities and coordinators of the Areco County *(Municipalidad de San Antonio de Areco*) and the Ricardo Güiraldes Gauchesco Museum (*Museo Gauchesco Ricardo Güiraldes*). Thanks to this, it was possible not only to contextualize the implementation of the management project but also to understand and respond to the needs of the community.

It is important to point out that the town of San Antonio de Areco began its touristification and local patrimonialization in the 1930s when part of its society and the provincial and local policy support created the Creole Park and Ricardo Güiraldes Gauchesco Museum (*Parque Criollo y Museo Gauchesco Ricardo Güiraldes*). The objective was to honor the *gaucho* figure and the deceased author of the book *Don Segundo Sombra,* Ricardo Güiraldes (1886–1927) (Pérez Winter, 2015).

In that same decade, the event that is currently considered one of the most important attractions of the town, the Traditional Festival, was established, and later, various provincial and national regulations placed San Antonio de Areco as the official national city of the celebration (Pérez Winter, 2016a, 2018).

In fact, the town has an important hotel and gastronomic infrastructure to support the implementation of tourism-heritage policies since 1938, which were accompanied by speeches and provincial and national representations that have configured it as the 'Cradle of Tradition' ("*Cuna de la Tradición*") (Pérez Winter, 2016a, 2018). These policies have fostered the development of very strong heritage awareness.

To be able to understand the patrimonialization and touristification processes of the town, it is pertinent to mention the influence of the Güiraldes family, who belonged to the landowner-conservative sector, and the descendants of the founding families of the city, who had a good political and economic position in the Buenos Aires Pampean Region as well as in the rest of the country (Pérez Winter, 2016b). The family's interest was focused on promoting values of love for the country by evoking and disseminating elements associated with a version of the regional tradition that they were shaping through several generations (Garavaglia, 2009).

In this way, they promoted actions that ended up consolidating the so-called tradition southern (*surera*) in the locality, a term that refers to the historical boundary as the southern limit of the Buenos Aires province, before the incorporation of the Patagonian territory. This tradition is characterized by claiming equestrian skills, dances, clothing, and crafts that are typical of the Pampean Region (Pérez Winter, 2016a, b).

Furthermore, the town was established as a city of great importance from very early times (eighteenth century) as a boundary point, with the creation of a fort to defend the territory (Garavaglia, 2009). From there, the development of the town positioned it as a relevant historical, economic, and cultural center in the province.

Likewise, this local development was promoted during the first half of the twentieth century thanks to the nationalist and conservative discourses, in which the true Argentine tradition was represented by the figure of the countryman. This situation brought about the valorization of folkloric expressions (music, dances, legends, myths, etc.), which were conceived by the national authorities as to the legitimate elements of the *criolla* (creole-like) identity, on which the idea of the country was being developed (Pérez Winter, 2016a, b).

As a consequence of this process, it is perceived that the local traditional identity is linked to the historical idiosyncrasy, which includes events such as the founding of the town, who were its first settlers, the defense of the sovereignty of the territory in the battles of Independence, (which started in 1810), the importance of the local territory in the national context and the permanence of traditional families and their descendants. The pride of history and the sense of belonging on the part of the local inhabitants can be perceived in daily life.

In 2009, this locality suffered an outflow of the Areco riverbed with the flooding of the historic center, impacting heritage monuments. Different institutions where large-scale cultural materials are deposited, such as the Ricardo Güiraldes Gauchesco Museum (*Museo Gauchesco Ricardo Güiraldes*), suffered severe damage. This situation generated among the local citizens a particular concern to safeguard the material cultural heritage, so much so that it motivated the creation of the National University of San Antonio de Areco (*Universidad Nacional de San Antonio de Areco, UNSAdA*), in which the Bachelor's degree in Heritage Management is offered. For all the above, the implementation of both the project and the management was facilitated in some way by a favorable local context. It is necessary to highlight the joint will to work between institutions, scientists, and the community to develop the project, through the establishment of a collaborative network, in which negotiation between the parties involved was essential to guarantee good practices.

Likewise, certain elements of the local context were used to apply specific strategies for the project to be successful and the active participation of the community called by the Secretary of Planification (*Secretaria de Planificación*). The visibility of the enhancement of cultural heritage took place during a week in the month of October, in which two local and relevant events were commemorated: the 81st anniversary of the creation of the Ricardo Güiraldes Gauchesco Museum (*Museo Gauchesco Ricardo Güiraldes*) and the 289th commemoration of the foundation of the town of San Antonio de Areco. In addition, the cannon was removed from the museum property to be reinstalled in front of the main plaza in the historic center. In this way, it was not only incorporated into the tourist circuit in the historical area but also had greater visibility due to the new location because national and international tourists came to this area, and it is an attraction to convene local festivals.

Integrated Management of Cultural Heritage Project

Management of cultural heritage involves the planning of strategies and actions in accordance with social needs and demands to enhance and value the revitalization of cultural assets. In this way, they receive a new social purpose and promote simultaneously a more adequate way to preserve them (Ballart Hernández & Tresserras, 2001; De Carli, 2018). However, it must be taken into account that during the management process, certain unexpected issues (explained further below) may arise in the implementation of the actions, leading us to rethink and reorganize the tasks carried out. In this case, these issues included the negotiations and challenges of decision-making and were part of the articulation process with the different actors involved during the enhancement and value revitalization of the cannon. Our goal in the management process was to make known the good practices of protection of cultural heritage based upon concrete and experiential strategies of research, conservation, and dissemination that would help promote heritage awareness in the community.

The starting assumption was that the cannon of the eighteenth-century constitutes a heritage object valuable for the community because it refers to the local history. However, there is a lack of knowledge regarding this object as well as the strategies for its protection.

Therefore, the management implied the planning of a series of actions linked to the articulation between social actors:

- Generation and dissemination of knowledge gained from scientific research on the cultural object
- Development and implementation of a plan for conservation and restoration which contemplated immediate intervention as well as long-term ones
- Providing guidance and collaboration with the community regarding enhancement and value revitalization activities
- Making known the legal framework and good practices for heritage protection
- Encouraging active participation of all the local inhabitants involved by means of experiential strategies to evocate and strengthen the collective memory

Our task was to record the process as anthropologists, without intervening in local political decisions and without making value judgments. In this way, management was constituted from a comprehensive view of heritage.

Cultural Heritage: Cultural Resource and Heritage Product

Over the process in which a cultural resource becomes a heritage product for consumption, it is frequent that challenges and unexpected situations arise, like direct and indirect consequences to the occasional or massive responses to the cultural offer proposed. In this case, it is essential to bear in mind the following sustainability strategies to make the project viable in the long term: a constant dialogue between social actors, monitoring protocols of control of the cultural heritage product, and the continuous dissemination of information generated in all the areas of the local community. Cultural resources present a particular dynamic related to the historical and social context so the permanency as a cultural product will directly rest on abovementioned strategies (De Carli, 2018; Palavecino, 2013).

Particularly noteworthy is that the primary function of heritage assets is not and cannot be exclusively economical. For this reason, the initiatives that include the cannon as an economic resource must be analyzed without altering its essence and the right of the community to access and enjoy it. As Bonfil Castro (2003) correctly pointed out, heritages have different values: first, the documental value of a certain period of the history of men and society; second, the value of identity and social cohesion of the group that creates it. Finally, of greater and irrefutable importance in the current times is the value of a tourist attraction that produces wealth and which should never replace the supremacy of the former.

The uses that a society makes of its heritage are linked to the transformation processes of the present that influence the assets considered heritage references. Each society creates, uses, recreates, and converts heritage into a resource according to the interests of the groups and the social, political, economic, and ideological contexts. Each society gives to cultural assets the social meanings that it considers at a certain time and as patrimonial referents these assets operate as cultural symbols and for change (Arévalo, 2010). Currently, heritage is a resource that is increasingly used socially, and as a social construction it is constantly being reformulated. As such, it has an exchange value that has made it a product of new consumption habits in society. It is an instrument of social and economic development, but also a means of identification, that is, it is a social-cultural resource that, as a heritage product, also allows economic development in local areas (Arévalo, 2010; Prats, 2005).

Cultural Heritage in the Management Process

Cultural heritage is a complex issue due to the multiple concepts and definitions that surround it. A common idea is that cultural heritage is something valuable that must be preserved and that shares the following attributes: it is dynamic, it has the potential to be conveyed (legacy, communication, dissemination, and education), it has an identity potentiality because it supposes relationships of belonging and relates to a specific socio-historical context, memory, and tradition (Acevedo, 2017; Acevedo et al., 2009; Mancini et al., 2016; Fontal, 2003). Heritage as a cultural product of human creativity is conceived, transmitted, modified, and optimized in accordance with each individual and his generation (Acevedo et al., 2009; De Carli, 2018; UNESCO, 1972).

Although cultural heritage is not a universal phenomenon, it is deeply linked to the context of the society that produces it (García Canclini, 1999; Manzini, 2011; Ramos, 2011) because it is generated by the local community in a given time and space and acquires meaning due to social common agreement (De Carli, 2018). Additionally, it is part of history materialization because it transmits a message through material support (Ballart Hernández & Tresserras, 2001).

In this sense, cultural heritage is part of the collective memory of a community. It is the memory linked to belonging to social groups and therefore, it is shared and contextualized (Halbwachs, 2004). The collective memory preserves the social heritage and serves as a reminder to keep it alive. It also implies a system of representations in constant dialectic between the past, the present, and the future (Arévalo, 2010).

The community is responsible for its care and as such requires to be investigated, protected, and preserved. Likewise, heritage as a local cultural product acts as a factor in human development and the community has the right to make sustainable use of its heritage resources (De Carli, 2018).

Thereby, the relevance to making visible interests and agencies involved in the process through communication, collaboration, and interdisciplinary work between researchers, public administration, and the general community for the protection of cultural heritage should be a shared duty.

In the course of the last decades, most American countries have modified their relationship and construction with respect to the recovery and preservation of their cultural heritage. This is a consequence of the processes of recognition and visibility of the original communities impacted by colonization and of the legitimization of the historical process of organization of the independent republics (Berberián, 2009; López et al., 2016).

Thus, the concern for the preservation of heritage, which in principle was of a local nature or responded to bilateral agreements between countries, is currently projected through resolutions and recommendations of official agencies, governments, and international organizations (Berberián, 2009; Pozzi Escot, 2014). The mentioned processes caused concern and led to legislative actions for the sanction of protection laws, which has had a direct impact on scientific and professional practice (Berberián, 2009; López et al., 2016; Ratto, 2013).

Initiatives to externalize recommendations through charters or conventions, which are ratified by different countries, have emerged from international organizations such as the United Nations Educational, Scientific and Cultural Organization (UNESCO), the Organization of American States (OAS), and the various academic meetings of specialists. These recommendations must be put into practice and disseminated to the community by the professionals based on open science and then be integrated and disseminated by the aware communities themselves.

In Argentina, we have the National Law No. 25,473 for the Protection of Archaeological and Paleontological Heritage ("*Protección del Patrimonio Arqueológico y Paleontológico*"), in addition to the different provincial laws in force. Within this framework, each province dictates its own regulations and policies regarding the protection of heritage, although they must be in accordance with national laws. However, work is being done to develop joint management policies between the different provinces in order to unify criteria for more effective protection.

In this context, national and international laws about cultural heritage contemplate the necessity to encourage an interdisciplinary investigation, conservation, assistance, and protection work along with the development of new actuation tools (ICOMOS, 1990).

The articulation between heritage, identity, and conservation is a collective right that often the state cannot guarantee in a concrete and effective way, being relegated to the expression of laws or regulations with almost no application (Ratto, 2013). That is why it is necessary to propose a change in the layout of management policies.

Thus, the creation of spaces for the exchange of knowledge and conservation in laboratories, universities, research institutes, museums, and other places where cultural knowledge is studied, preserved, and safeguarded, is of greater importance for patrimonialization processes linked to identity. It is also important that theoretical perspectives on these issues can be discussed in these spaces. Consequently, social interrelationships take a leading role because what is unknown cannot be protected and therefore it is of vital importance to generate scientific information that supports and serves for safeguarding, enhancement, and value revitalization and/or patrimonialization processes.

For this reason, the management plan for the enhancement and value revitalization of the cannon was guided by the three key points of the Project ArqueoLab-UBA: to do research, to conserve, and to disseminate. Investigation and archaeometric studies of heritage assets allowed the production of knowledge that worked as a base for the designing of a short and long-term conservation plan. In addition, they were communicated to the community using different experiential devices.

It is important to mention that our research group develops the guidelines for comprehensive management of cultural heritage following the guidelines proposed by Criado Boado (1996a, b), Pozzi Escot (2014), De Carli (2018), among others:

- Not only assets are managed, but we must operate with knowledge and intellectual and symbolic values; therefore, the management work must be respectful and must guarantee its preservation.
- It cannot focus on technical exhibitionism, it must always integrate the historical dimension that makes up the identity.
- It must articulate knowledge and its practical utility, which implies understanding all interventions on heritage as a unit. This will enable future profitability as a cultural resource and will be recognized as a heritage product.

This way of working includes, in our professional practice, conceiving and designing all projects and actions on heritage as an integrating cycle, in which all social actors contribute at all stages of the work. This makes them aware of the results derived from research through Participatory Science and allows them, in turn, to incorporate the historical significance of the treated heritage and recognize it in the identity of the local actors involved.

Regarding the link to science and society, our Project ArqueoLab-UBA proposes a dual work modality that refers on the one hand to Visible Thinking and on the other to Participative Science. They articulate together to promote critical and dialectical thought among citizens and academics.

Visible Thinking is a methodology that arose from Project Zero of the School of Education at Harvard University. It is based on certain ideas and principles that suggest that thinking occurs primarily in our brain, and is invisible to everybody and even to ourselves. Efficient thinkers make their thinking visible by externalizing it through speaking, writing, drawing, or another method, as well as documenting a process for later reflection (Ritchhart & Perkins, 2008).

Thereby, Project Zero developed several routines and strategies to promote an expressive and open mentality so that curiosity, attention, and imagination can be arranged for good thinking practice. Such characteristics do not refer so much to abilities owned by a person but rather to how a person invests those abilities in competencies in the social media where it also interrelates itself with others and with the social environment (Ritchhart & Perkins, 2008).

In this case, the strategies to make the thinking visible were applied to all the social actors involved during the negotiations, enhancement, and value revitalization of the cannon. This allowed us to know opinions, interests, and conditioning factors as well as to connect with and get feedback on expectations, motivations, and feelings.

Participative Science is a way to produce science that seeks to generate knowledge among scientists and interested citizens, working willingly and collaboratively. This work modality allows us to register concerns and gather systematic knowledge from community members as potential citizen scientists (Acevedo, 2018; Acevedo et al., 2018a, b, 2020). In this way, Participative Science convenes and assists in the investigation process, promoting dialogue and reflection (Acevedo et al., 2018a, b; Fernández & Torres, 2016; Torres & Fernández, 2016). "Citizen scientists" are those no professional scientists who participate willingly in the recollection, analysis, and dissemination of data in scientific projects (Torres & Fernández, 2016). Scientific problems usually arise from academics who seek the collaboration of citizens. However, sometimes "citizen scientists" are the ones who start a research project systematically and independently, searching later to connect with specialists in academia (Acevedo et al., 2018a, b; Fernández & Torres, 2016; Torres & Fernández, 2016).

On the basis of this experience, the aim is to link the diversity of necessities and knowledge to work together through the commitment of "citizen scientists," training them in specific activities, the work, and the handling of produced data, among many other things.

This way to make science is a didactical mechanism to disseminate knowledge built by means of shared research. Dissemination is a methodology that puts in action via graphical, oral, and virtual participative communication, making visible all the social actors involved in the communication act that is scientists, society, or citizen scientists (Acevedo, 2017). Allow for the engagement of mentioned practice increases productivity and scientific efficiency, possibilities the engagement of diverse social actors in the democratization of knowledge, and strengthens science-society relationship, thus replying to social demands (Fecher & Friesike, 2014).

The generation of knowledge through research, conservation, and dissemination projects for the enhancement and value revitalization of cultural heritage based on articulated social practices, builds and spreads the heritage consciousness of all actors involved. In this sense, the study of heritage from anthropological and archaeological investigations conveys and gives visibility to these processes based on the creation of exchange spaces that optimize the link between society and science.

Thanks to the communication between the community and researchers, this articulated space makes possible the initial dialogue about safeguarding cultural heritage in vulnerable conditions to, later on, enhance and value revitalize it for the patrimonialization process.

Patrimonializations begin with cultural referents selection considered part of the idiosyncrasy of the community. They propose certain objects for enhancement and value revitalization, and that includes the meaning of its cultural past and present.

In concordance with this, heritage activation is conveyed by the collective memory of the social group that was selected to legitimize an imaginary past, building his present memory. In our case study, patrimonial activation constitutes a dynamic involvement that recognizes places and people that took part in local historical events. Hence, the social actors are recognized and prompt the generation of a cultural link promoting social and economic planning.

Furthermore, this process comprises that certain discourses are appropriated by individuals that transform and resignify them in a historic continuity-discontinuity to give place to something new and particular. The present appropriation is linked to local identity, which at the same time, relates to the historic idiosyncrasy. Namely, the founding of the town, the territorial defense, and the traditional families, among others.

To summarize, the cultural material product of the collective memory is built upon the interaction of members of the community and institutions on the basis of a shared representation of the historical past. Thereby, it created an official discourse that may, sometimes, generate conflicts regarding representations and interactions, and also provoke conflicts around themselves (Acevedo et al., 2009; Mancini et al., 2016; Fontal, 2003).

The process of enhancement and value revitalization prompts the necessity of management plan and dialogue among all social actors involved. It also highlights the necessity to contemplate immaterial and material resources to be able to valorize cultural heritage for future generations.

Thus, cultural management cannot only and uniquely correspond to specialists of the past but to the local community that must be involved as well. For it is them who demand diversity of participation, cultural entertainment, expression, learning, and economic development spaces.

Actions and Devices for a Practical Proposal

Understanding that all heritage management must be integrated, two levels of work were developed. On one hand, the activity of our discipline and on the other hand, the strategies applied with and for the community. Mentioned levels imply a joint articulation with all the social actors involved.

In general terms, integral management presents the following actions:

- Social selection of cultural resources for safeguarding
- Social participation in elaborating a research, conservation, and dissemination plan that contemplates documentation and registration of the activities
- Intervention of the cultural asset from preventive conservation and restoration
- Dissemination and communication to increase the awareness of people
- Enhancement and value revitalization and patrimonialization activities
- Planning a sustainable use of the cultural asset like a cultural product for cultural consumption and its preservation

We have two interconnected devices for a workable proposal: one of them is our Cultural Lab as a workspace and the other is our Mobile ArqueoLab-UBA which has all the material resources to carry out the activities (Acevedo et al., 2021).

Regarding the first one, for several years, both in the public and private sectors, cultural labs have been developed as a way of connecting socially and seeking solutions to the needs and demands within the framework of current social transformations, in order to strengthen local development.

Usually, a laboratory is understood as a space equipped with the necessary means to carry out scientific research. Nonetheless, a cultural laboratory implies other characteristics especially when cultural and heritage assets are studied. Not only are they composed of their own materiality, but they are also endowed with meanings, symbolism, and belief. Our laboratory definition presents a new meaning, a new way of creating collectively. The Cultural Lab is understood as a space of analysis where hypotheses about the life history of cultural assets and their conservation are generated and contrasted. These assets are thought of as material media for information and symbolic meaning to disseminate to the general public and as part of the collective memory that identifies a society (Acevedo et al., 2018a, b, 2020, 2021).

Our Cultural Lab was created as a space where scientists and the community were brought together with the purpose of interacting, promoting, and facilitating local development and culture. Moreover, the lab is a medium of connection to advise, manage, and accompany different social groups, organizations, and institutions to establish a collective and collaborative network through innovative projects.

Our second interconnected workplace, the Mobile ArqueoLab-UBA, works as an implementation device and as a medium for remembrance and strengthening collective awareness. This motorized lab, scientific and social open was designed specifically for the investigation and dissemination of knowledge in educational areas as well as for the general public. The laboratory is versatile and it is equipped and conditioned to be able to carry out activities in any place where it is required.

Activities planned and implemented from the Mobile Lab have the objective of presenting in an educational way the different stages of an archaeological investigation through the organization of different workstations: introduction of the basic concepts of archaeology, excavation and fieldwork, preventive conservation, material analyses at the laboratory, digitalization of archaeological materials and closure of what has been learned using didactic material. All these workstations are designed so that participants can interact in the different tasks with scientists in a framework where learning is enhanced by a joint construction and dissemination of scientific knowledge and cultural heritage protection (Acevedo et al., 2021; Ávido et al., 2021).

During the commemoration festivities organized in October 2019, the mobile lab operated at the main "plaza" (i.e., square and park) of San Antonio de Areco (Fig. 3). A tent was set up to display the cannon, so visitors could know its history and actively participate in scientific activities alongside the professionals. Among the tasks carried out, we can mention digitalization of the cannon through a portable scanner, microscope observation of the cannon and diverse samples, visual material exposure, hand graphic material over, and cultural heritage talks, among many others. All these activities were destined to facilitate interaction with the people and respond to their concerns (Fig. 4) (Ávido et al., 2021).

Presented activities were conceived as a part of an intertwined work cycle where each one was interrelational with the preceding and the following so that as they are being developed and put into action, the knowledge displayed is reinforced.

Likewise, they were aimed at the local community as a whole (authorities, personnel of the museums and local archives, residents, inhabitants of the founding

Fig. 3 ArqueoLab-UBA Mobile Stand at the 289th. Commemoration of the Foundation of the San Antonio de Areco Town

Fig. 4 Community of San Antonio de Areco carrying out work on our Mobile ArqueoLab-UBA at the 289th. Commemoration of the town's foundation

families that took part in historical events, teachers, and students from all education levels, among many others and provincial, national, and international authorities.

Autoassesment of activities performed also constituted a reflection stage to improve the project in its workable instance according to the application context.

Moreover, it is important to note the participation at the festivities of different authorities from the National Commission of Monuments, Places and Historic Assets of the Government Secretariat of Culture of the Nation, Ministry of Education, Culture, Science, and Technology (*Comisión Nacional de Monumentos, de Lugares y de Bienes Históricos de la Secretaría de Gobierno de Cultura de la Nación, Ministerio de Educación, Cultura, Ciencia y Tecnología*), all professionals linked to cultural heritage protection that also gave a conference about heritage protection at the Gauchesco Ricardo Güiraldes Museum (*Museo Gauchesco Ricardo Güiraldes*) in the framework of a monthly series of talks called *Charlas de Pulpería*.

Cultural Management Challenges

As it has already been mentioned, despite a favorable local context for the development of our work proposal, different issues and challenges arose that led us to rethink and review the management plan. Reflecting on it, we understood that it was indispensable to redesign the plan in order to give satisfactory responses to particular necessities and demands.

Initially, the Secretary of Planification (*Secretaría de Planificación*) asked us to present a report arguing the importance of the intervention of scientists in the enhancement project. Said report was attached to a set of documents presented to the local Town Hall Council ("*Consejo Deliberante*") to justify the project approval. In this sense, it is outlined the relevance of complying with all the regulations and administrative procedures, respecting political decisions and bureaucratic acts time required.

At the same time, the final stage of the activities coincided with the October elections, a situation that generated a particular atmosphere between different political factions in the county.

Also, our actions as intermediaries in the communication between different social actors was crucial due to dissenting views and actions over cultural heritage that make the interaction between themselves very difficult. Such was the case with the personnel of the Ricardo Güiraldes Gauchesco Museum (*Museo Gauchescho Ricardo Güiraldes*) and people from the county.

Another situation in which we had to intervene was the negotiation for the manufacture of new support for the cannon to place it in its new location. Happily, the negotiation was a pleasant dialogue with the Credicoop Bank architects and the planning secretary. It was agreed that the new support was adequate for the design and functionality of the heritage asset, always appealing to good practices of heritage protection. Finally, the support was manufactured in wood and melted iron according to an original model of naval cannon support, instead of a modern design as the one suggested at the beginning. We participated also in the negotiation to enter reliable information in the QR code placed on the glass plate next to the cannon.

It is worth mentioning that the cannon support (a wagon axis more than 100 years old) was not the original one but served to expose the object during all the time in which it remained on the museum property (Fig. 2). It was destroyed during the cannon transportation to the main plaza. Prior to this, it was relevant to digitize and draw it when it was still in good condition, before its destruction. Immediately, we realized that it was necessary to generate a project for its recovery and restoration, which is underway.

At last, the cannon with the identification plate was placed on its new support in front of the Credicoop Bank, where it currently remains and was inaugurated in the course of the town's anniversary week in 2019, during a ceremony that was attended by the local authorities (Fig. 5).

One of the challenges of the management in this context was the change of local authorities one month after the enhancement and value revitalization of the cannon. With the new municipal government, QR code access was lost and so was all the information that was contextualized and provided.

Thanks to the warning of our 'scientists citizens', monitoring was performed at once and we succeeded in rectifying that problem by creating a new virtual space on the municipal website. From this situation forward, we realized the need to start a joint project for the creation of a digital repository of the Areco district heritage according to the National Law No. 26,899 of Institutional Digital Repositories of

Fig. 5 Cannon in his new location after the enhancement and value revitalization and the ArqueoLab-UBA working group with the authorities of San Antonio de Areco

Open Access (*Repositorios Digitales Institucionales de Acceso Abierto*) by opening a file at the Town Hall Council (*Consejo Deliberante*).

Later, both the QR code and the glass plate with the information were vandalized due to the lack of surveillance on the public street. The solution agreed to this problem was to make a new plate in a resistant material and to maintain control over the cannon, which is now a town monument.

These challenges led us to deeply reflect on the significance of developing an integrated and articulated workable proposal that considers the named contingencies by applying specific sustainability strategies based on the local context. Thus, the safest protection agents for cultural heritage are the town citizens involved in the patrimonialization process through Participatory Science activities.

Science Communication and Participatory Science to Preserve

The challenge of articulating public communication of science with participatory open science is of the utmost importance in the process of social appreciation of cultural heritage. Therefore, the act of communication, the awareness and a continuous monitoring of the assets after the process of patrimonialization and enhancement must be taken into account in the integral planification. In this way, all the actors can correctly communicate the information generated. However, this does not guarantee the emergence of unexpected situations due to the lack of fluidity of the information channels and the care of the heritage recovered by the citizens (Acevedo et al., 2021).

The simple act of communicating supposes a social relationship since it implies that every communicative process includes the existence and production of shared and agreed information (Conforti, 2013; Grimson, 2000). Hence, the actions of communicating require an effort of articulation of the interlocutors to generate a shared code, since it is impossible to think of communicating without an active community to interact with. An example of this situation was the preparation of the QR code. This information was agreed upon and built jointly between scientists and citizens, generating its dissemination by all social actors. The bond with the "citizen scientists" was active and heterogeneous since they have diverse capacities and knowledge about the sense of belonging and recognition of the local cultural heritage.

Noteworthy, doing science from local practices is a field under construction, where the roles of citizen scientists and citizen scientists are still being defined within the framework of the Open Science Paradigm.

Consequently, the activities of public communication of science from Participatory Science acquire a new meaning in the sharing of processes and in the validation and legitimization of scientific-academic-technological activities. The objective of this process is that citizens take ownership of it. The complex interactions, the multiple actors, and the cultural products can only be defined by making explicit, in each case, the frame of reference.

Concluding Remarks

Negotiations between the different social actors and institutions were constant throughout the enhancement process and were favored by cordial bonds and fluid communications. Our intervention was always guided by the objective of guaranteeing good practices, respecting political decisions, and improving communication between the actors to favor a good dialogue. Moreover, our participation as anthropologists/archaeologists, and not as heritage managers, respected the needs of the community and political decisions. In other cases, they only met partial agreement.

Understanding that contingencies and conflicts usually present naturally during any management, it was important to develop strategies and tools to resolve them in an optimal way for all sectors of the community that participated in the process, especially since it was a diverse audience. For this, it was necessary to know the context in which we were working, have a good dialogue, offer viable alternatives to provide solutions, among others.

At the same time, as cultural management is itself a process that constantly requires monitoring and adjustment of what is being done, we were deeply considering each decision as we were getting along and new situations arose. Thereby, the project became a workable proposal at the same time it was put into action.

Acknowledgments We thank the authorities of San Antonio de Areco, the staff of the Ricardo Güiraldes Gauchesco Museum (*Museo Gauchesco Ricardo Güiraldes*), the Urban and Rural Heritage Protection and Rehabilitation Program (*Programa de Protección y Rehabilitación del Patrimonio Urbano y Rural*), Dr. Alicia Tapia, Luis Lupini, Ariel Arellano, Andrea Vigil, and Santiago Amondaray. We appreciate the contribution of the companies Campisi Cranes, Zamin Lumber, and the Credicoop Bank in the enhancement and value revitalization process. Our gratitude goes to the entire community of San Antonio de Areco, province of Buenos Aires, Argentina, for the commitment and heritage awareness and to all those who came to our Cultural Lab and to those interested in the protection of cultural heritage for feeling it part of their own identity.

References

Acevedo, V. J. (2017). *Modelando La Vida. Puesta en valor de los saberes y costumbres en la producción, uso y consumo de la Alfarería Andina del Pasado y del Presente. Plan de Trabajo Ingreso CIC 17 Temas Estratégicos y Tecnológicos*. Unpublished manuscript.

Acevedo, V. J. (2018). *Habilidades del Pensamiento, percepción de los sentidos y pensamiento visible. Con especialidad en desarrollo de estrategias pedagógicas en educación e investigación en ciencias sociales*. Proyecto Zero, Harvard University. Unpublished manuscript.

Acevedo, V. J. (2019). Espacios interdisciplinarios integrales en los estudios arqueométricos. Proyecto ArqueoLab-UBA: Investigar, Conservar y Difundir. In *Libro de Resúmenes del XX Congreso Nacional de Arqueología Argentina* (p. 1144). Universidad Nacional de Córdoba.

Acevedo, V. J., Espinoza, A. L., López, M., & Mancini, C. (2009). *La feria de las Alasitas de Parque Avellaneda, Ciudad de Buenos Aires, y sus vinculaciones con la tradición andina de miniaturas. Temas de Patrimonio Cultural N° 24: Buenos Aires Boliviana, migración, construcciones identitarias y memorias* (pp. 249–264). Comisión Nacional para la Preservación del Patrimonio Histórico y Cultural.

Acevedo, V. J., Staropoli, L., Ávido, D., & Vitores, M. (2021). Proyecto ArqueoLab-UBA, un encuentro con la ciencia participativa. Sesión Temática C Prácticas vivenciales y dispositivos digitales para la construcción y comunicación de la ciencia abierta y participativa en la investigación y preservación del Patrimonio Cultural, Paleontológico y Natural. In *Actas del 3° Congreso de Ciencia Abierta y Ciudadana (CIACIAR)*. Universidad Nacional de La Plata. http://sedici.unlp.edu.ar/

Acevedo, V. J., Staropoli, L., Herrera, N. V., Ciarlo, N., & Landa, C. (2018a). ArqueoLab-UBA, Un encuentro con la Ciencia Participativa. In *Actas del II Congreso de Ciencia Abierta y Ciudadana*. Universidad Nacional de San Martín. www.2ciaciar.org

Acevedo, V. J., Staropoli, L., Herrera, N. V., Ciarlo, N., & Landa, C. (2018b). ArqueoLab-UBA, Laboratorio Móvil para la Ciencia Participativa. In *Actas del II Congreso de Ciencia Abierta y Ciudadana*. Universidad Nacional de San Martín. www.2ciaciar.org

Acevedo, V. J., Staropoli, L., Herrera, N. V., Ávido, D., Vitores, M., Soto, D., et al. (2020). Proceso de patrimonialización de un cañón del siglo XVIII en San Antonio de Areco, provincia de Buenos Aires, Argentina. *Cuadernos de Marte, 19*, 165–208. https://publicaciones.sociales.uba.ar/index.php/cuadernosdemarte/article/view/6281

Acevedo, V. J., & Staropoli, L. (2021). Desafíos en la gestión e investigación del patrimonio cultural. La patrimonialización del cañón EB 1789 en la localidad de San Antonio de Areco, Provincia de Buenos Aires. In *Libro de Resúmenes del IX Congreso de Arqueología de la Región Pampeana* (p. 49). Facultad de Humanidades, Universidad Nacional de Mar del Plata.

Archivo Histórico de la Dirección de Geodesia, Catastro y Tierras de la Provincia de Buenos Aires, Ministerio de Obras Públicas. (2022). http://catalogosuba.sisbi.uba.ar/vufind/Record/20160317044315782. Accessed 6 Apr 2022.

Arellano, A. (2017). *Informe sobre investigación. Donación de cañón del siglo XVIII al museo de Luján (1927)*. Archivo Histórico Municipal Juan H. Vieytes, Usina Vieja Museo Centro Cultural. Unpublished manuscript.

Arévalo, J. M. (2010). El patrimonio como representación colectiva. La intangibilidad de los bienes culturales. *Gazeta de Antropología, 26*(1), 1–14. http://hdl.handle.net/10481/6799

Ávido, D., Vitores, M., Acevedo, V. J., Staropoli, L., Herrera, N. V., Soto, D., et al. (2021). La digitalización como estrategia de comunicación. El caso de la patrimonialización del cañón EB 1789 en la localidad de San Antonio de Areco, Provincia de Buenos Aires. In *Libro de Resúmenes del IX Congreso de Arqueología de la Región Pampeana* (pp. 100–101). Facultad de Humanidades, Universidad Nacional de Mar del Plata.

Ballart Hernández, J., & Tresserras, J. J. I. (2001). *La Gestión del Patrimonio Cultural*. Editorial Ariel.

Berberián, E. (2009). *La protección del patrimonio cultural argentino. Arqueológico y paleontológico*. Editorial Brujas.

Bonfil Castro, R. (2003). *Ponencia presentada en su calidad de Presidente del ICOMOS Mexicano en el Congreso Iberoamericano de Patrimonio Cultural, Desarrollo y Turismo*. http://ilam.org/index.php/es/programas/ilam-patrimonio/gestion-patrimonio. Accessed 6 Apr 2022.

Ciarlo, N. C. (2017). Una síntesis histórica y propuesta para el estudio arqueológico de la artillería de las Armadas europeas del siglo XVIII. *Revista de Arqueología Histórica Argentina y Latinoamericana, 11*(1), 5–42.

Conforti, M. E. (2013). La comunicación pública de la ciencia y su importancia en el proceso de valoración del patrimonio arqueológico. In M. L. Endere, M. G. Chaparro, & C. I. Mariano (Eds.), *Temas de patrimonio cultural* (pp. 35–49). Universidad Nacional del Centro de la Provincia de Buenos Aires.

Criado Boado, F. (1996a). La arqueología del paisaje como programa de gestión integral del patrimonio arqueológico. *Boletín del Instituto Andaluz del Patrimonio Histórico, 4*(14), 15–19. https://dialnet.unirioja.es/revista/1025/A/1996

Criado Boado, F. (1996b). Hacia un Modelo integrado de investigación y gestión del patrimonio histórico: La cadena interpretativa como propuesta. *Boletín del Instituto Andaluz del Patrimonio Histórico, 4*(16), 73–78. https://dialnet.unirioja.es/revista/1025/A/1996

De Carli, G. (2018). *El patrimonio. Su definición, gestión y uso social. Protegiendo el Patrimonio latinoamericano*. Ediciones Ilam electrónica: Fundación ILAM. https://ilamdocs.org/documento/3505/

Fecher, A., & Friesike, M. (2014). Open science: One term, five schools of thought. In S. Bartling & M. Frieslie (Eds.), *Opening science. The evolving. Guide on how the internet is changing research, collaboration and scholarly publishing* (pp. 17–47). Springer. www.springer.com/gb/book/9783319000251

Fernández, A., & Torres, D. (2016). *Ciencia Ciudadana. ¿Qué es? Didáctica y TIC*. Blog de la Comunidad virtual de práctica "Docentes en Línea". https://blogs.ead.unlp.edu.ar/didacticaytic/2016/03/01/ciencia-ciudadana-que-es/

Fontal, O. (2003). *La educación patrimonial. Teoría y práctica en el aula, el museo e internet*. Trea.

Garavaglia, C. (2009). *San Antonio de Areco, 1680–1880. Un pueblo de la campaña, del Antiguo Régimen a la modernidad Argentina*. Protohistoria ediciones.

García Canclini, N. (1999). Los usos sociales del Patrimonio Cultural. In Aguilar Criado, E. (Coord.), *Patrimonio Etnológico. Nuevas perspectivas de estudio* (pp. 16–33). Junta de Andalucía, Instituto Andaluz del Patrimonio Histórico.

Grimson, A. (2000). *Fronteras, naciones e identidades: la periferia como centro*. Ediciones CICCUS – La Crujía: Madrid.

Halbwachs, M. (2004). *La memoria colectiva*. Universidad de Zaragoza.

ICOMOS. (1990). *Charter for the protection and management of archaeological heritage*. https://icomos.org

López, M. A., Mancini, C. E., & Acevedo, V. J. (2016). ¿Es posible una arqueología sin excavación? El caso de la Quebrada de Humahuaca en el contexto latinoamericano. *Latin American Research Review, 51*(4), 102–119.

Mancini, C. E., Acevedo, V. J., & López, M. A. (2016). "Narrativas sobre el sitio arqueológico Peñas Blancas, Departamento de Humahuaca, Provincia de Jujuy". *Territorios, memoria e identidades*. Ponencia presentada en *las IV Jornadas Multidisciplinarias* (pp. 79–92). IMHICIHU- Instituto Multidisciplinario de Historia y Ciencias Humanas.

Manzini, L. (2011). El significado cultural del patrimonio. *Estudios del Patrimonio Cultural, 6*, 27–42.

Palavecino, V. A. (2013). Naturaleza y cultura; patrimonio y turismo. Algunas precisiones sobre su relación. In M. L. Endere, M. G. Chaparro, & C. I. Mariano (Eds.), *Temas de Patrimonio Cultural* (pp. 83–100). Universidad Nacional del Centro de la Provincia de Buenos Aires. e-Book: ISBN 978-950-658-343-9.

Pérez Winter, C. (2015). La gestión turístico-patrimonial de los centros históricos. Dos "pueblos" de la pampa bonaerense (Argentina). In O. Niglio (Ed.), *Historic towns between east and west* (pp. 225–239). Ermes edizione scientifiche.

Pérez Winter, C. (2016a). *Patrimonio y proceso de patrimonialización en dos pueblos de la provincia de Buenos Aires* [Unpublished doctoral thesis]. Facultad de Filosofía y Letras de la Universidad de Buenos Aires.

Pérez Winter, C. (2016b). "Tierra de Gauchos" o "Gaucholandia"? Autenticidad en San Antonio de Areco (Argentina). *Paso, Revista de Turismo y Patrimonio Cultural, 14*(5), 1281–1296.

Pérez Winter, C. (2018). San Antonio de Areco y la celebración de la Tradición. *REIA, Revista de Estudos e Investigações Antropológicas, 5*(1), 208–216.

Pozzi Escot, D. (2014). *Pachacamac: conservación en arquitectura de tierra*. Ministerio de Cultura de Perú.

Prats, L. L. (2005). Concepto y gestión del patrimonio cultural. *Cuadernos de Antropología Social, 21*, 17–35.

Ramos, A. (2011). Perspectivas antropológicas sobre la memoria en contextos de diversidad y desigualdad. *Alteridades, 21*(42), 131–148.

Ratto, N. (2013). *Patrimonio Arqueológico y megaproyectos mineros en Argentina: Turismo, desarrollo y sociedad*. Aspha.

Ritchhart, R., & Perkins, D. (2008). Making thinking visible. *Educational Leadership, 65*(5), 57–61.

Torres, D., & Fernández, A. (2016). *Formas de participación en Ciencia Ciudadana. Didáctica y TIC*. Blog de la Comunidad virtual de práctica "Docentes en Línea". https://blogs.ead.unlp.edu.ar/didacticaytic/2016/03/29/formas-de-participacion-en-ciencia-ciudadana/

UNESCO. (1972). *Convención sobre la Protección del Patrimonio Cultural y Natural Mundial*. https://whc.unesco.org/archive/convention-es.pdf

The Incorporation of the Cultural Landscape as a Heritage Category in South American Legislation. New Perspectives for the Protection of Pampean Heritage Sites

Norma Elizabeth Levrand, Nahir Meline Cantar, and María Luz Endere

Abstract The heritage category of cultural landscape has been recognized internationally by charters and recommendations adopted by international organizations in the last decades of the twenieth century. However, these documents are not binding on States. The UNESCO Convention Concerning the Protection of the World Cultural and Natural Heritage of 1972 included this category in the Operational Guidelines for its implementation in 1992. In this chapter, we analyze the incorporation of this category in the internal legislation of South American countries. From the comparison of secondary sources and the micro-comparison of elements in each norm, the flexibility of the legal systems to safeguard the cultural heritage through these new categories is examined. The studied elements are organized around two legal dimensions: the first refers to the conceptualization of the studied category, and the second relates to the devices with which such protection is provided. The results show, on the one hand, the lack of regulations for the protection of cultural heritage in certain South American countries, while on the other hand, the gradual incorporation of the category analyzed into national legislation. However, in those countries that recognize the cultural landscape, the legal development of both conceptual and guardianship is still incipient. The challenges of the cultural landscape

N. E. Levrand (✉)
CONICET, INES, Universidad Nacional de Entre Ríos (UNER), Entre Ríos, Argentina
e-mail: levrand.norma@uader.edu.ar

N. M. Cantar
CONICET, ILHAM-Facultad de Arquitectura, Urbanismo y Diseño, Universidad Nacional de Mar del Plata (UNMDP), Mar del Plata, Buenos Aires, Argentina
e-mail: nahircantar@conicet.gov.ar

M. L. Endere
INCUAPA, UNCPBA-CONICET. Universidad Nacional del Centro de la Provincia de Buenos Aires (UNCPBA), Facultad de Ciencias Sociales, Olavarría, Buenos Aires, Argentina
e-mail: mendere@soc.unicen.edu.ar

G. F. Bonnat et al. (eds.), *Current Research in Archaeology of South American Pampas*, The Latin American Studies Book Series,
https://doi.org/10.1007/978-3-031-55194-9_13

approach are analyzed concerning some examples of Pampean archaeological heritage.

Keywords Cultural landscapes · Cultural heritage · South America · Legislation · Heritage categories

Introduction

The scope and content of what is considered cultural heritage have varied considerably over time and are under constant discussion. These debates, which have not been devoid of conflicts and contradictions, have hovered not only about what is considered heritage but also about the reasons why it should be and who has the power to define it (de la Torre, 2013; Jokilehto, 2016).

The United Nations Educational, Scientific and Cultural Organization (hereinafter UNESCO) has adopted the results of some of these deliberations in The Operational Guidelines for the Implementation of the World Heritage Convention (UNESCO, 1972). Although cultural landscapes were internationally recognized both from a conceptual[1] and from a heritage management point of view (Rigol Savio, 2004), their inclusion in these guidelines in 1992 finished consolidating them as a heritage category.

The conceptualization of cultural landscapes from the heritage point of view has given rise to numerous expert debates in the framework of international meetings. The conclusions of these meetings tend to permeate national borders—with a greater or lesser degree of success—and have an impact on national heritage regulations. As Brumann and Gfeller (2022) indicate, most of these debates have taken place in countries from the Global North. In the Latin American sphere, the countries have fostered some debates on new heritage categories, both in the academic/scientific field and by social and community organizations (see Martorell Carreño, 2010; Rigol Savio, 2004; Urquijo, 2020). Likewise, recent constitutional reforms allow us to rethink the traditional classifications of tangible and intangible heritage in the light of integrating categories.

Academic debates and those produced in community spheres generate new demands on governmental agencies involved in heritage protection in order to safeguard assets that do not fit into the categories of monumental or artistic heritage. In this sense, the cultural landscape appears as an integrating category of immovable, movable, and immaterial assets.

However, there are difficulties in legislating the cultural landscapes. Some that can be mentioned are: the complexity of the goods included; the different domain regimes, and the role of the communities in their management and preservation. Therefore, the aim of this chapter is to consider how the legal systems of the region

[1] Cultural landscape was used for the first time by the geography scholar Carl Sauer in 1925.

have resolved these difficulties. This objective is developed from the comparative analysis of the legislation. This is without losing sight of, on the one hand, the incorporation of this category in the norms issued by UNESCO and, on the other, the nomination of sites for the World Heritage List by the States. Both aspects are also part of this analysis.

To carry out this study, a qualitative approach was used and the documentary analysis technique was mainly employed (Duverger, 1981) for the interpretation of the national regulations of the countries. To this end, this study includes legal sources available in digital legal repositories of an official nature or provided by officials from some of the countries, as well as specialized literature, which was previously compiled (Saltalamacchia, 2005).

According to the abovementioned objective, this chapter is organized into three parts. In the first one, cultural landscapes are defined and their incorporation into the documents emanating from UNESCO and other international organizations is analyzed. The second part analyzes the legal framework for the protection of cultural heritage in Argentina, Bolivia, Brazil, Chile, Colombia, Ecuador, Paraguay, Peru, Uruguay, and Venezuela.[2] Finally, from the comparison of the national legal systems mentioned, the modalities and characteristics under which the cultural landscape category was incorporated in the region and the challenges that it presents—both from the legal point of view and from its implementation in practice—are discussed. Some considerations concerning the implementation of this category for safeguarding archaeological heritage sites in the Pampean region are also made.

Cultural Landscapes as a Category of Cultural Heritage

The notion of cultural landscape emerged in various disciplines and was conceptualized at the beginning of the twentieth century. However, its reception into legal instruments occurred dispersed and fragmented during the past century (Aguilar Bellamy, 2006; Díaz-Otero, 2009; Fernández Rodríguez, 2007). One of the inaugural documents in the region was the Convention for the Protection of Flora, Fauna, and Natural Scenic Beauty of the countries of the Americas, which was approved by the Organization of American States (OAS) in 1940. Beyond the document's name, the landscape is referred to as a kind of "cultural heritage." Thus, in the Preamble of this Convention, it is established that its objective is to "protect and conserve landscapes of incomparable beauty, extraordinary geological formations, regions and natural objects of aesthetic interest or historical, scientific value, and places where conditions exist primitives."

[2] Guyana, Suriname, and Trinidad and Tobago have not been included in the study due to the lack of sufficient information to carry out the analysis.

Another relevant antecedent was the consideration, by the Franceschini Commission,[3] of the category of environmental, cultural assets, such as "orographic zones that constitute landscapes, natural or transformed by the work of man, and the delimited zones that constitute installed urban and non-urban structures, which, presenting particular merit for their values of civilization, must be conserved for the enjoyment of the community" (cited in Hernández, 1996, p. 256). The environmental and cultural assets present two categories: landscape and urban assets.

In the context of UNESCO, after decades of discussions, this category was approved at the 16th. session of the World Heritage Committee, held in New Mexico in 1992. In this way, it is legally adopted within the framework of the World Heritage Convention, adopting an innovative perspective by recognizing the interaction between nature and culture. To include this category in the Operational Guidelines for the Implementation of the World Heritage Convention, a specialized meeting was held in Petit Pierre, France, in 1992, organized by the World Heritage Center in coordination with ICOMOS, IUCN, and other international organizations (Rigol Savio, 2004; see also von Droste et al., 1995; Rössler, 2007; Brumann & Gfeller, 2022).

Subsequently, the debate about the category of cultural landscapes intensified within the framework of the Global Strategy for a More Representative, Balanced, and Credible List adopted by the World Heritage Committee (main body of the Convention on World Cultural and Natural Heritage) in 1994. At that time, documents were produced to broaden the definition of world heritage. The role of the International Council on Monuments and Sites (ICOMOS hereinafter), an organization that acts as a consultative body of the World Heritage Committee for the implementation of the World Heritage Convention (UNESCO, 1972), in the preparation of these documents was preponderant.

In the World Heritage Convention framework, cultural landscapes are considered: "cultural properties and represent the 'combined works of nature and man' designated in Article No. 1 of the Convention. They are illustrative of the evolution of human society and settlement over time, under the influence of the physical constraints and/or opportunities presented by their natural environment and of successive social, economic, and cultural forces, both external and internal.

They should be selected on the basis of both their Outstanding Universal Value and their representativity in terms of a clearly defined geo-cultural region. They should also be selected for their capacity to illustrate the essential and distinct cultural elements of such regions.

The term 'cultural landscape' embraces a diversity of manifestations of the interaction between humankind and the natural environment.

Cultural landscapes often reflect specific techniques of sustainable land use, considering the characteristics and limits of the natural environment they are established in, and may reflect a specific spiritual relationship to nature. Protection of

[3] It is a commission of the Italian parliament that worked on protecting and valuing cultural heritage between 1964 and 1968. It bears the name of its president, deputy Francesco Franceschini.

cultural landscapes can contribute to current techniques of sustainable land use and maintain or enhance natural values in the landscape. The continued existence of traditional forms of land use supports biological diversity in many regions of the world. The protection of traditional cultural landscapes is therefore helpful in maintaining biological diversity" (WHC, 2021, Paragraph 47).

These landscapes fall into three main categories: landscapes intentionally conceived and created by man; landscapes that have evolved organically, and associative cultural landscapes. The first refers to gardens and parks "created for aesthetic reasons, which are often (but not always) associated with buildings or religious or monumental complexes." The second considers those places that originated "as a result of an originally social, economic, administrative and/or religious requirement and have reached its current form by association and as a response to its natural environment." They are subdivided into two categories: on the one hand, the relict landscapes, with mainly material characteristics, are those that have undergone an evolutionary process but it is stopped in the present; on the other, living landscapes, in which their "active social function in contemporary society, closely linked to the traditional way of life" is still preserved and for which their evolutionary process continues. Finally, associative cultural landscapes are linked to religious, artistic, or cultural events, in which the material aspects may be insignificant or absent (UNESCO, 2003, Paragraph 47 bis, Operational Guidelines for the Implementation of the World Heritage Convention).

In addition to incorporating cultural landscapes in the Practical Guidelines, at an international level, the idea of cultural landscapes is consolidated in the Recommendation on the Historic Urban Landscape (UNESCO, 2011). It should be remembered that this term had been included in the Vienna Memorandum (UNESCO, 2005), a product of the Conference "World Heritage and Contemporary Architecture. Management of historic urban landscapes". It had a previous precedent, the European Landscape Convention (Council of Europe, 2000), which conceives landscape as "any part of the territory as perceived by the population, whose character is the result of the action and interaction of natural and/or human factors" (Art.1), a definition that presents a renewed view of the landscape by highlighting the role of community perception.

The UNESCO Recommendation of Paris of 2011 proposes a methodological approach that provides "a broad conceptualization of heritage, which includes the traditional categories of monuments and historical centers and other components, natural and cultural, tangible and intangible (…). In this way, the strict notion of a historic area or center is overcome to include the surroundings, which implies, for example, the conservation of visual impacts of interventions that can sometimes be carried out frankly far from the historic areas" (Conti, 2015, p. 5). The urban landscape is defined in this Recommendation as the urban area resulting from a historical stratification of cultural and natural values and attributes, which transcends the notion of "ensemble" or "historical center" to encompass the general urban context and its geographical environment (point I.8) The Recommendation considers the context, which includes "other features of the site (…); its urbanized environment (…); its infrastructures (…); its open spaces and gardens, the configuration of land uses and their spatial organization, visual perceptions, and relationships, and all other elements of the urban structure. It also incorporates social and cultural uses and values, economic processes, and intangible heritage aspects in its relationship with diversity and identity (Art. 9). However, it should be noted that the historic urban landscape is not presented as a legal heritage category but as a "notion," which could transform by its content into a methodology of analysis and management of heritage for urban areas.

This proposal, which refers to a new way of approaching the management of historic cities, is the most recent methodology recognized by UNESCO for the treatment of world heritage sites (Bandarin & Van Oers, 2012, p. 200). In this sense, it has been stated that "one of the main characteristics of this new approach is the intensity of interaction between the population that develops its activities in the city and the physical environment chosen for its location. The values of balance and coexistence between contemporary architecture, historic buildings, and open spaces also appear" and its "coexistence transmits a single idea of the city as a whole" (Rey Pérez, 2017, p. 21).

Other international charters and recommendations that reinforce the idea of cultural landscape are the "Declaration of Florence. Heritage and landscape as human values" (ICOMOS, 2014) and the Principles concerning Rural Landscapes as Heritage (ICOMOS, 2017).

The Ibero-American Charter of Cultural Landscape (II Meeting of Cultural Landscapes 2012) incorporated this recommendation at a regional level. Also, the Latin American Landscape Initiative is a declaration adopted in 2012 by associations of landscape experts from more than ten countries in this region. This declaration provides a set of ethical principles for the recognition, management, and protection of cultural landscapes and recognizes a series of challenges to address them in the region.

It should be clarified that unlike the UNESCO Conventions that become part of their normative framework once ratified by the Member States, the declarations or recommendations are not binding on the States. Therefore, they have more moral than legal authority. The same happens with the charters, principles, and declarations of ICOMOS. These documents generally establish principles, definitions, and rules of conduct that professionals and specialists must take as good practice guides (Luxen, 2004). Regarding the ICOMOS documents, it has been pointed out that its self-regulatory and collective approach provides flexibility, adaptation to changes, and moral credibility to the extent that they result from broad consultation with professionals from all over the world (Jameson, 2014).

Therefore, although these charters and recommendations are not binding for the countries' legal systems, it must be recognized that these documents constitute a stimulus for the processes of asset declaration in recent decades (Prats, 2012). They are the result of the internationalization of the debates around heritage and the evidence of the attempts of professionals to adopt conservation criteria to an expanding reality (de la Torre, 2013; Labadi & Logan, 2016). In this regard, Rotman (2015, p. 18) states that UNESCO "suggests guidelines regarding the heritage policies of the States, operates as a space for the generation of conceptions, recommendations, lines of action and management on the subject, as well as an area of control of the performance of nations [and] sets agenda on heritage."

It should be noted that the documents issued by these international organizations have been recognized as soft law, that is, as "instruments that do not have legal force or legally binding character." However, they have, at the same time, "a certain legal relevance" (Levrand & Endere, 2020, p. 3), given their impact on the internal regulations of the countries.

Besides the influence of the standards issued by various international organizations, community participation generated significant changes in regulatory processes. In this sense, Mallarach and Verschuuren (2019) show the relevance that bottom-up approaches, based on rights and with broad participation of communities, have in the effective regulation of the management of natural sites. The situation of cultural landscapes does not differ, in this sense, from that presented by the authors.

The Incorporation of the Heritage Category Cultural Landscape in the Internal Regulation of South American Countries

The introduction of new heritage categories manifests the expansion of the concept of cultural heritage, which is mentioned by multiple investigations (Aguirre Arias, 2007; Ariño Villarroya, 2002; Cornu, 2001; Endere, 2021; Gatti & Muriel, 2006). Likewise, the consolidation of a reference from the legal device evidences the modalities by which the law, as a mechanism of social regulation, intervenes in constructing meaning about this object. In this way, the institutionalization of new heritage categories establishes them as an object of intervention and management by the State through the implementation of a set of legal and administrative devices to manage these assets. These devices include the definition of the goods that make up the category, the protection mechanisms, and the institutions in charge.

One of the undeniable legal devices is the definition of a heritage category. These definitions usually incorporate knowledge from non-legal disciplines. In some cases, as mentioned in the previous section regarding the notion of the cultural landscape, the legal consecration can only be carried out after the stabilization of the concept has taken place.

Another legal device that can be identified in legal texts is the inclusion of cultural landscapes in an inventory or registry. Although the purpose of recognizing the goods that make up cultural heritage is usually indicated, other aspects can also be considered. Inventories or registries typically require the preparation of a file that forms a passage point that translates, through the clauses it contains, the various meanings of the patrimonial categories, sifts their elements, and thus purifies the registered assets.

The declaration of those assets that make up cultural heritage is a device used in the countries studied by the legislative or executive body of government. Unlike inventories or registries, which are the responsibility of the enforcement authority, declarations require not only a technical but also a political consensus to classify the property as cultural heritage.

Cultural landscapes are categories made up of an assortment of goods, which may belong to the State, to natural or legal persons. For this reason, the domain regime and its limitations or restrictions are essential elements of the regulations that protect these assets.

In the absence of specific provisions on domain restrictions, an intervention or authorization device is usually incorporated. Before implementing substantial modifications to the legal regime of the goods that are included in the registry or inventory or being declared as cultural assets, the enforcement agency reserves the right to know, advise, suggest changes, and even authorize these modifications. The foundation of this legal device is the preservation of the cultural values of the property.

Argentina

In Argentina, there is no comprehensive regulation of cultural landscapes. However, court rulings and legislative projects have recognized them as a category of cultural heritage protection (Levrand, 2021). Argentina is a federal country where the provinces recognize jurisdictions not expressly delegated to the Nation. In cultural heritage issues, provincial and the Nation States share powers according to the scope of the declared assets.

In 2015, through the reform of National Law No. 12,665, creating the "National Commission for Historical Monuments, Places, and Assets," the cultural landscape category was incorporated into national legislation (Levrand & Endere, 2020). A relevant precedent was the recognition of cultural landscapes by the National Commission itself, in an internal provision whose purpose was to establish the structure of the National Registry of Historic and Historic-Artistic Assets. This provision mentions the urban and natural landscape as a manifestation of the "Asset of Historical Interest" category. It clarifies that it includes urban-architectural ensembles and sequences, parks and gardens, etc.

However, beyond the mention of this category, no specific provisions are included in the law that allow special protection of cultural landscapes. Thus, the general rules establish the possibility that these cultural assets are recognized and declared as such (Art. 4, National Law No. 12,665). On the other hand, Article 5 of the cited law establishes the prior intervention of the enforcement authority in ownership transfers or other modifications of legal status. Beyond these norms, the national regulation does not provide for the preparation of management plans, the modalities of communities participation in the management of these assets, or certain limitations to ownership based on their heritage value.

Despite the absence of comprehensive regulation of the cultural landscape category, expert organizations' development of management tools is relevant. Thus, in 2011 the Argentine Center of Landscape Architects (an association that is part of the International Federation of Landscape Architects) drew up the Argentine Landscape Charter. It is defined as "a social construction from a cultural tradition where both human intentionality and nature's biological and physical processes are deployed together."

Argentina currently has two assets inscribed in this category on the World Cultural and Natural Heritage List established by the 1972 UNESCO Convention: on the one hand, in 2003, the "Quebrada de Humahuaca" (i.e., a natural fluvial canyon with steep sides) was inscribed as an evolving cultural landscape; on the other hand, in 2014, the Qhapaq Ñan, Andean Main Road, was registered jointly with other Latin American countries.[4]

Bolivia

The Plurinational State of Bolivia is organized as a unitary, decentralized, and autonomous state. Cultural heritage is governed by the Bolivian National Law No. 530, enacted in 2014, which allows declaring cultural landscapes as cultural assets. Article 4 of this law establishes a set of definitions. It incorporates the presumption of quality of cultural heritage for those assets that "due to their historical, architectural, ecclesiastical, religious, ethnographic, documentary, paleontological, archaeological, artistic, scientific, technological value, landscape and social, may be considered as Cultural Heritage."

The legislation explicitly recognizes cultural landscapes and defines them as "the result of the development of human activities and their meanings over time in a specific territory, reflecting the mutual relationship between nature and culture" (Art. 4, Bolivian National Law No. 530). In this way, the law adheres to the criteria that delimited this category in the international arena.

The declaration of cultural landscapes enables measures to register, protect, strengthen, and disseminate cultural expression in the State. Likewise, the norm establishes that the declared assets must enter the Plurinational System for the Registration of Bolivian Cultural Heritage. Under the regulations of the Bolivian National Law No. 530, this Registry is nourished by the information provided by the autonomous territorial entities, the national legislative body, and the updates reported by the owners and custodians of assets.

[4]The Qhapac Ñan or Andean Road System is a vast road network of 30,000 km built over several centuries by the Incas. This site was jointly nominated by Argentina, Bolivia, Chile, Colombia, Ecuador, and Peru and was inscribed on the World Heritage List in 2014. Each of the countries chose components that make up the network and are testimonies of the network's infrastructure, highway, and social and political function.

Article 54 of the Bolivian National Law No. 530 prohibits new construction in the surroundings of cultural landscapes unless authorized by the competent entity. This limitation to the domain must be complemented with a precise indication of the scope of the declaration of cultural landscape or road system and the properties that are affected by this declaration.

Bolivian law also provides for the intervention of the competent authority in the material or legal modifications proposed to the declared assets. In such cases, the express authorization of the enforcement authority is required.

It is relevant that the law itself establishes duties for the owners of assets declared in Article 19. Thus, they are guarantors of the same and "are obliged to prevent any risk that affects their integrity and conservation"(Article 19, I). This article establishes some duties for the owners or custodians, such as facilitating preventive defense actions, allowing inspections, and assisting in the execution of construction, restoration or revaluation works.

Brazil

The Federative Republic of Brazil is a federal country. The Federal Constitution of 1988 recognizes cultural heritage in its Article No. 216, which includes among cultural heritage assets "urban complexes and places of historical, scenic, artistic, archaeological, paleontological, ecological and scientific value."

The Decree-Law No. 25 of 1937 established the organization of protecting historical and artistic heritage. It recognizes those movable or immovable properties that have archaeological, ethnographic, bibliographic, or artistic value as historical or artistic heritage. Likewise, Article 1 equates the previous ones with "Natural monuments and sites and landscapes that are important to conserve and protect due to the remarkable aspect with which they have been endowed by nature or managed by human action."

The Brazilian Decree-Law No. 25 established a registry book of archaeological, ethnographic, and landscape assets managed by the Institute of National Historical and Artistic Heritage (IPHAN). However, it was not until 2009 that cultural landscape protection became an autonomous heritage category. In that year, IPHAN issued Resolution 127, which defines and establishes the legal regime of cultural landscapes.

Article 1 of the cited Decree-Law defines the cultural landscape as: "a particular portion of the national territory, representative of the interaction process between man and the natural environment, to which life and human science have imprinted marks or attributed values." This general definition tries not to limit, initially, the goods recognized in this category. However, an administrative procedure—directed by a technical body of the IPHAN—is established. This procedure aims to obtain the official declaration (known as "chancela") of the Brazilian cultural landscape. As it can be seen, this declaration requires a translation of the elements that compose it and a purification carried out by the experts who work at the state level.

One of the requirements to start the declaration procedure is presenting an agreement whose objective is the shared management of the territory considered cultural landscape. This pact can involve public power, civil society, and private initiative. As Weissheimer (2012) indicates, the novelty of this joint preservation mechanism between the different agents that act on the territory configures one of the most arduous and complex stages of the procedure. This pact may include declarations of private real estate as cultural heritage, records of intangible heritage, and even the improvement of urban planning or development plans compatible with preserving the cultural landscape.

Even though, in 2012 the Rio de Janeiro site -consisting of Rio de Janeiro landscapes between the mountains and the sea- was inscribed on the World Heritage List, the national procedures for the declaration of cultural landscapes were interrupted in 2013, and they have not been taken up ever since. Pereira (2020) indicates that although this interruption only affects the administrative procedure, it generates frustration with the expectations of both the communities and the local governments. Although the IPHAN initiated some procedures between 2009 and 2013, none of them were completed, providing evidence of administrative liability in this agency.

Chile

The Republic of Chile is a unitary country. The Chilean National Law No. 17,288 related to national monuments does not accept the category of cultural landscapes. However, it participates in Qhapac Ñan and has declared the Rapa Nui National Park a cultural landscape (see Silva Pérez & Fernández Salinas, 2015). Pía Acevedo Méndez (2018, p. 210) indicates that considering the industrial landscapes, "Chile has two declarations: the Humberstone and Santa Laura Saltpeter Works, declared in 2005, and the Sewell Mining Camp, declared in 2006 as a heritage of mankind".

The site of the Saltpeter Offices has an extension of 631 hectares, and its buffer zone covers 12,055 hectares. It was declared a National Monument under the Historic Monument category. However, it has been pointed out that urgent measures are required, including launching the priority intervention plan. In addition, "it is of crucial importance to protect the desert landscape, as a cultural landscape, both in its geographical dimension and related to the vestiges of the exploitation and transport of saltpeter in general" (National Council of Monuments, 2013, p. 17).

It should be noted that the National Monuments Council has the power to protect material remains, spaces, or objects of importance to local and/or national identity through one of its six types of declarations: historical monument, typical and picturesque area, sanctuary of nature, public monument, archaeological monument, and paleontological monument. According to Acevedo Méndez (2018), by establishing only the legal device of declaration, a management plan is not conceived to safeguard and protect these assets.

Colombia

The Republic of Colombia is a unitary country whose constitution establishes the duty of the State to protect cultural heritage (art. 72). The Colombian General Law of Culture of 1997 was modified in 2008 (Law No. 1185), incorporating a broad concept of cultural heritage. Thus, the new Article 4 establishes that it is made up of material goods, immaterial manifestations, and representations of culture, including the cultural landscape.

The application of the law is limited to those declared sites. According to their respective competencies, the declaration procedure may involve the national, territorial, or indigenous authorities and the community councils of the Afro-descendant communities (Art. 4, subsection b, Law No. 397). In all cases, the intervention of expert bodies that rule on the interest of the property is foreseen. However, the legal text does not mention the cultural landscape as one of the possible declaration categories, nor does it include a definition of it.

In 2011, the Coffee Cultural Landscape was inscribed on the World Cultural and Natural Heritage List. Despite this, in 2015, Diana Zuluaga Varón indicated that "the reference to landscape in the Colombian legal system is lagging and incidental in some provisions whose main objective is the regulation of environmental issues (…) or urban development" (Zuluaga Varón, 2015, p. 35).

In 2019, the Single Regulatory Decree of the Culture Sector No. 1080 was modified, incorporating relevant provisions for the management of landscapes. According to this, the National System of Cultural Heritage comprises competent public bodies and the owners of declared immovable or movable assets and bearers of cultural manifestations, among others. Likewise, the declaration as Assets of Cultural Interest of cultural landscapes was contemplated, requiring the description and georeferenced location, approval of the management plan, and the determination of the obligations of the owners or possessors, among other provisions.

The mentioned modification incorporated Title III to Decree 1080 relative to Cultural Landscapes. Article 2.4.3.1 defines cultural landscapes as "the territories that are the product of the interrelation between social groups, communities or collectivities with their territory or nature, of historical, economic, social, political, cultural or spiritual processes, which illustrate the forms of occupation and management of the territory." Therefore, "they are factors of identity, belonging or citizenship, they contain goods, manifestations, products and all those elements that are expressions of cultural identity and that are representative of a clearly defined region and illustrate the elements essential and distinctive cultural; through the valuation and sustainable management of these places, the enjoyment of cultural rights is effectively made possible." Likewise, it enumerates and defines the following classes of cultural landscapes: landscape clearly defined, conceived, and created intentionally by man; organically evolved; associative; places of memory and sociocultural complexes.

The area affected by the declaration of cultural landscapes is recognized as a buffer zone. Diverse management instruments are proposed, which must be developed by the different authorities with competence over the elements that make up the cultural landscape, gathered in inter-institutional committees. This legal figure seeks to remedy the dispersion of state action, which is usually disaggregated into various administrative bodies that do not act in a coordinated manner (Levrand, 2020). The design and development of the management instrument aim at the cultural landscape conservation, protection, and safeguarding. In this sense, it must foresee the interventions that will require authorization from the application authority and the criteria to grant said authorization.

Ecuador

The Republic of Ecuador is a unitary state. The National Constitution of 2008 recognizes as manifestations of cultural heritage the paths and landscapes that constitute identity references for peoples or that have historical, artistic, archaeological, ethnographic, or paleontological value (Art. 279).

The Organic Law of Culture of Ecuador defines tangible heritage (Art. 51), stating that they are "the material, movable and immovable elements that have been produced by the various cultures of the country and that have a historical, artistic, scientific or symbolic significance for the identity of a community and the country.

Tangible cultural heritage can be archaeological, artistic, technological, architectural, industrial, contemporary, funerary, railway, underwater, documentary, bibliographic, film, photographic, urban, rural, river, and maritime cultural landscapes, gardens, routes, paths, and itineraries and, in general, all those elements whose relevance is inscribed in the indicated definition".

Article 54 of this Law lists the assets that belong to the Ecuadorian cultural heritage, considering the landscape as a preservation unit. Thus, in the case of paleontological sites, archaeological sites, and underwater heritage, the law requires the definition of a natural and cultural environment that endows them with landscape unity. On the other hand, the law provides for the declaration as an intangible cultural heritage the "expressions of food and gastronomic heritage, including landscapes and territories of agro-biodiversity heritage" (Art. 80, sub. e). However, beyond these mentions, the law does not establish a definition of cultural landscape.

The declarations of cultural landscapes require the approval of comprehensive management, conservation, protection, and safeguard plans (Art. 58). The Regulatory Decree of the Organic Law of Culture, of 2017, establishes the possibility of establishing associations or consortiums for the management of archaeological or paleontological areas, including pre-Hispanic routes or paths (Art. 69, Executive Decree 142, 8). This norm establishes that the investment of public resources in these areas (even when they are privately owned) must be for mutual benefit and seek alliances for their management and sustainability. Thus, the participation of peoples, communities, and nationalities in the management and conservation of these sites is provided for (Art. 70, Executive Decree 1428).

Paraguay

The Republic of Paraguay is a unitary country. In Article 81, the National Constitution incorporates the ideas of space of historical value and physical environments by recognizing the right to cultural heritage. This right is regulated by Law No. 5621 of 2016. Article 5 of the law establishes the classification of cultural assets that make up the heritage, incorporating: the territorial space necessary for the maintenance and development of indigenous cultural forms (Art. 5, inc. d) and the complexes and sites (Art. 5, inc. k). The latter includes some categories that can be considered within this definition without being identified as cultural landscapes.

On the one hand, "the complexes, urban or rural, are provided with traditional, natural, historical, environmental, artistic, architectural or urban values. The sets are made up of the centers, cities, and towns, and include the special character of certain towns and cities or their areas, as well as the temples with their surrounding squares, streets, and buildings located in certain traditional towns" (Art. 5, inc. k). Archaeological areas and their buffer zones can also be considered cultural landscapes.

The law establishes different mechanisms for the protection of declared assets. First, the National Cultural Heritage Registry is created (Art. 38). Secondly, promotion measures are granted for immovable goods, such as the tax exemption for both, national and municipal taxes or the performance of property conservation tasks in the event of the holders' economic incapacity. Thirdly, general limitations to the domain are provided, such as the power of the municipalities to establish a special regime for cultural heritage assets in the regulatory plans or the intervention of the enforcement authority for modifications to declared assets.

Although the National System of Cultural Information of Paraguay lists more than 20 sites in the category "cultural landscape" (see http://sicpy.gov.py/busquedas/?categorias-41), most of them are natural sites or cultural landscapes intentionally created by man (such as squares, equestrian centers, among others).

Peru

The Republic of Peru is a unitary state. The National Constitution recognizes that archaeological sites and remains and expressly declared places are part of the cultural heritage (Art. 21). Although cultural landscapes are not mentioned in it, the General Law of Cultural Heritage of the Peruvian Nation (Law No. 28,296), when classifying cultural heritage assets, incorporates, as immovable material assets, the environments and monumental complexes, the material evidence resulting from urban and/or rural life and human activity and the landscape environment (Art. 1, Law No. 28,296).

This law creates a National Registry of Real Estate that is part of the Nation's cultural heritage. Also, in Art. 20 establishes restrictions on ownership, such as the prohibition of dismembering the asset and the necessary intervention of the enforcement authority to modify or restore these assets.

Given the absence of specific measures in the said norm, the Supreme Decree No. 002-2011-MC regulated the procedure for declaring cultural landscapes in Peru. The Directorate makes this declaration of Cultural Landscape of the Ministry of Culture based on the information included in the technical sheet.

Article 6 of this Supreme Decree defines cultural landscapes as "cultural assets that represent the joint works of man and nature, illustrate the evolution of human society and its settlements over time, conditioned by limitations and/or physical opportunities presented by its natural environment and by the successive social, economic and cultural forces, both external and internal, integrating the following dimensions: diachronic (...) territorial dynamics (...) perceptive...". It also establishes the categories of cultural landscapes maintaining the classification of the Operational Guidelines for the Implementation of the World Heritage Convention of UNESCO.

As indicated, one of the leading legal provisions of the standard is the declaration of cultural landscapes. One of the high points of the technical file for the declaration is the delimitation of the cultural landscape. It implies the identification of significant cultural and territorial elements for the configuration of the cultural landscape, which are independent of the existing political and administrative limits. In the delimitation, the elements that define the uniqueness of the cultural landscape, the trends and processes of change over time, and the main existing and foreseeable conflicts must be highlighted.

Once declared, the Directorate of Cultural Landscape must prepare the management plan for the property. The Plan is defined as "a technical instrument that contains programs and projects aimed at sustainable development in order not to alter the exceptional value of the landscape" (Art. 20, Supreme Decree No. 002-2011-MC). Thus conceived, the management plan must be designed by a multidisciplinary group of experts who, in addition, must define strategies for citizen participation in its formulation.

Although none of the 13 sites inscribed by Peru are in the "cultural landscape"category on the World Heritage List, some have been inscribed as mixed sites (such as the Historic Sanctuary of Machu Picchu or the Nazca lines and hieroglyphs and Palpa) can fit into the category. Peru also had a leading role in the nominating process of the Qhapac Ñan as a world heritage site.

Uruguay

The unitary Republic of Uruguay approved, in 1971, the National Law No. 14,040, by which it created the Historical, Artistic, and Cultural Heritage Commission of the Nation. The law provides for the declaration of Historical Monument of those movable or immovable property linked to relevant events, to the national historical evolution, to notable figures in the life of the country, or to what is representative of the culture of a national era (Art. 5). The declared goods cannot be modified or altered without the authorization of the competent authority. This law does not mention the category of cultural landscapes.

Despite the absence of the category in national legislation, Uruguay has inscribed the cultural landscape of Fray Bentos on the World Heritage List. This cultural landscape has a Management Commission created through a framework agreement between the local government and four national ministries: the Ministry of Housing, Territorial Planning and the Environment, the Ministry of Education and Culture, the Ministry of Tourism, and the Ministry of Industry, Energy and Mining. Representatives from the municipality of Río Negro, the Commission for Cultural Heritage of the Nation, and the National Directorate for Territorial Planning also participate. As indicated by Levratto et al. (2018), after the site's registration, other agents were incorporated into the Management Commission, such as educational institutions, the local community, and commercial entrepreneurs. In this way, the incidence of the UNESCO recommendations in the management of heritage sites has made it possible to expand and involve citizens in management.

The Site Commission (such as its current name) is organized around a site manager, ultimately responsible for the management, and an executive board made up of technicians and experts, and the plenary of the Site Commission. Consequently, the absence of a specific regulation of cultural landscapes has not hindered the design of a particular management device that is normatively based on the declaration as a World Heritage.

Venezuela

The Bolivarian Republic of Venezuela, a decentralized federal state, enacted the Law for the Protection and Defense of Cultural Heritage in 1993. Article 6 sets out the assets that make up the cultural heritage, which includes: the environmental or landscape environment -rural or urban- required by the cultural, movable, or immovable assets for their visualization or adequate contemplation (Art. 6 inc. 12); the archaeological and paleontological heritage wherever they are found (Art. 6 inc. 13); and any other asset of cultural interest that deserves to be declared as such (Art. 6 inc. 14).

When determining the assets that are part of the cultural heritage, the enforcement authority has the power to establish conservation plans and ensure their execution (Art. 10 inc. 2). In addition to the declaration device, in the case of national monuments, the intervention of the applicable authority is implemented in the legal and factual modifications. Article 23 of the law determines the consideration of a protected area or environment of national monuments, which must be regulated by law.

Article 31 of the law refers to the declaration of historic towns, centers, and sites. It indicates that those that possess typical, traditional, natural, environmental, artistic, architectural, or archaeological values can be considered as such. In this category, the intervention of the applicable authority is foreseen in the modification proposals and the planning and development programs, including the specific urban regulations.

The Incorporation of the Cultural Landscape Category as an Opportunity for the Safeguarding of Pampean Heritage Sites

The Argentine pampas region in a broad sense includes the territory of the current Argentine provinces of Buenos Aires, Córdoba, Entre Ríos, La Pampa, Santa Fe, San Luis, and northern Río Negro province. As indicated, Argentine legislation does not establish a definition of cultural landscapes. However, the category was accepted in the reform of National Law No. 12,665 (2015). Despite this incorporation, at the time of writing these lines, only the site "Cuenca del Río Pinturas" in the Province of Santa Cruz, which is located in the Patagonian region, is recognized as a cultural landscape.

However, some properties that are in other categories can be considered cultural landscapes. For example, some assets which have been declared national historic sites such as the Pehuen-Có Provincial Geological, Paleontological and Archaeological Reserve (Buenos Aires province); the entrance of the Arroyo de la China (Entre Ríos province); the Treru Lauquen Camp (La Pampa province); the Chacras Camp (San Luis province) or the Fort of Sancti Spiritu (Santa Fe province). In other cases, the declaration was made in the category of historical monuments, such as "Fortín Cuatreros," Fort Barragán or "Estancia del Virrey del Pino" (Buenos Aires province); the "Comandancia de Fronteras del Ejército de Línea" (the Frontier Command of the Line Army) (Córdoba province); the Ruins of Cayastá (Santa Fe province) or the "Fortín Beslile" (Río Negro province). Finally, in some cases, they were declared under categories introduced in the 2015 law reform, such as "Pueblo Liebig" (Entre Ríos province), which was declared an asset of national industrial interest.

The declaration of these sites allows their inclusion in the List of Declared Assets, which functions as a national inventory of cultural assets. However, the national administrative body will not take into account the criteria for the conservation and restoration of cultural landscapes when supervising the interventions or actions proposed for these cultural places.

Besides, the National Law No. 24,743 on the protection of archaeological and paleontological heritage created the National Registry of Archaeological Sites, Collections, and Objects (RENYCOA-INAPL). This registry allows knowing the state of conservation of archaeological assets and managing archaeological concessions and surveys. However, the implementation of the registry does not incorporate other legal provisions aimed at the conservation of archaeological sites as cultural landscapes.

The existence of lists or registers of cultural heritage could be enriched by considering the elements that identify cultural landscapes. In this sense, the legislation of Peru, by pointing out that the territorial and significant elements to identify the landscape are independent of the existing political and administrative limits, offers a relevant criterion to consider in Argentina, which is a federal country.

The domain regime must also be considered when evaluating the challenges of the category of cultural landscapes. It should be remembered that in Argentina the declaration as a national monument, national historic site, or other category established in National Law No. 12,665 does not imply a domain restriction. On the other hand, the buffer zones of the sites are not regulated, despite being stated as a category of cultural property in the aforementioned law. This prevents considering, as the regulations of Ecuador, Colombia or Paraguay do, limitations to ownership in these areas. Regarding the management plans, they are not a requirement for the declaration of monuments or cultural assets, nor for the prospecting of archaeological sites. In Argentina, the agency that has the major experience in developing management plans is the National Parks Administration. Many of the landscapes found within national parks are managed with comprehensive criteria. It should also be noted that, unlike countries such as Brazil, which include in the legislation a pact for shared management of the territories, in Argentina, the agreements with indigenous communities that inhabit these lands do not yet constitute a priority issue on legal instruments (Endere, 2020).

All these issues require a deeper analysis that exceeds the limits of this chapter, however, it is interesting to highlight the possibilities offered by the cultural landscape category to protect some sites which extend throughout the territory. It allows the planning and implementation of cultural heritage safeguard strategies that are not limited to protecting points but areas in the territory and consider them comprehensively in all their complexity (see Endere & Prado, 2009).

Discussion and Concluding Remarks

Based on the analysis carried out, it is observed that although the category of cultural landscapes was debated and incorporated into international regulation in the last decades of the twenieth century, its admission to the legislation of South American countries has been gradual. However, there are nuances between them that deserve consideration.

Some countries have incorporated the category of cultural landscapes into their domestic legislation, developing their definitions and specific regulatory instruments. Sometimes, such definitions are similar to those set within the UNESCO framework, as is the case of Bolivia. Argentina is also within this group; however, the national legislation does not include provisions that allow special protection of this heritage category, presenting a legal loophole.[5]

In other countries, this category was incorporated through regulations of the laws that govern cultural heritage in general. These regulations come from the Executive Branch or the enforcement agencies. As various authors have indicated (Galbrath & Zaring, 2014), non-binding norms such as recommendations or charters issued by different international organizations are usually implemented within the States by executive bodies. Thus, national administrations are permeable to the international level in this regard. As was considered in the first part of this chapter, the cultural landscapes category is found in many non-binding documents. This new category is thus admitted at the national level in countries such as Ecuador, Colombia, and Peru.

Brazil has incorporated the category of cultural landscape through a resolution issued by the IPHAN, which is an autonomous body that depends on the Ministry of Culture. Similar is the case of Colombia, where the cultural landscape category was incorporated by decree of the President of the Republic, and Peru, where it was incorporated by decree of the Ministry's Directorate of Cultural Landscape. It is worth noting that the legal definitions of these last two countries have several similarities with that of UNESCO. Ecuador does not incorporate the category of cultural landscape; however, it has a broad definition of cultural heritage that could include it. Likewise, the Organic Law of Culture of Ecuador admits the declaration of cultural assets under the cultural landscape category. It establishes the approval of comprehensive management, conservation, protection, and safeguard plans. Venezuela and Paraguay also have broad definitions of cultural heritage that could include cultural landscapes. For example, the former protects landscape environments in its Law for the Protection and Defense of Cultural Heritage. Meanwhile, the latter protects spaces of historical value and their physical surroundings in the National Constitution and in its specific law to protect cultural heritage.

Finally, there are those countries that have not incorporated the category into national legal regulations although have a history of applying the concept that could be considered predecessors to the rules. For example, Chile's case, which participates in the nomination of the Qhapac Ñan and has declared the Rapa Nui National Park a cultural landscape. Also, Uruguay has prepared the nomination for the Fray Bentos site, developing a management plan based on the initiative of the national administration.

Regarding the legal provisions that enable sustainable management of sites, it should be noted that this aspect is regulated in a few countries in the region. Thus, the standards of Peru, Ecuador, Colombia, and Brazil stand out. The framework of this research exceeds the analysis of the effective application of site management standards or the evaluation of their effectiveness in preserving cultural landscapes. As indicated, the existence of management plans or agreements between various

[5]Those cases not expressly provided by the law are considered legal loopholes, mainly when the law provides only a general directive that administration regulation must complement.

agents to protect cultural landscapes goes beyond the legal requirement. It can still be found in countries that do not provide for it, as in the case of Uruguay.

Among the legal provisions, the declaration or inventory of cultural landscapes is only one of the elements that enable their recognition and preservation. It is also a significant challenge, both for the regulation and for the administrations, to establish management plans that consider the different interests that converge in the territory and facilitate agreements for its sustainable management.

Finally, the impacts generated by this type of declaration and its implementation in communities cannot be left out, an issue that has been debated at the regional level (Marcone Flores, 2019) and internationally (Brumann & Gfeller, 2022; Pocock & Lilley, 2017; Wardana, 2020).

Academic discussions about heritage categories, translated into charters and recommendations, eventually result in a legal translation. However, the indirect effects of recognizing comprehensive categories that incorporate religious and spiritual values and traditional knowledge should not be lost sight of. Among the latter, communities' claim in the participatory management of cultural sites and landscapes is a challenge that requires greater attention from legal operators, scientists, governments, and other actors involved. However, the opportunities offered by the implementation of the cultural landscape category to protect the archaeological heritage should not be underestimated, particularly when it has considerable spatial dimensions, as in some heritage places located in the Pampas region.

References

Acevedo Méndez, P. (2018). Vestigios industriales en Tierra del Fuego, Chile: elementos de un paisaje cultural en el fin del mundo. In Comisión del Patrimonio Cultural de la Nación. Intendencia de Río Negro (Ed.), *Gestión del patrimonio. Paisajes culturales y participación ciudadana* (pp. 207–220). Universidad CLAEH.

Aguilar Bellamy, A. (2006). Algunas consideraciones teóricas en torno al paisaje como ámbito de intervención institucional. *Gaceta Ecológica, 79*, 5–20.

Aguirre Arias, B. (2007). Del concepto de bien histórico-artístico al de patrimonio cultural. El patrimonio urbano y arquitectónico en Chile. *Revista Electrónica du&p (Diseño Urbano y Paisaje), 4*(12), 1–16.

Ariño Villarroya, A. (2002). La expansión del patrimonio cultural. *Revista de Occidente, 250*, 129–150.

Bandarin, F., & Van Oers, R. (2012). *The historic urban landscape: Managing heritage in an urban century*. Wiley.

Brumann, C., & Gfeller, A. (2022). Cultural landscapes and the UNESCO world heritage list: Perpetuating European dominance. *International Journal Of Heritage Studies, 28*(2), 147–162. https://doi.org/10.1080/13527258.2021.1941197

Consejo Nacional de Monumentos. (2013). *Sitios de Patrimonio Mundial en Chile*. Ministerio de Educación.

Conti, A. (2015). *La conservación y la gestión de las ciudades históricas desde la perspectiva del Paisaje Urbano Histórico*. Resource document of Encuentro Internacional "El Paisaje Urbano Histórico como herramienta del desarrollo urbano sostenible". Centro de Investigaciones Científicas. Accessed 5 Apr 2020.

Cornu, M. (2001). La formation du droit du patrimoine culturel. In M. Cornu & J. Fromageau (Eds.), *Genèse du droit de l'envirnnement* (Vol. V. I, pp. 41–57). L'Harmattan.
Council of Europe. (2000). *The European landscape convention*. Council of Europe. https://rm.coe.int/european-landscape-convention-book-text-feb-2008-en/16802f80c6. Accessed 4 Apr 2022.
de la Torre, M. (2013). Values and heritage conservation. *Heritage and Society, 60*(2), 155–166.
Díaz-Otero, A. (2009). El valor jurídico del paisaje en el derecho público gallego. *Anuario de derecho parlamentario, 21*, 139–165.
Duverger, M. (1981). *Métodos de las Ciencias Sociales*. Editorial Ariel.
Endere, M. (2020). Latin America: Indigenous peoples' rights. In C. Smith (Ed.), *Encyclopedia of global archaeology*. Springer. https://doi.org/10.1007/978-3-319-51726-1
Endere, M. (2021). Enfoques sobre el patrimonio arqueológico y su impacto en el sistema legal de los países sudamericanos. *Complutum, 32*, 657–671.
Endere, M., & Prado, J. (2009). Criterios de selección valorización y zonificación de yacimientos arqueológicos y paleontológicos. In M. Endere & J. Prado (Eds.), *Patrimonio, Ciencia y Comunidad. Su abordaje en los Partidos de Azul, Tandil y Olavarría* (pp. 49–66). UNCPBA.
Fernández Rodríguez, C. (2007). *La protección del paisaje. Un estudio de derecho español y comparado*. Marcial Pons.
Galbrath, J., & Zaring, D. (2014). Soft law as foreign relations law. *Cornell Law Review, 99*, 735–794.
Gatti, G., & Muriel, D. (2006). El patrimonio, en el quicio de lo viejo y lo nuevo. La producción de la identidad en la sociedad del conocimiento. In G. Gatti & I. Martínez de Albeniz (Eds.), *La producción de la identidad en la sociedad del conocimiento. Cultura experta e identidad en el País Vasco* (pp. 25–67). Azkoaga. Cuadernos de Ciencias Sociales y Económicas.
Hernández, F. (1996). La conservación integral del patrimonio. *Complutum Extra, 6*(II), 251–260.
ICOMOS. (2014). *Declaración de Florencia. El patrimonio y el paisaje como valores humanos*. ICOMOS. https://www.icomos.org/images/DOCUMENTS/Secretariat/2015/GA_2014_results/GA2014_Symposium_FlorenceDeclaration_EN_final_20150318.pdf. Accessed 23 June 2019.
ICOMOS. (2017). *Principios concernientes a los Paisajes rurales como Patrimonio*. ICOMOS. http://www.revistas.unam.mx/index.php/mecedupaz/article/view/63951/56841. Accessed 23 June 2019.
Jameson, J. (2014). International conventions and charters and archaeology presentation. In C. Smith (Ed.), *Encyclopedia of global archaeology. Section: Legislation* (pp. 3948–3954). Springer.
Jokilehto, J. (2016). Valores patrimoniales y valoración. *Conversaciones con, 2*, 20–32.
Labadi, S., & Logan, W. (2016). Approaches to urban heritage, development and sustainability. In S. Labadi & W. Logan (Eds.), *Urban heritage, development and sustainability* (pp. 1–20). Routledge.
Levrand, N. (2020). Federalismo de Concertación y nuevas formas de gobierno de espacios interjurisdiccionales. El caso del sitio Ramsar Delta del Paraná. *Derecho y Ciencias Sociales, 23*, 93–110.
Levrand, N. (2021). El paisaje como un objeto de regulación jurídica en Argentina: Matriz de análisis legal y jurisprudencial. *Justicia y Derecho, 4*(3), 1–17.
Levrand, N., & Endere, M. L. (2020). Nuevas categorías patrimoniales. La incidencia del "soft law" en la reciente reforma a la Ley de patrimonio histórico y artístico de Argentina. *Revista Direito GV, 16*(2), 1–31.
Levratto, G., Delgrosso, M., Capoleoni, M., & Fiorelli, C. (2018). Los desafíos de una gestión participativa. Paisaje industrial Fray Bentos. In Comisión del Patrimonio Cultural de la Nación e Intendencia de Río Negro (Ed.), *Gestión del Patrimonio. Paisajes culturales y participación ciudadana* (pp. 191–205). Universidad Claeh.
Luxen, J. L. (2004). Reflections on the use of heritage charters and conventions. *The GCI Newsletter, 19*(2). http://www.getty.edu/conservation/publications_resources/newsletters/19_2/feature.html. Accessed 21 Apr 2021.

Mallarach, J. M., & Verschuuren, B. (2019). Changing concepts and values in natural heritage conservation: A view through IUCN and UNESCO policies. In E. Avrami, S. Macdonald, R. Mason, & D. Myers (Eds.), *Values in heritage management: Emerging approaches and research directions* (pp. 141–157). The Getty Conservation Institute.

Marcone Flores, G. C. (2019). Superando las dicotomías: el Qhapaq Ñan como ejemplo del patrimonio como proceso social. *Chungara Revista de Antropología Chilena, 51*(3), 457–469.

Martorell Carreño, A. (2010). *Itinerarios culturales y patrimonio mundial.* Universidad de San Martín de Porres.

Pereira, D. C. (2020). A chancela da Paisagem Cultural Brasileira: 10 anos de caminhos e descaminhos de uma política de cultura com compromisso social. *Revista Memória em Rede, 12*(22), 173–197.

Pocock, C., & Lilley, I. (2017). Who benefits? World heritage and indigenous people. *Heritage & Society, 10*(2), 171–190.

Prats, L. (2012). El patrimonio en tiempos de crisis. *Revista Andaluza de Antropología, 2*, 68–85.

Rey Pérez, J. (2017). *La aplicación de la recomendación sobre el paisaje urbano histórico (PHU) en Cuenca—Ecuador. Una nueva aproximación al patrimonio cultural y natural.* Universidad de Cuenca.

Rigol Savio, I. (2004). *Módulo 4. Gestión de Paisajes culturales. Programa de desarrollo de capacidades para el Caribe (CCBP).* UNESCO.

Rössler, M. (2007). World heritage cultural landscapes: A UNESCO flagship programme 1992–2006. *Landscape Research, 31*(4), 333–353.

Rotman, B. (2015). Procesos patrimoniales: redefiniciones, dinámica y tensiones en la contemporaneidad. *Que Haceres, Revista del Departamento de Ciencias Antropológicas FFyL-UBA, 2*, 11–26.

Saltalamacchia, H. R. (2005). *Del proyecto al análisis: aportes a una investigación cualitativa socialmente útil. Tomo Tercero.* El Artesano.

Silva Pérez, R., & Fernández Salinas, V. (2015). Los paisajes culturales de Unesco desde la perspectiva de América Latina y el Caribe. *Conceptualizaciones, situaciones y potencialidades. Revista INVI, 30*(85), 181–212.

UNESCO. (1972). Convention concerning the Protection of the World Cultural and Natural Heritage, Paris, 16 November, 1972. Entry into force: 17 December 1975. https://www.unesco.org/en/legal-affairs/convention-concerning-protection-world-cultural-and-naturalheritage Accesed 19 March 2024.

UNESCO. (2003). Convention for the safeguarding of the Intangible Cultural Heritage, Paris, 17 October 2003. Entry into force: 20 April 2006. https://www.unesco.org/en/legal-affairs/convention-safeguarding-intangible-cultural-heritage Accesed 19 March 2024.

UNESCO. (2005). *Memorándum de Viena Patrimonio Mundial y la Arquitectura Contemporánea Gestión del Paisaje Histórico Urbano.* UNESCO. http://www.journals.unam.mx/index.php/mecedupaz/article/view/57157/50716. Accessed 19 May 2020.

UNESCO. (2011). *Recomendación sobre el Paisaje Urbano Histórico.* UNESCO. http://portal.unesco.org/es/ev.php-URL_ID=48857URL_DO=DO_TOPICURL_SECTION=201.html. Accessed 23 May 2019.

Urquijo, P. (2020). Paisaje cultural, un enfoque pertinente. In P. Urquijo and A. Boni (Coord.), *Huellas en el paisaje. Geografía, historia y ambiente en las Américas* (pp. 17–37). Centro de Investigaciones en Geografía Ambiental.

von Droste, B., Plachter, H., & Rossler, M. (1995). *Cultural landscapes of universal value: Components of a global strategy.* Gustav Fischer.

Wardana, A. (2020). Neoliberalizing cultural landscapes: Bali's agrarian heritage. *Critical Asian Studies, 52*(2), 270–285.

Weissheimer, M. R. (2012). Paisagem Cultural Brasileira: Do Conceito À Prática. *Fórum Patrimônio, 5*(29), 1–17.

World Heritage Commission. (2021). *The operational guidelines for the implementation of the world heritage convention*. UNESCO. https://whc.unesco.org/en/guidelines/. Accessed 5 Apr 2022.

Zuluaga Varón, D. C. (2015). *El derecho al paisaje en Colombia: consideraciones para la definición de su contenido, alcance y límites*. Universidad Externado de Colombia.

Archaeological Impact Studies: A Reflection on an Experience in the Salado River, Buenos Aires Province, Argentina

Olivia Sokol, Virginia Salerno, Paula Escosteguy, María Magdalena Frère, and María Isabel González

Abstract Archaeological impact assessment is a fundamental tool for heritage conservation in the context of activities involving earthwork. In Latin America, this professional practice strengthened during the last decades of the twentieth century. Its implementation involves tensions and agreements between different actors: inhabitants, landowners, companies, government agencies, international credit organizations, and diverse professionals. This has provoked a series of disciplinary debates regarding the role of archaeology in development projects. In this chapter, we present the case of the Salado River Integrated Master Plan (PMI), a large-scale project covering a total area of 170,000 km^2, that has been underway for more than two decades in the province of Buenos Aires, Argentina, in a territory where an extensive settlement process —more than 2000 years long— took place. In the late 1990s, the PMI was designed to protect the environmental and economic value of the basin, improve the drainage network, and mitigate the effects of floods and droughts. Its execution was planned in stages involving different infrastructure developments that resulted in significant alterations to the river landscape, affecting the archaeological heritage of the area. We present our own experience, both before and after carrying out archaeological impact studies, as well as the methodology applied, and the results obtained. From this case study, we highlight the importance of integrating the resulting information with that of preexisting research projects. We also reflect on the achievements and difficulties of heritage management.

O. Sokol (✉) · V. Salerno · P. Escosteguy
CONICET; Instituto de arqueología, Facultad de Filosofía y Letras, Universidad de Buenos Aires, Buenos Aires, Argentina
e-mail: olivia.sokol@conicet.gov.ar; vmasalerno@conicet.gov.ar; pdescosteguy@uba.ar

M. M. Frère · M. I. González
Instituto de arqueología, Facultad de Filosofía y Letras, Universidad de Buenos Aires, Buenos Aires, Argentina
e-mail: magdafrere@filo.uba.ar; igonzale@filo.uba.ar

G. F. Bonnat et al. (eds.), *Current Research in Archaeology of South American Pampas*, The Latin American Studies Book Series,
https://doi.org/10.1007/978-3-031-55194-9_14

Keywords Heritage · Landscape · Professional practice · Social actors · Development projects · Hydraulic works

Introduction

Development projects, particularly those involving earthwork, require archaeological impact assessments (AIA). These assessments are essential tools for heritage conservation (Criado Boado, 1999; Guráieb & Frère, 2012; Ratto, 2009, 2013). These types of tasks entail different problems related to the practice, both from a professional point of view and from the possible conflicts that mobilize large-scale works or that involve modifications in areas of great environmental and archaeological sensitivity.

In the first part of this chapter, we reflect on the development of impact studies in Latin America, in general, and in Argentina, in particular. In our region, this professional practice consolidated at the end of the twentieth century with the introduction of legal actions that required the execution of environmental impact assessments (EIA) within sustainable development projects. Its consolidation, based on the formulation of specific legal frameworks, has provoked a series of disciplinary debates regarding the role of archaeology in development projects and heritage management. From this regional framework, we address the particularities of the development of impact archaeology in Argentina, where its implementation has posed multiple challenges regarding the link with academic archaeology and the participatory processes that are mobilized through the implementation of dissimilar heritage management approaches.

In the second part, we present the experience developed by our research team in the context of a large-scale project in Argentina. This covers approximately 170,000 km^2 and has been in execution for more than two decades in the Salado River basin in Buenos Aires province. At the end of 1990, the Salado River Integrated Master Plan (PMI by its acronym in Spanish) was designed to protect the environmental and economic value of the basin, improve the drainage network, and mitigate the effects of floods and droughts (Dirección Provincial de Obra Hidráulica, 2017). Its execution was planned in stages involving different infrastructure works, with great modifications in the river landscape that affected the archaeological heritage of the area.

Archaeological research in the Salado River Depression, initiated at the end of the 1980s, covers the Lezama, Chascomús, General Paz, General Belgrano, San Miguel del Monte, Lobos, Roque Pérez, and Navarro districts (i.e., counties). We know from these studies that in this territory there was an extensive settlement process, for more than 2000 years, by populations whose economy was based on hunting, gathering, and fishing. In addition, they were expert potters who used local clays to make pots of different sizes and shapes (González, 2005; González & Frère, 2019).

After warning in various instances and to different authorities about the impact generated by the works involved in expanding the river's capacity, our research group was called to participate in the archaeological impact studies only in sector IV. The objectives of this work are to present our experience before and after the implementation of archaeological impact studies; also to publicize the applied methodology and the results obtained with the work carried out; and finally, to reflect on the challenges faced by archaeology and, in particular, impact archaeology in our country, considering our role as professionals given the need to work with the different actors involved and the implications of the landscape changes that the large public works entail.

Archaeological Impact Assessment: A Regional Overview

In Latin America, impact archaeology consolidated as a field of professional practice in the last decades of the twentieth century. This was fostered by the introduction of legal actions that required the execution of EIA within sustainable development projects. On average, by the decade beginning in 2000, almost all the countries in the region had environmental laws that, in many cases, included strategic environmental assessments (SEA) or sectorial or regional legislation resources (CEPAL, 2013). In this regard, García Leyton (2004) points out that the institutionalization process of these studies is linked to the need to satisfy the requirements demanded to grant credits by international financial organizations (e.g., International Development Bank, World Bank, European Bank of Investments, International Bank for Reconstruction and Development, among others). For these reasons, in most cases, these studies were implemented with the aim of complying with international standards in relation to the development of projects in a particular place or public work. From the concrete results, the specificity of these ways of environmental assessment, linked to specific projects (which start late and finish early) and without a criterion for their formulation and evaluation, has been recognized as a weakness. Over time, the SEA has positioned itself as an overcoming resource because it incorporates the principles of sustainability from policies, plans and programs to the formulation of projects (Shepherd & Ortolano, 1996).

These legal frameworks also defined the scope and limits of AIA. Depending on the political-institutional trajectories of each country, the evolution of the regulations opened a wide range of situations and forms of implementation. Thus, for example, in some countries, such as Chile and Brazil, this field of archaeology application represents most of the professional practice (Carrasco, 2006; Menezes Ferreira, 2010; Salazar et al., 2011). While in other countries, such as Ecuador, Colombia, and Peru, there was a growing expansion of fieldwork for archaeologists in areas distanced from traditional scientific research (Echeverría-Almeida, 2009; Londoño, 2016; Tantaleán & Gonzales Panta, 2016). In some countries, like

Uruguay and Argentina, impact archaeology is being developed with a reduced workflow, little presence of private consultants and few specific training spaces. In Argentina in particular, as will be described in the following sections, some impact archaeology experiences have developed within mining and hydroelectric mega-projects that endured over time and involved archaeological interventions on a large spatial scale (e.g., González, 1982; Madero & Reigadas, 2002; Ratto, 2013). In addition, some of these experiences are articulated with academic training and research spaces (Beovide & Caporale, 2009; Guráieb & Frère, 2012; Ratto et al., 2019; Verdesio, 2019).

In this work, without setting aside diversity, we focus on some common aspects to contextualize and reflect on the challenges that impact archeology faces today at a regional level. At first, we consider it relevant to remark that the inclusion of archaeological heritage within the environmental assessment system is based on an "environment" notion that includes sociocultural elements. This began to be forged in the 1970s with the perspectives that revised the restricted idea of cultural heritage and focused on physical assets (Millán, 2004). Since then, the debates and conceptual shifts have been reflected in the speeches spread mainly by international organizations through declarations that argued about the importance of sustaining and promoting cultural diversity (e.g., ICOMOS, 1990; III Earth Summit, 1992; UNESCO, 1972). Thus, cultural heritage management positioned itself at the center of both environmental and heritage problems, which was expressed through the institutionalization of heritage based on specific policies, legislation, and management plans (Yúdice, 2003).

From this interpretational framework, cultural heritage was conceived in limited and static terms, as a possible resource to be integrated into the productive fabric (Lombardo, 2009). Heritage policies were intertwined with contradictory multiculturalist discourses of "diversity recognition" and "sustainable development" (Ortiz, 2002; Wright, 2004).[1] In Latin America, these processes developed alongside the emergence of new historically subalternized sectors of civil society, especially indigenous peoples (Gnecco & Ayala Rocabado, 2011) and Afro-descendants (Sampaio Guimarães, 2016), whose demands are often articulated with heritage issues. In this way, the archaeological heritage management included a framework of increasing diversification of relations between the state, archaeologists, and local and indigenous actors, with new spaces for participation (Ayala Rocabado, 2017).

In turn, the consolidation of AIA generated multiple debates regarding the possibilities and limitations of this way of practicing archaeology in close connection with commercial logic and development projects (Ayala Rocabado, 2017; Cáceres Roque & Westfall, 2004; Londoño, 2013; Pellini, 2017; Piazzini Suárez, 2001; Villa, 2003). Thus, for example, the way in which the intersection between state,

[1] Multicultural policies contributed to generating tools for disciplining cultural diversity, however, their formulation responds to the confluence of various factors, among which we underline the historical struggle that indigenous peoples have been developing regarding their autonomy and self-determination. In this sense, these legal frameworks also enable spaces for political action and organizational strengthening (Wright, 2004).

market, and territory configured fragmented practices between research and management that reproduced modern dichotomies—state-market, academy-service provision—has been discussed (Carrasco, 2006). In some countries of the region, changes in study programs, both undergraduate and postgraduate, were also observed with a tendency to shorten their duration and emphasize technical aspects. Different authors have argued that these changes reduce the possibility of promoting critical thinking regarding the link between heritage management, the study of the past, and social conflicts (Gnecco & Schmidt Días, 2017). From our point of view, this field of professional application is crossed by a series of tensions and conditioning factors linked to the social, political, and economic system in which it is inserted. In addition, in most cases, archaeologists usually have a limited capacity to influence decision-making processes on development projects. Despite these limitations, we consider it necessary to include the problems that affect these areas of professional practice in both undergraduate and graduate training. Also, we stress the importance of promoting spaces for discussion and exchange where the experiences of the so-called impact archaeology and academic archaeology can be integrated.

Finally, another aspect discussed in relation to AIA in the Latin American sphere refers to the approaches to heritage management that it mobilizes. Based on specific experiences, it has been observed that these approaches prioritize an essentialist view of heritage referents (Menezes Ferreira, 2010) and generate participatory dynamics that foster the disciplining of cultural diversity (Ayala Rocabado, 2017; Jofré, 2015). On the other hand, there is consensus on the importance of interdisciplinary and comprehensive approaches that consider the different levels of conflict involved in heritage management linked to impact archaeology, whether with local communities, academy, state agencies, and private companies (Beovide & Caporale, 2009; Menezes Ferreira, 2010; Villa, 2003; Zarankin & Pellini, 2012).

Archaeological Impact Assessment in Argentina

In Argentina, the first impact studies were developed within the framework of rescue archaeology in sites that would remain underwater after the construction of large hydroelectric dams (Sanguinetti de Bórmida, 1981; Gradín, 1984; Ceruti, 1981; Rodríguez & Rodríguez, 1985; Rodríguez, 1978; González, 1982). From these experiences, the need to have a legal framework for heritage safeguarding became visible. It was only in 2003 that the National Law 25,743/2003 (National Law for the Protection of Archaeological and Paleontological Heritage) and its Decree-law 1022/2004 were sanctioned, thus declaring archaeological heritage as public domain. Since then, the State exercises its tutelage and arbitrates all the necessary steps for its preservation, study, and dissemination (Endere & Rolandi, 2007). This includes actions for its management, registration, transport, research permissions, or report/registration of offenders to the heritage. It should be clarified that each province is the authority that applies the law in its jurisdiction and therefore can dictate its own regulations regarding heritage in its territory (Guráieb & Frère, 2012). Since the enactment of the law, different realities and levels of conflict have risen depending on the types of territorial intervention projects, the provinces where they were implemented, the involvement of local actors, and the presence of social movements, indigenous communities, and environmentalists.

Unlike what happens in other countries in the region, archaeology professionals in Argentina are generally employed by the state and devoted to research and academic teaching, with impact activities being a complementary economic reinforcement. For these reasons, to date, most of the impact activities have been carried out by members of the research groups that work in the areas under intervention. This scenario allows feeding back the investigation results with the impact studies. In recent years, archaeologists who do not have a scientific-academic insertion have begun to be hired by private consultancy firms. Besides, few worker cooperatives formed by archaeologists have been registered. These areas of insertion provide the possibility of obtaining new positions in our profession. Faced with this, the debates were updated on the importance of having public records of the professionals involved and accessible repositories with the results of the studies that allow the integration of impact archaeology into academic research (e.g., Carballo Marina et al., 1999; Franco, 2017).

Regarding impact work, we differentiate between larger and smaller projects. The first group includes projects that either affect large areas or occupy smaller areas but have a high archaeological, paleontological, or sociocultural sensitivity (Lisboa & Oliva, 2020). Many of them, with public, private, or mixed financing and execution, are linked to the use of non-conventional energies such as mining developments, hydraulic works, construction of gas pipelines, dams, and neighborhoods, exploitation of hydrocarbon deposits, road works, high tension lines, dredging, among others (Funes & Ferreira, 2015; Jofré, 2015; Lucero et al., 2020; Madero & Reigadas, 2002; Matera & López, 2015; Núñez Camelino, 2015; Ratto, 2009; Ratto & Orgaz, 2002; among others). Other cases of AIA are grouped into smaller projects, linked to minor work surfaces. In recent years, these projects have increased considerably in Argentina, linked to urban and public service works, installation of wind turbines in wind farms, photovoltaic solar parks, gated communities, and landscape interventions in protected areas (Ambasch et al., 2019; Lisboa & Oliva, 2020; Loponte, 2012; Naturata, 2016; Onetto et al., 2009; Ramos et al., 2013; Weissel et al., 2019, among others).

Some papers that refer to AIA offer exclusively archaeological references, since they describe the remains and their finding condition, and do not discuss the conflicts caused by those works. On the other hand, at meetings, conferences, and congresses, the discussion arises about the need for management that considers the institutions and social actors directly or indirectly involved. In addition, the professionals reflect on the actions aimed at conserving and publicizing the archaeological impact studies, as well as agreeing on theoretical-methodological outlines that will guide these studies in their different undertaking stages (Jofré, 2015; Lucero et al., 2020; Ratto, 2013; Ratto & Carniglia, 2019; Weissel et al., 2019). These spaces for academic exchange have increased visibility. In this sense, we highlight the journal

Práctica Arqueológica of the Association of Professional Archaeologists of the Argentine Republic (AAPRA for its acronym in Spanish), created in 2017, which encourages the presentation and discussion of experiences and results related to impact studies on archaeological assets in different contexts.

As a result of these archaeological impact studies in Argentina, several points emerge to be remarked: they constitute a new work field within the archaeological discipline, they lead to the mediation of different state institutions in different stages; they promote the elaboration and promulgation of regulations, and they produce data in areas that were not previously researched, improving and complementing existing information (Ratto, 2009). In addition, it has been observed that these activities can modify behavior patterns at the community level. This is how, for example, indigenous communities interested in preserving the sites of their ancestors began to actively report works that did not have an AIA (Jofré, 2015; Picoy, 2020). In this framework, heritage management linked to these forms of intervention acquires centrality as expressed in the multiple actions carried out: among others, we mention workshops to generate participatory processes in local areas; activation of archaeological sites for touristic benefit; dissemination of proposals in graphic, audio-visual and digital media (Aichino et al., 2012; Athor, 2016; Bianchi Villelli & Fernández Do Río, 2019; Corbalán, 2011; Corbalán et al., 2017; Lucero et al., 2020; Matera et al., 2019; Núñez Camelino, 2020; Onetto et al., 2009; Weissel et al., 2019 among others). Finally, in recent years, the topics related to this area of disciplinary exercise began to be included in the university curriculum. An example of this is the first specialization career oriented to the evaluation and study of archaeological impact of our country, which has been created by the Faculty of Philosophy and Letters of the University of Buenos Aires (Ratto et al., 2019).

In this sense, cultural agency, that is, the capacity for action of social groups through their cultural practices and worldviews– is constantly negotiated in the relationships between the local and the global, in coordination with other actors such as the state, market, cultural industries, tourism sector, non-governmental organizations, and transnational organizations (Yúdice, 2003). The set of experiences mentioned here highlights the need to contemplate the different levels of interrelation, articulation, and negotiation between actors to achieve the goals proposed. However, it also shows that it remains a challenge for Argentina.

Next, we develop our impact work experience in a hydraulic megaproject in the Salado River basin, Buenos Aires province.

The Framework of the Integrated Master Plan (PMI) in the Salado River Basin

The Salado River is the main fluvial corridor of Buenos Aires province. It is a plain course that suffers periodically and frequently from floods and prolonged droughts. This condition brought significant negative consequences for production in the agricultural sector, especially in the form of flooding and economic losses but also social impacts (Dirección Provincial de Obra Hidráulica, 2017). For these reasons, the provincial authorities called an international consultant financed by the World Bank to carry out the pre-feasibility studies. As a result, in 1999, they presented the PMI, which included the basin plus the surrounding territory, completing a total area of 170,000 km^2 within the province of Buenos Aires (D.P.H., n.d.). The objective of this megaproject was to reduce flooding impacts and to improve the basin's economic growth along with the preservation and development of its environmental value (Halcrow, 1999a, b, c). Its implementation was planned in stages and the effective start occurred 5 years after the consultant's presentation (Fig. 1).

The PMI execution was estimated to remove, in total, around 200 million m^3 of excavation to be deposited in approximately 500 km of land in the filling enclosures (Halcrow, 1999b). The works immediately near the river included the excavation of its banks (using a system of dredges and backhoes), while the resulting sediments were used to increase the surfaces above the flood level and to expand productive areas. The distribution of the land was carried out in filling enclosures that were arranged considering the morphology of the terrain—concentrated in marginal lowlands—and the presence of wire fences and buildings. In this way, different work sectors were established: (a) canalization area on both banks of the river; (b) 200-m area defined as a biological corridor; and (c) sector of possible location of the enclosures. On the other hand, complementary actions to the work were established, the

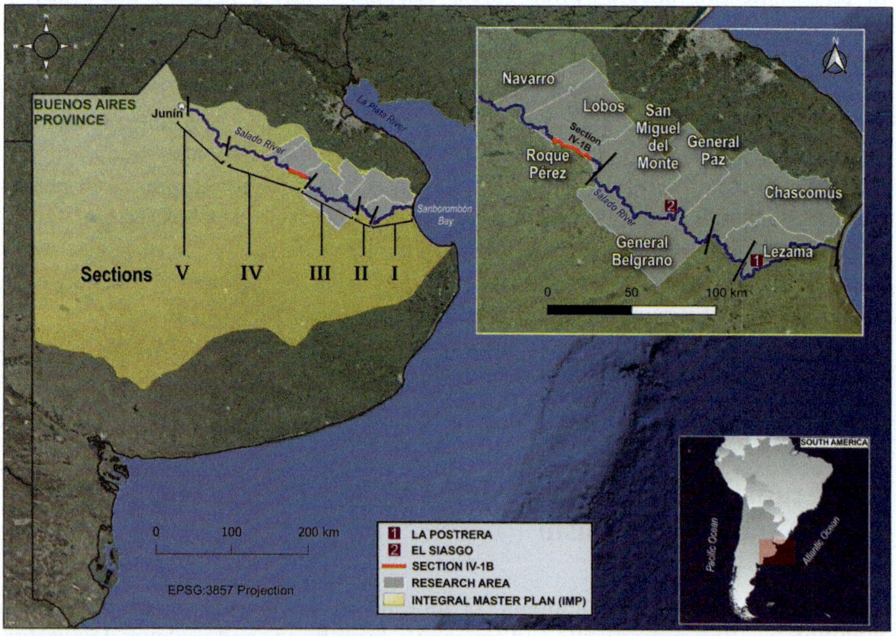

Fig. 1 Total area and sections of the PMI, research permit sector, and sites affected by the work in 2004 and 2016

development of projects to protect urban centers against flooding, improvement and drainage of rural routes and roads, repair of bridges, development of sewers, and creation of spas and recreational spaces, among others.

Throughout its different implementation stages, this plan included various instances of participation with local actors, the academic sector, and non-governmental organizations linked to issues relevant to the basin (Dirección Provincial de Obra Hidráulica, 2017; Halcrow, 1999a; UT Salado, 2019). Regarding the inhabitants' involvement, a basin committee was formed with officials from the different stakeholders involved. In addition, over time, public hearings were held with the participation of officials, specialists, and civil organizations. The reports of these instances detail that there were no conflicts regarding the works and that the main issues addressed referred to concerns of the owners about the landscape transformations that resulted from the location of spas and areas of sediment removal and filling of flooded sectors.

The Archaeological Dimension in the Framework of the PMI

A work of this magnitude is highly relevant for archaeological studies since human groups existed in Buenos Aires province from the early Holocene. The remains found that document populations date from at least 10,000 years BP to historical times (Prates et al., 2020). Particularly, the Salado River basin, affected by this work, was a scenario in which hunter-gatherer-fisher groups settled between 2400 and 400 years BP and perhaps earlier, since dating on human remains yielded antiquity of ca. 3500 years BP (Escosteguy et al., 2017; Frère et al., 2016; González, 2005).

By the end of the 1980s, systematic archaeological research began in the sector. The studies cover the lower and middle basins of the river up to its mouth. Thus, in recent years, solid information has been acquired on the populations that inhabited the area, which allowed the systematization of a cultural development model integrated with the rest of the Pampean region during the late Holocene (González, 2005; González & Frère, 2019). From these studies, it was determined that the hunter-gatherer-fisher and potter groups that occupied the Salado River wetland had a variety of both animal and plant resources. Among them, small- and medium-sized fauna, the tala forest (*Celtis tala*) and the abundant mud stand out. At the same time, they developed extensive interaction and exchange networks, for example, to obtain rocks not available locally (González, 2005; González et al., 2021; González & Frère, 2019).

The archaeological sites are in landscape units (dunes, watersheds) that have remained emerged at times of maximum flooding of the Salado River and the lakes (González, 2005; González & Frère, 2019). Research has allowed us to propose that lake bodies are highly sensitive to climatic variations and periods of drought and flooding. Therefore, the river could have constituted a safer environment (González & Frère, 2019).

From the beginning, our archaeology research group has carried out different types of activities, around the archaeological heritage, with various actors: on the one hand, within the framework of institutional relations with museums, public administration officials, civil society organizations, schools, and higher education institutions; and on the other hand, with actors linked to our research work in experimental archaeology, such as farmers, fishermen and potters, as well as landowners, rural residents, collectors and other local referents. The latter correlates with the promotion of a work line in public archaeology that we have been developing for several years (Salerno, 2014, 2018; Salerno & González, 2014).

The PMI works began in 2004 at the Salado River mouth in Samborombón Bay and were projected in stages toward western districts. Sections I, II, and III developed without archaeological impact studies. In different moments of these first years of execution, we recorded the landscape transformations and the impacts that the work had in sectors with archaeological sensitivity. Thus, in 2004, we observed alterations at the La Postrera site, in which we had worked 2 years before. Among the modifications, we recognized the cutting of "tala" forests, large removals of sediment, and roads widening. In addition, on that occasion, we recorded ceramic material on the surface as a product of soil alterations. In 2016, our team returned to the site and continued to document the dramatic changes generated in the landscape that included, among other things, the removal of ravines (Fig. 2).

Another sector where we recorded impacts was the El Siasgo stream, at its mouth in the Salado River. Here, in May 2015, during our archaeological surveys, we observed the work of large machinery and the dredger, as well as vehicles passing through the area. We walked the path of the workshop, where we recovered surface or half-buried material and recorded embankments and pipes parallel to the river ravine. Along the internal paths of the work, we found ceramic remains and lithic material (Fig. 3).

As a result of this evidence, formal complaints were submitted to the intervening municipal and provincial agencies (González et al., 2017). At that time, we identified communication difficulties and lack of articulation between the different agencies, as well as the lack of control in the EIA, especially in the AIA. This situation was communicated at the International Symposium on Cultural Heritage and Local Community (González et al., 2017), where the damage caused in the different sectors of the work throughout those years was exposed to the academic community.

Finally, in 2018, the company in charge of the EIA contacted us to carry out the AIA in Section IV-Stage 1B (TIV-1B), a territory of the Lobos and Roque Pérez districts (counties). Therefore, between February and March 2019, we carried out field and office activities, including informal interviews with residents, municipal officials, owners, and producers of the land affected by the works. These interviews sought to reveal these actors' knowledge regarding the presence of archaeological materials, as well as the expectations and concerns about the work and the landscape transformations involved. The information that these local actors can provide to the AIA processes is extremely useful because they live there and, hence, they know the dynamics of the lakes and the river, their rises and flood levels, and the impact of torrential rain or drought periods, among other things. In this way, we can

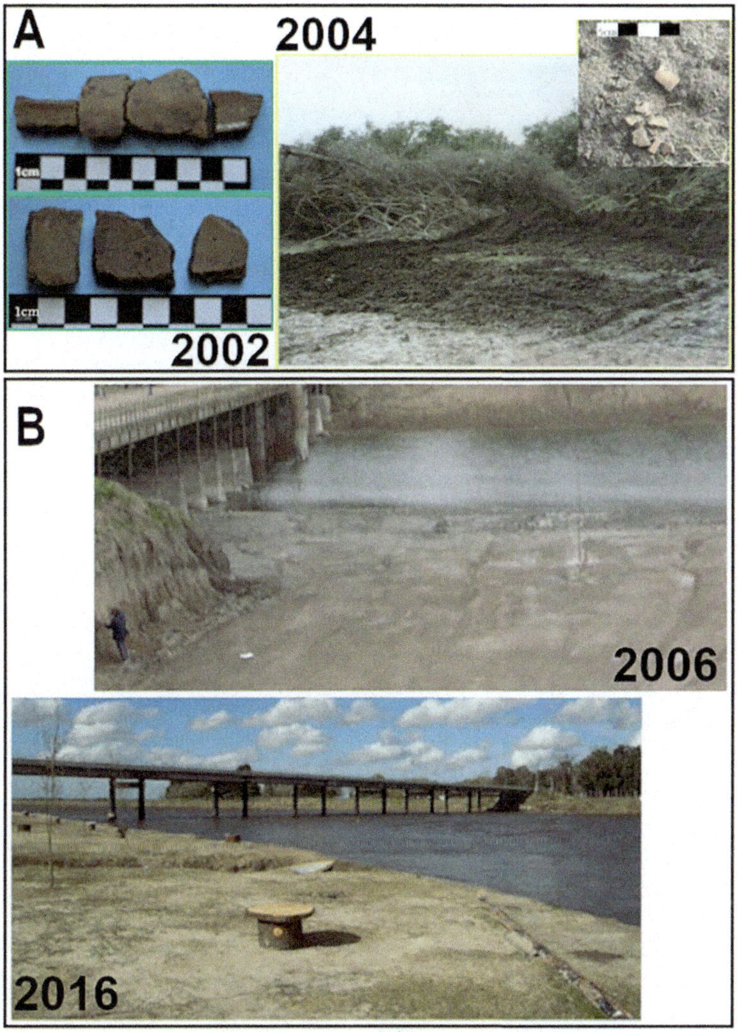

Fig. 2 La Postrera site. (**a**) Materials found in 2002 and a photo of the sector taken during May 2004, showing fragments on the surface and soil movement caused by machinery. (**b**) Photos from 2006 and 2016 showing landscape transformations

recognize the field of interpersonal relationships and the potential emergence of conflicts. In addition, these instances allow knowing different ways of inhabiting and signifying space, as well as the existence of positions regarding potential memory and heritage processes linked to archaeological materials.

The baseline study included surveying the area directly and indirectly affected by the work through pedestrian surveys on both banks of the Salado River. Although we had archaeological information on Laguna de Lobos, approximately 15 km from

Fig. 3 El Siasgo in 2015: surveys during the execution of the work. (**a**) Machinery working at the confluence of the Salado River and the El Siasgo stream. (**b**) Lithic artifacts on the surface. C and D: Surface collection along paths of the work

the river (Frére & González de Bonaveri, 1993; Escosteguy et al., 2017), we did not have a systematic survey of the river coast. Therefore, the possibility of carrying out these tasks contributed to the knowledge of the area that we had been developing in our research project.

The methodology applied in this archaeological impact study was the full-coverage prospection modality (Barreiro Martínez, 2000; Barreiro Martínez & Criado Boado, 2005). During the observations made in the transects, we controlled and recorded variables related to the description of possible archaeological evidence and other contextual evidence related to environmental aspects that define archaeological visibility. Regarding the excavation method, we made test pits at high points of the land, close to the water, where the low visibility due to the vegetation did not allow for a visual examination. We carefully checked the soil profiles and removed the sediment to locate possible artifacts or traces of past human activity and, finally, we covered them. All the information was georeferenced on a map, and we created a chart with the surveyed points (Fig. 4). This allowed us to visualize the degree of archaeological sensitivity for the sector. In total, 75 test pits were carried out, 43 on the margins of the Roque Pérez district and 32 on the margins of the Lobos district. We did not find archaeological material in any case, but we found a quartzite flake on the surface. Based on these activities, we wrote the baseline reports that permitted the company to start channeling the section. We must clarify that due to the characteristics of the work, through the dredging of the river, it was impossible to hold a continuous presence in the field during the soil movements. However, on

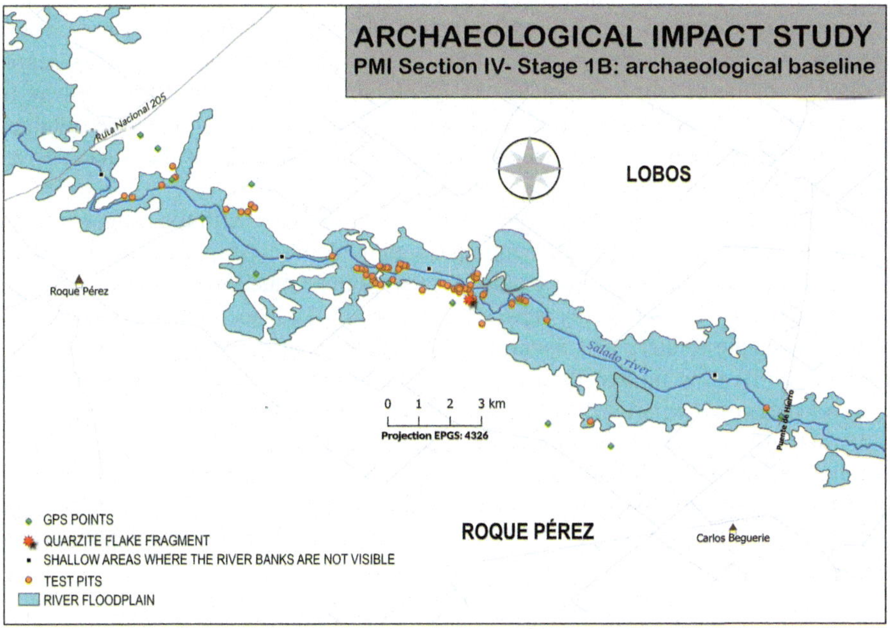

Fig. 4 Baseline Section IV-Stage 1B

several occasions, we had the opportunity to see and walk through different sectors of the dredging as we continued performing tasks (baselines of backfill sectors) for at least two more years.

In conclusion, the results of the test pits proved to be sterile with respect to their specific informative potential. We have established as of low archaeological risk both the channeling sector and the 200-m area of the biological corridor—because it is a low-lying and floodable land area—since both exhibited a very low expectation of containing archaeological findings. Similarly, we identified a low archaeological sensitivity in the affected sectors as supporting points and workshops. However, in the report, we made explicit the need to train the work staff, to take rescue actions when finding material, and carry out mitigation measures. The latter included the enhancement of the findings, if any, for their study, conservation, exhibition, and dissemination to the community. Based on this, the recommendations for the corresponding actions in the event of finding cultural goods were published in the environmental management plan (*Plan de Gestión Ambiental*) for the sector (UT Salado, 2019). Within the management proposals, it was established that baselines should be made for all the filling enclosures.

On the other hand, conducting interviews with local actors, as mentioned previously, confirmed our results of low archaeological sensitivity, since, for the area, all the interviewed actors indicated that they had no knowledge of pre-Hispanic cultural goods in that section of the river. In turn, these interviews allowed us to identify different positions and issues regarding the work that changed over time, and their realization. Mainly, at the beginning of the project, we registered several doubts about the effectiveness of the work to prevent flooding. Concern was also expressed among some rural residents about whether the filling enclosures would become a fertile land for agricultural production. In other cases, the work represented a source of employment for the inhabitants of both districts and a way of economic activation (mainly from local officials' perspectives). Also, we obtained information about the knowledge that the inhabitants had about the Salado River wetland, its hydrography, fauna, and flora.

During the various fieldwork carried out simultaneously with the progress of the work, we registered different perceptions about the landscape modifications and the impacts of some activities. For example, local potters stated that the sectors of the river where they collected clay were affected, becoming inaccessible in some cases. Similarly, some hunters and fishermen pointed out that the changes in the river's course and its flow limited their recreational and extractive activities. Such is the case of fishermen who used to frequent the river to catch various species and now choose to go to nearby lakes, where the waters are calmer. The hunters mentioned a similar situation: they highlighted certain changes in coypus' (*Myocastor coypus*) distribution, a prey that they currently prefer to obtain in streams, marshes, and lakes. The landscape changes over the years in the study area which can be seen in Fig. 5.

Fig. 5 Google Earth satellite images of Section IV-1B. The aerial photograph was taken from the YouTube channel of the Dirección Provincial de Obra Hidráulica (D.P.H., 2021)

Discussion

The PMI of the Salado River basin is an emblematic project for Buenos Aires province, not only because of its size but also because of the duration of its implementation, which has been running for almost two decades. The proposed structural measures have a great impact on the living conditions of the area's population. For these reasons, its implementation had been long-awaited, and virtually from the beginning, this mega-work had a social license. Even though consulting firms and international financing organizations participated in its formulation, the environmental impact assessment feasibility report did not consider archaeological problems. This study only mentioned that the execution of specific sections would require additional assessments, because "there is a great interest among researchers in the study of archaeological sites in the Salado River basin associated with lake bodies, though there are no sites protected by any regulation" (Halcrow, 1999c, p. 29). For these reasons, our intervention, at first, could only document the destruction and alteration of archaeological contexts simultaneously with the development of the works or, in subsequent instances, to their implementation. Based on our claims, the subsequent development of the projected, bided, and executed sections included AIA within the framework of EIA. From this instance, we were able to generate concrete actions to document the presence or not of archaeological materials and, if necessary, prevent their destruction.

As previously mentioned, the impact studies were carried out simultaneously with our research work in the area, which articulates the search for pre-Hispanic archaeological evidence with the study of the ways in which archaeology forms part of the present in local areas. The background and previous research knowledge were crucial to outlining the AIA work plan. Throughout three decades, the integration of various material, written and oral sources related directly or indirectly to the population processes we study allowed us to consider multiple dimensions. In this way, we generated a broad perspective of the archaeological problems addressed.

This approach contemplates the diverse population trajectories in the Salado River basin, which in part are used in the current representations made on archaeological objects. In this way, we could think about the microregion's territory as a space in which historical processes marked by the transformation and permanence of practices and narratives took place. Among them, we remark on the experiences gone through in the river that are nourished by long-lasting historical meanings from activities such as fishing and the use of clay to make pottery (Salerno, 2018). These activities mobilize processes of territorial appropriation and identification (Benedetti, 2011; Brubaker & Cooper, 2001). In particular, the reproduction of pre-Hispanic pottery techniques was recognized as a practice that conceives memorial processes. These can be interpreted from the notion of embodied memory (sensu Connerton, 1989) because through bodily actions, the "know-how" with local clays relative to the indigenous past is updated in the present (Salerno & Cañardo, 2022).

This background has led us to problematize the space of the microregion as a historical, political, and strategic construction (Lefebvre & Enders, 1976) and its territoriality as the expression of the way in which people use the land, organize space, and assign meanings to the place within multiple conflict links (Benedetti, 2011). For these reasons, we underline the importance of the landscape modifications involved in the PMI and their impact on the different actors' relations with the space, the way of experiencing the river, and the activities that they unfold there. Given the situation of the socio-environmental emergency in which the PMI was developed, these impacts were considered as minor problems and have not been particularly stressed by the people interviewed. However, in a long-term perspective, we consider that these are significant impacts because they imply the interruption or modification of practices that allow updating the links between past and present from the reuse of places, objects, and extractive activities or creation processes, as is the case with pottery and fishing.

According to the debates that have been taking place in the region regarding the implementation of the AIA, we share the importance of contemplating conflicts caused by the works in their different stages of implementation and of generating truly participatory instances that include the perspectives of institutions and social actors involved. Another aspect that our experience led us to rethink is how development projects promote a naturalistic view of the environment (Saldi et al., 2019). This emphasizes the notion of space as a physical container of natural resources to be exploited or conserved without considering the social and historical processes that constitute it.

Conclusions

There is a tendency to expand economic development through infrastructure works. Within this framework, EIA are carried out as key elements to preserve the environment and especially the cultural heritage. The experience presented here led us to reflect on our professional role and on how a policy that links archaeological studies and sustainable development is currently defined. In fluctuating economic scenarios, common in Argentina, there is a real need to reconcile and link different lines of activity that generally remain isolated, such as basic research—committed to the production of new knowledge about the past –, applied research –oriented to the development of cultural heritage management conventions and procedures– and the provision of specialized technical services (Cáceres Roque & Westfall, 2004).

Often, archaeologists participate in these lines of activities, particularly in technical instances aimed at recording the presence/absence of archaeological sites and materials as well as establishing recommendations regarding conservation and mitigation. These interventions are even more limited when the archaeological remains have little visibility in the landscape, as is the case of the Salado River basin. This condition caused the archaeological dimension to be underestimated within the planning and execution of EIA, to the point that we found ourselves working isolated from the rest of the professionals involved in the assessment. By limiting our interventions to technical instances, the potential of archaeological contributions is lost, especially regarding the knowledge about how current residents are linked to archaeological materials and the space they inhabit, sometimes building meanings between the past and the present.

In turn, in the face of previous impact situations on the microregion heritage, the archaeological work team itself achieved instances that allowed linking the research project with the investment project. In sections I, II, and III of the PMI, the absence of archaeologists caused the loss of this type of information. From the moment we began to intervene, the results of the AIA were addressed considering our scientific background. On the one hand, the realized baseline helped us get to know a sector that up to the moment had not been prospected or assessed within the framework of our research project, with external financing. Thus, we verified the absence of archaeological materials and directed the study toward other sectors of the microregion. On the other hand, the residents' perspectives on the impact the work had on their daily practices allowed us to deepen our knowledge of the dynamic processes of territorial identification and appropriation. In this sense, large-scale transformations such as those brought about by the implementation of the PMI made it possible to see how relationships and practices that can be modified converge in the landscape. This leads us to consider the way in which, as archaeologists, we theorize about the past based on the intersection of different ways of feeling and inhabiting the territory that the different social groups have, both past and present.

In the EIA, the right to live in a healthy environment as well as to access and participate in the processes that directly involve people's lives are put into action. A solid and high-quality impact study that considers workshops and interviews with the community, and which includes an approach to the environment that considers its complexity, arbitrates the means to ensure that the projects are carried out with the guiding principle of sustainable development. This, in accordance with socially responsible actions and initiatives, would allow economic resources to be directed toward issues, aspects, or needs considered to be priorities for these communities (Criado Boado, 1999). Thus, residents, companies, professionals involved, and government authorities will benefit from the results of realizing these investment projects. In the same way, the quality of the information produced through the AIA, in addition to contributing to the protection of the past and the local memories, can result in a broader knowledge of the regional pre-Hispanic past.

Acknowledgments We thank Engineer Roberto Molina for providing us with videos and aerial photographs, and Professor Alejandro Fernández for the translation. We also thank the two anonymous reviewers for their constructive comments. This research was funded by the Agencia Nacional de Promoción Científica y Tecnológica, grants PICT 2015-0272 and PICT 2016-0368, and by the Secretaría de Ciencia y Técnica, Universidad de Buenos Aires, grant UBACyT 2018-20020170100525BA. The authors are the sole responsible for the arguments discussed in this chapter.

References

Aichino, L., De Carli, M. C., Zabala, M. E., & Fabra, M. (2012). Procesos de activación y valoración del patrimonio arqueológico a través de la Cartografía Social. *Revista EXT, 3*(27), 1–27.

Ambasch, M., Andueza, P., & Villaroel, A. (2019). *Informe de Estudio de Impacto Arqueológico (EIArq). Proyecto Solar Fotovoltaico Distribuido de la provincia de Jujuy "Parque Solar Fotovoltaico Lavayén"*. Departamento de San Pedro.

Athor, J. (2016). Museos costeros, historias de pasiones. In J. Athor & C. Celsi (Eds.), *La Costa Atlántica de Buenos Aires. Naturaleza y patrimonio cultural* (pp. 591–638). Vazquez Mazzini editores, Fundación Azara.

Ayala Rocabado, P. (2017). El multiculturalismo neoliberal y la arqueología de contrato en el norte de Chile. In C. Gnecco & A. Schmidt Dias (Eds.), *Crítica de la razón arqueológica: arqueología de contrato y capitalismo* (pp. 143–162). Instituto Colombiano de Antropología e Historia.

Barreiro Martínez, D. (2000). *Propuesta para una Metodología de Evaluación de Impacto Arqueológico*. Universidade de Santiago de Compostela, Spain.

Barreiro Martínez, D., & Criado Boado, F. (2005). Evaluación de impacto ambiental y arqueología: perspectivas. In *Primeras jornadas de Patrimonio Arqueológico en la Comunidad de Madrid* (pp. 115–132).

Benedetti, A. (2011). Territorio: concepto integrador de la geografía contemporánea. In P. Souto (Ed.), *Territorio, lugar, paisaje. Prácticas y conceptos básicos de geografía* (pp. 11–82). Facultad de Filosofía y Letras.

Beovide, L., & Caporale, M. (2009). La Arqueología de Contrato en el marco de la gestión integral del patrimonio arqueológico de la región metropolitana de Montevideo. *Revista de Arqueología Americana, 27*, 7–29.

Bianchi Villelli, M., & Fernández Do Río, S. (2019). Arqueología de rescate en Bariloche. Límites y posibilidades en la práctica arqueológica. *Revista Práctica Arqueológica, 2*, 14–34.

Brubaker, R., & Cooper, F. (2001). *Más allá de "Identidad"* (Apuntes de Investigación del CECyP, 7) (pp. 30–67).

Cáceres Roque, I., & Westfall, C. (2004). Trampas y amarras: ¿es posible hacer arqueología en el sistema de evaluación de impacto ambiental? *Chungara Revista de Antropología Chilena, 36*(Special Supplement, espec. Arica), 483–488.

Carballo Marina, F., Espinosa, S., & Belardi, J. B. (1999). Estudios del impacto ambiental (EsIA) sobre el recurso arqueológico en la provincia de Santa Cruz. *Propuestas teórico metodológicas. Arqueología, 9*, 215–237.

Carrasco, C. (2006). La práctica arqueológica y la actual construcción de conocimiento en Chile. *Boletín de la Sociedad Chilena de Arqueología, 39*, 35–50.

CEPAL. (2013). Acceso a la información, participación y justicia en temas ambientales en América Latina y el Caribe. Situación actual, perspectivas y ejemplos de buenas prácticas. *Serie Medio Ambiente y Desarrollo, 151*, 13–39.

Ceruti, C. (1981). *Proyecto Investigaciones Arqueológicas en el área del Parana Medio - Margen Entrerriana*. Resumen de Actividades (period March 1978-September 1981) [MS].

Connerton, P. (1989). *How societies remember*. Cambridge University Press.

Corbalán, M. (2011). Pertenecer no tiene sus privilegios. Reflexiones sobre la práctica de los estudios de impacto arqueológico en el valle de Tafí, Tucumán. *Mundo de antes, 6*(7), 53–69.

Corbalán, M., Di Tullio, E., Rodríguez, M. E., & Díaz, O. (2017). Encuentros y Desencuentros en la gestión del patrimonio arqueológico en la provincia de Tucumán. *Andes, 1*(28), 00–00.

Criado Boado, F. (1999). Tendencias en la conservación del patrimonio cultural: demandas tecnológicas y científicas. *Trabajos de Prehistoria, 56*(1), 13–26.

D.P.H. (2021, June 5). *Clip: Presentación Banco Mundial (Mayo 2021)*. https://www.youtube.com/watch?v=Moxo1K9qWMM. Accessed 12 Apr 2022.

D.P.H. (n.d.). *Cuenca del Río Salado, Provincia de Buenos Aires*. https://www.gba.gob.ar/hidraulica/interior_bonaerense/cuenca_salado. Accessed 12 Apr 2022.

Dirección Provincial de Obra Hidráulica. (2017). *Evaluación de impacto ambiental y social. Ampliación de capacidad del río Salado Superior-Tramo IV, Etapa 1B. Apoyo a la Gestión Integral de la Cuenca del Río Salado y Ejecución de Obras Contempladas en el Tramo IV, Etapa 1B del Plan de Manejo Integral*. Buenos Aires.

Echeverría-Almeida, J. (2009). La arqueología de contrato en Ecuador. Breves reflexiones. *Revista de Arqueología Americana, 27*, 37–48.

Endere, M. L., & Rolandi, D. (2007). Legislación y gestión del patrimonio arqueológico. breve reseña de lo acontecido en los últimos 70 años. *Relaciones de la Sociedad Argentina de Antropología, XXXII*, 33–54.

Escosteguy, P. D., Scabuzzo, C., & González, M. I. (2017). Análisis bioarqueológico de los restos del arroyo El Siasgo, (supuesto Homo caputinclinatus de Ameghino 1910). *Revista Argentina de Antropología Biológica, 19*(2), 1–14. https://doi.org/10.17139/raab.2017.0019.02.04

Franco, N. (2017). La relación entre los estudios de impacto ambiental e investigaciones arqueológicas entre los años 2006 y 2016 en la Provincia de Santa Cruz (Argentina). *Revista Práctica Arqueológica, 1*(1), 31–45.

Frère, M. M., & González de Bonaveri, M. I. (1993). Utilización de ambientes lagunares: laguna de Lobos -Provincia de Buenos Aires. *Arqueología, 3*, 171–197.

Frère, M. M., González, M. I., & Greco, C. (2016). Continuity in the use of shallow sites of the Salado River Basin in the Pampean Region, Argentina. *Radiocarbon, 58*(4), 921–933.

Funes, M. A., & Ferreira, M. E. (2015). Obra Altas Cumbres: corrección y mitigación del impacto arqueológico en una obra vial de la Provincia de Córdoba. In S. Matera, M. Kergaravat, & G. Spengler (Eds.), *Estudios de impacto ambiental y la protección del patrimonio arqueológico* (pp. 11–36). Imprenta digital.

Gradin, C. (1984). *Investigaciones Arqueológicas en Casa de Piedra, Santa Rosa*. Ministerio de Educación y Cultura.

García Leyton, L. A. (2004). *Aplicación del Análisis Multicriterio en la evaluación de Impactos Ambientales* (Unpublished doctoral thesis). Universidad Politécnica de Catalunya.

Gnecco, C., & Ayala Rocabado, P. (Eds.). (2011). *Indigenous peoples and archaeology in Latin America*. Left Coast.

Gnecco, C., & Schmidt Días, A. (2017). Sobre Arqueología de Contrato. In C. Gnecco & A. Schmidt Días (Eds.), *Crítica de la razón arqueológica: arqueología de contrato y capitalismo* (pp. 9–28). Instituto Colombiano de Antropología e Historia.

González, R. (1982). Generalidades y objetivos. In R. L. Wilson & G. Loyola (Eds.), *Arqueología de rescate. Ponencias presentadas en la Primera Conferencia de Arqueología de Rescate del Nuevo Mundo* (pp. 102–139). The Preservation Press.

González, M. I. (2005). *Arqueología de alfareros, cazadores y pescadores pampeanos (Colección.)*. Sociedad Argentina de Antropología.

González, M. I., & Frère, M. M. (2019). Río Salado: Espacio de interacción de cazadores-recolectores-pescadores (provincia de Buenos Aires, Argentina). *Revista del Museo de la Plata, 4*, 611–632.

González, M. I., Frère, M. M., Salerno, V., & Escosteguy, P. D. (2017). Estimaciones del impacto arqueológico de las obras hidráulicas en la cuenca del Río Salado. In *Simposio internacional de Patrimonio Cultural y Comunidad local. Buenas prácticas de gestión*. Olavarría.

González, M. I., Escosteguy, P. D., Salemme, M. C., Frère, M. M., Weitzel, C., & Vecchi, R. (2021). Assessing strategies for coypu hunting and use in the Salado River Depression (Buenos Aires Province). In *Ancient hunting strategies in Southern South America* (pp. 59–81). Springer. https://doi.org/10.1007/978-3-030-61187-3_3

Guráieb, A. G., & Frère, M. M. (2012). *Caminos y encrucijadas en la gestión del patrimonio arqueológico argentino* (2nd ed.). Facultad de Filosofía y Letras.

Halcrow. (1999a). *Plan Maestro Integral de la Cuenca del Río Salado. Volumen Principal y 17 Anexos*. Halcrow.

Halcrow. (1999b). *Plan Maestro Integral de la Cuenca del Río Salado. Evaluación de Impacto Ambiental. Volumen Principal y 8 Apéndices*. Halcrow.

Halcrow. (1999c). *Subregión A1: Estudio de Factibilidad Informe de Evaluación de Impacto Ambiental*. Halcrow.
ICOMOS (1990). *Carta Internacional para la Gestión del Patrimonio Arqueológico*. https://www.icomos.org/images/DOCUMENTS/Charters/arch_sp.pdf
III Earth Summit. (1992). *Río earth summit*. Río de Janeiro.
Jofré, I. C. (2015). Mega-mining, contract archaeology, and local responses to the global order in Argentina on JSTOR. *International Journal of Historical Archaeology, 19*(4), 764–774.
Lefebvre, H., & Enders, M. (1976). Reflections on the politics of space. *Antipode, 8*(2), 30–37.
Lisboa, L., & Oliva, F. (2020). Vista de Nuevas perspectivas acerca de la regulación de los estudios de impacto arqueológicos y paleontológicos en la Provincia de Buenos Aires. *Revista Práctica Arqueológica, 3*(2), 30–38.
Lombardo, E. (2009). Patrimonio y nuevas formas de colonialidad. In B. Manasse & P. Arena (Eds.), *Arqueología, tierras y territorios: conflictos e intereses* (pp. 203–213). Editorial Lucrecia.
Londoño, W. (2013). Arqueología para el desarrollo y arqueología del desarrollo: una visión desde Colombia. In A. Herrera (Ed.), *Arqueología y desarrollo en América del Sur* (pp. 147–165). Universidad de los Andes.
Londoño, W. (2016). Arqueología por contrato y nuevos contratos arqueológicos. *Jamgwa Pagna, 15*(1), 117–128.
Loponte, D. (2012). *Análisis de sensibilidad arqueológica y paleontológica del área metropolitana concesión AySA*. Biblioteca Aysa.
Lucero, G., Sacchi, M., & Saleta, M. J. (2020). Negativo por negativo, positivo: una reflexión sobre el uso de las líneas de base en las evaluaciones de impacto arqueológico en la Argentina. *Revista Práctica Arqueológica, 3*(2), 39–52.
Madero, C., & Reigadas, M. C. (2002). Estudio de caso, evaluación y mitigación de impacto arqueológico. El trabajo del arqueólogo en el ámbito de las relaciones institucionales y empresariales. In *Cuadernos de la Facultad de Humanidades y Ciencias Sociales, Universidad Nacional de Jujuy, 18* (pp. 203–213). San Salvador de Jujuy, Argentina.
Matera, S., & López, L. (2015). Estudios de impacto arqueológico en el Norte de Santa Cruz: los alcances del registro arqueológico en zonas con diferentes niveles de modificación antrópica. In *Estudios de impacto ambiental y la protección del patrimonio arqueológico* (pp. 77–104). La Imprenta Digital.
Matera, S., López, L., Pallo, M. C., & Cirigliano, N. A. (2019). Estudios de impacto e investigación arqueológica: una experiencia complementaria en Tapi Aike (sur de la Provincia de Santa Cruz, Argentina). *Revista Práctica Arqueológica, 2*(2), 35–51.
Menezes Ferreira, L. (2010). Arqueología comunitaria, arqueología de contrato y educación patrimonial en Brasil. *Jangwa Pana, 9*(1), 95–102. https://doi.org/10.21676/16574923.23
Millán, S. (2004). Cultura y patrimonio intangible: contribuciones de la antropología. *Patrimonio Cultural y Turismo, 9*, 55–70.
Naturata, S. A. (2016). *Estudio de Impacto Ambiental Parque Eólico Kosten*. Área Pampa de Castillo-Kosten. Comodoro Rivadavia. https://ambiente.chubut.gov.ar/wp-content/uploads/2016/07/EsIA-Parque-E%C3%B3lico-Kosten-.pdf
Núñez Camelino, M. (2015). Las experiencias de protección del patrimonio arqueológico en la provincia de Corrientes. In S. Matera, M. Kergaravat, & G. Spenlger (Eds.), *Estudios de impacto ambiental y la protección del patrimonio arqueológico* (pp. 37–55). Imprenta digital.
Núñez Camelino, M. (2020). Los estudios de impacto y rescate arqueológico y su importancia en la protección del patrimonio de la ciudad de Corrientes. *Revista Práctica Arqueológica, 3*(2), 19–29.
Onetto, M., Funes, M. L., & Murgo, A. (2009). Integrando a la comunidad en la protección del patrimonio: el caso de Cueva de las Manos, Perito Moreno, provincia de Santa Cruz. *Novedades de Antropología, 19*(64), 7–11.
Ortiz, R. (2002). Globalización y esfera pública. Entre lo nacional y lo transnacional. *Signo y Pensamiento, XXI*(41), 68–81.

Pellini, J. R. (Ed.). (2017). *Arqueología Comercial. Dinero, alienación y anestesia*. JAS Arqueología.

Piazzini Suárez, C. E. (2001). Arqueología, legislación y desarrollo. Balance de los noventa y perspectivas desde la coyuntura actual. In D. Patiño Castaño (Ed.), *Simposio Arqueología, Sociedad y Estado* (pp. 23–40). Universidad del Cauca.

Picoy, C. (2020). *De Sitio Punta Canal a Territorio Indigena Punta Querandí. Patrimonios y memorias en disputas* (Unpublished graduation thesis). Universidad de Buenos Aires.

Prates, L., Politis, G. G., & Pérez, S. I. (2020). Rapid radiation of humans in South America after the last glacial maximum: A radiocarbon-based study. *PLoS One, 15*(7), e0236023. https://doi.org/10.1371/JOURNAL.PONE.0236023

Ramos, M., Helfer, V., Salerno, V., Darigo, M., Dottori, C., Raies, A., et al. (2013). Estudio de impacto arqueológico por la obra del monumento de Vuelta de Obligado, San Pedro, Provincia de Buenos Aires. In M. Ramos, V. Helfer, V. Pernicone, F. Bognanni, C. Landa, V. Aldazábal, & M. Fernández (Eds.), *Arqueometría Argentina. Estudios pluridisciplinarios* (pp. 37–51). PROARHEP, UNLu, Aspha, Docuprint.

Ratto, N. (2009). Aportes de la arqueología de contrato al campo de la investigación: estudios de casos en Patagonia y Noroeste de la Argentina. *Revista de Arqueología Americana, 27*, 49–70.

Ratto, N. (2013). *Patrimonio Arqueológico y Megaproyectos Mineros en Argentina. Turismo, desarrollo y sociedad*. Aspha ediciones.

Ratto, N., & Carniglia, D. (2019). Práctica arqueológica y la construcción de puentes comunicacionales entre disciplinas. *Práctica Arqueológica, 1*(3), 28–44.

Ratto, N., & Orgaz, M. (2002). Arqueología e impacto arqueológico: el caso del sistema de producción agrícola en las Quebradas de Villavill, Carapunko y Las Pampitas (Departamento Andagalá, Catamarca). *Mundo de antes, 3*, 43–63.

Ratto, N., Guráieb, A. G., Franco, N., & Basile, M. (2019). La especialización en evaluación y estudio de impacto arqueológico de la Facultad de Filosofía y Letras de la Universidad de Buenos Aires. *Revista Práctica Arqueológica, 2*(2), 78–83.

Rodríguez, J. (1978). El proyecto Antropológico-Ecológico Salto Grande y el manejo de los recursos culturales. In *VI Congreso Nacional de Arqueología Argentina. San Juan, Argentina.*

Rodríguez, J., & Rodríguez, A. (1985). *Proyecto Antropológico – Ecológico Salto Grande (Primer Informe)*. Concordia.

Salazar, D., Jackson, D., & Troncoso, A. (2011). Entre dos tierras: Reflexionando sobre la arqueología chilena en el siglo veintiuno. In P. Ayala & F. Vilches (Eds.), *Teoría arqueológica en Chile. Reflexionando en torno a nuestro quehacer disciplinario* (pp. 41–70). Universidad de Chile, Universidad Católica de Chile y Quillqa.

Saldi, L., Maferra, L., & Barrientos Salinas, J. A. (2019). Ontologías en disputa. Diálogos entre la antropología y la arqueología para la problematización de paisajes regionales. *Antípoda. Revista de Antropología y Arqueología, 37*, 3–26. https://doi.org/10.7440/antipoda37.2019.01

Salerno, V. (2014). *Trabajo arqueológico y representaciones del pasado en la provincia de Buenos Aires*. Editorial de la Facultad de Filosofía y Letras, UBA.

Salerno, V. (2018). Testimonios que nos da la tierra. Apropiación de objetos arqueológicos en la provincia de Buenos Aires, Argentina. *Antipoda. Revista de Antropología y Arqueología, 31*(abril-junio), 89–107. https://doi.org/10.7440/antipoda31.2018.05

Salerno, V., & Cañardo, L. (2022). El "otro lado" de la historia: Patrimonio arqueológico y procesos de memoria. *Latin American Antiquity, 33*, 1–15. https://doi.org/10.1017/laq.2022.1

Salerno, V., & González, M. I. (2014). Conocimiento en relación. Reflexiones sobre el trabajo de campo arqueológico en el curso medio e inferior del Río Salado bonaerense. *Revista del Museo de Antropología Universidad Nacional de Córdoba, 7*(1), 25–38.

Sampaio Guimarães, R. (2016). Entre memorias de Portugal y de África: las políticas patrimoniales en la región portuaria de Río de Janeiro, Brasil. *Quadernos, 44*, 51–66.

Sanguinetti de Bórmida, A. (1981). Síntesis del desarrollo cultural de la cuenca inferior del río Limay en el área de El Chocón y su ubicación en el cuadro de la prehistoria de la Patagonia. *Trabajos de Prehistoria, 1*, 171–179.

Shepherd, A., & Ortolano, L. (1996). Strategic environmental assessment for sustainable urban development. *Environmental Impact Assessment Review, 16*(4–6), 321–335. https://doi.org/10.1016/S0195-9255(96)00071-6

Tantaleán, H., & Gonzales Panta, A. (2016). La arqueología de contrato en el Perú: surgimiento, caracterización y perspectivas. In J. R. Pellini (Ed.), *Arqueología de Contrato en Latinoamérica* (pp. 177–198). JAS Editores.

UNESCO. (1972). *Convención sobre la protección del Patrimonio Mundial, cultural y natural*. 17° General Conference, Paris, November 16th.

UT Salado. (2019). *Plan de Gestión Ambiental Ampliación de la capacidad del Río Salado Superior*. Tramo IV/Etapa 1B. LPI N01/17. Provincia de Buenos Aires.

Verdesio, G. (2019). La arqueología y la explotación de la riqueza de la tierra en Uruguay en el marco del desarrollismo y la globalización neoliberal. In P. Aparicio Resco (Ed.), *Arqueología y Neoliberalismo* (pp. 233–242). JAS Arqueología Editorial.

Villa, F. A. (2003). Arqueología de rescate y desarrollo sostenible en Colombia: crítica a un vínculo instrumental. In C. Gnecco & E. Piazzini (Eds.), *Arqueología al desnudo: Reflexiones sobre la práctica disciplinaria* (pp. 327–344). Universidad del Cauca.

Weissel, M. N., Weissel, A. R., & Rodríguez Basulto, B. A. (2019). Arqueología y patrimonio cultural en las obras de Aguas y Saneamientos Argentinos de la Cuenca Matanza Riachuelo, Provincia de Buenos Aires. *Cuadernos del INAPL Series Especiales, 8*(2), 324–337.

Wright, S. (2004). La politización de la "cultura". In M. Boivin, A. Rosato, & V. Arribas (Eds.), *Constructores de Otredad. Una introducción a la antropología social y cultural* (Vol. 14, pp. 128–141). Antropofagia.

Yúdice, G. (2003). *El recurso de la cultura. Usos de la cultura en la era global*. Gedisa.

Zarankin, A., & Pellini, J. R. (2012). Arqueologia e Companhia: Reflexões sobre a introdução de uma lógica de mercado na prática arqueológica brasileira. *Revista de Arqueología, 25*(2), 44–60. https://doi.org/10.24885/SAB.V25I2.354

Archaeological Heritage Management of Baradero's Historical Area (Buenos Aires Province): Salvage and Socialization for Public Use

Aniela R. Traba, Verónica Noya, and Ulises A. Camino

Abstract The archaeological site Anchorena 942 (the San Martín theater) is located within the foundation area of Baradero's city, in the same place where the indigenous town (*pueblo de indios*) was settled between 1615 and 1775 A.D., originating the Spanish settlement in colonial times. Faced with the fortuitous finding of old structures beneath the theater's stage, the first stage of an archaeological rescue was initiated, accompanied by proposals for the musealization in situ. The excavations revealed an exceptional record linked to traditions of indigenous, European, and Creole societies, exposing the town's historical depth and multicultural origin. The discovery enhanced the interest to participate in various local community social actors (official and ecclesiastical authorities, neighbors, among others), revealing the need to investigate and preserve this kind of culturally and historically significant places. These scenarios create opportunities to elaborate proposals for the preservation and socialization of cultural heritage at different scales, articulating significant memories and historical and archaeological spaces, new and already known, which contribute to the construction of collective memory. Our objective here is to present this experience by focusing on the challenges of generating strategic and multidisciplinary planning. It aims to integrate the social construction of knowledge with different social actors' interests and demands through research and heritage management. To conclude, two main subjects are discussed: their inclusion

A. R. Traba (✉)
Cooperativa de Trabajo Arqueocoop ltda. – Dto. de Cs. Sociales, División Historia, Universidad Nacional de Luján, Luján, Argentina

V. Noya
Dirección de Cultura, Secretaría de Cultura, Turismo, Educación y Deportes, Municipalidad de Baradero, Baradero, Argentina
e-mail: cultura@baradero.gob.ar

U. A. Camino
Centro de Arqueología Urbana, Instituto de Arte Americano, Facultad de Arquitectura, Diseño y Urbanismo, Universidad de Buenos Aires – Cooperativa de Trabajo Arqueocoop ltda, Buenos Aires, Argentina
e-mail: camino.4381@umsa.edu.ar

G. F. Bonnat et al. (eds.), *Current Research in Archaeology of South American Pampas*, The Latin American Studies Book Series,
https://doi.org/10.1007/978-3-031-55194-9_15

in comprehensive long-term projects aimed at public use for educational, cultural, and tourist purposes, and the scope and possible modalities of action for contract archaeology in Argentina.

Keywords Urban archaeology · Contract archaeology · Historical archaeology of Argentina · Archaeology cooperative · Heritage strategic planning · Value-based management

Baradero and Its Historical Area

Baradero is the head city of the homonymous district located in the northeast of Buenos Aires Province (Argentina), on the western bank of the Baradero River, a secondary tributary of the Paraná River (Fig. 1). Founded in 1615, it is considered the oldest city in the province and is also characterized by having the first agricultural colonies developed from the arrival of European immigrants to the country in the mid-nineteenth century (especially Swiss, German, and French settlers). It also has a port built at the beginning of the twentieth century.

Since 2010, the district of Baradero has had a Historical Heritage Department, currently depending on the Municipality Department of Culture, Education, Tourism and Sports. The archaeological research developed in the area since the 1990s is directed by Ph.D. Alicia Tapia. Since its creation, the HH Department counted on her full collaboration for archaeological works and other heritage-related activities. All the localities of the district (Baradero, Alsina, Irineo Portela, Santa Coloma) are characterized by a strong historical and cultural heritage.

The indigenous town (*"pueblo de indios")* founded between 1615 and 1775 A.D. originated the Spanish settlement of Santiago de Baradero in colonial times. The first archaeological findings related to the functioning of this seventeenth century reduction were at the Cementerio Indígena site (Indigenous Cemetery), located on a ravine of the Baradero river, 800 m from the main square of the city (Mitre Square). There, the archaeologist Salvador Debenedetti exhumed 14 burials in 1910 (Debenedetti, 1911).

The discovery of an early nineteenth century plan of Baradero at the Dirección Provincial de Geodesia, allowed the drawing of a first plan of the archaeological potential of the city, superimposing its ancient layout with modern satellite images. The first archaeological excavation with an urban approach took place in an old house built in the last third of the nineteenth century (Anchorena 994 site). This house, which had suffered the collapse of its roof, was located in the block in front of the main square, close to the constructions related to the church of the early nineteenth century. Two campaigns were carried out between 2014 and 2015, excavating under the beams that formerly supported a floor of pitch-pine wood. A stratigraphic layer characterized by fragments of "*adobes,*" bricks, tiles, and "*musleras*" roof tiles (handmade) was detected at this site. Within the adobes (sun-dried mud bricks), fragments of indigenous and Hispanic-Indigenous ceramics were rescued, which

Fig. 1 Location map with the historical sites mentioned in this chapter

brings us closer to a relative dating for the manufacture of adobes around the eighteenth century (Camino & Noya, 2015). We assume that these remains come from the demolition of a previous construction in the same place, or in the same block, and that they were used to level the ground. The characteristics of the construction artifacts found are size of the bricks, type, thickness, and paste of the tiles, as well as the ceramic fragments found within the adobe components, which are coincident with Baradero's colonial period.

These initial works continued in 2017 with an open-area excavation campaign in the backyard of the Asociación Española de Socorros Mutuos (Spanish Mutual Aid Association), located in the same block. The results of this intervention consisted of the finding of several subway structures, such as remains of canals, walls, and foundations, accompanied by evidence of successive soil removal activities and embankments with sediment, debris remains, and discarded domestic material culture (Camino & Tapia, 2017). Among these fill components, two levels were distinguished. Based on the analysis of the ware and glass assemblages, their deposition could be ascribed to ca. 1840–1860 the first one, and ca. 1900–1915 the second one (Traba, 2018). Finally, the finding of a brick floor with an east–west orientation could be linked to one of the ancient churches of the city, dated to the first half of the nineteenth century (Camino et al., 2022).

Intervention in the San Martin Theater

The San Martin Theater is located in the city of Baradero (District II, Section A, Block 17, Lot 19), facing 942 Anchorena Street, at 30 m.a.s.l. Its geographic coordinates are 33°48′27″ S, 59°30′4″ W (WGS84); 6259559.04, 5546192.92 (POSGAR 94 – Argentina 5); IGN Chart 6296 (1:50.000). This theater was built in 1954 by the Santiago Apostol Parish in a parcel of its property, adjacent to the church and in front of the main square (the Mitre Square), that is, within the foundational city block. Since then, it has been used as a theater and cinema, becoming an important cultural reference for the local community. The space also fulfilled other social functions such as being a hall for school events and end-of-year diploma ceremonies. The cinema ceased its activities around 1990, while the theater remained open until 2013 when it was closed due to the need for structural repairs.

The building works required for the reopening of the theater began to be developed at the end of 2018. At the time of lifting the wooden boards that constituted its stage, a filling of rubble was observed with the presence of certain ancient objects, such as a large metal clock marker and the remains of a baptismal font, among others. Thanks to the predisposition of the parish priest, the Historical Heritage Department was called in to supervise the works. The discovery of more archaeological objects, features, and construction materials posed the need to communicate those findings to the archaeologists who had previously worked in the area and, at the same time, receive some advice from them. It should be noted that the Municipality of Baradero already has specific legislative provisions aimed at the preservation of its cultural heritage. In this regard, it is also worth mentioning Ordinance 4.213 for the creation of the Historical Center of the city, in force since 2009.

After an initial professional assessment of the archaeological potential of the site, the Secretary of Culture of the Municipality, collaborating with the Santiago Apostle Parish, decided to carry out the systematic recovery of the deposited material goods; thus, the Cooperativa de Trabajo Arqueocoop Ltda. (an archaeological professional cooperative) was hired to accomplish an archaeological rescue within a period of three to six months.

From Archaeological Rescue to Heritage Intervention

Systematic excavations on the area free of construction inside the theater were carried out in 2019 with a total area of ca. 80 m^2. The exposed deposits were disturbed by successive anthropic actions on the site. The archaeological context of intervention was organized in three sectors (1–3 from north to south), differentiated by a notorious disruption between the deposits. The methodology consisted of an open-area excavation by stratigraphic units (sensu Harris, 1991) and its registration was made in descriptive charts. In the case of particular features or findings, sediment and material samples were taken for dating and/or characterization, which are available for future research.

We also worked with a general survey of historical sources that included plans, photographs, documents, and occasional oral testimonies. Given the stipulated time limitations, this task was carried out in an unsystematic manner by the authors, with the special collaboration of Alexis Urquiza (collaborator of the Historical Heritage Area of the Santiago Apostle Parish).

In this first stage of excavation, a total of 91 stratigraphic units were identified throughout the exposed surface. Some of these units presented an important palimpsest of anthropic interventions; especially the sediments of Sector 3 West have suffered diverse removals and redeposits, causing the mixture of materials belonging to different temporalities.

So far, about 50% of the materials recovered in excavation have been processed. We base inventory on the identification of general macroscopic criteria (class of material, functionality, segment of the piece, surface treatments, inscriptions, taxon at class level, and other particularities) in order to obtain a first quantification of the findings and diagnostic information on their cultural and temporal ascription. As for the material recovered, the following were obtained: vegetal, malacological, osseous, vitreous, ceramic, metallic, lithic, plastic, and textile remains, mostly representative of daily and construction activities. The sample reached 4498 elements.

As for the archaeological context excavated, four major construction phases were identified, chronologically ascribed to the nineteenth and twentieth centuries. It is worth noting the presence of artifacts of indigenous tradition (plain and decorated ceramic sherds, with and without red paint, and a lithic flake) mixed with modern fill deposits, which raises the hypothesis that their presence is due to the removal of older deposits of the site. The construction phases and preliminary interpretations are briefly described below.

San Martín Theater (1954–Present)

The theater was built between 1953 and 1954 on the initiative of Father Eugenio Betoño of the Santiago Apóstol Parish, with the intention of creating a cultural space that would also generate economic support for educational work. For this purpose, it was proposed to re-functionalize a piece of land adjacent to the church, which was occupied by the parish house. To this end, a large part of that building was demolished, particularly the side rooms.

As shown in the construction plan approved by the Municipality (Obras de ampliación Casa Parroquial de Baradero, 1953), the perimeter structure and the rear rooms of the previous building would remain standing, while those at the front of the building would be modified, including an upper floor; likewise, on the first floor, there was the creation of a slope for the future hall and other rooms related to the stage and dressing rooms. Several of these structures would have been modified again until today. Since its opening in 1954, the building has functioned as a cinema-theater, becoming over the years a cultural reference of the city, highly valued in the local community memory.

The context corresponding to this stage presents poorly consolidated and negative anthropic fillings of trenches located especially near the walls, primarily linked to the adaptation of the soil for the support of the stage and possibly for the removal/closure of drainage pipes of the previous building. The archaeological material includes demolition debris with construction pieces (tiles, bricks) and a variety of modern objects from both the construction itself and the subsequent use of the theater (including sequins, clothing, bottles of alcoholic beverages, cables, plastic objects, a 1954 coin and soda caps from the 1978 World Cup).

Parish House (1893–1953)

The building that was located on the current site of the theater served as the Parish House from 1894 to 1953. According to the construction sketch approved for the beginning of its construction in 1893 (Casa Parroquial, 1893), the sidewall of the church, two rear enclosures connected to it and, on the other side, the wall of the adjacent neighbour Don Pablo Morixe (currently the land of the Asociación Española de Socorros Mutuos [AESM]) would be reused. In addition, a series of enclosures identified as "rooms" were built: two toward the front and two on the right side, continued by another for "kitchen" and a smaller one for "latrine"; the space in turn had an internal "L" shaped courtyard, where a cistern was located (see Fig. 2a).

The archaeological context presents some features related to this building, such as two drains (Sector 1 and 3); a flat surface demarcated by bricks, and remains of red tiles in a horizontal position associated with the location of the latrine (Sector 3); a small sidewalk of French red tiles (Sector 1, Fig. 2.a.1) that would communicate the exit of a closed opening to the SE with the inner courtyard; and at least two

Fig. 2 (continued) San Martín Theater is superimposed; (a.1–4): Views of stratigraphic units and constructive elements; (**b**) above. General view of the site with its historic façade; (**b**) below. Stratigraphic section at the end of excavation, NE-SW profile; (**c**) kaolin tube pipe; (**d**) *pas de calais* tiles; (**e**) English annular ware bowl and shell-edge dish fragment; (**f**) stoneware bottles and jars; (**g**) adobe fragment; (**h**) Hispanic-Indigineous ceramic remains; (**i**) Guaranitic-style ceramic rim; (**j**) European majolica base; (**k**) fragmented siliceous rock flake; (**l**) turned hispanic pottery with green glaze. (Source: (a.4) Municipality of Baradero Press; (**b**) above, by Pablo Berninger; (**b**) below, by Lucía Suárez Paz)

Fig. 2 (**a**) Site plan of the parish house approved by the Municipality of Baradero, ca. 1893; in hatching, the pre-existing walls are indicated; in shading, the initial excavation plan inside the

large earthen deposits (Sector 1 and 3) composed of silty-clayey sediments mixed with demolition debris and other various artifacts, as well as archaeofaunal remains (UE 13 and UE 17). Among the finds linked to these units and of chronology between the end of the nineteenth century and the first decades of the twentieth century, we find, for example, a kaolin pipe tube from Glasgow, Scotland (Fig. 2c); French red tiles and *pas de Calais* tiles (Fig. 2d); a patent medicine glass flask and a torpedo-type glass bottle; a painted ceramic marble; an industrial porcelain insulator; metalized wrapping paper; and mother-of-pearl buttons, among others.

Finally, there are two elements whose stratigraphic reference was interrupted by modern cutting events: a rectangular cavity (see the stratigraphic section in Fig. 2b) and a vaulted cesspit (reference 5 in Fig. 2a); in both cases, the preliminary reconstruction of their work would indicate the association of these structures to the parish house.

Ancient Rooms (1883–1893)

As already mentioned in the 1893 construction plan (Casa Parroquial, 1893), that new building incorporated at the back a previous structure with two rooms directly connected to the church. When the renovations of the theater stage began, and after removing a false wall that was used as scenery, another hidden and preserved wall was uncovered: the façade of the ancient rooms (Fig. 2b, above).

This Italianizing style building would have been built around 1883, when an extension of the church was made to place the main altar where it is currently located, replacing the former sacristy space (Barbich, 2007). According to historian Alejandro Barbich (2007), the land in question would have been ceded by the municipality to the church and a new sacristy would have been built there, a function that this structure continues to partially maintain to this day.

Following the same line, the historian Micucci Tarsetti (1998) reproduces 1885 budgets for the construction of a parish house, which in the end was never built. It is worth noting that in the description of the building to be constructed (with almost identical characteristics to those carried out in 1893), the master mason Don Juan Formica mentions that the rooms to be built would have "*the same shape and construction of the front of the sacristy*" (Micucci Tarsetti, 1998, p. 114, translated).

In addition to this front, remains of brick masonry with an east–west orientation were identified in the excavation area (Sector 1, Fig. 2.a.1), and at least one fill component in Sector 3 (UE 19). This fill presents silty-clayey sediments with demolition debris, archaeofaunal remains, and artifacts of chronology linked to the mid-late nineteenth century. Among them, fragments of creamware, pearlware, and whiteware, with the presence of decorative techniques and motifs such as flow blue, polychrome floral painting, ring-shaped earthenware, shell-edge rims (Fig. 2e), stoneware bottles with brown glaze (Fig. 2f), and glass bottles blown in molds.

Additionally, two other units were excavated in Sector 1 that could correspond to this same occupation phase, although it is maintained as a hypothesis given that the field work has not yet been completed, they correspond to a brick sidewalk (some of them partially vitrified) that runs along the east side towards the façade (Fig. 2.a.2) and a fill deposit with objects more related to the late nineteenth century, such as a coin of the Argentine Republic of 1884 (UE 44).

Contexts ca. 1800–1880

Alejandro Barbich (2007) points out that according to documents in the municipal archives, the site where the theater is located had been acquired in 1825 and then transferred to the church sacristan, Don Domingo Troncoso. In 1835, it was acquired by another owner and became a wasteland, until it finally went into the Municipality's hands, sometime between 1835 and 1882 (Barbich, 2007).

Historical and bibliographic sources referring to the first half of the nineteenth century and the colonial period are scarce, having been surveyed by the research project in the Foundational Area of the Franciscan Reduction Santiago del Baradero (FFyL, UBA), directed by Dr. Alicia Tapia (Camino et al., 2022; Camino & Tapia, 2017; Tapia et al., 2015). Among them, a reference was found to an old church (prior to the current one built in 1858–1861), which would be located toward the center of the block next to the priest's house, and which appears in the only available plan, dating from 1815 (Camino et al., 2022).

On the other hand, the works carried out in the adjoining property of the Asociación Española de Socorros Mutuos (AESM) resulted in the discovery of an old brick floor at a depth of 1.06 m (IGM datum). Immediately above this floor, a brick fragment was recovered and recently dated at the Instituto de Geofísica, Laboratorio de Paleomagnetismo of the Universidad Nacional Autónoma de México, yielding a geomagnetic dating of 1818–1854 A.D., with a confidence level of 65% (Camino et al., 2022). This data, together with the east–west orientation, allows researchers to hypothesize that it is part of the structure of the church built at the beginning of the nineteenth century. Likewise, the entire block is identified by Camino and Tapia (2017) as part of the original lands of the indigenous reduction of the Franciscan mission, settled in 1615.

Among the stratigraphic units of the site identified as evidence prior to the historic façade, three stand out: a tamped earth surface, on which a few adobe fragments were found (UE 70, Sector 2) (Fig. 2g); a deposit with remains of consumer waste that could be part of a garbage dump or landfill (UE 23-sondeo 1, Sector 3) (Fig. 2.a.3); and an "L" structure of fired brick foundations with mud mortar (UE 24, Sector 2) (Fig. 2.a.4). These deposits have only been partially excavated, leaving their continuation for future work; a cross-section of the excavation is shown in Fig. 2b, below.

The indicated foundations, both for the characteristics of its factory (handmade bricks laid in mud), and for its stratigraphic position, could be ascribed to a chronology between the late eighteenth century and the first decades of the nineteenth century. It remains for future interventions to test its stratigraphic and architectural association with the brick floor of the AESM site found about 8 m to the west and a few centimeters deeper. On the other hand, the test pit in UE 23, in an area of 1 × 0.50 m (between 0.70 and 1 m deep) resulted in the discovery of a large concentration of archaeofaunal remains (large, medium, and small mammals, including cattle and hairy "armadillos" (*Dasypodidae*)-, fish and birds), nine fragments of earthenware, three fragments of glass bottles (square and cylindrical), and more than a hundred sherds of Hispanic-indigenous pottery, both plain and decorated (modeled and painted) (Fig. 2h). Particularly remarkable is a fragment of a vessel rim, ca. 38 cm in diameter, with corrugated decoration typical of Guaranitic traditions (Fig. 2i) (Tapia et al., 2021).

It is worth noting the presence of numerous diagnostic artifacts of chronologies between the late eighteenth and early nineteenth centuries, as well as Hispanic and indigenous traditions, some found intermixed in later fill deposits. These include a fragment of a European majolica dish (Fig. 2j); numerous sherds of Hispanic-indigenous ceramics; a lithic flake in siliceous rock (Fig. 2k); and sherds of turned Hispanic pottery with internal green glaze (olive jar type) (Fig. 2l). Given the evidence of successive anthropic interventions in the terrain, particularly with digging activities (ditches, pits) and terracing, it is more likely to support the hypothesis that these objects come from the substrate itself at a greater depth. In the same sense, the contracting document for the construction of the parish house (Parroquia Santiago Apóstol, 1896) states that "*the earth that is removed from the wells and foundations will be levelled inside the rooms, so that when the wooden floors are placed, they will have the same level as that of the church*" (p. 2, translated).

This interpretation reinforces the potential of the site to preserve the archaeological record generated by the installation of the indigenous reduction that gave rise to the town between the seventeenth and eighteenth centuries. It also provides clues to link the findings made and to be made in this site with those identified in the surrounding land of the AESM on the foundation area of the city.

Patrimonialization and Contract Archaeology

During the last decades, the so-called contract archaeologies (Guráieb & Frère, 2008) gained relevance in Argentina, marking some differences with respect to the development of research born purely from academic fields. These differences lie mainly in the type of agents involved and their capacity to influence the objectives of archaeological work, which revolve around the management of material heritage and the actions required for its conservation. Faced with the imminent destruction of material goods (protected by law), various actors (companies or state agencies for example) require archaeological work, and fundamentally its methodology to mitigate heritage damage. This situation particularized each case, which in turn is subject to multiple negotiations that vary according to the ways in which consensus is built and according to the heritage valuations reached in that dispute.

This *applied archaeology*, as Barreiro Martínez (2006) points out, is "deeply immersed in social action" (p. 7, translated) and translates into different management models within the public sphere. Such models for heritage management start from different perspectives on the link between archaeology, communities, and the state or other private organizations. In this regard, two classes of hegemonic models can be distinguished fundamentally. The first is a "conventional" model of cultural property management, oriented to the conservation of heritage sites defined almost exclusively by expert knowledge (Endere et al., 2021). The second is inspired by the recommendations of the Burra Charter and its modifications (Australia ICOMOS, 2013), which are oriented to the values granted from the significance that a good or manifestation has for certain interest groups (Endere et al., 2021). The values-based heritage management thus implies expanding the range of material and immaterial "values" and the actors considered, seeking a balance between past, present, and future (Avrami et al., 2019).

This second type of approach is based on the concept of heritage as a social, subjective, and multivocal construction (Endere et al., 2021). For this reason, any patrimonialization project necessarily implies a reference social diagnosis as a first step, from which to identify the meanings or values that are constructed and put into play by different actors in their social, cultural, or even political bond with a patrimonial asset.[1] It is understood that these values are diverse, dynamic, and often conflictive, so it is important to include transversal, internal, and external views in the search for an integrating identification process (Mason, 2002). The actors, for their part, defined as stakeholders, are those groups, persons, or organizations with an interest or concern in the project (Avrami et al., 2019). They are linked directly or indirectly to the property/place in question and are the main agents that give it meaning. This sociocultural diagnosis is known as cultural significance assessment (Endere et al., 2021; Mason, 2002), and includes experts as one more of those stakeholders.

From this theoretical framework, we seek to build value-oriented strategic planning (Endere et al., 2021), that is, a heritage management plan that weighs the cultural significance assessment, the relationship with and between stakeholders, as well as a projection of sustainability for the enhancement project based on the search for consensus. Despite using these methodologies and languages that largely come from the business world[2] and are recommended by international organizations such as ICOMOS and UNESCO, our goal is to create spaces for this type of project to be supported by the local community itself. That the search for a social license for the "success" of the project is transmuted into a process of appropriation that favors

[1] The logical process proposed by the Burra Charter consists of: understanding the place, assessing cultural significance, identifying all factors and issues, developing policy, preparing a management plan, implementing it and finally, monitoring the results and reviewing the plan (Australia ICOMOS, 2013).

[2] The concepts of strategic planning originate in the military and business spheres, and were adopted at the end of the twentieth century by the areas of territorial planning, urban management and sectoral economic development (Kuper et al., 2010).

that same community. In this sense, we point out that the involvement of the key actors is not limited only to their "consultation" in the framework of a social diagnosis, but to the search for their active and significant participation throughout the patrimonialization process (Kuper et al., 2010; Mason, 2002), which is what will ultimately give it meaning. As stated by Avrami et al., "broader involvement, especially of local communities, in deciding what and how to conserve is critical to socially responsive and socially relevant practice" (2019, p. 6).

Before moving forward, let us return to some of the concerns that challenge contract archaeology and that, in turn, constitute the main foundation axes of its critical researchers (e.g., Haber, 2017, Londoño, 2016, Zorzin, 2017). Namely: 1. Is it possible to carry out scientific research in this area? 2. How participatory or communicative can our work be when pragmatic issues such as means and budgets limit the task "outside" the archaeological intervention? 3. How to overcome the idea of a commercial archaeology that is solely responsible for delivering a service or building cultural resources for its commodification?

We have already advanced on the first point in the previous sections, concluding that a research agenda can be maintained and data generated (minimally descriptive) to be taken and deepened in other instances and/or by research teams, thus contributing to the development of scientific knowledge (see Bárcena & Schávelzon, 1991; Izeta et al., 2014; Ratto, 2009). Regarding the second, we understand that participatory and communicational *instances* must be integrated into our practice as part of the interventions and not as "extra" services (Clark & Horning, 2019; Söderström, 2018). Become work habits more than tasks, which implies contemplating in advance times or moments to carry them out. Finally, to respond to the third challenge—and here we do address the *participatory* process—we appeal to strategic planning, seeking to shape a proposal and define tools that can be carried out by other actors (e.g., authorities) or even other researchers in the territory and the long term, beyond a specific contract. This proposal will be oriented toward the integral management of cultural heritage, "which not only implies the management of archaeological practice (…), but the management of everything that this practice entails as a social practice" (Barreiro Martínez, 2006, p. 214, translated).

Work Strategies When Time Is a Tyrant

Taking into account some issues of time and budget, the first stage of the archaeological rescue incorporated some actions oriented on three axes: the work team, social communication, and strategic planning (sensu Kuper et al., 2010).

A first approach to the cultural significance assessment was developed in parallel to field work. Social diagnosis was carried out through qualitative methods such as the rapid ethnographic assessment procedures ("REAP") (Mason, 2002; Ross, 2018) that allow an initial identification of stakeholders in addition to being sensitive to the social values of the place. Thus, an unstructured ethnographic approach was applied, which consisted of informal interviews (group and directed) with the different actors and participant observation during collaborative work spaces and instances of dialogue with the community. In this sense, the San Martín theater itself functions as an access point (sensu Ross, 2018) to the opinions of part of the population involved, as it is a social center in which members of the parish, mainly but not exclusively, congregate. In addition, other indirect methods were the consultation of bibliography and data from previous interviews carried out by other researchers, and the review of testimonies on social networks.

Given the archaeological relevance of the finding, but above all its heritage potential expressed in the willingness of the actors involved to generate a medium/long-term project, it became essential to encourage the integration of different sectors of the local population in the intervention process.

It is worth mentioning that the intervention was carried out by archaeologists associated in a work cooperative, which implies a work philosophy intrinsically founded in by the principles of horizontality and social responsibility. The employment of cooperativism values seeks to strengthen both the work team and its links with other actors already or potentially involved in the project. We understand that a horizontal, supportive, and open-to-dialogue work environment contributes to generating spaces and opportunities to share the construction of knowledge and tools or means for research, management, and scientific and social development.

In this sense, the work team was formed with open doors to the participation of different actors, both professionals and amateurs interested in the project, which in turn will also promote its social appropriation. This collaborative proposal (Clark & Horning, 2019), although limited given the characteristics of the contract, allowed the construction of an initial instance of integration between local people's interests and knowledge and the required training on archaeological practice and heritage issues. This articulation took place during the fieldwork, laboratory tasks and in the planning and development of heritage-linked activities such as workshops and exhibitions. At the same time, reciprocal cooperation links were generated with the researcher in charge of the study area and interdisciplinary work was promoted, collaborating with other teams and professionals from related disciplines (such as architecture and museology) from other institutions (Instituto de Arqueología, Universidad de Buenos Aires; Universidad Nacional de San Martín). This was especially oriented to the site enhancement and its management plan for public use.

Local community, both civil and political authorities' involvement and legitimation were sought, based on the conviction of the relevance and potential contribution of the project to the preservation and sustainable development of Baradero's historical heritage. At this point, it should be noted that there is a social memory linked to other previous heritage projects, with positive and negative perceptions, which invites to problematize the relationship of the local community with heritage resources and their activation (sensu Prats, 2000) (e.g., (Pineau & Charlin, 2007).[3]

[3] The survey in question has been updated in 2018 but the analysis of its results has not yet been published (Alicia Tapia, pers. comm., 2019).

As a contribution to this process of patrimonialization, actions oriented to communication and participation were defined and organized by the researcher in charge, the Direction of Culture and the Heritage Area of the Santiago Apostle Parish; they were communicated through the invitation to the local media (radio and electronic newspapers). The activities included two informative talks, in which the actors involved and the different work stages were presented, and which served to address the concerns of the local community with respect to the heritage projects and the development projections that are planned. Two participatory workshops were also held, open to the public. In them, we communicate and share the work through site visits and laboratory activities with the archaeological materials extracted, providing an open space for direct interaction with the heritage assets and the work team. Parallel to the excavation tasks, some visits to the site were made by high school students in which perspectives on the historical heritage of the city were also addressed. These visits had a positive predisposition on the part of the students, although they could not be developed frequently (some difficulties were the permissions for the educational outings or the transfer for the most distant establishments). The limited timeframe of the intervention was not favorable for a more comprehensive work with schools and teachers in the classrooms, although we tried to provide a first approach. Other sporadic activities were also carried out by the Parish and the Department of Culture, such as exhibitions of objects in the framework of catholic festivities.

This series of meetings revealed the interest in the history of the place in much of the population near the site, as well as the predisposition to participate in **dialogic activities** in which the researchers themselves also have the opportunity to enrich themselves with local knowledge and interpretations (Fig. 3).

During the year 2020 and within the context of the COVID-19 sanitary emergency, all on-site work ceased. Although the archaeological intervention had concluded, the researchers collaborated remotely in accompanying the municipal Culture Department. Some attempts were made for virtual activities linked to the history of the theater but there was practically no response through the media used (Facebook and Instagram). On the other hand, a call for secondary school students was proposed, in which they would present stories and infographics linked to the historical heritage of Baradero. This activity was organized in the anniversary month of the city together with some teachers through the Education Department, to be incorporated into the school curriculum; the works would be published in turn as a special section in a specialized scientific magazine.[4] In this case, even though the activity was successful, the challenges faced by teachers and students during the 2020 school year, together with the specificities of virtuality and the lack of prior joint preparation, made it difficult for students to participate more widely.

In sum, the positive experiences resulting from this intervention process reinforces the certainty about the importance of this type of participatory activity as part

---

[4] See section "Convocatoria Patrimonio Joven—Baradero", in *Urbania. Revista latinoamericana de arqueología e historia de las ciudades*, 9 (2020). https://revistaurbania.wordpress.com/archivo/n9-2020/

Fig. 3 Record of the activities developed with the community: informative talks (**a**), school visits (**b**), guided tours and laboratory workshop (**c**–**e**), exhibition of archaeological objects and collection of the Municipal Historical Museum (**f**). (Source: Baradero Te Informa (**a**, **c**, **d**, **e**); Pablo Berninger (**f**))

of the development of the project itself and of the construction of social license and appropriation. The difficulties observed so far indicate the need to plan these activities in a more integrated manner with the different local sectors (municipal and parish, in principle), thus providing better continuity and fluidity to the dialogue and proposals. Likewise, it is important to adjust the channels for this communication according to the reality of each territory. In this case, the best ones to reach were face-to-face workshops, the local media, and other institutionalized proposals (e.g., through the Education area). Finally, it should be noted that these tasks did not require a considerable amount of time to be carried out, which would have been detrimental to the deadlines established for field work. Instead, they were possible to coordinate in parallel thanks to the articulation with the areas and persons in charge involved.

Through this process and the background, we identified a series of values associated with the site that needed to be adressed in a statement of significance (sensu Australia ICOMOS, 2013; Mason, 2002). The *scientific value* of the find stands out, especially from archaeologist researchers and other actors interested in cultural heritage. Closely linked to this, its *historical value* is nevertheless recognized by all stakeholders, although some and others value more or less certain moments of that "history." *Educational value* is also introduced here, as this is a space that materially reveals these stories. On the other hand, its *social value* is associated with the function of the theater as a community center, linked not only to the activities promoted by the church but also as a space for meeting, sociability, and entertainment.[5] The development of religious activities of the parish (such as workshops and catechism.) endows the place with a *spiritual value* for the local Catholic community, while an *identity value* is identified more closely linked to the modern history of theater (in its last 50 years). Finally, a *tourist value* appears as potential, especially for local authorities, although its formulation is highly dependent on interest in the educational value of the site and oriented to the local population.

Strategic Planning and Heritage

Based on all of the above, it was proposed to initiate a multifunctional enhancement of the site. This includes both the preservation of the archaeological heritage and the creation of a museum space, as well as the continuation of the cultural activities of the theater itself. The objective is to articulate the heritage value of the site and the theater with the religious context with which both elements are integrated. From this point of view and in agreement with different local authorities, we began to plan the creation of a site museum for the exhibition of the excavations carried out. A museological space with a historical-religious theme for the Parish collection of heritage goods and for the archaeological objects recovered. The project includes the reopening of the theater hall that is used for artistic purposes as well as other sociocultural events.

We understand that this site constitutes a space that could condense and integrate several layers of meaning of the local memory, from the modern history with the theater as a cultural reference, the history linked to the settlers and immigrants that promoted the remodeling of the church and the urban growth of the town in the second half of the nineteenth century, to the colonial processes and indigenous occupation that were the founding context of the town of Baradero. In this line, we propose that it is fundamental for the interpretative discourse of the site (sensu

[5] The theater and its entrance hall was usually used in addition to theatrical performances for social events, such as graduation diplomas ceremonies for students from schools.

Ballart Hernández & Tresserras, 2001) to weave a connecting thread between the present and this complex and multi-ethnic past.

Keeping this in mind, the archaeological work team focused on generating all possible contributions from professional practice within the stipulated time frame, seeking to make available its evaluation and recommendations on the various dimensions involved in a long-term enhancement process. We understand that strategic planning implies the search for consensus on the objectives and courses of action of a given project through participation, and thus, cooperating to achieve a good implementation (Clark & Horning, 2019; Kuper et al., 2010). It is therefore essential in the first instance to identify the main challenges to be faced in the formulation and implementation of a project of these characteristics, as can be seen from a SWOT analysis.

We detected several contingency factors for the development of the project, such as the strong dependence on the state administration which in turn usually translates into deficiencies in funding sources. Related to this, there is also a strong dependence on the ecclesiastical authorities for decision-making as owners of the land in question and also a dependence on the state authorities as the main administrator of the archaeological assets.[6] These factors have an impact on the possibilities of carrying out the projects, the collaboration in infrastructure, and the contribution of human resources, but also leave the heritage projects at the mercy of the ups and downs of the political situation at any given time.

Another aspect to consider is the commitment -or lack thereof- of the local population and also of the municipal authorities themselves, to comply with the recommendations for prevention and mitigation of impacts on the archaeological heritage in general and the sites in particular, within the framework of the legislation in force. In this sense, we also add the background on the low interest or distrust of the local community in the proposals for the patrimonialization of archaeological and other historical sites, which challenge us regarding the processes of appropriation and genuinely collective construction of historical heritage and the articulation with the scientific and state sphere.[7]

On the other hand, we emphasize that a project such as the one presented here intrinsically implies the need to work in an interdisciplinary manner. This is not only in terms of the technical issues of research, preservation, and exhibition, but

---

[6] In Argentina, the vast majority of heritage projects (as well as archaeological research projects) are developed from the state sphere, as an institutional and funding framework. Unlike the legislation of other countries, every good (movable and immovable) considered archaeological is in the "public" domain and is under the guardianship of the State, even when it is located on private lands. In spite of this, Law 25.743 on the Protection of Archaeological and Paleontological Heritage contemplates mechanisms for a "guardianship" or "tutelage" by private individuals as long as they are legally registered, and the care and access for research and dissemination are guaranteed.

[7] For an analysis of the heritage issues related to the case of the Indigenous Cemetery, for example, see Pineau & Charlin (2007) and Tapia (2003), where the heritage process related to this iconic archaeological site of the city is developed.

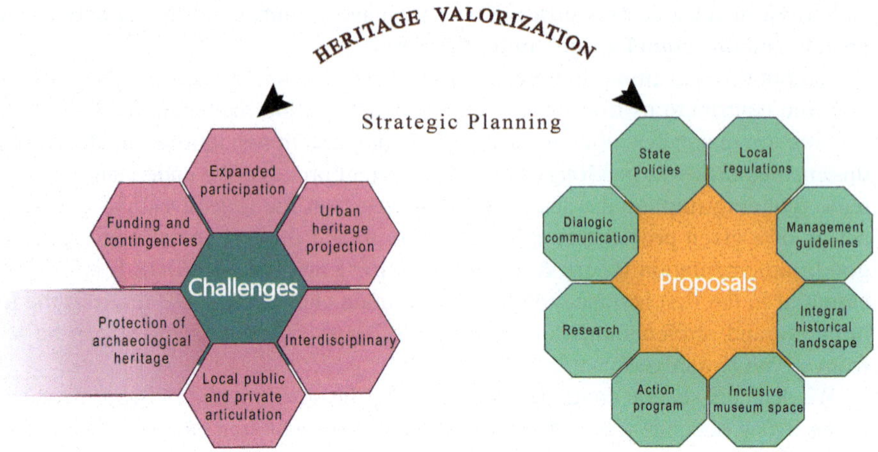

Fig. 4 Diagram on the strategic planning for the process of patrimonialization of the San Martin Theater

also with regard to the insertion of this site within a broader heritage and urban framework, which entails its own challenges for its management.

Based on the challenges posed, we summarized in the next sections a series of proposals to continue with the development of the work on three axes: the link with stakeholders, recommendations for a management plan for the public use of the site and guidelines for the urban projection of the historical heritage of the city (Fig. 4). These formulations were included as part of the final report of the intervention delivered to the Municipality of Baradero as an input for future interventions, **heritage management state policies** and/or **research** projects that would not necessarily be carried out by the same work team.

Site Management for Public Use

In this axis, we include what concerns the preservation and enhancement of the site itself, as a starting point for its linkage to public use and within the framework of the heritage management of the Historic Center of Baradero. Therefore, we emphasize the need to elaborate a master plan for the realization of the site museum, which is conceived in short and medium-term stages to allow obtaining sufficient resources for its conclusion and guarantee its immediate partial functionalities. That is to say, to contemplate the opening of the theater for diverse activities during the periods in which technical tasks related to the patrimonial intervention are not being developed, thus establishing priorities and alternatives in the use of the space. On the other hand, to foresee the possibility of setting up a temporary exhibition of archaeological materials from the site at the Municipal Historical Museum during the time that the museographic setting for the San Martín Theater is completed.

On the other hand, basic recommendations were established for the preservation of the site until the next stage of archaeological intervention: keep the excavation covered to avoid the transit of people that could deteriorate the structures and contexts (either by burying them or other preservation and coverage measures under the supervision of archaeologists); do not carry out building modifications without prior planning with the professionals in charge of the master plan; guarantee the supervision of the archaeological materials recovered so far and in custody by the parish; periodically monitor the state of conservation of the site in the face of risk and deterioration factors such as rain, animal activity, humidity, and microorganisms. Notify all authorities, personnel and collaborators who work at or frequent the site of these best practices guidelines for the management of the archaeological site. More generally, create and implement an emergency and risk prevention plan for any potential disaster impacting the heritage property or the building.

For professionals, it is also recommended to maintain the work methodology in the next stage of excavation to generate homogeneous scientific information and to ensure that the archaeological material stays in the locality, with a final deposit at the site provided by the Department of Historical Heritage.

It will also be important to generate multilateral agreements between the Municipality and the Diocese of Zárate—Campana[8] to guarantee the realization of the valorization project and the subsequent public access to the heritage property in the long term. Related to this, it is also suggested to diversify the funding sources, both public and private (subsidies and public programs, donations, corporate social responsibility projects, etc.).

Regarding other aspects of management, it is recommended that a specific legal figure be created to guarantee the site's comprehensive heritage protection, including the regulation of its management and administration plan. In addition, the objectives of its enhancement for public use, specifically oriented to be incorporated into the educational, cultural and tourist programs of the locality, should be established; hence, the relevance of articulating these areas from the administrative management.

It is important to point out that a first step in this direction has already been taken, since within the framework of the archaeological intervention project we collaborated in the articulation with universities (Instituto de Arquitectura y Urbanismo of the Universidad Nacional de San Martín) that have specialists in heritage projects. This team was then hired through a university agreement by the Municipality of Baradero for the design of a Master Plan from an architectural, urban, and heritage perspective to frame the enhancement of the San Martín Theater.

Finally, we conceive public access to heritage sites as a reflexive resource that in its interpretive discourse invites to highlight the historical tensions and dimensions of meaning that such heritage represents (Londoño, 2016; Santana Talavera, 2002). Understanding that the selection of moments and manifestations of culture and history is part of the heritage activation process (Santana Talavera, 2002), this should

[8]The Santiago Apostle Parish is under the jurisdiction of the aforementioned ecclesiastical authority.

tend to the enrichment of the collective memory and ideally favor the appropriation by the community of the diversity of cultural referents in the construction of its historical identity (Tapia, 2003). Therefore, we suggest stimulating such reflexivity of the social imaginary, particularly with regard to the indigenous past. Thus, promoting the visibility of the multiculturalism that characterizes the identity roots of the locality, which "only exists in an underlying way in the documentary sources but not in the historical consciousness of today's youth" (Tapia, 2003, p. 7, translated), and which has been systematically concealed by the official history since the conformation of the Argentine nation-state in the nineteenth century. In this sense, it is also proposed to link the site with other spaces of patrimonial significance and memory (such as the Asociación Española de Socorros Mutuos, the church it self, the historical museum, or the Cementerio Indígena) that feed back into the construction of intertwined and diverse historical narratives. Along the same lines, it will also be important to work on the local perception of the site and its history(ies) through oral memory, which in turn will enrich the development of a museographic script.

It is hoped that these **management guidelines** will serve as a starting point for the application of specific protocols throughout the historical district. The measures to be presented aim at the prevention of potential negative impacts on heritage assets and the promotion of positive ones through planned management. We understand that the planning of these integrative cultural resource management policies is essential to "contain the different value instances of the components of Cultural Heritage"(Ratto, 2013, p. 11, translated) and that much of their value lies in their analytical, preventive and participatory preparation.

Ties with Key Stakeholders

It is usual for heritage activation projects (sensu Prats, 2000) and the associated regulations, very especially in urban contexts, to overlook the consultation instances and -even more so— the spaces for direct community participation, partially based on the complexities of population density and heterogeneity. This can also affect the degree of legitimacy of the project (or social license) for its development and sustainability in the long term, understanding by this the dynamic capacity to continually seek the balance in development (in its environmental, economic, and social aspects) (Söderström, 2016).

We start from the idea that a *process of enhancement* requires working on the integration and participation in the project of different actors, even if the proposal arises in the first instance from an institutional framework as in the case of the San Martin Theater. We understand that without the collective, there is no heritage. For this reason, and also following the guidelines on international heritage management (Australia UNESCO, 2013; Clark & Horning, 2019), we believe that it is necessary for official organisms to consider as a state policy to provide mechanisms so that the population can have access to information and active participation during the development process of the proposals, even in those cases in which a properly "community" management of heritage is not contemplated. In this way, it will also be possible to seek to disarticulate potential conflict situations with actors with conflicting interests, exploring the opening of dialogue for consensus building.

In order to organize and begin to plan these mechanisms, it is important to draw up a map of stakeholders to organize the social panorama. This will help in the creation of possible processes of participation, both of interested agents (positively or negatively) and of those not interested but directly and indirectly involved (Clark & Horning, 2019). Based on the first diagnosis in this sense, we define here four groups as priorities. At the same time, we propose an **action program** that complements those already carried out in the first stage of work and that can be executed during the process of elaboration of the management plan and later during its implementation. The objective of these formulations is to arouse or enhance the positive interest of these key stakeholders in the development of the valorization project. The four groups defined are the local population, owners of the historical district, elementary, middle and high school institutions (basic primary and secondary education in our system), and municipal bodies.

Local Population

Although it consists of a very heterogeneous group of actors and also includes the others we define below, we do not want to neglect to mention the need for a general approach to the link with the local inhabitants. This broader view encourages the generation of open and nonspecific spaces for dialogue and the exchange of visions in their greatest diversity.

In this sense, we highlight the positive experiences resulting from the workshops and exhibitions already carried out, in which there was direct interaction between technicians, authorities, enthusiasts, adults, and children, both among themselves and with the heritage assets. These participatory instances fulfill several objectives among which are the communication of the work stages and those responsible, the encouragement to value the site, the integration of the interest of the local population and their knowledge, the training on archaeological practice and heritage issues, and the communication of the population's demands and concerns to professionals and/or officials. In short, the promotion of an initial appropriation of the project (in terms of taking a position on it) and a general approach to the status of the link between the population and the heritage projects.

Just as we encourage the continuation of these activities, we also propose as desirable the deepening and development of new ones. Among them to encourage the opening of communication channels about the project through radio, local press, and social networks. We also encourage generating programs for oral history promotion, especially aimed at integration activities with older adults. Other thematic participative spaces can also target the deficiencies in schooling by promoting educational and recreational activities for young audiences using axes addressing heritage, reflexivity, research, and the revaluation of local history. It is worth mentioning that although all the proposals presented here were conceived about the project in question, they constitute valuable tools also for the broader management of historical heritage in Baradero.

Municipal Organisms

In general, it will be important to engage the authorities in the development of strategic planning understood as a state policy, with application not only for this case but also for other future ones. In relation to the other actors, it will be important for the Department of Culture to promote coordination strategies between the different government areas (education, public works, tourism) and with other institutions. In this sense, it is also relevant to make visible the potential of heritage projects as a factor of local social and economic development based on public management.

We consider that it would be of interest to carry out a detailed tourism research oriented to the diagnosis of the enhancement of the site in its potential as a tourism product, to be incorporated into the offers of rural tourism and historical towns that the district of Baradero already has.[9] Likewise, to encourage the involvement of the agents of the area in the participative activities to be developed in the site as well as the elaboration of others, thoughts from the local tourist policy and achieved from spaces of joint planning with the Culture area.

Finally, the elaboration of regulations for the protection and promotion of Baradero's historical heritage in general, and of this site in particular, could guarantee or at least reciprocally encourage the assumption of responsibility and specific management mechanisms for the corresponding municipal competences.

Educational Institutions

We have pointed out as one of the main objectives of the enhancement of the site (in addition to the preservation of the heritage) is the creation of an educational cultural space for the local community. In this regard, we reiterate that the complementarity between the different areas of government is essential in terms of strategic planning so the articulation of the area of culture with educational institutions is essential. To this end, it is recommended in the first instance to call authorities and teachers to informative meetings about the project and to consider the contents that could be addressed from the school curriculum based on the interaction with the archaeological site and the historical narratives. From there, a program of activities in the formal educational environment could be organized to involve the children and the youth, for example, through guided visits to the site (during and after the field work) or in coordination with periodic visits to the Municipal Historical Museum and/or

[9] One of the closest archaeological tourism proposals to have as a reference in this evaluation is the natural historic park Vuelta de Obligado (San Pedro, Province of Buenos Aires), which also has an open-site museum (historic battle field) for public use (Hernández de Lara et al., 2014).

the Cementerio Indígena. In addition, we point out to take into account the possibility of conducting visits and/or small traveling exhibitions in educational institutions with limited resources such as an island or rural schools, since their mobility to the site is much difficult.

Private Owners in the Historical Center of Baradero

Finally, we highlight as stakeholders the owners of properties belonging to the historical center of Baradero, primarily in its foundational block. It is within this group in particular that one of the greatest points of tension concerning preservation heritage policies and private interests is to be found. Therefore, it is recommended that these owners be particularly invited to community participation activities and informative meetings, seeking to raise awareness of the heritage value of archaeological sites as well as to properly communicate the current legislative framework on these matters. It is not uncommon for conflict situations to arise from partial ignorance of these legislations or, more specifically, of the possible mechanisms of action and strategies, both on the part of individuals and of the state and professionals. In this regard, the preparation of informative documents may be added to these actions, making their distribution effective through additional communication channels such as brochures, official websites, and social networks. Along the same line, it is desirable to integrate the private owners of the foundational block into instances of interchange and dialogue on the planning of the enhancement of the site and the projection of a heritage management plan for the area that also favors their property.

Urban Projection of the Historical Heritage of the City

The urban spaces considered within the current historical center of Baradero, or even more so a specific archaeological site such as the San Martín Theater, consist of an arbitrary —intentional— segmentation of a broader territorial system, which also has its historicity, and socio-cultural and environmental diversity as a *landscape* (sensu Criado Boado, 1997). We should consider the articulation inside and outside these "units" with different places, assets, or heritage manifestations (tangible and intangible) of significance for the local population and that provide meaning to the interpretation and historical and cultural experience. To this end, an *Integral Strategic Heritage Plan* should be drawn up for the city, incorporating on a broad scale the associations between different historical landmarks, places, and heritage assets, as well as memories (resignifications of the past), as in the historic urban landscape approach (Conti, 2015). Thematic axes organizing the urban heritage landscape can be arranged in circuits aimed primarily at education, reflexivity, appreciation, and appropriation of the history, culture, and natural environment of the locality through its tour and experience. The integrating proposal is applied in a double sense, both to understand/construct the past landscapes, and act/construct in the present ones, in close connection with the demands of our current society.

The adoption of the holistic or integrative perspective to address the management of cultural resources and heritage policies implies that research should be basic and applied, by not circumscribing itself only to the objects but also to the implementation of a management plan oriented to the sustainable development of heritage assets and the local population (Conti, 2015; Criado Boado, 1996; Molinari et al., 2000; Ratto, 2013; Söderström, 2018). Thus, to ensure respect for the different dimensions of heritage value that manifest the cultural and natural diversity of the city, we recommend the execution of a comprehensive urban design, incorporating land-use planning guidelines.

Urban heritage conservation requires policies and mechanisms to balance conservation and sustainability in the short and long term, which implies the inclusion of citizen participation tools that promote social diagnosis and inter - and intra- cultural dialogue through learning. This participation facilitates negotiation processes between actors and groups that may have conflicting interests, reinforcing their common ground. These policies should also incorporate knowledge and planning tools through the training of the main parties involved (communities, civil organizations, professionals, authorities, and institutions) to characterize inventory and set objectives, strategies, and action plans. These action plans begin with the evaluation, interpretation, management, and conservation of the social, environmental, and cultural values of urban heritage.

In this sense, the need to update the municipal regulatory framework to overcome the deficiencies in the provincial heritage laws in force today is indicated as a priority. This update should be specifically oriented to the establishment of protection measures for archaeological remains in case of casual findings, those that would not be expected due to the absence of superficial evidence, as well as for *potential* heritage assets, for example, through impact studies prior to the beginning of engineering works. It would be desirable that for this purpose a system of review and monitoring of works would operate based on the definition of *heritage zoning*, in which the Undersecretariat of Public and Private Works and Planning, and the Department of Historical Heritage in the case of Baradero, share competences.

Along the same line it is recommended to generate municipal legislation that includes the creation of a specific protection figure for the block or foundational area, as well as its own heritage management protocol (for which, in turn, it would be necessary to update the current ordinances which define the historical center).

Below, we present an approach to this problem through the definition of areas of known and potential archaeological and architectural sensitivity within the urban area of the city.[10] Such sensitivity was inferred from the available historical cartography (working on the georeferencing of the urban plan of Baradero from the years 1815 (Manso, 1815), 1872 (Plano de Baradero, 1872) and 1908 (Barbich, 1908);

[10] Other heritage-sensitive areas, such as those linked to riverside areas with pre-Hispanic archaeological remains, are not taken into consideration in this work.

and on the processing of archaeological findings made so far, the figures of heritage protection in force (Cementerio Indígena and Historical Center); and the information gathered in previous research (Camino et al., 2022; Camino & Noya, 2015; Camino & Tapia, 2017). This information was uploaded to a Geographic Information System (QGIS v3.10) through which five areas with potential for protection and/or specific intervention protocols were delimited (Fig. 5, above).

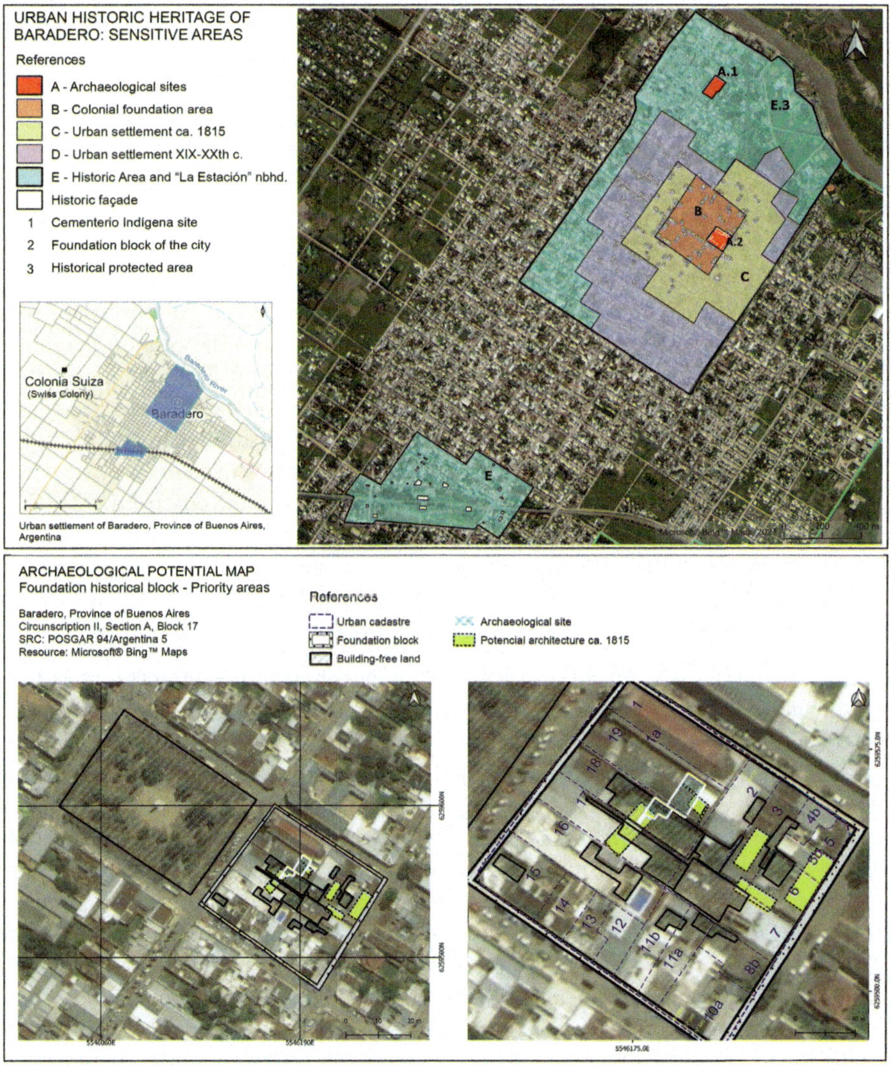

Fig. 5 Cartography for the protection of Baradero's historical heritage. Above, map of sensitive areas of urban historic heritage; below, indication of priority areas for protection and research within the foundational block

Each of these areas has particular historical and archaeological implications and can be characterized as follows:

- Area A: archaeological contexts of known heritage value, requiring the highest level of recognition and heritage safeguarding. It currently includes the Cementerio Indígena (with the declaration of Heritage Interest and Historical Site (Ord. 3.186/05)) and the foundational block.
- Area B: greatest archaeological potential to preserve vestiges of the initial settlement of the seventeenth Century indigenous reduction, the foundational area of the city (Camino et al., 2022; Camino & Tapia, 2017). It requires a high level of safeguarding that contemplates the character of an unexamined sensitive area and its potential immediate link with the enhancement of the foundational block.
- Area C: delimited by the 1815 urban plan, from which its potential to evidence material remains from the colonial period and the beginning of the Republican period is inferred. Its patrimonial protection should be due to its character as an area not archaeologically explored and buffer zone of the nodal center of the initial settlement of the city.
- Area D: urban layout occupied around 1870, with greater potential for preserving remains of nineteenth century architecture and material culture. Also, adds the historical standing buildings of *Barrio La Estación* (The Station neighborhood), one of the three population centers of the city at the beginning of the twentieth century and located around the railroad station inaugurated in 1885. It requires a heritage protection approach that draws on historical and archaeological research to determine specific management measures in each case.
- Area E: limits of the area declared as historical center of the city (Ord. 4213/09), wich details the historic facades and incorporates the maximum area built on its periphery, around the Railway Station. It is recommended that its patrimonial protection be linked to the monitoring of the surveyed historic fronts and to the potential fortuitous archaeological findings in the case of building works.

In addition to these definitions, a map of archaeological potential (Fig. 5 below) was made specifically for the foundational block. It shows the known archaeological sites, the constructive structures represented in the oldest known historical cartography (Manso, 1815) and the municipal cadastre. The overlap of this information, and without prejudice to potential unexpected findings, generates a couple of *priority areas with high archaeological sensitivity* for research and heritage protection. Those are indicated as priority areas free of construction within lots 18 and 19 (corroborated archaeological contexts), and lots 3, 6, and 17 (with immediate potential for findings); not to mention the Mitre Square itself, immediately adjacent to the NW. Regarding these lands, it is recommended the implementation of a *public research promotion program*, agreed with the owners, oriented to generate archaeological, architectural, and historical knowledge directly linked to the enhancement of the foundational block in general, and the San Martín Theater in particular.

Although these sensitivity maps are presented here with an instrumental nature (Mason, 2002) and fundamentally oriented to the scientific and historical value of these areas, they are valid as a basic approach. This in turn, should be evaluated by the municipal agencies involved together with professionals in an interdisciplinary manner (archaeologists, architects, paleontologists, conservation specialists, etc.) and other local stakeholders, in order to identify additional cultural significances and develop management plans addressing specific protection and other heritage issues (Conti, 2015). Among the measures that could be considered are complementing current regulations, defining priority archaeological research/protection areas, requiring archaeological impact studies to grant building permits, and monitoring during activities involving soil movements in the most sensitive sectors. Also, establishing periodic monitoring of the state of conservation of standing historic architecture, and establishing technical and aesthetic parameters (styles, volumes, heights, etc.) for new architectural works in the most sensible areas that are harmonious with the existing heritage landscape. Finally, there should be established concrete mechanisms for the implementation of heritage protection regulations by the private sector and the state, and encourage the articulation between heritage enhancement processes and strategic plans for sustainable development. In this sense, promote historical and archaeological research as well as heritage education and generate protocols for citizen participation.

An *integral management* model implies constituting a coherent "chain of interpretive assessments," covering the diverse dimensions of the historical document, as well as the planning of this process from the identification of the elements, their research, diagnosis, evaluation, revaluation, interpretation, and socialization, to their inclusion as a heritage property in the cultural market (Criado Boado, 1996, p. 77, translated). This process will involve the integration of technical and scientific information to generate a master plan for urban heritage that contemplates and foresees both the positive effects to be enhanced (e.g., educational, socio-cultural, and tourist purposes), and the negative effects to be prevented (affectation of heritage assets, gentrification, real estate speculation, centralization, heritage impact, environmental deterioration, etc.). In that, there is a need to develop a projection of the historical urban heritage of Baradero, paying particular attention to work on promoting respect, representation, appreciation, and appropriation of the cultural diversity (indigenous, European, African, and Creole) that makes up its historical landscape and to generate specific legislation and **regulations** for the evaluation, valuation, and protection of heritage. In addition, integrate the historical-cultural heritage with the historical-natural heritage (paleontological and ecological), producing programs that support the association and appreciation of the historical landscape and the wetland landscape (typical of the region) as a socio-environmental unit, past and present.

Among other management guidelines, we propose generating heritage education programs for all audiences and other specific ones with amateurs; promoting socially responsible sustainable development policies that include programs for universal **accessibility and inclusion** of the local population; and generating long-term policies and programs whose development is organized in progressive stages that can be implemented in shorter periods of time.

Discussion and Final Considerations

This paper displays the initial organization of the project designed for integrated management of the archaeological heritage of the foundational area of Baradero in general and of the San Martín Theater in particular. It was proposed here to promote access to information and participation of the local community in the planning of the enhancement, favoring the linkage and joint construction of the social value of this heritage locus, even if the management/administration of the site will be developed in a mixed private-state way.

We understand that the plan for public use of the site must be based on the universal right to culture and oriented to its sustainable development, which implies guaranteeing the protection, research, access, and transmission of the archaeological heritage to future generations while generating other social benefits; that is, seeking to favor positive implications for both the receiving community and the preservation of the heritage assets themselves (Moreno Melgarejo & Sariego López, 2017).

As an educational space, it will allow us to know both the archaeological work and the material culture of the city's past, adding a narrative about the history of its development from the original occupations, through the European colonization and up to the last century's urban and population growth. This script can be incorporated into both the educational curriculum and the visiting program for the general public. The conditioning of the site will foresee an architectural system that guarantees both its preservation and its visitation.

The religious museological space will channel the work of recovery and registration of the patrimonial heritage of the Parish, which is being carried out with the encouragement of several local collaborators. In this regard, advice will be provided for a museographic setting that articulates with the archaeological material record found at the site.

The theater will be reopened with the necessary remodeling, incorporating the archaeological context in situ without detriment to its normal operation. To this end, work is already underway in agreement with the Facultad de Arquitectura, Diseño y Urbanismo (Universidad de Buenos Aires) and the Universidad Nacional de San Martín. Theater's enhancement will also respond to social demand, since it is a cultural reference in the memory and the present of the *Baraderense* community.

The virtue of strategic planning in heritage management is to articulate the development of scientific knowledge with the imperatives of heritage preservation and social development policies. It invites to generate and promote workflows between research projects, the concerns of local populations and organizations, and the management of state administrators. In this sense, the relationships between scientific research and public, private, or mixed policies for the revaluation and management of archaeological heritage are both evident and necessary. Whether it stems from purely scientific projects or from circumstantial and daily constraints, they call and challenge both the state and the professionals. This challenge is to cultivate these opportunities to build value chains between the interests and needs of each field of action, having as a principle the social commitment to improve the quality of life, material and spiritual, of the communities.

Experiences such as those initiated in the case of the archaeological rescue of the San Martín Theater highlight the need to transcend the dichotomies between public and private, basic science vs. applied science, or even "research archaeology and management archaeology" as stated by Criado Boado (1996, p. 74, translated) in the 1990s. To understand and propose that there is a professional field of development in which archaeology can contribute a committed and socially responsible look, and not merely a instrumental one as argued by some colleagues detractors of contract archaeology (Haber, 2017; Londoño, 2016). Even within the contractual scope of archaeology of management, and embedded in the same hegemonic economic system as all workers -social scientist included-, alternative spaces such as work cooperatives are oriented toward taking a socially committed attitude toward their work and acting from the ethics of solidarity (Zorzin, 2017). From this framework, the goal of archaeological practice is not indiscriminate and uncritical profit, but to provide proposals aimed at expanding the right of the population to heritage -often alienated from the community or reified-, the right to enjoy and cultivate their own culture and memory. These alternative spaces, from different positions and even theoretical/epistemological guidelines, are developing in very dissimilar territories in the face of very similar circumstances (development projects, neoliberal extractivism, heritage legislation policies, etc.) (Zorzin, 2017). In this sense, as Criado Boado (1996) stated, it is better to face the problem by contributing to the design of "integrated models of management" of historical heritage, which allows, in addition to incorporating social demands, not to dissociate the cultural industry and its administration from the construction of knowledge.

Maintaining our social commitment implies a concern for providing contributions to people's daily lives, but also for making our professional practice transparent, inasmuch as any enhancement always involves a variety of interpretations and valuations of the assets, elements, or memories that make the construction of knowledge and discourses viable. It is then our ethical responsibility to highlight this interpretative chain and the double hermeneutics that implies that the archaeological practice will seek to reconstruct (interpret) "the" original value and recognize the current value, thus generating an evaluation and revaluation that produces heritage assets (Criado Boado, 1996). It is clear that this construction of discourses and narratives, beyond the discipline's own competencies as a field of professional training, involves and actively interpellates the communities. Thus, we bet on the collective construction of archaeological heritage, which will ultimately grant and guarantee its legitimacy as such in the long term.

The case presented here, although still at an early stage, shows the potential of specific archaeological interventions derived from cultural resource management "contracts" to strengthen (or even generate) broader scientific research projects. It also shows the value of their incorporation in heritage protection policies through the development of archaeological baselines, especially for contexts characterized by a high and continuous anthropization such as urban spaces.

In this regard, it is worth noting the contributions that these baselines can make to enrich scientific knowledge about the archaeological record by contributing quality information. As stated by Ratto (2009), the so-called contract archaeology allows to survey areas that do not have previous baseline information, or that are not included in research projects; to identify the incidence of taphonomic processes in different seasonal periods; to act when unexpected archaeological assets are found, enabling their rescue in context or other appropriate protection measures; to provide information for the implementation of preventive, mitigating or corrective measures of archaeological impacts, among others. For this positive feedback with scientific research to be generated, however, it is essential that the information is disclosed, not only in the public but also in the academic sphere, and most especially when there are other projects and/or researchers linked to the area (Franco, 2017; Guráieb & Frère, 2008).

The expansion of scientific knowledge about the past is not the only objective that can be achieved with these interventions. Also is the viability to contribute to the revaluation of heritage and its activation as a cultural resource through sustainable development projects (Molinari et al., 2000; Ratto, 2013; Söderström, 2018), which not only ensure the sustainability of that heritage itself but also provide added value to the local population from educational, cultural, tourist, or other proposals.

It does not escape our reflection, nevertheless, that this kind of archaeological intervention will not provide in all cases the possibility of carrying out those heritage processes in a complete or finished way (either due to time constraints, budget, or other factors). We assume that not all the proposals presented here, for example, will be carried out by the same work team or at the same time, and they may even be developed during different stages of the same project. In this sense, the emphasis has been placed on *planning*, following models based on research and participation as principles. This will make it easier to generate control factors over heritage projects so that they effectively meet agreed objectives that tend to forge broader benefits in the long term.

In those cases where it is not possible to monitor these programs within the framework of each project, our contribution could be to build specific guidelines so that they can be taken up, developed, or improved in other areas, in other timeframes, or by other actors in the future. Establishing these working principles to ensure that our intervention can be capitalized to some extent by both researchers and local communities is part of our ethical commitment and our disciplinary challenge to contribute to today's society.

Acknowledgements To the parish priest Lucas Martínez, to the municipal authorities in office, City Major Fernanda Antonijevic and Secretary of Culture, Tourism and Sports Marcos Barlatay. To Ph.D. Alicia Tapia, researcher in the area, thanks to whom it was possible to carry out the intervention project presented. We also thank Alexis Urquiza, Luciano Rey, Tomás Cricri and María Emilia Do Campo for their participation in the fieldwork, archives and social activities, as well as to the community of the Santiago Apóstol Parish, who kindly opened its doors to us. Finally, we appreciate the contributions of the evaluators to enrich this manuscript.

References

Australia ICOMOS. (2013). *The Burra Charter: The Australia ICOMOS charter for places of cultural significance, 2013*. Australia ICOMOS.

Avrami, E., Macdonald, S., Mason, R., & Myers, D. (2019). Introducción. In *Values in heritage management: Emerging approaches and research directions* (pp. 1–8). The Getty Conservation Institute.

Ballart Hernández, J., & Tresserras, J. (2001). *Gestión del patrimonio cultural*. Ariel.

Barbich, M. J. (1908). *Plano Topográfico del Pueblo de Baradero [Plano]*. Municipalidad de Baradero.

Barbich, J. A. (2007). *Historia de Santiago del Baradero*. Amerindia.

Bárcena, R., & Schávelzon, D. (1991). *El cabildo de Mendoza. Arqueología e historia para su recuperación*. Municipalidad de Mendoza.

Barreiro Martínez, D. (2006). Conocimiento y acción en la Arqueología Aplicada. *Complutum, 17*, 205–219.

Camino, U., & Noya, V. (2015). Primeras excavaciones arqueológicas urbanas en Casco Histórico de Baradero. In *XV Congreso de la Historia de los Pueblos*.

Camino, U., & Tapia, A. (2017). *Asociación Española de Socorros Mutuos, Baradero*. Informe de campaña arqueológica [Informe de campaña]. https://www.baradero.gob.ar/sites/default/files/descargas/informe_de_campana_arqueologica-_aesm_-2017.pdf

Camino, U., Noya, V., Ronco, F., & Tapia, A. (2022). Evolución del patrón urbano a partir de un pueblo de indios (siglo XVII), Santiago del Baradero, Provincia de Buenos Aires, Argentina. *Canto Rodado. Revista especializada en patrimonio cultural, 15,* 73–89.

Casa Parroquial. (1893). [Plano]. Archivo Histórico Municipal.

Clark, B., & Horning, A. (2019). Introduction to a global dialogue on collaborative archaeology. *Archaeologies: Journal of the World Archaeological Congress, 15*(3), 343–351.

Conti, A. L. (2015). *La conservación y la gestión de las ciudades históricas desde la perspectiva del Paisaje Urbano Histórico*. Encuentro Internacional «El Paisaje Urbano Histórico como herramienta del desarrollo urbano sostenible», Quito, Ecuador

Criado Boado, F. (1996). Hacia un modelo integrado de investigación y gestión del Patrimonio Histórico: La cadena interpretativa como propuesta. *Boletín de Instituto Andaluz del Patrimonio Histórico, 4*(16), 73–78.

Criado Boado, F. (1997). *Del Terreno al Espacio: Planteamientos y Perspectivas para la Arqueología del Paisaje*. Grupo de Investigación en Arqueología del Paisaje, Universidad de Santiago de Compostela, Spain.

Debenedetti, S. (1911). Noticia sobre un cementerio indígena en Baradero. *Revista de la UBA*, XIII.

Endere, M. L., Chaparro, M. G., & Giacomasso, M. V. (2021). Grupos de interés, valores y significados del sitio arqueológico Intihuasi, provincia de San Luis, Argentina. Un análisis orientado a su gestión patrimonial. *Arqueología, 27*(1), 69–90.

Franco, N. (2017). La relación entre los estudios de impacto ambiental e investigaciones arqueológicas entre los años 2006 y 2016 en la Provincia de Santa Cruz (Argentina). *Práctica Arqueológica, 1*(1), 31–45.

Guráieb, A. G., & Frère, M. (2008). *Caminos y encrucijadas en la gestión del patrimonio arqueológico argentino*. Facultad de Filosofía y Letras.

Haber, A. (2017). Contratiempo. Arqueología de contrato o una trinchera en la batalla por los muertos. In C. Gnecco & A. Schmidt Dias (Eds.), *Crítica de la razón arqueológica: Arqueología de contrato y capitalismo* (pp. 79–99). Instituto Colombiano de Antropología e Historia.

Harris, E. (1991). *Principios de estratigrafía arqueológica*. Crítica.

Hernández de Lara, O., Motta, A. P., & Umaño, M. (2014). Patrimonio arqueológico y comunidad: Confluencias y problemáticas en el sitio Vuelta de Obligado (Argentina). *Revista Ph, 85*, 142–159.

Izeta, A., Pautassi, E., Costa, T., Robledo, A., & Bilinski, C. (2014). Excavaciones arqueológicas en el área fundacional de la ciudad de Córdoba, Argentina: Dean Funes 67. *Revista de Arqueología Histórica Argentina y Latinoamericana, 8*(1), 45–69.
Kuper, D., Ramírez, L., & Troncoso, C. (2010). *Política turística y planificación: ¿de las estrategias centralizadas a las estrategias participativas?* Actas del XI Coloquio Internacional de Geocrítica.
Londoño, W. (2016). Arqueología por contrato y nuevos contratos arqueológicos. *Revista Jangwa Pana, 15*(1), 117–128.
Manso, J. M. (1815). *Plano de Baradero [Plano].* Archivo de la Dirección Provincial de Geodesia.
Mason, R. (2002). Assessing values in conservation planning: Methodological issues and choices. In M. de la Torre (Ed.), *Assessing the values of cultural heritage* (pp. 5–30). The Getty Conservation Institute.
Micucci Tarsetti, A. (1998). *Historia del partido de Baradero: Vol. I.* Instituto de Investigaciones Históricas «Domingo Oro»..
Molinari, R., Ferraro, L., Paradela, H., Castaño, A., & Caracotche, S. (2000). 2001 Odisea del manejo: Conservación del Patrimonio Arqueológico y Perspectiva Holística. In *II Congreso Virtual de Antropología y Arqueología.*
Moreno Melgarejo, A., & Sariego López, I. (2017). Relaciones entre turismo y arqueología: El turismo arqueológico, una tipología turística propia. *Pasos, Revista de Turismo y Patrimonio Cultural, 15*(1), 163–180.
Obras de ampliación Casa Parroquial de Baradero. (1953). [Plano]. Archivo de la Parroquia Santiago Apóstol.
Pineau, V., & Charlin, J. (2007). *Poniendo en valor la memoria colectiva.* Recuperación del patrimonio arqueológico de la comunidad de Baradero [Manuscript].
Plano de Baradero. (1872). [Plano]. Archivo Histórico Municipal.
Prats, L. (2000). El concepto de Patrimonio Cultural. *Cuadernos de Antropología Social, 11.* https://doi.org/10.34096/cas.i11.4709
Ratto, N. (2009). Aportes de la arqueología de contrato al campo de la investigación: Estudios de casos en Patagonia y Noroeste de la Argentina. *Revista de Arqueología Americana, 27*, 49–70.
Ratto, N. (2013). *Patrimonio arqueológico y megaproyectos mineros en Argentina: Turismo, desarrollo y sociedad.* Aspha.
Ross, S. (2018). Strategies for municipal participatory governance and implementing UN-habitat's new urban agenda: Improving consultation and participation in urban planning decision-making processes through rapid ethnographic assessment procedure. *Canadian Bar Review, 96*, 294–323. SSRN Electronic Journal 10.2139/ssrn.3008945
Santana Talavera, A. (2002). Mirar y leer: Autenticidad y patrimonio cultural para el consumo turístico. In *Sexto Encuentro Nacional de Turismo con Base Local, Campo Grande, Brasil.*
Söderström, U. (2016). Archaeology and urban sustainability: Can the past provide a key to the future? *The 17th International Planning History Society Conference*, Delft.
Söderström, U. (2018). *Contract archaeology and sustainable development: Between policy and practice* [Thesis]. Department of Cultural Sciences, Linnaeus University.
Tapia, A. (2003). Patrimonio arqueológico local: Uso social y globalización. Un caso de estudio en Baradero, provincia de Buenos Aires. *Revista de la Escuela de Antropología, VIII*, 191–201.
Tapia, A., Nespolo, E., & Noya, V. (2015). *Santiago del Baradero. Origen Evolución de un pueblo de indios.* Pangea Ediciones de la Tierra.
Tapia, A., Traba, A., & Camino, U. (2021). Investigaciones en el área fundacional de Baradero (Provincia de Buenos Aires): Análisis de un conjunto de cerámica hispano-indígena. In *X Congreso de Arqueología de la Región Pampeana Argentina.* Facultad de Humanidades, Universidad Nacional de Mar del Plata.
Traba, A. (2018). *Sitio Asociación Española de Socorros Mutuos. Informe preliminar del análisis de materiales vítreos y cerámicas de pasta dura* [Internal report]. Proyecto de Investigación en el Área Fundacional de la Reducción Franciscana Santiago del Baradero (FFyL, UBA), direction: Dr. Alicia Tapia.

U. A. Camino Centro de Arqueología Urbana, Instituto de Arte Americano, Facultad de Arquitectura, Diseño y Urbanismo, Universidad de Buenos Aires – Cooperativa de Trabajo Arqueocoop ltda, Buenos Aires, Argentina.

Zorzin, N. (2017). Arqueología distópica: La implementación de la lógica del capital en el manejo del patrimonio. In C. Gnecco & A. Schmidt Dias (Eds.), *Crítica de la razón arqueológica: Arqueología de contrato y capitalismo* (pp. 163–188). Instituto Colombiano de Antropología e Historia.

Index

G. F. Bonnat et al. (eds.), *Current Research in Archaeology of South American Pampas*, The Latin American Studies Book Series,
https://doi.org/10.1007/978-3-031-55194-9

The manufacturer's authorised representative in the EU is Springer Nature Customer Service Centre GmbH, Europaplatz 3, 69115 Heidelberg, Germany. If you have any concerns regarding our products, please contact ProductSafety@springernature.com

Printed and bound by CPI Group (UK) Ltd, Croydon, CR0 4YY
30/09/2026
02225571-0016